HEAT AND MASS TRANSFER
IN BOUNDARY LAYERS

VOLUME 2

INTERNATIONAL CENTRE FOR HEAT AND MASS TRANSFER

P.O. Box 522, Beograd, Yugoslavia

SCIENTIFIC COUNCIL

E. A. BRUN, President, Académie des Sciences, Paris
E. R. G. ECKERT, Vice-President, University of Minnesota, Minneapolis
M. A. STYRIKOVICH, Vice-President, USSR Academy of Sciences, Moscow
D. VELIČKOVIĆ, Vice-President, Serbian Academy of Sciences, Beograd
K. O. BEATTY, Jr., North Carolina State University, Raleigh
F. BOŠNJAKOVIĆ, Technische Hochschule, Stuttgart
A. FORTIER, Faculté des Sciences, Paris
J. GINOUX, von Karman Institute for Fluid Dynamics, Rhode Saint Genese
U. GRIGULL, Technische Universität, München
W. B. HALL, University of Manchester, Manchester
S. S. KUTATELADZE, Thermal Physics Institute, Novosibirsk
P. M. C. LACEY, University of Exeter, Exeter
M. LEDINEGG, Technische Hochschule, Wien
A. V. LUIKOV, BSSR Academy of Sciences, Minsk
D. MALIĆ, University of Beograd, Beograd
T. MIZUSHINA, Kyoto University, Kyoto
Y. MORI, Tokyo Institute of Technology, Tokyo
L. NAPOLITANO, Istituto di Aerodinamica, Napoli
N. NISHIWAKI, University of Tokyo, Tokyo
W. M. ROHSENOW, Massachusetts Institute of Technology, Cambridge
J. T. ROGERS, Carleton University, Ottawa
O. SAUNDERS, University of London, London
K. STEPHAN, Technische Universität, Berlin
M. VERON, Conservatoire National des Arts et Métiers, Paris
K. VORONJEC, Serbian Academy of Sciences, Beograd
J. W. WESTWATER, University of Illinois, Urbana
N. ZUBER, Georgia Institute of Technology, Atlanta

ORGANIZATION COMMITTEE

T. F. IRVINE, Jr., Chairman, State University of New York, Stony Brook
N. AFGAN, Vice-Chairman, University Beograd, Beograd
J. GOSSE, Ecole Nationale Supérieure d'Eléctricité et de Mécanique, Nancy
E. HAHNE, Technische Universität, München
J. P. HARTNETT, University of Illinois, Chicago
T. W. HOFFMAN, McMaster University, Hamilton
D. B. SPALDING, Imperial College, London
A. I. LEONT'EV, Institute of High Temperatures, Moscow
D. A. DE VRIES, Technische Hogeschool, Eindhoven

SECRETARY GENERAL

Z. ZARIĆ, University of Beograd, Beograd

HEAT AND MASS TRANSFER IN BOUNDARY LAYERS

VOLUME 2

PROCEEDINGS
OF THE
INTERNATIONAL SUMMER SCHOOL
HEAT AND MASS TRANSFER IN TURBULENT BOUNDARY LAYERS
HERCEG NOVI, SEPTEMBER, 1968

AND

SELECTED PAPERS AND ABSTRACTS
OF THE
INTERNATIONAL SEMINAR
HEAT AND MASS TRANSFER IN FLOWS WITH SEPARATED REGIONS
HERCEG NOVI, SEPTEMBER, 1969

Edited by

N. AFGAN,
Z. ZARIĆ
and
P. ANASTASIJEVIĆ

Boris Kidrič Institute
Beograd

PERGAMON PRESS
OXFORD · NEW YORK · TORONTO
SYDNEY · BRAUNSCHWEIG

Pergamon Press Ltd., Headington Hill Hall, Oxford
Pergamon Press Inc., Maxwell House, Fairview Park, Elmsford,
New York 10523
Pergamon of Canada Ltd., 207 Queen's Quay West, Toronto 1
Pergamon Press (Aust.) Pty. Ltd., 19a Boundary Street,
Rushcutters Bay, N.S.W. 2011, Australia
Vieweg & Sohn GmbH, Burgplatz 1, Braunschweig

Copyright © 1972 Pergamon Press Inc.
All Rights Reserved. No part of this publication may be reproduced, stored in a retrieval system, or transmitted, in any form or by any means, electronic, mechanical, photocopying, recording or otherwise, without the prior permission of Pergamon Press Inc.
First edition 1972
Library of Congress Catalog Card № 72-85858

Printed in Yugoslavia by »Radiša Timotić« Belgrade

08 017140 0

CONTENTS

VOLUME 1

	Page
FOREWORD	

A. INTERNATIONAL SUMMER SCHOOL ON HEAT AND MASS TRANSFER IN TURBULENT BOUNDARY LAYERS

A brief review of theories of the turbulent boundary layer
   by D. B. SPALDING — 1

On the turbulent boundary layer with vanishing viscosity
   by S. S. KUTATELADZE — 9

Calculation of constant χ of the wall turbulence
   by M. A. GOLDSHTIK, S. S. KUTATELADZE — 39

Principle of maximum stability of the averaged turbulent fluxes
   by M. A. GOLDSHTIK — 42

Methods of calculation of the turbulent boundary layer by limiting laws
   by A. I. LEONT'EV — 51

Some results of investigation of the turbulent boundary layer on a permeable surface
   by B. P. MIRONOV — 71

Radiation-convection heat transfer in a blow-off region of the boundary layer
   by N. A. RUBTSOV — 87

Heat transfer in transpired turbulent boundary layers
   by J. P. HARTNETT and V. M. K. SASTRI — 97

Heat and mass transfer of moist capillary-porous bodies with laminar and turbulent heated rarefied gas flows
   by L. L. VASIL'EV — 125

Studies in mass transfer between hot jet of air and moist surfaces
   by R. N. KUMAR — 141

Heat and mass transfer in a turbulent boundary layer on a penetrable surface
   by B. I. FEDOROV — 151

Turbulent boundary layer with suction and heating to the wall
   by E. VEROLLET, L. FULACHIER, R. DUMAS and A. FAVRE — 157

The effect of the slope of a flat plate on the transfer of mass on its surface
   by G. POPOVIĆ — 169

Gaseous film cooling at various degrees of hot-gas acceleration and turbulence levels
   by L. W. CARLSON and E. TALMOR — 177

Asymptotical theory of turbulent boundary layer
   by A. FORTIER — 199

	Page
Measurement of the temperature profiles in a turbulent flow of air in a tube (incompressible fluid)	
by L. TACCOEN	217
Heat transfer in the entrance region of an annular channel	
by V. JOVAŠEVIĆ	227
A method for prediction of turbulent boundary layer development based on the shear-work-integral	
by M. P. ESCUDIER and W. B. NICOLL	239
Calculation of heat transfer in the turbulent boundary layer with injection and pressure gradient	
by M. R. HEAD and F. A. DVORAK	257
A general method for predicting friction, heat transfer and mass transfer in two-dimensional boundary layers	
by D. B. SPALDING	269
Some applications of the general theory of friction, heat transfer and mass transfer in two-dimensional boundary layers	
by J. H. WHITELAW	289
On the prediction of laminarisation	
by B. E. LAUNDER and W. P. JONES	307
Predictions of transpiration cooling in turbulent boundary layer	
by V. K. JONSSON and R. J. BAKER	321
Study of combined free and forced-convection turbulent boundary layer with variable fluid properties and chemical reaction	
by F. C. LOCKWOOD and P. H. ONG	339
Calculation of local characteristics of a turbulent boundary layer with a positive pressure gradient	
by G. N. PUSTYNTSEV	351
Transfer of scalar substances in turbulent shear flows	
by B. A. KOLOVANDIN	359
Disturbed wall flows	
by J .MATHIEU	379
Fully-developed flow in rectangular ducts of non-uniform surface texture: An experimental investigation	
by K. HANJALIĆ and B. E. LAUNDER	393
Mixing-length predictions of flow between planes of dissimilar surface texture	
by K. HANJALIĆ and B. E. LAUNDER	411
Experimental study of convective mass transfer at a rough plate surface with cavitation in a boundary layer	
by V. P. POPOV and L. K. GLEB	419
Flow field past a single roughness element in channel of rectangular cross-section	
by S. OKA and Ž. KOSTIĆ	425
Some questions of the vortex flow theory	
by M. A. GOLDSHTIK	437
Mass transfer between vortices and main stream	
by M. M. ZDRAVKOVIĆ	447
Fluid flow and convection heat transfer in boundary layer on smooth, cylindrical surface of a tube in a tube bank in cross flow	
by Ž. KOSTIĆ and S. OKA	451

	Page
Nonlinear stability theory of the boundary layer by V. V. STRUMINSKII	459
Investigation of turbulent flames by V. C. BAEV and C. P. TRETJAKOV	475
Study of diffusion flame stability by V. C. BAEV and V. A. YASAKOV	489

VOLUME 2

Heat transfer within close-to-critical single-phase region by B. S. PETUKHOV	495
The effect of free convection on heat transfer and flow resistance in turbulent flow in horizontal tubes by A. F. POLYAKOV	517
Turbulent free convection heat transfer under supercritical conditions by E. HAHNE	523
Turbulent mixing in subchannel analysis by I. DEVOLD	529
Measurements close to the wall in a turbulent boundary layer by Z. ZARIĆ	555
Study of turbulence at a wall by stroboscopic vizualization method by E. M. KHABAKHPASHEVA	573
Heat transfer from hot wire and hot film anemometer probes by M. R. DAVIS	577
Measurement of the velocity near a smooth wall and the determination of the local wall shear stress in a turbulent flow by NGUYEN VAN THINH	585
Calibration of a DISA hot-wire anemometer and measurements in a circular channel for confirmation of the calibration by B. KJELLSTRÖM and S. HEDBERG	593
The theory of electrochemical technique of investigating turbulence at a wall by V. YE. NAKORYAKOV	619
A method of temperature measurement close to a wall, in a turbulent boundary layer by J. P. MAYE	629
Heat and mass transfer in a boiling boundary layer by M. A. STYRIKOVICH	641
Investigations of two-phase boundary layer characteristics by various methods by E. I. NEVSTRUYEVA	651
Investigation of pool boiling heat transfer of water and freon-113 in a wide range of temperature heads by S. A. KOVALEV, V. M. ZHUKOV. YA. A. KUZMA-KICHTA, and V. P. OGORODNIKOV	657
Critical heat fluxes in boiling liquid metals by G. I. BOBROVICH	663
Electrochemical technique of flow studies by A. P. BURDUKOV	669
On boiling crisis at the core stream of vapor-water mixture by I. G. DRUKER	677

	Page
Post-burnout heat transfer to mist flow by W. M. ROHSENOW and E. FEDOROVICH	683
Relationship between the structure of two-phase layer near a heating wall and the mechanisms of thermal exchange by R. SEMERIA	701
The general equations of two-phase systems applied to flashing flows by J. M. DELHAYE, A. FIORE, PH. VERNIER and B. BROISE	715
Maximum superheating of liquids in flashing and boiling by M. STEFANOVIĆ	743
The mechanism of nucleate boiling heat transfer by LJ. JOVANOVIĆ and M. STEFANOVIĆ	753
Boiling heat transfer of the binary mixtures by N. H. AFGAN	761
List of participants	773

B. INTERNATIONAL SEMINAR ON HEAT AND MASS TRANSFER IN FLOWS WITH SEPARATED REGIONS AND MEASUREMENT TECHNIQUES

SELECTED PAPERS

Dynamics and thermodynamics of separated flows by H. H. KORST	781
Generalized turbulence transport equations by C. W. HIRT	827
Heat transfer from single tubes and banks of tubes in crossflow by Z. ZHUKAUSKAS	843
Gas concentration measurements in boundary layer by E. A. BRUN	867
Pressure measurements in highly turbulent flows by M. BARAT	887
The use of electrochemical techniques to study flow fields and mass transfer rates by T. J. HANRATTY	919
Optical measurement of fluid velocity utilizing the Doppler effect by R. J. GOLDSTEIN	941
Turbulence measurement by optical methods by A. M. TROHAN	953
ABSTRACTS	967
Seminar Committee	1007
List of participants	1007
Author index	1013

HEAT TRANSFER WITHIN CLOSE-TO-CRITICAL SINGLE-PHASE REGION

B. S. PETUKHOV

Institute for High Temperatures, USSR Academy of Sciences, Moscow, USSR

The problem of heat transfer within a single-phase medium under close-to-critical parameters are of great concern in science and practice. This problem is closely associated with the development of a theory of unisothermic turbulent flows of fluid featuring variable physical properties and the development of the methods concering the design of high-tension heat transfer systems.

When speaking of the heat transfer within a single-phase medium under to close-to-critical parameters (or, in short, within a supercritical region) we mean the processes pertaining to heat transfer occuring under pressures above critical and under close-to-critical or pseudo-critical temperatures (i.e. under the temperature corresponding to the maximum of specific heat at constant pressure).

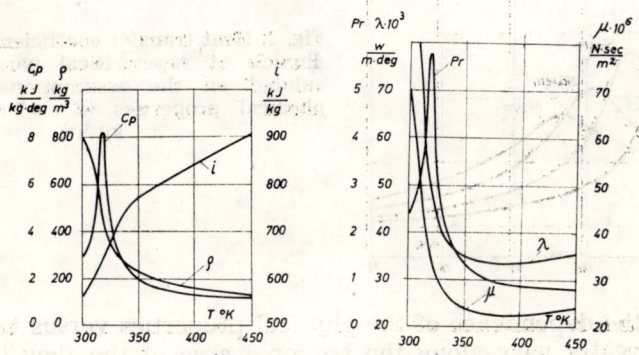

Fig. 1. Physical properties of carbone dioxide at $P=100$ bars

Specific features of the heat transfer occurable within super critical region lie in the fact that physical properties of the medium within this region are strongly and peculiarly changed depending upon the temperature involved and are substantially dependent upon the pressure. A picture of the character of variations of the physical properties can be seen from Fig. 1 wherein are the data concerning carbon dioxide at $P=100$ bars.

If the process of heat transfer occurs at rather small (within the limit of infinitesimal) temperature differences, it is no matter how would the physical properties of the fluid vary versus temperature; the calculations of such a process may be made on the assumption of the constancy of physical properties. Therefore in the case of turbulent flow of fluid through the tubes with $T_w - T_b \to 0$ and for suppercritical region the relationship, for example, [1] holds true of

$$Nu = \frac{\frac{\xi}{8} Re Pr}{k_1 + k_2 \sqrt{\frac{\xi}{8}} (Pr^{2/3} - 1)}, \qquad (1)$$

where

$$\xi = (1{,}82 \lg Re - 1{,}64)^{-2},$$
$$k_1 = 1 + 3{,}4 \xi \quad k_2 = 11{,}7 + 1{,}8 Pr^{-1/3} \qquad (2)$$

Approximately we may assume $k_1 \approx 1.07$ and $k_2 \approx 12.7$. Figure 2 represents the heat transfer coefficient calculated from the abovementioned Eq. (1) versus T_b with different pressures $P > P_{cr}$. The character of variation of α is specified, of course, by a peculiar dependance of carbone dioxide physical properties versus T and P.

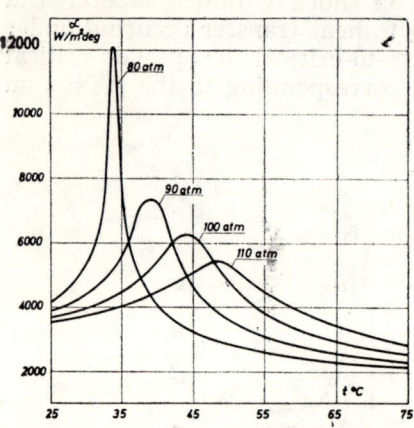

Fig. 2. Heat transfer coefficient for carbone dioxide at supercritical parameters, calculated on the assumption of constant physical properties ($d = 6.7$ mm, $G = 100$ kg/hr)

Due to the dependence of the physical properties versus temperature, these will greatly vary along the sectional area of the flow in question. Thus, for example, if $T_b < T_m < T_w$ (here T_m is a pseudocritical temperature), the specific heat C_p as removing from the wall tends to increase rapidly, after that it goes through the maximum and then will be decreased sharply. The density with the fluid being heated will increase in the direction from the wall toward the axis of the tube involved. Variations of physical properties with respect to the radius affects greatly the flow of fluid and the entire heat transfer process.

Theoretical Analysis

A theoretical analysis of heat transfer and friction at supercritical parameters of the state with due account of the dependence of physical properties of the fluid versus temperature was for the first time conduced by American investigators Deissler [2] and Goldman [3, 4]. Both works were concerned with the study of heat transfer with turbulent flow of uncompressible fluid through a tube far from the inlet and at constant heat flux on the wall (q_w=const.). Calculations were carried out through the use of the following equations:

$$q = \lambda \left(1 + \beta\, Pr\, \frac{\varepsilon_\sigma}{\nu}\right) \frac{\partial T}{\partial r}, \tag{3}$$

$$\sigma = -\mu \left(1 + \frac{\varepsilon_v}{\nu}\right) \frac{\partial w_x}{\partial r}. \tag{4}$$

The distribution of q and σ along the radius was assumed either constant (by Deissler) or linear (by Goldman). The Eqs. (3) and (4) written down preliminarily in a dimensionless form were solved by the method of subsequent approximations. The coefficient $\beta = \varepsilon_q/\varepsilon_\sigma$ was assumed equal unity. The difference between the methods resorted to by Deissler and by Goldman lies mostly in the technique of calculations of the eddy diffusivity of momentum ε_σ at variable properties of the medium involved. To determine ε_σ/ν in the case of variable properties Deissler resorts to the same dependences as in the case of constant properties, considering kinematic viscosity coefficient which appears in these dependences as a variable quantity. The calculation method proposed by Goldman is based upon a hypothesis that turbulent characteristics at each point of the flow featured by variable properties are dependent upon the values of the properties at this particular point and do not depend upon variations occurring in the neighbourhood of this point. Resting upon this hypothesis, Goldman has come to the conclusion that to calculate ε_σ/ν with variable properties it is possible to use the same dependences ε_σ/ν versus η, as with constant properties the only thing required is to substitute $\eta = \dfrac{v^* y}{\nu}$

$\left(\text{here } v^* = \sqrt{\dfrac{\sigma_w}{\rho}}\right)$ by a generalized variable $\eta^+ = \displaystyle\int_0^y \dfrac{1}{\nu}\sqrt{\dfrac{\sigma_w}{\rho}}\, dy$.

Deissler calculated heat transfer and friction for the case of water heating at the pressure of 344 bars within the temperature range from 204 up to 650°C. The results thus obtained were generalized by resorting to the method of reference temperature which appear to be a complex and unmonotonous function T_w/T_b and T_w. The curve of this function is represented in [2]. Goldman likewise carried out the calculations of heat transfer and friction for the case of water heating at the pressure of 344 bars but within a wider temperature range, i.e. from 260 to 840°C. The data obtained were represented in the form of the following equation:

$$\frac{q_w \cdot d^{0,2}}{(\rho w)^{0,8}} = f(T_w, T_b), \qquad (5)$$

which is based on the assumption that the dependence of heat transfer versus ρw and d with variable properties of the medium is the same as with constant properties. The function $f(T_w, T_b)$ is plotted in Fig. 3. If the physical properties are constant, f would be a linear function versus T_w. Thus, deviation from a linear dependence characterises the effect of variable physical properties.

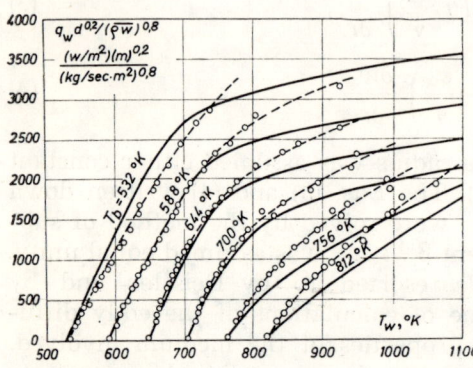

Fig. 3. Results of theoretical calculation by Goldman for water at $P = 344$ bars (solid lines) in comparison with experimental data (broken line; designated by circles are average experimental data). Limits of measurement: $d = 1.27$ to 1.9 mm, $q_w = (0.32$ to $9.5) \cdot 10^6$ W/m²; $_0\overline{W}_x = (2$ to $4) \cdot 10^3$ kg/sec·m²

Using Goldman's method Tanaka, Nishiwaki and Hirata [5] carried out calculations of heat transfer for carbone dioxide at $P=78.5$ bar both with heating and cooling down of the fluid involved. The results they obtained are represented in Fig. 4. As is seen from the figure, the dependence of the heat transfer coefficient α versus t_b with $t_b \approx t_m$ passes

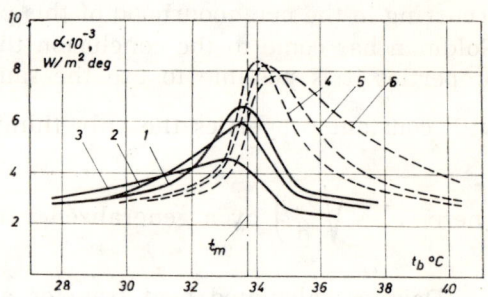

Fig. 4. Dependence of α versus t_b for carbone dioxide with heating and cooling down of fluid ($P=78.5$ bar, $G=3.9 \cdot 10^{-2}$ kg/sec., $d=10$ mm) with $q_w \cdot 10^{-4}$ W/m²; 1) 0.70; 2) 1.4; 3) 2.8; 4) —0.70;; 5) —1.4; 6) —2.8.

through the maximum which is due to the availability of corresponding maximums $C_p(T)$ and $Pr(T)$ (see Fig. 1). The maximum value of α in the case of cooling ($q_w < 0$) is larger than in the case of heating ($q_w > 0$) and decreases with the increase of q_w. This may be explained by the fact that thermal conductivity of the fluid within a viscous sublayer in the case of cooling is two times as high as in the case of fluid heating and decreases with the increases of q_w and T_w.

Petukhov and Popov [1] have developed another method for calculating heat transfer and friction with a turbulent flow of the uncompressible fluid featured by variable physical properties, and flowing through the tubes. This method provides the possibility to accomplish calculation of the flow and heat transfer far from the inlet to the tube in question with q_w=const. The calculations of heat transfer are based upon the analytical expressions for the number Nu, the distribution of temperature and of velocity:

$$\frac{1}{Nu_w} = 2\frac{c_{pw}}{\overline{c_p}} \int_R^1 \frac{\left(\int_0^R \frac{\rho w_x}{\overline{\rho w_x}} R\, dR\right)^2}{\frac{\lambda}{\lambda_w}\frac{c_{pw}}{\overline{c_p}}\left(1+\beta\, Pr\, \frac{\varepsilon_\sigma}{\nu}\right) R}\, dR, \qquad (6)$$

$$T_w - T = \frac{q_w d}{\lambda_w}\int_R^1 \frac{\int_0^R \frac{\rho w_x}{\overline{\rho w_x}} R\, dR}{\frac{\lambda}{\lambda_w}\left(1+\beta\, Pr\, \frac{\varepsilon_\sigma}{\nu}\right) R}\, dR, \qquad (7)$$

$$w_x = \frac{\sigma_w d}{2\mu_w}\int_R^1 \frac{R}{\frac{\mu}{\mu_w}\left(1+\frac{\varepsilon_\sigma}{\nu}\right)}\, dR, \qquad (8)$$

where

$$Nu_w = \frac{q_w d}{\lambda_w (T_w - T_b)},$$

$$\overline{c_p} = \frac{\int_{T_b}^{T_w} c_p\, dT}{T_w - T_b} = \frac{h_w - h_b}{T_w - T_b},$$

$$R = \frac{r}{r_0}.$$

The expressions (6), (7), and (8) are obtained from the equations of energy and motion (written down upon approaching of the boundary layer) with the following assumptions:

1) variation of axial component of mass velocity along the tube axis is small, i.e.

$$\partial(\rho w_x)\, \partial x \approx 0.$$

2) pressure P is constant across the section of the tube;
3) longitudinal gradient of the enthalpy $\partial h/\partial x$ is constant across the section of the tube;

4) effect of mass forces and dissipation of kinetic energy is negligible. By having recourse of the method of successive approximations it is possible through the use of the Eqs. (6), (7), and (8) to calculate first the velocity distribution and temperature and then heat transfer.

Resorting to this method Popov carried out calculations of heat transfer and friction within close-to-critical single-phase region. The eddy diffusivity of momentum ε_σ was determined from Reichardt's equations (7) generalized for the case of variable physical properties by introducing Goldman's variable. The value of β was assumed equal unity. The calculations were carried out for the case of heating carbone dioxide at $P/P_{cr}=1.33$ ($P=98$ bar); $0.94 \leqslant T_b/T_m \leqslant 1.24$ and $0.97 \leqslant T_w/T_m \leqslant 1.24$. The data obtained were represented in the form of interpolation equations for heat transfer and friction.

Recently, Shiralkar and Griffith [36] by resorting to Goldman's method carried out calculations of heat transfer and friction for the case of heating water at $P=228$ bars and carbone dioxide heated at $P=75.8$ bar and 79.3 bar and compared the data of calculations of heat transfer with experimental data for regimes with deteriorated heat transfer.

The analysis of the results of theoretical calculations and experimental data shows that the dependence of the Nusselt number versus Re and Pr with variable physical properties tends to be approximately the same as with the constant physical properties. Then, the relationship Nu_b/Nu_{ob} will depend only upon the character of variations of physical properties of the fluid with temperature. For the given fluid Nu_b/Nu_{ob} will be the function of T_b, T_w and P or T_b/T_m, T_w/T_m and P/P_{cr}, which is more convenient to compare the data pertaining to different pressures and fluids.

Represented in Figs. 3, 5 and 6 are some data concerning theoretical calculations of heat transfer and friction in comparison with experimen-

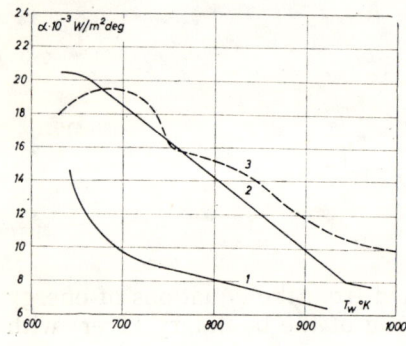

Fig. 5. Dependence of α versus T_w for water at $P=345$ bars, $T_w/T_b=1.25$; $\rho \overline{W}_x=2150$ kg/m^2 sec and $d=9.4$ mm;
1. Deissler's theoretical calculation;
2. Goldman's theoretical calculation;
3. Empirical equation by Swenson and al.

tal data or empirical equations. The character of dependence Nu_b/Nu_{ob} versus T_w/T_m can be clearly seen from Fig. 6. With $T_b/T_m<1$, heat transfer first tends to rise with the increase of $\dfrac{T_w}{T_m}$, with $\dfrac{T_w}{T_m} \approx 1$, it reaches its maximum, and then tends to decrease. If $T_b/T_m \leqslant 1$, heat transfer tends to decrease with the increse of T_w/T_m (see Fig. 5).

As it is seen from Fig. 5 borrowed from the work [8] the results of theoretical calculation carried out by Deissler present lower values of heat transfer as compared with experimental data (approximately in two times). Bringer and Smith [9] having calculated heat transfer for the case of heating carbone dioxide according to Deissler's method and compared it with the results of their own measurements, found out the difference up to 25%. Melik-Pashaev [10] likewise calculating heat transfer according to Deissler's method but using Prandtl expression for the mixing length of occurred within the turbulent core and approximately taking into account the effect of density fluctuations has obtained better agreement between the culculated and experimental data. The results of calculations carried out by Goldman (cf. Figs. 3 and 5) are in good agreement with the experiment where $T_w - T_b \gtrsim 100°C$. With the increase of $T_w - T_b$ the divergence tends to rise thereby reaching 25% with $T_w - T_b = 250$ to 300°C. The results of theoretical calculations carried out by Popov as is shown in the work [6] are in good agreement with the data concerning heat transfer to carbone dioxide with an accuracy of $\pm 20\%$. Good agreement between the design and experimental data obtained by Popov is represented in Fig. 6.

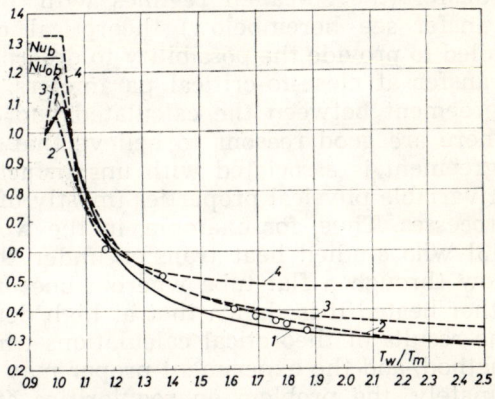

Fig. 6. Dependence of Nu_b/Nu_{ob} versus T_w/T_m for carbone dioxide at P=98 bars, $T_b=303°K$, $G=100$ kg/hr and $d=4.08$ mm: 1 — Popov's theoretical calculation; 2, 3 and 4 — empirical equations (10), (11) and (9) respectively; circles — experimental data

Hsu and Smith [11] undertook the attempt to approximately take the account of the efect of density fluctuations upon turbulent momentum diffusivity. The calculations carried out for carbone dioxide have shown that by introducing such a correction it is possible to improve somewhat the agreement between the calculated and experimental values of heat transfer. However, the calculations carried out for hydrogen [12] with account of the correction for density fluctuations found to be in worse agreement with the experiment than calculations without this correction. Thus, the problem of taking account of density fluctuations as well as fluctuations of other physical properties has not been yet solved in the positive.

Due to strong variation of density with temperature it is likely expectable that free convection substantially affects the heat transfer with forced flow occurrable within a supercritical region. Theoretically this problem was analized by Hsu and Smith [11] with resort to the case of upward turbulent flow of carbone dioxide through a vertical tube under the conditions of heating. As it was expectable, the effect of free convection upon heat transfer is as greater than larger the Grashof number Gr and lower the number Re. For example, with $Gr=10^8$ and $Re=10^4$ heat transfer tends to two times increase due to free convection, while with the same values of Gr and $Re=10^5$ free convection is of no effect upon heat transfer. The here-in-described analysis does not completely cover the problem concerning the effect of free convection upon forced turbulent flow. This problem has been considered but insufficiently.

The analysis of the results of theoretical calculations shows that with not so great variations of physical properties across the section of the flow involved the latter provide the possibility to describe satisfactorily regularities of the heat transfer within a supercritical region (however, not for all possible regimes). A certain disagreement between the calculated and experimental data observed in this case is likely associated not only with imperfection of theoretical calculation methods but with a known indefinity to appreciate physical properties of the fluid under close-to-critical environmental conditions.*

In the case of strong variation of physical properties across the section of the flow as well as for certain specific regimes of flow and heat transfer (the so-called regimes with deteriorated and improved heat transfer, see hereinbelow) theoretical calculation methods now in use failed to provide the possibility to describe satisfactory regularities of heat transfer at close-to-critical parameters. In this case a considerable disagreement between the calculated and experimental data is observable. There are good reasons to believe that the hereinabove-mentioned disagreement is associated with unsatisfactory taking account of the effect of variable physical properties (mostly of density) on turbulent diffusivity processes. Thus, for example, in the works by Hall, Jackson, and Khan [13] who studied heat transfer under the conditions of carbone dioxide flow through a flat tube wherein one of the walls was cooled and the other heated it is shown that at high density gradients within a sublayer the results of theoretical calculations carried out according to Goldman's method, and the experiment proper may differ by a factor of two. Unfortunately, the problem on regularities of turbulent diffusivity occurable with variable physical properties has not been practically considered. Therefore experimental and theoretical results concerned will be of great interest.

Experimental Data and Empirical Equations

By the present time a large number of experimental works concerning heat transfer within supercritical region is available. The overwhelming majority of works deals with heat transfer to turbulent flow of water and carbone dioxide through the circular tubes under the con-

---

* Thus, for example, according to certain data the thermal conductivity within pseudo-critical temperature features a marked maximum, whereas according to other data it varies monotonously. The present work relies on the second conception. However, as it is described in [13] taking account of thermal conductivity maximums observable in some papers within a very narrow temperature range differs but negligibly with the results of theoretical calculations.

ditions of heating with q_w=const. An uncomplete list of the papers in question is represented in Table 1. Some works are dealt with heat transfer to hydrogen [12, 28], oxygen [24] and freon-12 [30], as well as under the conditions of heating. The experimental data on heat transfer under the conditions of cooling do not practically exist.

As one can judge by experimental data, with small temperature a viscous-inertial regime of flow and heat transfer and a viscous-inertialficantly the heat transfer processes occurring within a supercritical region feature no specific characteristics and can be described well by known dependences for constant properties.

With sufficiently high temperature differences one should distinguish a viscousinterial regime of flow and heat transfer and a viscous-inertial-gravitational regime. The former is characterized by the absence of a marked efect of Archimedes' forces, i.e. free convection, upon a forced turbulent flow. The latter is characterized by the presence of such effect.

With definite combination of regime parameters both with a viscous-inertial and viscous-inertial-gravitational flows, a sharped decrease in heat transfer is observed on some sections of the tube involved. Such regimes which are of particular concern in practice are often called the regimes with deteriorated heat transfer. In other cases, with this or other combination of regime parameters an unusual and for the time being difficult-to-explain increase in heat transfer is observable. Therefore, such regimes are often called the improved heat transfer regimes.

Isolation of regimes with improved and deteriorated heat transfer to certain specific cases, from my point of view, is not due to the fact that the latter feature some specific characteristics, but due to insufficient knowledge of the mechanism of turbulent diffusivity processes with variable physical properties of the fluid involved. The last point fails to provide the possibility to describe all possible regimes of flow and heat transfer by unified dependences and makes us consider the regimes with deteriorated and improved heat transfer as some specific cases. As we acquire the required knowledge, the abovementioned isolation will become unnecessary. One may expect that in so far as the regimes with deteriorated heat transfer are concerned this will occur recently.

The viscous-inertial regimes are frequently observed in the heat trasfer systems operating under the conditions of supercritical parameters. There are known a few empirical equations for calculation of heat transfer at such regimes. One of the first equations was proposed by Miropol'sky and Shitsman [16]. Their equation is written as follows:

$$Nu_b = 0.023 \ Re_b^{0.8} \ Pr_{min}^{0.8}, \qquad (9)$$

where Pr_{min}-stands for the least value of Prandtl number calculated with an average bulk temperature of the fluid (Pr_b) and the temperature of the wall (Pr_w).

Krasnoshchekov and Protopopov [25] on the basis of the experimental data they obtained for CO_2 and of the experimental data obtained by other investigators for CO_2 and H_2O proposed the equation of the following form:

$$Nu_b = Nu_{ob} \left(\frac{\rho_w}{\rho_b}\right)^{0.3} \left(\frac{\overline{C_p}}{C_{p}}\right)^n \qquad (10)$$

where Nu_{ob} symbolises Nusselt number with constant physical properties that is derived from the equation (1); $\overline{C}_p$ designates average integral heat capacity within the temperature range from T_b up to T_w; and n is the exponent dependent upon T_w/T_m and T_b/T_m. This relationship is represented in Fig. 7. The Eq (10) describes experimental data with a maximum

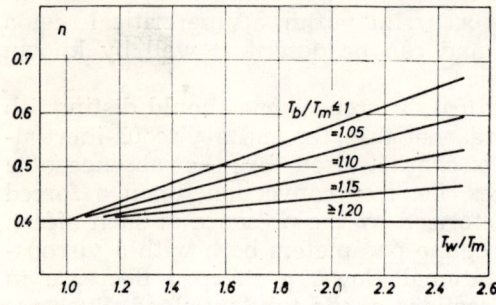

Fig. 7. Dependence of n versus T_w/T_m and T_b/T_m

possible error of $\pm 15\%$ and holds true of within the following measurement limits of characteristic parameters:

$$1.01 \leqslant P/P_{cr} \leqslant 1.33;$$
$$0.6 \leqslant T_b/T_m \leqslant 1.2;$$
$$0.6 \leqslant T_w/T_m \leqslant 2.6;$$
$$2.10^4 \leqslant Re_b \leqslant 8.10^5;$$
$$0.85 \leqslant Pr_b \leqslant 55;$$
$$0.09 \leqslant \rho_w/\rho_b \leqslant 1.0;$$
$$0.02 < \overline{C}_p/C_{pb} < 4.0;$$
$$2.3.10^4 \leqslant q_w \leqslant 2.6.10^6 \ W/m^2 \text{ and } \frac{1}{d} > 15.$$

Swenson, Carver and Kakarala [8] generalized their experimental data concerning heat transfer to water through the use of the following equation:

$$Nu_w = 0.00459 \ Re_w^{0.023} \ \overline{Pr}_w^{0.613} \times$$
$$\times \left(\frac{\rho_w}{\rho_b}\right)^{0.231}, \tag{11}$$

where $\overline{Pr}_w = \dfrac{\overline{C}_p \mu_w}{\lambda_w}.$

The Eq. (11) describes the experimental data obtained by the aforementioned investigators with a root-mean-square deviation of $\pm 10\%$ and covers the measurement limits of characteristic parameters tabulated in Table 1.

Apart from the abovementioned equations there are known some other empirical equations, such as Bishon equation, for heat transfer to water [8] and Khess-Kuntz equation [3] for heat transfer to hydrogen.

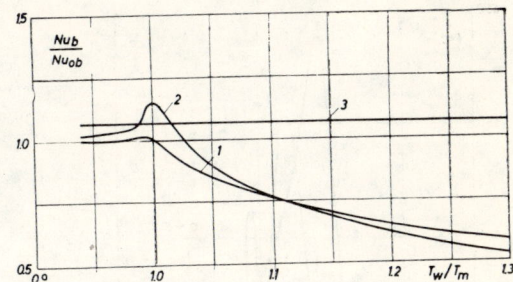

Fig. 8. Dependence of Nu_b/Nu_{ob} versus T_w/T_m for water at $P = 235.4$ bar, $T_b = 573°K$, $\rho \overline{W}_x = 2150$ kg/m² · sec and $d = 9.4$ mm: 1, 2 and 3 — empirical equations (10), (11) and (9) respectively

In Figs. 5, 6 and 8 the empirical equations are compared between each other, with the results of theoretical calculations and directly with experimental data of Fig. 6. The Eq. (9) provides more or less satisfactory results only with the values of $T_w/T_m < 1.05$ for water and with $T_w/T_m < 1.4$ for carbone dioxide. The region of high values of T_w/T_m this equation fails to desribe. The disadvantage of the Eq. (9) likewise lies in the fact that it does not satisfy the requirements of a limiting transition to constant physical properties. The Eqs. (10) and (11) as it has been already stated hereinabove, are well suited to satisfactory describe experimental data and simultaneously are in good agreement with each other and with the results of theoretical calculations. Therefore the Eqs. (10) and (11) may be utilized to calculate heat transfer for the case of heating water and carbone dioxide under the conditions where free convection is of no effect.

The question on the effect of free convection upon the heat transfer with forced turbulent flow has not been yet considered properly. However, the data available are of the evidence that under certain conditions this effect may be of particular concern. Thus, for example, according to the data [32] with flow of water through a uniformly heated horizontal tube with $P = 245$ bars, $\rho \overline{W}_x = 374$ kg/m² · sec and $q_w = 0.45 \cdot 10^6$ W/m² the nonuniformity of the distribution of the wall temperature round the circumference was of the order of 220°C. In the case of water flowing through the vertical or inclined tubes with small value of $\rho \overline{W}_x$ it is expectable that free convection may substantially effect upon the heat transfer in question.

In a number of experimental works [29, 33, 15, 18, 22] there are found to exist the regimes with deteriorated heat transfer. These regimes are the most observable in the Shitsman's experiments [18] for water with upward flow of fluid and pressure $P = 226$ to 245 bars. In the case of heating the fluid with the temperature of the flow through given across the section of $t_b < t_m$, the temperature of the wall $t_w > t_m$ and with a certain combination of the both mass velocity and heat flux a sharply impaired heat transfer and correspondingly (for the case $q_w = $ const.) a rapidly increased wall temperature are occurrable. Obsereved in this case the values of heat transfer coefficient or wall temperature may be described but qualitatively by the herein-above-stated Eqs. (10) and (11).

Figure 9 represents variation of t_w and t_b along the length of the tube with upward flow of fluid for a few regimes with normal and deteriorated heat transfer [18]. With $q_w = 221 \cdot 10^3$ and $281 \cdot 10^3$ W/m² (see curves 1

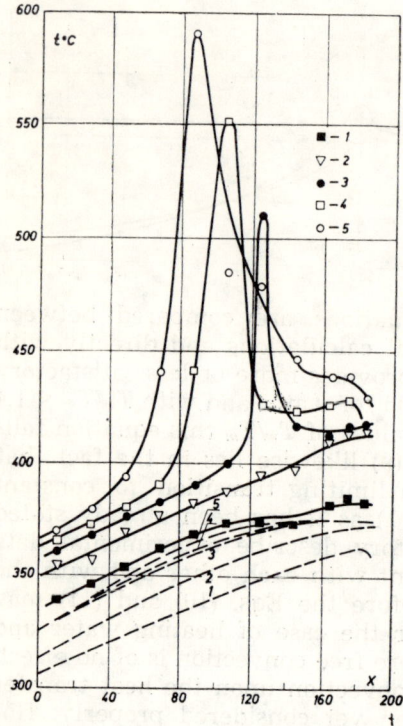

Fig. 9. Distribution of t_w (solid lines) and t_b (broken lines) along the entire length of the tube for water at $P = 226$ bars, $\rho W_x = 430$ kg/m² · sec and $q_w \cdot 10^{-3}$ W/m²; 1) 221; 2) 281; 3) 300; 4) 337; 5) 386

and 2) a standard dependence of t_w versus x is observed. However, even with $q_w = 300 \cdot 10^3$ W/m² (see curve 3) the wall temperature within the section of the tube featuring the length of (10 to 15) d tends to increases sharply and becomes above the expected temperature by approximately 100°C. With further increase of the heat flux (see curves 4 and 5) the temperature of the wall at the point of maximum increases, the maximum t_w tends to shift toward the direction to the tube inlet, while the high temperature region of the wall is expanded covering a considerable section of the tube in question. The heat transfer coefficient taken at the points of maximum t_w tends to decrease by a factor of five as compared with its values for the regimes corresponding to the curves 1 and 2.

The regimes with deteriorated heat transfer are dangerous for the screened tubes of boiler units operating under conditions of supercritical parameters. According to the data by Styrikovich, Margulova and Miropol'sky [38] the deteriorated heat transfer is observed with $\rho \overline{W} <$ $< 2 \cdot 10^{-3} q_w$, where $[\rho W] = $ kg/m² sec, while $[q_w] = $ Kcal/m² hr.

The far from being complete data pertaining to deteriorated regimes of heat transfer are obtained for water and carbone dioxide only. Figure 10 represents limiting values of heat flux q_w for the case of heat transfer to water, i.e. the values at which deteriorated heat transfer originates depending upon the average bulk enthalpy through that crosssection of the flow starting from which (as viewed from the tube inlet) a rapid increase of the wall temperature is observable*. Each particular value of an average mass velocity is its respective straight line, in other words, the relation q_w^{lim} versus h_b may approximately be considered as linear. It is of particular interest that q_w^{lim} with present values of h_b and ρW_x does not depend upon pressure, if anything, within the limits of variation of P from 226 to 343 bars. The region which is lower the respective straight lines represented in Fig. 10 corresponds to normal heat transfer regimes (from beginning of heating up to the section under consideration) which, naturally, may be in fact the outlet section, the region which is above the aforementioned straight lines, corresponds to the regimes with deteriorated heat transfer.

With the origination of the regimes with deteriorated heat transfer observable within a certain range of heat fluxes and other regime para-

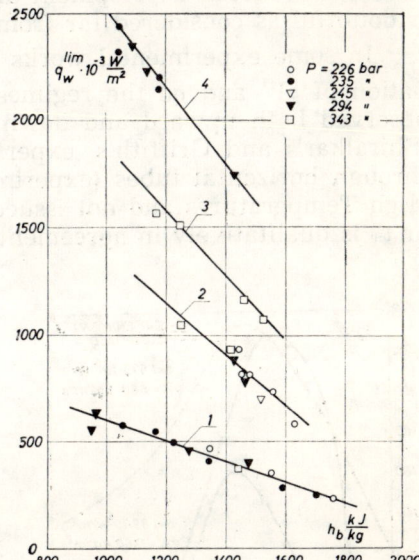

Fig. 10. Limiting values of heat flux for water flowing downward a 8 mm dia. tube with values of ρW_x in kg/m² · sec:
1) 430; 2) 700; 3) 950; 4) 1500

meters are sharp variations of pressure, inlet fluid temperature and wall temperature. Thus, for example, according to data [18] with $P=245$ bars the range of pressure variations amounts to 25 bars.

The occurrence of the regimes with deteriorated heat transfer is associated with a specific feature of distribution of physical properties within the flow of fluid and with a corresponding variation of the flow hydrodynamics, and, in particular, of the turbulent diffusivity processes. In this case a great influence upon the character of motion may be caused by free convection which arises due to great density gradients.

Hall in his recently issued work [39] has made an assumption that the occurrence of regimes with deteriorated heat transfer is associated

---

* The processing of the data [18] represented in Fig. 10 is carried out by Silin.

with the effect of free convection upon the fluid sublayer with upward turbulent motion through a heated vertical tube. He supposed that with certain values of Re and Gr under the action of upward forces the velocity distribution within a sublayer becomes so that the velocity gradient and the shear stress on the outer boundary of this sublayer turns to zero. Under such conditions, from Hall's point of view, the turbulence does not penetrate into the sublayer and the flow of fluid within said sublayer remains laminar. The depth of the sublayer along the flow increases, the film heat transfer coefficient α decreases, while t_w tends to increases. With certain values of Re (before the sublayer) which corresponds to the critical value, the flow in the sublayer becomes turbulent, α tends to increase and t_w, to drop. So, from Hall's point of view, maximums t_w occurs for the regimes with deteriorated heat transfer. In so far as Hall's analysis is concerned, it is of great doubt that a true laminar sublayer with a turbulent flow within a core do exist. The process model under consideration appears to be trustful in the vicinity of the tube inlet (i.e. with simultaneous development of velocity and temperature profiles) and is doubtful as considered far from the tube inlet.

In some experimental works with $t_b < t_m$, $t_w > t_m$ and certain combination of $\overline{\rho W}$ and q_w the regimes with deteriorated heat transfer were observed both upward and downward motions through vertical tubes (Shiralkar's and Griffith's experiments [36]), as well as with motion through horizontal tubes (experiments carried out at the Institute for High Temperatures and not issued yet). The character of variations of α or t_w is qualitatively in agreement with the results of theoretical calcula-

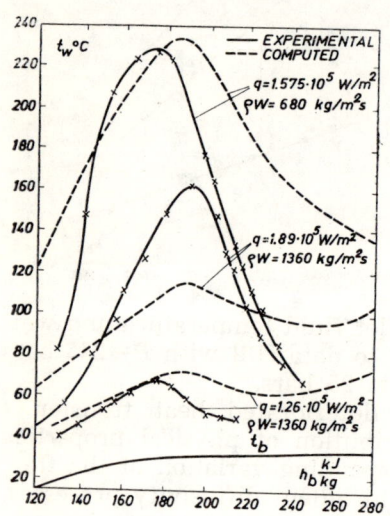

Fig. 11. Comparison of calculated and experimental values of wall temperature for experiments with carbone dioxide

tions carried out without taking account of the effect of free convection for the analogous conditions as compared with the experiment. This is clearly seen from Fig. 11 wherein there are compared the results of the experiment and calculations carried out by Shiralkar and Griffith.

The deterioration of the heat transfer observed in the abovementioned experimental works and approximately predicted by calculations may be explained as follows. Across the section of the tube 1 (see Fig. 12),

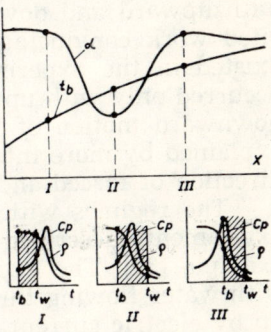

Fig. 12. Concerning onset of deteriorated heat transfer regime

where $t_b < t_m$ and $t_w < t_m$; the fluid density in the flow core and within the wall is of a high value, and, consequently, ρW_x and $\rho C_p \overline{W_y \, 't'}$ in the core and within the close-to-wall region are sufficiently high as well. Therefore across the section l the heat transfer coefficient α is likewise high. Across the section of the tube II downstream the flow, where t_b as before is less than t_m, while t_w has become greater than t_m, the density of the fluid within the close-to-wall layer tends to drop, ρW_x and $\rho C_p \overline{W_y \, 't'}$ decrease which results in decreased α. α tends to decrease more rapidly with the increase of t_c. Across the next section (section III), where $t_b \approx t_m$, while $t_w > t_m$, the fluid specific heat in the flow core increases rapidly, whereas the density remains sufficiently high. Therefore the turbulent heat diffusivity $\rho C_p \overline{W_y \, 't'}$ and, consequently α tend to increase. If the tube is of a sufficiently great length, there occurs across one of the next-to-come section downstream the flow, that t_b is greater than t_m and the heat transfer coefficient will start to drop. Due to the change of α the temperature of the wall involved will likewise tend to change. Across the section where α features its minimum, t_w passes over the maximum. It goes without saying that such character of variations of α or t_w may be observed with sufficiently high values of q_w only. If q_w is small, $t_w - t_b$ will be small across the section where $t_b \approx t_w \approx t_m$ heat transfer coefficient will pass over the maximum.

It follows that the regimes with deteriorated heat transfer may occur even with the absence of substantial efect of free convection upon the forced flow. In other words, the presence of buoyancy upstream forces is not an obligatory condition for the occurrence of the aforementioned regimes. It does not of course follow that buoyancy forces are of no effect upon the origination and the character of the regimes with deteriorated heat transfer. Moreover, under certain conditions this effect may be of a particular concern. Taking account of the effect of buoyancy forces in a theoretical calculation or empirical equation would probably make it possible to obtain better agreement between calculated and experimental values of heat transfer or temperature of the wall in question. The fact that buoyancy forces are of great concern with the occurrence of the regimes with deteriorated heat transfer is dealt with in the experiments by Jackson and Evans-Lutterord [37] recently carried out at University of Manchester. The present authors measured heat transfer to carbone dioxide at a pressure somewhat above critical in a 19 mm dia. vertical tube with upward and downward motion of the fluid in question. The aforecited work convincingly proved that under the same conditions which existed in the experiment, the regimes with deteriorated heat transfer occurred only with upward motion of the fluid and are not available with downward motion of the fluid involved. The latter, probably, may be explained by more intense mixing of the fluid with the mutually opposed direction of forced and free convection within the wall in question.

The regimes with improved heat transfer are found to exist in the experiments described by Goldman [4] and later on in other works for example [35]. The experiments described by Goldman were carried out with water flowing through a 1.58 mm dia. and 203 mm long tube heated up by electric current. Represented in Fig. 13 is a typical graph showing

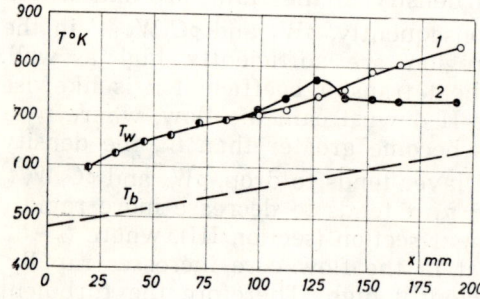

Fig. 13. Distribution of T_w (solid lines) and T_b (broken lines) along the entire length of the tube with the occurrence of an deteriorated heat transfer regime: 1) $q_w = 3.54 \cdot 10^6$ W/m^2; 2) $q_w = 3.67 \cdot 10^6$ W/m^2. Experiments with water at $P=345$ bars, $\overline{\rho W_x} = 2 \cdot 10^3$ kg/sec. m^2 and $d=1.58$ mm

distribution of the fluid and wall temperatures along the length of the tube at $P=345$ bars and $\overline{\rho W_x}=2.10^3$ kg/sec. m^2. With $q_w=3.54.10^6$ W/m^2 (curve 1) the temperature of the wall increases along the entire length of the tube and corresponds to the relations typical for a normal heat transfer regime. However, with a slight increase of the heat flux up to $q_w=3.67.10^6$ W/m^2 (provided the inlet values of P, $\overline{\rho W_x}$ and T_b are constant), the temperature of the wall at a distance from the inlet $x \approx 125$ mm, after a slight local increase tends to drop rapidly and on the length of the left section of the tube amounting to about 50d remains constant. This constant value of T_w, for the particular case equal 733°K, is somewhat above the value of $T_m=691$°K. The heat transfer on the section of the tube with a constant value of T_w is rather higher, while T_w is lower as compared with the values expectable for the case of a normal heat transfer regime.

The occurrence of the regimes with deteriorated heat transfer is always accompanied by the appearance of a strong sound like a squawk or squael, with an oscillating frequency of the order of 1400 to 2200 c/s.

The regimes with improved heat transfer occur only at high heat fluxes (usually above 3.10^6 W/m^2) and with $T_b < T_m$ after that the wall temperature reaches the value somewhat lower or frequently higher than T_m. The values of heat fluxes corresponding to the onset of improved heat transfer, under otherwise equal conditions, tend to increase with the increase of P and $\overline{\rho W_x}$ and tend to drop with the inlet fluid temperature rise. Upon onsetting of the regime with improved heat transfer further increase of heat flux results in the increased wall temperature on the section adjacent to the outlet cross section, the wall temperature within the abovementioned section remining approximately constant on the entire length.

The fact of improved heat transfer is probably associated with a rather marked variations of the physical properties of the medium involved (mostly of ρ and C_p) across the section of the flow at high heat fluxes. Under the conditions wherein the regimes with improved heat transfer are observed, the density of the medium on the wall is several times less and the specific heat is several times as high as in the flow core. The fluid particles which happen to get from the flow core into the hot wall feature due to a turbulent diffusivity a comparatively high thermal conductivity and low specific heat. With larger temperature differences between the sublayer and the particles which got into it from the flow core, these particles as is noted by Goldman [4] are rapidly heated up and expand in a burstlike manner. This process results in a more intense mixing of the fluid within a sublayer, and, consequently, in an increased heat transfer. The specific features of the heat transfer mentioned hereinabove with variable physical properties of the medium are not taken into account by the present theory. Therefore, the theory fails to satisfactory describe the regimes with improved heat transfer.

Some investigators interpret the fact of improvement and deterioration of heat transfer by the analogy with boiling of a subcooled fluid. The first fact is likened to the heat transfer burnout with boiling of the subcooled fluid, whereas the second fact is likened to bubbling of the undercooled fluid. This analogy is based on the following two considerations. The supercritical region may be conditionally subdivided into two areas whose boundary may serve the line of maximums of heat capacity on the isobars. The area corresponding to $T < T_m$ is likewise conditionally understood to feature the properties of a liquid phase, while the area corresponding to $T > T_m$, to feature the properties of a vapour phase. The fact that both of the abovedescribed phenomena occur in the case where $T_b < T_m$, while T_w is usually $\gtrsim T_m$ even with sufficiently high heat fluxes* (i.e. with a substantial difference between the density within the flow core and that on the wall) has given rise to an attempt to explain these phenomena by the occurrence of the processes which are analogous to those occurring with boiling of the subcooled fluid. However, these assumptions need thorough checking and experimental and theoretical substantiation.

---

* The first at relatively small heat fluxes and the second with rather high heat fluxes.

TABLE 1

Specifications of experimental data with respect to heat transfer for water and carbone dioxide at supercritical parameters

Authors		$\dfrac{P}{P_{kr}}$	$\dfrac{T_w}{T_m}$	$\dfrac{T_b}{T_m}$	$Re_b \cdot 10^{-3}$	Pr_b	$q_w \cdot 10^{-6} \dfrac{w}{m^2}$
1		2	3	4	5	6	7
			Water				
Armand Tarasova Kon'kov	[14]	1.01—1.16	0.91—1.12	0.87—1.09	35—180	0.89—25	0.18—0.73
Vikhrev Barulin Kon'kov	[15]	1.1 —1.2	0.6 —1.4	0.5 —1.05	18—800	0.9 —8.7	0.23—1.25
Miropolsky Shitsman	[16, 17]	1.02—1.25	0.81—1.17	0.97—1.12	13—570	0.84—55	0.23—2.5
Shitsman	[18]	1.02—1.11	1.04—1.57	0.85—1.25	—	—	0.27—1.1
Dickinson Welch	[19]	1.09—1.4	0.69—1.35	0.59—1.22	58—860	0.75—16	0.87—1.7
Chalfant Randall	[20, 21]	1.56	0.74—1.6	0.74—1.25	64—1020	0.82—3.3	3.1 —9.5
Swenson Carver Kakarala	[8]	1.03—1.87	0.56—2.1	0.54—1.9	27—680	0.8 —9.4	0.2 —1.8
Bischop Sandberg	[8]	1.03—1.25	0.89—1.35	0.87—1.22	36—530	0.89—16	0.32—3.6
Jamagata Nishikawa Hasegawa Fujii	[22]	1.11	0.84—1.33	0.6 —1.25	12—400	0.9 —15.4	0.12—0.93

1	2	3	4	5	6	7
		Carbone Dioxide				
Krasnoschekov Protopopov [23, 24, 25]	1.06—1.46	0.98—2.6	0.96—1.7	65—500	0.84—40	0.02—2.5
Melik—Pashaev Kobel'kov [10]	1.2 —5.7	1.4 —2.9	0.91—1.2	150—650	1.0 —8.1	to 9.5
Bringer Smith [9]	1.13	0.99—1.06	0.97—1.02	38—270	2.6 —7.4	0.02—0.31
Koppel Smith [26]	1.00—1.03	—	0.95—1.06	30—300	0.9 —11	0.06—0.63
Wood Smith [27]	1.004	0.97—1.05	0.97	910—950	0.85—10.5	0.01—0.25
Hall Jackson Khan [13]	1.51	0.84—1.32	0.84—1.32	52—400	2.5 —10.5	0.002—0.25
Shiralkar Griffith [36]	1.025—1.07	1.07—1.6	0.84—1.13	250—835	—	0.05—0.45
Jackson Evan— Lutterord [37]	1.025	0.95—1.22	0.94—0.99	113—202	—	0.01—0.07

NOMENCLATURE

α	heat transfer coefficient
$\beta = \varepsilon_q/\varepsilon_\sigma$	the ratio of the eddy diffusivity of heat to the eddy diffusivity of momentum
C_p	specific heat with constant pressure
$d = 2r_o$	tube diameter
ε_q	eddy diffusivity of heat
ε_σ	eddy diffusivity of momentum
η	universal co-ordinate
η^+	Goldman's variable
Gr	Grashoff number
h	enthalpy
λ	thermal conductivity
μ	dynamical viscosity
Nu	Nusselt number
ξ	friction factor
Pr	Prandtl number
P	pressure
q	heat flux
Re	Reynolds number
$R = r/r_o$	dimensionless radial co-ordinate
r	radius-oriented co-ordinate
ρ	density
$\overline{\rho W_x}$	average over tube section mass velocity
σ	shear stress
T	temperature in °K
t	temperature in °C
W_x	velocity vector axial component

SUBSCRIPTS

b	properties with average bulk temperature
cr	critical
m	pseudocritical
o	theoretical calculation or constant properties
W	values measured on the wall or properties at the wall temperature

REFERENCES

1. B. S. Petukhov, V. N. Popov, High Temperatures, **1**, 1 (1963).
2. R. G. Deissler, Trans. ASME, **76**, 1 (1954).
3. K. Goldman, Chemical Engineering Progress Simposium. Series Nuclear Engineering, Part I, 50, № 11.

4. K. Goldman, Internat. heat transfer conference, Colorado, USA, 1961.
5. H. Tanaka, N. Nishiwaki, M. Hirata, JSME, Semi-international symposium, Abstracts of papers, Tokyo, Sept. 4—8, 1967.
6. V. N. Popov *Teplo i massoperenos*, v. 1, Izd. Nauka i Tekhnika, Minsk, 1965.
7. H. Re chardt, Arch. Ges. Wärmetechnik, Heft 6/7, 129—142 (1951).
8. H. S. Swenson, J. R. Carver, C. R. Kakarala, Trans. ASME, ser. C, 4 (1965).
9. R. P. Bringer, J. M. Smith, AIChE J. **3**, 1 (1957).
10. Melik-N. I. Pashaev. High Temperatures, **4**, 6 (1966).
11. I. U. Hsu, D. Smith, Trans. ASME, ser. C, 2, 94—104 (1961).
12. E. J. Szetela, ARS J. **32**, 8, 1289 (1962).
13. W. B. Hall, J. D. Jackson, S. A. Khan, Proc. 3rd Int. Heat Transfer Conf. 1, 257—266, 1966.
14. A. A. Armand, N. V. Tarasova, A. S. Kon'kov, »*Teploobmen pri Vysokikh Teplovykh Nagruzkakh i Drugikh Spetsialnykh Usloviyakh*«, Armand A. A., GEI, 1959.
15. Yu. V. Vikhrev, Yu. D. Barulin, A. S. Kon'kov, Teploenergetika, 9, pp. 80—82 (1967)
16. Z. L. Miropol'sky, M. E. Shitsman, ZhTF, **27**, IO (1957).
17. Z. L. Miropol'sky, M. E. Shitsman, *Issledovaniya teplootdachi k paru i vode kipyashchei v trubakh privysokikh davleniyakh*, Atom-izdat, 1958.
18. M. E. Shitsman, H gh Temperatures, 1, 2, pp. 267—275 (1963).
19. N. L. Dickinson, C. P. Welch, Trans. ASME, **80**, 746 (1958).
20. A. I. Chalfant, PWAC=109, June, 1954.
21. D. G. Randall, NDA2=51, Nov. 1956, TID=7529, Pt. 3, Nov. 1957.
22. K. Jamagata, K. Nish kawa, S. Hasegawa, T. Fujii, JSME, 1967, Semi-international symposium, Papers, Heat- and Mass Transfer, Thermal Stress, II, Tokyo, 1967.
23. B. S. Petukhov, E. A. Krasnoshchekov, V. S. Protopopov, Int. Develop. Heat Transfer, Part III, Rep. 67, Inter. Heat Transfer Conf. Colorado, USA, 1961.
24. E. A. Kransnoshchekov, V. S. Protopopov, Van-Fan, I. V. Kuraeva, *Teplo i massoperenos*, v. I, Nauka i tekhnika, Minsk, 1965.
25. E. A. Krasnoshchekov, V. S. Protopopov, High Temperatures, **4**, 3, pp. 389—393 (1966).
26. L. B. Koppel, J. M. Smith, Int. Develop. Heat Transfer, part III, Rep. 69, Int. Heat Transfer Conf. Colorado, USA, 1961.
27. R. D. Wood, J. M. Smith, AIChE J. **10**, 2 (1964).
28. R. C. Hendriks, R. W. Graham, Y. Y. Hsu, A. A Mederios, ARS J. **32**, 2, 244 (1962).
29. W. B. Powell, Jet Propulsion, **27**, 7, 776 (1957).
30. J. P. Holman, S. N. Rea, C. E. Howard. Int. J. Heat Mass Transfer, **8**, 8, 1095 (1965).
31. H. L. Hess, H. R. Kunz, J. Heat Transfer, 1 (1965).
32. M. E. Shitsman, Teploenergetika, 7 (1966).
33. K. R. Schmidt, Mitt. der Vereinigung der Grosskesselbesitzer, H. 63, Dezember, 1959.
34. Yu. V. V khrev, V. A. Lokshin, Teploenergetika, 12 (1964).
35. N. L. Kafengauz, DAN, **173**, 3, pp. 557—559 (1967).
36. B. S. Shiralkar, P. Griffith, Report № 70332—51, Massachusetts Institute of Technology, March 1, 1968.
37. J. D. Jackson and K. Evans-Lutterord, Research Report N E. 2, University of Manchester, March 1968.
38. M. A. Styrikovich, T. Kh. Margulova, Z. L. Miropol'sky, Teploenergetika, 6 (1967).
39. W. B. Hall, Research Report N. E. 1, University of Manchester, Dept. of Nuclear Eng., 1968.

THE EFFECT OF FREE CONVECTION ON HEAT TRANSFER AND FLOW RESISTANCE IN TURBULENT FLOW HORIZONTAL TUBES

A. F. POLYAKOV

Institute for High Temperatures USSR Academy of Sciences, Moscow, USSR

The effect of free convection on local heat transfer and friction with turbulent flow of water in horizontal tubes and at constant heat flux on the wall was investigated experimentally. The experimental data cover the region of values of Re from $8.5 \cdot 10^3$ to $35 \cdot 10^3$ and the values of Gr from $5 \cdot 10^7$ to $5 \cdot 10^9$. It was ascertained that gravitational forces substantially effect the local heat transfer.

The effect of free convection on heat transfer in horizontal tubes especially manifests itself with application to viscous flow [1]. However even with developed turbulent flow $(Re > 8 \cdot 10^3)$ free convection may considerably effect the local heat transfer.

Specific features of viscous-inertial-gravitational flow in horizontal tubes have been investigated in the works [2, 3, 4]. Kirschbaum [2] measured the field of temperature within the flow of fluid and showed that free convection leads to a considerable asymetry in distribution of temperature within a vertical diametral plane with Re up to $12 \cdot 10^3$. The work [3] represents interesting data on the deterioration of stability of viscous-gravitational flow, however due to a small difference of the temperatures between the fluid and the wall a noticeable effect of free convection upon heat transfer with turbulent flow has not been found out. At supercritical parameters of the state, as is stated in the work [4], the effect of free convection is manifested the most and leads to a great difference of the temperatures between the upper and lower generatrices.

We investigated the local heat transfer and friction with viscous-inertial-gravitational flow of water in horizontal tubes and at constant heat flux on the wall. The experimental data were obtained at the described installation in [1] through the use of two test sections involved. The first test section is the round tube made of stainless steel, grade IXI8H9T with 18.8 mm inner diameter and is described in the same work. The second one differed in that it was manufactured from the tube with 50 mm inner dia. and featured no entry isothermic section. The length of the heated part of the first test section was 100 calibres, whereas that of the second, 80 calibres. In both cases heating was effected by an alternating current passed through the tube wall. The local heat flux on the wall along the entire perimeter of the tube involved differed from the average by not more than 5%. Apart from taking measurements of the local heat transfer and friction along the entire length of the test section under consideration some experiments were likewise dealt with taking measurements of temperature distribution within the flow of fluid in the horizontal and vertical diametral planes with $x/d=74$.

The effect of free convection results in deformation of the temperature field within the flow and in redistribution of the local heat transfer around the tube perimeter. Shown in Fig. 1 is an example of fluid temperature distribution across the tube section with $Re=18.2 \cdot 10^3$.

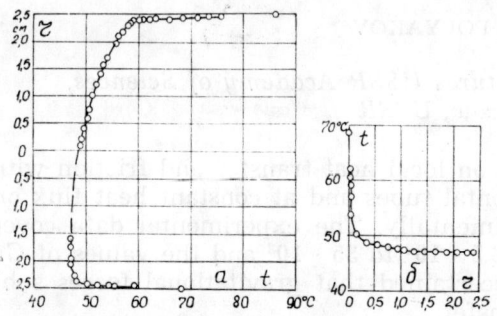

Fig. 1. Temperature distribution within fluid flow
(a) within a vertical diametral plane
(b) within a horizontal diametral plane with $x/d=74$, $Re=18.2 \cdot 10^3$, $Gr=4 \cdot 10^9$; $Pr=3.6$

Figure 2 represents variation of a tube wall temperature on the upper ($\varphi=0$), lower ($\varphi=\pi$) generatrices and with $\varphi=\pi/2$, $\pi/3$ and a bulk temperature of the fluid along the length for various values of Re and Gr. Specific variation of the temperature in close vicinity to the inlet with the points of maximum and minimum is due to formation of a laminar boundary layer and its subsequent transition to a turbulent layer. It is worth noting that at the inlet to the heated test section the degree of turbulence of the flow was sufficiently high, since the chamber featured a turbulizing lattice provided at a distance of $2d$ from the tube inlet. As is seen the effect of free convection starts to manifest itself with comparatively low values of x/d. The transition from a laminar boundary layer to a turbulent one occurs on different generatrices but not simultaneously. The farther from the upper generatrix the higher values of x/d are required for this transition and it occurs no sharp. This fact may be probably explained by attenuation of the turbulence under the conditions of stable stratification of the density within the region adjacent to the upper generatrix and by more intense growth of a laminar boundary layer. With the increase of x/d the effect of free convection is intensified and then the local heat transfer tends to stabilize. Since the degree of flow turbulence at the inlet was not measured, the data concerning the area of stabilization of the local heat transfer may be interpreted as representing the qualitative picture of the process only. Further we shall consider the data pertaining to heat transfer with $x/d>70$.

Figure 3 shows variation of dimensionless temperature of the wall ($T_c=1/Nu$) with respect to the angle read off from the upper generatrix.

In some experiments Nu number varied around the perimeter of the tube more than 3.5 times. For example, with $Re = 10.6 \cdot 10^3$ and $Gr = 9.0 \cdot 10^8$ Nu number on the upper generatrix equals 26, while on the lower generatrix it equals 100. Substantial difference between heat transfer occurring on the upper and lower generatrices is observable with rather higher values of Re. So with $Re = 17.5 \cdot 10^3$ and $Gr = 3.2 \cdot 10^9$ the ratio $Nu/Nu_0 = 2.1$.

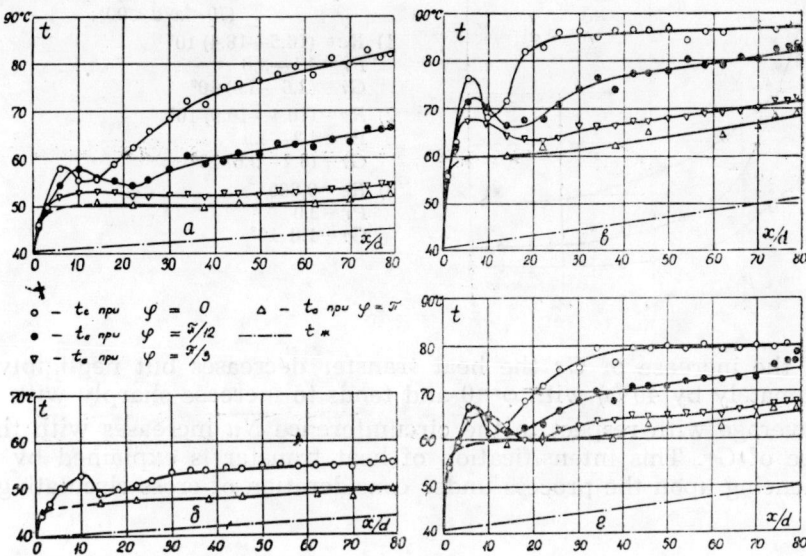

Fig. 2. Variation of wall temperature with $\varphi = 0$, $\pi/2$, $\pi/3$, π and bult temperature of the fluid along the tube length:
  a) $Re = (10.4—10.9) 10^3$, $Gr = (8.2—9.6) 10^8$, $Pr = 4.3—4.0$;
  б) $Re = (12.5—13.0) 10^3$ $Gr = (7.8—8.8) 10^8$;
  b) $Re = (16.5—18.4) 10^3$, $Gr = (2.6—3.9) 10^9$;
  г) $Re = (18.8—20.6) 10^3$, $Gr = (2.8—3.6) 10^9$.

The dependence of the ratio of Nu with $\varphi = 0$, $\pi/2$, π to the Nu_T with purely turbulent (i.e. viscous-inertial) flow versus Re and Gr is represented in Fig. 4. Nu was calculated from the equation obtained in [5]:

$$Nu = \frac{\xi/8 \, Re \cdot Pr}{1.07 + 12.7 \sqrt{\xi/8} \, (Pr^{2/3} - 1)}$$

As is seen from the figure, the effect of Gr upon the local heat transfer is substantial and it considerably depends upon Re. With increase of Re the effect of free convection is reduced rapidly. The character of the dependance of Nu versus Gr is different on various generatrices. So if on the lower generatrix Nu increases, it is decreased with the increase of Gr on the upper generatrix. In this case a relative increase of heat transfer on the lower generatrix is substantially lower than its decrease on the upper generatrix. The decrease of heat transfer in the vicinity to the upper generatrix and the increase of heat transfer close to the lower generatrix occur and with a viscous-gravitational flow (cf. [1]). However in this case

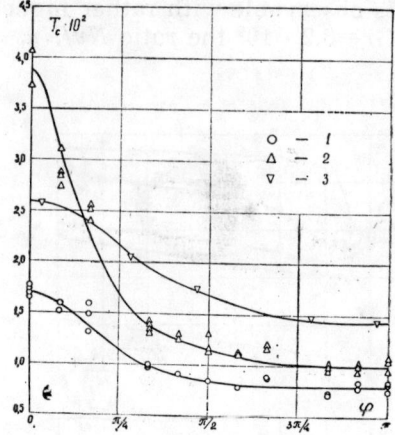

Fig. 3. Variation of dimens'onless wall temperature around tube perimeter
$(70 < x/d < 90)$:
1) $Re = (16.5-18.4) \cdot 10^3$
   $Pr = 4.1-3.6$
   $Gr = (2.6-3.9) \cdot 10^9$
2) $Re = (10.4-10.9) \cdot 10^3$
   $Pr = 4.2-4.0$
   $Gr = (8.2-9.6) \cdot 10^8$
3) $Re = 9.860$
   $Pr = 3.0$
   $Gr = 1.8 \cdot 10^8$

with the increase of Gr the heat transfer decreases but negligibly (approximately by 40%) with $\varphi = 0$ and tends to increase sharply with $\varphi = \pi$. An average with respect to the circumference Nu increases with the increase of Gr. This intensification of heat transfer is explained by great influencing upon the process under consideration of cross circulating cur-

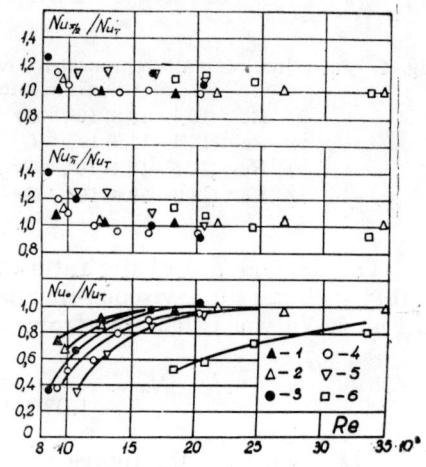

Fig. 4. Dependence of the ratio Nu/Nu_T on the upper ($\varphi=0$), lower ($\varphi=\pi$) an middle ($\varphi=\pi/2$) generatrices versus Re:
1) $Gr = (0.9-1.1) \cdot 10^8$
2) $Gr = (1.6-1.8) \cdot 10^8$
3) $Gr = (4.7-5.5) \cdot 10^8$
4) $Gr = (6.2-8.3) \cdot 10^8$
5) $Gr = (0.9-1.1) \cdot 10^9$
6) $Gr = (3.6-4.1) \cdot 10^9$

rents. While the region adjacent to the upper generatrix is free from the effect of circulating currents. In the case of a viscous-inertial-gravitational flow the nature of the effect of free convection on the local heat transfer in horizontal tubes, as has been stated above, differs greatly from the herein-considered one, and may be explained probably as follows. In the vicinity of the lower generatrix an additional effect of free convection intensifies but slightly the turbulent heat transfer from the wall involved. In close proximity to the upper generatrix under the conditions of stable density stratification a turbulent transfer is made substantially difficult due to opposing gravitational forces. We may suppose that a somewhat attenuation of turbulence occurs in this case. This may be judged indirectly according to the data represented in [6].

Figure 5 illustrates the effect of Gr upon average with respect to circumference Nu and the friction factor in the dependence upon Re. Within the considered range of the numbers Gr with $Re > 25 \cdot 10^3$, $\overline{Nu} = Nu_T$, while with $Re < 25 \cdot 10^3$, $\overline{Nu} < Nu_T$, however this deviation is within the limits of 20%. With higher numbers of Gr and $Re < 25 \cdot 10^3$ the flow resistance tends to decrease that is in agreement with the results obtained for an average with respect to circumference number of Nu.

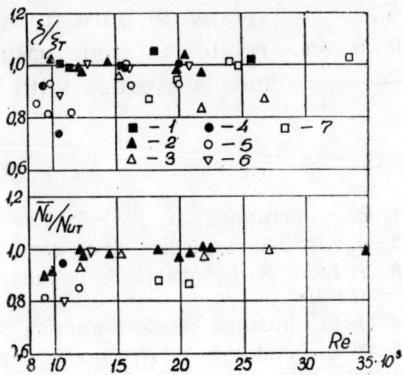

Fig. 5. Dependence of the ratio $\overline{Nu}/Nu_T$ and ξ/ξ_T versus Re:
1) $Gr = 0$
2) $Gr = (0.9—2.2) \cdot 10^8$
3) $Gr = (2.5—4.0) \cdot 10^8$
4) $Gr = (4.7—5.5) \cdot 10^8$
5) $Gr = (6.2—8.3) \cdot 10^8$
6) $Gr = (0.9—1.1) \cdot 10^9$
7) $Gr = (3.6—4.1) \cdot 10^9$

The herein-considered experimental data were obtained with $8.5 \cdot 10^3 < Re < 40 \cdot 10^3$; $5 \cdot 10^7 < Gr < 5 \cdot 10^9$; $2.5 < Pr < 6.5$ for the case of fluid heating. the maximum ration between the absolute viscosity coefficients with a bulk temperature of the fluid and with local temperature of the wall amounted to 1.8.

NOMENCLATURE

d	inner tube diameter
q	heat flux
t	temperature
W	mean velocity
x	distance from initial point of heating
β	volumetric expansion coefficient
λ	thermal conductivity
ν	kinematic viscosity
ξ	friction factor
φ	angle read off from the upper generatrix

$Nu = \dfrac{q_w d}{\lambda(t_w - t_f)}$ Nusselt number $= \dfrac{1}{T_w}$

$Gr = $ Grashoff number $= \dfrac{\beta g q_w d^4}{\nu^2 \lambda}$

Pr Prandtl number

$Re = \dfrac{wd}{\nu}$ Reynolds number

SUBSCRIPTS

f relates to the bulk temperature of the fluid flowing through a given tube cross sectional area
w relates to inner tube surface
T relates to pure turbulent flow
$0, \pi, \pi/2$ relates to angle value
—— means average with respect to circumference.

REFERENCES

1. B. S. Petukhov, A. F. Polyakov, Teplophizika Visokih Temperatur, 1, 2, (1967).
2. E. Kirschbaum, Neues über den Wärmeaustausch, c. 1, t. 28, 361, 1951.
3. Y. Mori, K. Futagami, S. Tokuda, M. Nakamura, Int. J. Heat Mass Transfer, 9, 5 (1966).
4. M. E. Shitsman. Teploenergetika, 7 (1966).
5. B. S. Petukhov, V. N. Popov, Teplophizika Visokih Temperatur, 1 (1963).
6. L. Prandtl. »Hydroaeromechanik«, M., 1952.

TURBULENT FREE CONVECTION HEAT TRANSFER UNDER SUPERCRITICAL CONDITIONS

E. HAHNE

Technische Hochschule München, Institut A für Thermodynamik, Germany

1. Introduction

High intensity cooling as required in gas turbines, nuclear reactors and rocket motors has reactivated the idea of using liquid coolants in the near critical state. At the critical point the specific heat c_p and the coefficient of thermal expansion β assume infinite values, while the kinematic viscosity ν reaches the smallest possible value for a liquid. In the vicinity of the critical point these values are still so extreme that the dimensionless numbers

$$Gr = g\beta\delta^3\Delta t/\nu^2 \quad \text{and} \quad Pr = \nu\rho c_p/k$$

which are formed with these properties also attain extremely high values. The product of the Grashof- and Prandtl-number, the Rayleigh-number $Ra = Gr \cdot Pr$, has been found the characteristic parameter of the flow behaviour in free convection with transition from laminar to turbulent flow to occur around $Ra = 30\,000$.

It was the aim of this investigation to study heat transfer and flow patterns in free convection turbulent flow as it exists in the so-called supercritical region, at pressures above the critical. The experiments were performed in a pressure vessel, with carbon-dioxide in a horizontal, cylindrical gap.

2. Experimental Set Up

The test arrangement is shown in Fig. 1. It consists of a circular borosilicate plate, the heating-plate, with an electrically conducting surface. Bonded to its non-conducting side is a pyrex glass disc, the so-called measuring plate with diametric groves on each side. Into these groves the thermocouples were cemented with their junctions in the center of the disc. A second measuring plate was glued to the upper window to form the upper boundary of the test gap. The gap itself had a height of $\delta = 1.5$ mm and a diameter of 47 mm. It was sealed around its circumference so that fluid exchange with the outside was prevented.

Cooling of the upper gap boundary was provided by a copper coil fastened to the top of the vessel and rinsed with tap water.

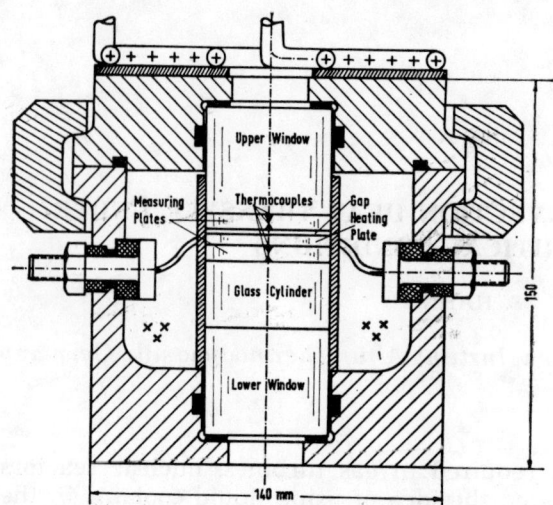

Fig. 1. Section view of the test chamber.

The pressure vessel was mounted over an optical bench. The light coming from a carbon arc lamp was directed through the chamber in such a way that schlieren-images on a screen gave the view normal to the heating plate. For good contrast, colored filters were used.

3. Experimental Procedure

All experimental data were obtained under steady state conditions. At constant pressures of either 75.8 bar (1100 psi), 89.6 bar (1300 psi) or 103.4 bar (1500 psi) the heat input was changed in steps. With the cooling water temperature being constant, a certain temperature distribution in the glass plates was connected to each heat input. From this distribution the surface temperature of the heating-plate t_h and the downward heat loss could be calculated. The temperature of the upper boundary t_c was determined directly.

4. Experimental Results

The experimental results are presented in Fig. 2. The effective thermal conductivity k_e defined by

$$k_e = \Phi \cdot \delta / A(t_h - t_c)$$

is plotted against the mean temperature $t_m = (t_h + t_c)/2$. The three different pressures are indicated as parameters. The coinciding feature in Fig. 2 is the appearance of a maximum in the effective thermal conductivity. This maximum is more pronounced, the closer the pressure is to the critical

(p_{cr}=73.84 bar, 1071 psi). Comparison with the properties of CO_2 shows that the peaks in the effective thermal conductivity occur for the various pressures approximately at the values of the respective pseudocritical temperature.

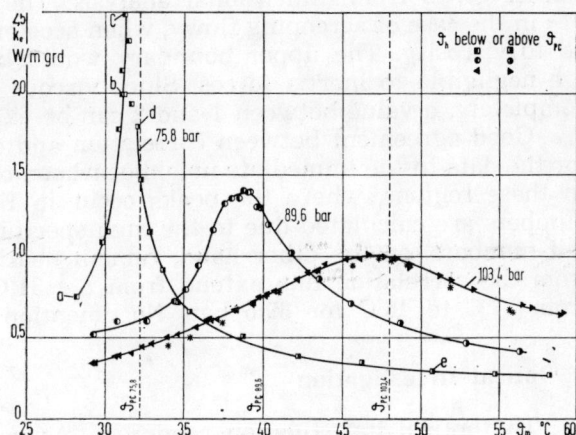

Fig. 2. Effective thermal conductivity k_e vs. mean temperature t_m at various pressures.

A correlation of all experimental data is presented in Fig. 3. In double logarithmic scale the ratio of the effective thermal conductivity k_e to the thermal conductivity k of CO_2 at the respective mean temperature t_m is plotted against a Rayleigh-number multiple. The ratio of thermal conductivities can be considered a Nusselt-number by putting $k_e = \delta \cdot h$.

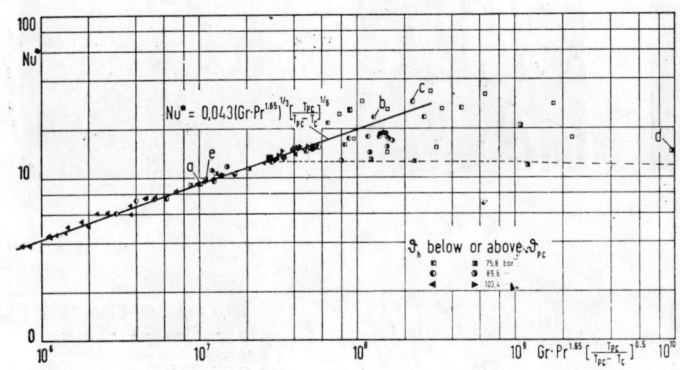

Fig. 3. Correlation of experimental data.

The correlation

$$Nu = 0.043\,(Gr \cdot Pr^{1.65})^{1/3} \cdot T_{pc}/(T_{pc} - T_c)^{1/6}$$

follows a consideration given in the book Grober/Erk/Grigull, and is based on fluid properties at mean temperature. It was set up in a form common to free convection heat transfer in enclosures.

The main difference to correlations for experiments far away from the critical point is the term giving the temperature ratio. The exponent $1/3$ of the Rayleigh-number multiple is an empirical value which was found to be characteristic for turbulent flow in free convection. The exponent 1.65 of the Prandtl-number is also empirical but theoretical boundaries can be given from dimensional analysis. The lower boundary would be »1« in the case of »creeping flow«, when acceleration terms can be neglected to viscosity. The upper boundary would be »2« when viscous forces are negligible to inertia forces. Since viscous forces cannot be neglected completely, a value between 1 and 2 can be expected.

Good agreement between correlation and experiment is found except for the data in the immediate neighborurhood of the pseudo-critical points. In these regions, where the peaks occur in Fig. 2, very high Rayleigh-numbers are calculated due to the high specific heat. The respective Nusselt-numbers, on the other hand, remain nearly constant. The deviations from the correlation line extend from $t_m = 31°C$ to $36°C$ for 75.8 bar, and from $37°C$ to $46°C$ for 89.6 bar. No deviation was found for 103.4 bar.

5. Optical Investigation

The optical investigation rendered flow patterns which showed also characteristic changes with Rayleigh-number variations. In Fig. 4, a series of photographs is presented which was taken at a constant pressure of 75.8 bar at various mean temperatures. The letters correspond to those of Figs. 2 and 3. The photograph 4a is an enlargement of the first picture in the series; all consecutive pictures are also enlargements to the same scale.

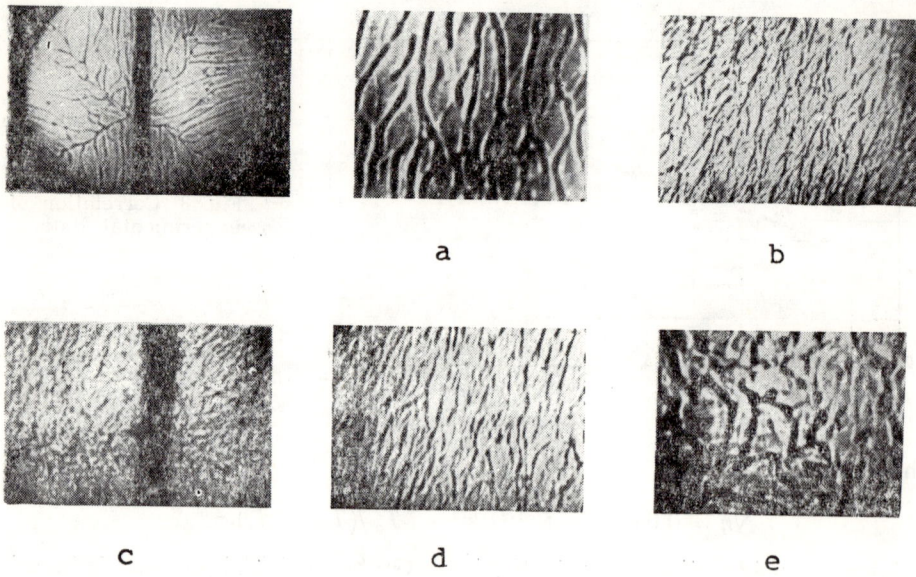

Fig. 4. Steady state flow patterns in supercritical carbon-dioxide at 75.8 bar and various temperatures. (Black l ne: thermocouples).

Streaks of hot fluid show as black lines in the pictures. They move parallel to the heating plate from the edge to the center of the disc. These streaks are not stable, but they appear and vanish and they oscillate sideways. An X-shaped pattern is formed for Rayleigh-numbers which correspond to temperatures below the pseudo-critical. This pattern is very sensitive to disturbances such as tiny leaks around the circumference and inclinations of the gap. With an increase in temperature i.e. in Rayleigh-number, the long streaks become smaller and their motion becomes faster and more irregular. In the vicinity of the pseudo-critical temperature, for very high Rayleigh-numbers, the streaks break up into spindle-like structures which become shorter and thinner the closer the pseudo-critical temperature is approached.

For temperatures higher than the pseudo-critical, the spindles thicken again but remain in a highly irregular motion and do not form longer streaks again. This flow behaviour, demonstrated for 75.8 bar, is also characteristic for pressures of 89.6 and 103.4 bar, although in these latter cases, the »granulation« of the spindles is not as fine as for the former.

6. Conclusions

Concluding it may be stated:

1. The effective thermal conductivity k_e shows the same behaviour as the specific heat; it assumes a maximum near the respective pseudo-critical temperature which is more pronounced, the closer the critical point is approached.
2. Comparison of experimental data with Nusselt-equations gives good agreement except in the immediate vicinity of the pseudo-critical point.
3. The immediate vicinity of the pseudo-critical point can be associated with a flow pattern with spindle-like structures of an especially fine »granulation«. Below and above this region, the spindles are longer and thicker.

TURBULENT MIXING IN SUBCHANNEL ANALYSIS

IVAR DEVOLD

AB Atomenergi, Sweden

1. Subchannel Analysis

INTRODUCTION

Determination of thermal and hydraulic characteristics of a coolant flowing through an irregular flow passage e.g. rod bundles with varying rate of heat input is a complicated mathematical problem. Non-linearity of the basic equations describing the phenomena makes a straight-forward analytical solution of the equations practically impossible. On the basis of some simplifying assumptions it is possible to rewrite these differential equations in terms of differences. Utilising the large capacity of digital computers it is possible to perform stepwise iterative calculations to obtain solutions to the difference equations.

An irregular passage as shown in Fig. 1 is considered here. The thick lines in the figure represent heating surfaces. It is an extreme example

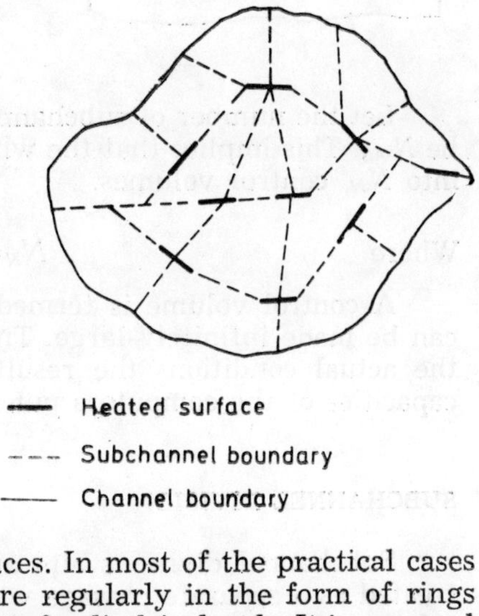

Fig. 1. An irregular flow passage.

— Heated surface
--- Subchannel boundary
— Channel boundary

of irregular distribution of heating surfaces. In most of the practical cases the heating surfaces are distributed more regularly in the form of rings of cylindrical rods, parallel plates or rows of cylindrical rods. It is assumed that the cross section of the flow passages remain constant for all axial locations. However, in practice it should be possible to take into account discrete variations in the cross section of the flow passage axially.

Using a standard lumping procedure in a mixed flow analysis the thermal and hydraulic characteristics of a flow passage are determined in the following way. An equivalent diameter of the flow passage is determined from the given geometrical data. The equivalent diameter is used as the length parameter in the correlations for determining the magnitude of different characteristics like heat transfer coefficient, friction factor, burnout heat flux etc.

In subchannel analysis the whole flow passage is subdivided into a number of smaller passages. These smaller flow passages are referred to as subchannels. Dotted lines in Fig. 1 show how the whole channel is divided into subchannels. In the axial direction all the subchannels are divided into a finite number of discrete steps. Figure 2 shows a sketch of the $(n-1)^{th}$, n^{th} and $(n+1)^{th}$ axial steps.

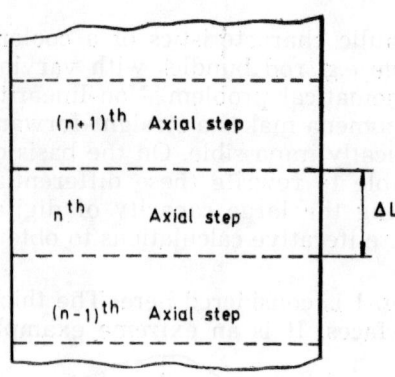

Fig. 2. Axial finite calculation steps.

Let the number of subchannels be N_{SC} and the number of axial steps be N_{AS}. This implies that the whole volume of the flow passage is divided into N_N control volumes.

Where $$N_N = N_{AS} \cdot N_{SC}. \tag{1}$$

A control volume is termed a node. In theory the number of nodes can be made infinitely large. The larger the number of nodes the nearer the actual conditions the results of calculations will be. However, the capacities of the computers put a practical limit on the number of nodes.

SUBCHANNEL DIVISION

Subchannel divisions aim at simpler geometries. Each subchannel is treated as a separate passage where heat transfer and pressure drop correlations for circular pipes are applicable directly or in a modified form.

The hydraulic diameter D_E is used as the length parameter. The hydraulic diameter is given by

$$D_E = \frac{4 A_K}{P_{SCW}}. \qquad (2)$$

Subchannel divisions can be done in many different ways. Five basic types, applicable to rod bundles, are discussed here.

1. Coaxial ring
2. Polygon
3. Similar boundary conditions sector
4. Sectoral
5. Square pattern

The geometrical parameters which are normally needed in the thermal and hydraulic calculations are

1. Wetted perimeter $\qquad P_{SCW}$
2. Heated perimeter $\qquad P_{SCH}$
3. Gap perimeter (pitch/diameter) p/D
4. Equivalent diameter $\qquad D_E$
5. Flow area of the K-th channel A_K

In the coaxial ring subdivisions the rings of rods, the outer wall of the channel and the central feature form the boundaries of different subchannels as shown in Fig. 3. In the same figure an example of a rod cluster divided into 4 subchannels is shown. The subchannels are in the form of coaxial annular passages. With such a subdivision the above mentioned five geometrical parameters can be simply evaluated from such basic geometrical data as

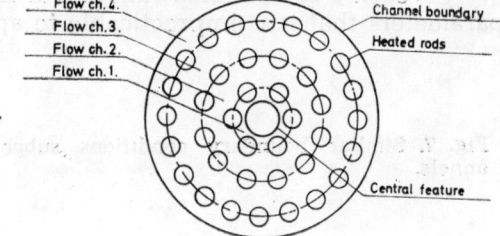

Fig. 3. Coaxial annular subchannels.

1. Central feature diameter
2. Outer wall diameter
3. Rod ring diameters
4. Number of rods in each ring

In the polygonal subdivisions the centers of adjacent rods in the same ring are joined to form polygons. Figure 4 is a sketch of such a subdivision.

Here again the five geometrical parameters mentioned above can be simply evaluated from the basic geometrical data. Intuitively, it seems that a polygonal subdivision is a better form of subdivision than a coaxial

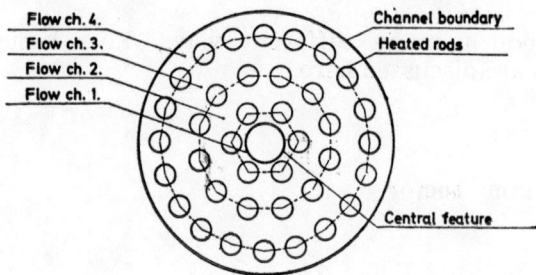

Fig. 4. Polygonal subchannels.

subdivision. The subchannel boundaries in a polygonal subdivision coincide with the minimum-gap line between two adjacent rods in a ring (see Figs 5 and 6).

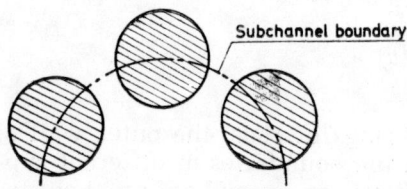

Fig. 5. Coaxial-annular or circular subchannels.

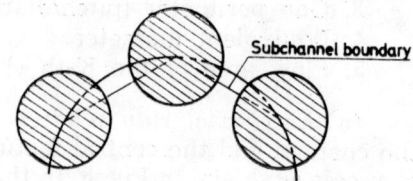

Fig. 6. Polygonal subchannels.

An even better subchannel boundary in the gaps between two adjacent rods would have been the no-shear boundary. *No-shear boundary* is the surface along which the shear stress is zero. A no-shear boundary, however, is such a complicated function of flow, heat transfer and geometrical parameters that it is impracticable to apply such a boundary (see Fig. 7).

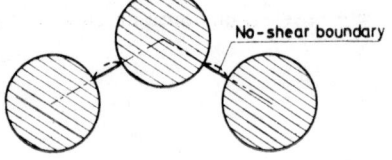

Fig. 7. Similar boundary conditions subchannels.

In the sectoral subdivision the whole rod cluster is divided into units smaller than the ones in the previously mentioned subdivisions. Fig. 8 shows a rod cluster divided into a number of sectors. The sectors which have similar boundary conditions are given the same group number. The total number of subchannels are 62 in the example shown in Fig. 8. The subchannels with identical boundary conditions are isolated into separate groups. So, one is left with 18 groups of subchannels, as against 62 individual subchannels. Each group of actual subchannels is treated as one flow passage or as one subchannel in a subchannel analysis. This finer subdivision is achieved at the expense of simplicity in calculating the geometrical parameters. In a sectoral subdivision the geometrical parameters must be calculated separately and fed as input data into a computer programme.

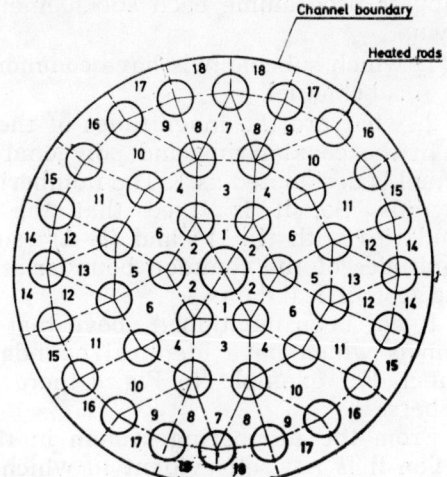

Fig. 8. Sectoral subchannels.

With a *square lattice* the subdivisions are done in a very regular pattern. Fig. 9 shows such a subdivision. Since the geometrical pattern is so regular all the five geometrical parameters can be very simply calculated from the following basic geometrical data

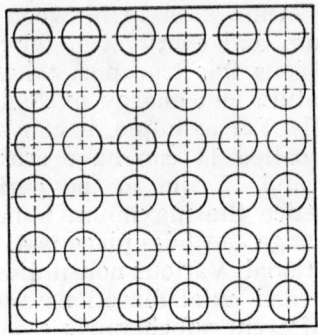

Fig. 9. Quadratic subchannel divisions.

1. no. of rods in each subchannel
2. rod diameter
3. rod pitch.

SUBCHANNEL IDENTIFICATION

One of the basic differences between a lumped-channel or mixed flow analysis and a subchannel analysis is that in a subchannel analysis each flow passage is treated separately and that the heat transfer and flow parameters in one individual channel, say n, are allowed to be affected by the heat transfer and flow parameters in the channels which have common boundaries with the channel n. This necessitates that in a digital computer programme each subchannel should be identified so that it is known,
1. which subchannels have common boundaries with every individual subchannel,
2. what are the magnitudes of the common boundaries.

In the coaxial ring and polygonal subdivisions the innermost subchannel is designated as 1. The numbering of the channels proceed radially outwards. So, it is fixed that the subchannel i will have common boundaries with $(i+1)^{th}$ and $(i-1)^{th}$ subchannels. It is readily seen that magnitudes of the common boundaries are straight forward functions of basic geometrical data.

It has been mentioned above that in a sectoral subdivision the subchannels which have identical boundary conditions are given the same number. For example, in Fig. 8 there are 62 subchannels but only 18 numbers.

From the numbering system in the sectoral method of subchannel division it is not self-evident to which group of subchannels one chosen subchannel have common boundaries with. Such information along with the information about how many subchannels there are in each individual group must be provided separately in the programme in this case.

CROSS FLOW OF MASS, MOMENTUM AND ENERGY

The information regarding which subchannels have common boundaries with each other indirectly determines as to between which subchannels cross flow of mass, momentum and energy can take place. In the coaxial ring and polygonal subdivisions this transport takes place in the radial direction only. Mass, momentum and energy exchange can take place with more than two subchannels. For example, it can be seen that in Fig. 8 the group of subchannels called 11 can have cross flow exchange in the radial direction with 6 and 15 and in the peripheral direction with 12 and 10. In a square lattice the maximum number of boundaries through which cross flow exchange can take place is 4.

In a rod cluster the flow is not undirectional. Various non-uniformities like non-uniform heat input in the radial and axial directions and non-uniform resistance to mass flow in the different subchannels create this disorder. At an axial level the pressure is not uniform at every point in the cross section. Channel 11 in Fig. 8, for example, may have a pressure different from channels 6, 10, 12 and 15. To balance these gradients there will be a continuous mass flow to or from channel 11. This mass flow is generally referred to as cross flow of mass.

DEFINITION OF HEAT TRANSFER COEFFICIENTS

In performing a subchannel analysis, at least three different heat transfer coefficients can be defined. These are:

»True heat transfer coefficient«

$$q = \alpha_\varphi A_H (T_{s\cdot\varphi} - T_{l\cdot m\cdot\varphi}) \qquad (3)$$

»Mean heat transfer coefficient«

$$q = \alpha_u A_H (T_{s\cdot u} - T_b) \qquad (4)$$

»Apparent heat transfer coefficient«

$$q = \alpha'_\varphi A_H (T_{s\cdot\varphi} - T_b). \qquad (5)$$

The mean heat transfer coefficient, α_u, which is the average heat transfer coefficient around the rod periphery, can be obtained from a suitable correlation like for instance:

$$\alpha_u = 0.0197 \frac{\lambda}{D_E} Re_b^{0.82} Pr_b^{0.4}. \qquad (6)$$

The true heat transfer coefficient for a subchannel we then can obtain from the following equation:

$$\alpha_\varphi = 0.0197 \frac{\lambda}{Y_E} \left(\frac{Y_E u_{l\cdot m\cdot\varphi}}{\nu}\right)^{0.82} Pr_{l\cdot m\cdot\varphi}^{0.4}. \qquad (7)$$

The resistance to cross flow of mass, momentum and energy between the subchannels is a complicated function of various parameters of flow, geometry and heat input. In the following chapter an attempt is made to summarize some theories applicable to turbulent flow in subchannel analysis, and to make a comparison between them.

2. Theory of Turbulent Flow as Applied in Subchannel Analysis

GENERAL THEORY OF TURBULENT FLOW

In this section the general theory of turbulent flow in a channel is summarized.

Assumptions:

1. Fully developed turbulent flow.
2. Incompressible fluid.
3. Thermal properties of fluid are constant.
4. Steady state flow in the X-direction (axial direction) of a coolant channel.

The equation of momentum:

$$\rho\left[\frac{\delta u}{\delta t}+\frac{\delta(u^2)}{\delta x}+\frac{\delta(uv)}{\delta y}+\frac{\delta(uw)}{\delta z}\right]=-\frac{\delta p}{\delta x}+\mu\nabla^2 u. \qquad (8)$$

The velocity components of the turbulent flow in the three directions of a rectangular coordinate system can each be divided into an average velocity and a fluctuating velocity as shown in the following three equations:

$$u=\bar{u}+u' \qquad (9)$$

$$v=\bar{v}+v' \qquad (10)$$

$$w=\bar{w}+w' \qquad (11)$$

In the same way the pressure component is divided into an average and a fluctuating component:

$$p=\bar{p}+p' \qquad (12)$$

Timeaveraging the fluctuations is, of course, equal to zero:

$$\overline{u'}=\overline{v'}=\overline{w'}=\overline{p'}=0 \qquad (13)$$

Introducing the equation of continuity:

$$\frac{\delta\bar{u}}{\delta x}+\frac{\delta\bar{v}}{\delta y}+\frac{\delta\bar{w}}{\delta z}=0 \qquad (14)$$

and combining the above equations, we obtain:

$$\delta\left[\frac{\delta\bar{u}}{\delta t}+\bar{u}\frac{\delta\bar{u}}{\delta x}+\bar{v}\frac{\delta\bar{u}}{\delta y}+\bar{w}\frac{\delta\bar{u}}{\delta z}\right]=-\frac{\delta\bar{p}}{\delta x}+\mu\nabla^2\bar{u}$$
$$-\rho\left[\frac{\delta(\overline{u'^2})}{\delta x}+\frac{\delta(\overline{u'v'})}{\delta y}+\frac{\delta(\overline{u'w'})}{\delta z}\right]. \qquad (15)$$

Since we are studying stationary flow in the *X*-direction only, we can further reduce Eq. 15 to:

$$\mu\nabla^2\bar{u}-\rho\left[\frac{\delta(\overline{u'^2})}{\delta x}+\frac{\delta(\overline{u'v'})}{\delta y}+\frac{\delta(\overline{u'w'})}{\delta z}\right]=\frac{\delta p}{\delta x}. \qquad (16)$$

The following definitions are then introduced:

$$\tau_x=-\rho\overline{u'^2}=A_\tau\frac{\delta\bar{u}}{\delta x} \qquad (17)$$

$$\tau_{xy}=-\rho\overline{u'v'}=A_\tau\frac{\delta\bar{u}}{\delta y} \qquad (18)$$

$$\tau_{xz}=-\rho\overline{u'w'}=A_\tau\frac{\delta\bar{u}}{\delta z}. \qquad (19)$$

The stress terms defined in Eqs. (17—19) are called apparent or virtual stresses of turbulent flow. According to the theory of Boussinesq it is assumed that the apparent shearing stresses are proportional to the derivative of the mean velocities with respect to the different coordinate directions. This assumption is expressed on the right hand side of Eqs. (17—19). Introducing Eqs. (17—19) into Eq. (16), we obtain:

$$\rho\left[(\mu+A_\tau)_y\frac{\delta^2\overline{u}}{\delta y^2}+(\mu+A_\tau)_z\frac{\delta^2\overline{u}}{\delta z^2}\right]=\frac{\delta p}{\delta x}. \quad (20)$$

Finally assuming that the channel consists of a circular tube, it is most convenient to transform Eq. (20) into cylindrical coordinates and we then obtain the following expression:

$$(\nu+\varepsilon_r\cdot m)\frac{\delta^2\overline{u}}{\delta r^2}+(\nu+\varepsilon_r\cdot m)\frac{1}{r}\frac{\delta\overline{u}}{\delta r}+(\nu+\varepsilon_\varphi\cdot m)\frac{1}{r^2}\frac{\delta^2\overline{u}}{\delta\varphi^2}=\frac{1}{\rho}\frac{\delta p}{\delta x}. \quad (21)$$

In Eq. (21) the following definition has been introduced:

$$\frac{A_\tau}{\rho}=\varepsilon. \quad (22)$$

The equation of energy:

The energy balance can be superimposed on the momentum equation, and the final equation of energy takes the following form:

$$(a+\varepsilon_r\cdot h)\frac{\delta^2\overline{T}}{\delta r^2}+(a+\varepsilon_r\cdot h)\frac{1}{\rho}\frac{\delta\overline{T}}{\delta r}+$$

$$+(a+\varepsilon_\varphi\cdot h)\frac{1}{r^2}\frac{\delta^2\overline{T}}{\delta\varphi^2}=\overline{u}\frac{\delta\overline{T}}{\delta x}, \quad (23)$$

where $a=\dfrac{\lambda}{\rho C_p}$.

Equations (21) and (23) are the main equations for the special turbulent theory in the following chapter.

SPECIAL THEORIES OF TURBULENT FLOW

Promising attempts of analytically solving the mixing process between subchannels are: a) Kattshee and Reynolds: HECTIC-II, b) Nijsing et al.: Analysis of fluid flow and heat transfer in a triangular array of parallel heat generating rods, c) Rapier: Turbulent mixing in a fluid flowing in a passage of constant crossection, and d) Buleev: Theoretical model of the mechanism of turbulent flow. Only the two first of these are dealt with here.

a) HECTIC-III

The original programme HECTIC-II has been further developed at AB Atomenergi, and is referred to as HECTIC-III. In HECTIC-II the thermal properties of the coolant are assumed constant. In HECTIC-III these properties are variable with temperature and pressure.

Fluid flow calculations

In HECTIC the flow passages can be divided into any kind of subchannels. Symmetry is used to reduce the configuration to the smallest possible symmetrical part. This part is then subdivided into arbitrarily small parts. Examples of subchannel division for triangular arrays are given in Fig. 10. For the k-th subchanel the momentum equation is written:

$$A_k \Delta p_k = f_k L_k A_k \frac{\rho_{av} u_k^2}{2 D_{EK}} + A_{fsk} C_{ds} \frac{\rho_{av} u_k^2}{2} + $$
$$+ 2 \left(\frac{\rho_{av}}{\rho_{out}} - \frac{\rho_{av}}{\rho_{in}} \right) \frac{\rho_{av} u_k^3 A_k}{2} + \sum_{i=1}^{N} F_M \rho \varepsilon_{mik} C_{ik} L_k (u_k - u_i). \quad (25)$$

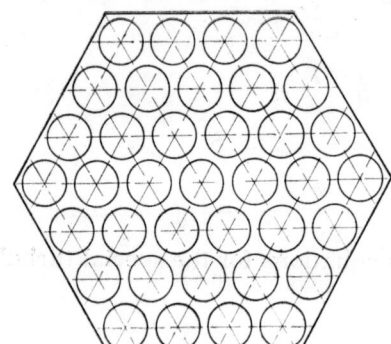

Fig. 10. Sectoral subchannels in a hexagonal channel.

On the right hand side of Eq. (25) the first term is due to wall friction, the second term to drag forces against spacers, the third term to acceleration and the last term is due to eddy diffusion (or mixing) between the subchannels under consideration.

Heat generation calculation

The differential equation for determining the coolant temperature rise is obtained by an energy balance. (The subchannel division regarded is the same as for the momentum equation.)

$$m c_p \frac{dT_k}{dx} = \sum_{j=1}^{M} a_j b_{kj} (T_j - T_k) + \sum_{i=1}^{N} F_H \rho c_p \varepsilon_{hik} C_{ik} (T_i - T_k). \quad (26)$$

Here the first term on the right hand side represents convection from the heated wall surfaces of the subchannel and the second term is representing turbulent convection in passages between the subchannels.

Semiempirical correlations used in HECTIC

In HECTIC the eddy diffusivity is computed in each subchannel as if it is equal to the eddy diffusivity in the center of a circular tube with a Reynolds number equal to that of the subchannel i.e.

$$\frac{\varepsilon}{\nu} = \frac{Re}{20} \sqrt{\frac{f}{8}}. \quad (27)$$

For the mixing between subchannel i and k the eddy diffusivity is thus given by

$$\frac{\varepsilon_{ik}}{\nu} = \left(\frac{Re_i \sqrt{2f_i}}{80} + \frac{Re_K \sqrt{2f_K}}{80} \right) / 2. \quad (28)$$

The Reynolds number in subchannel i is:

$$Re_i = \frac{\rho u_i D_{Ei}}{\mu} \quad (29)$$

and the friction coefficient is found from:

$$f_i = 0.18 \, Re_i^{-0.2}. \quad (30)$$

The eddy diffusivities for heat and momentum are assumed to be equal. However, the magnitude of ε_h and ε_m, can be artificially altered in HECTIC by means of the input data of the programme.

b) **The Theory of Nijsing**

The theory of Nijsing assumes constant thermal properties of the coolant. The theory applies to a triangular array of fuel rods only, as shown in Fig. 11. If this subchannel is further divided into smaller subchannels normal to the wall as shown in Fig. 12, turbulent mixing will only occur in the peripheral direction.

By applying such subdivisions the two first terms in Eq. (21) and (23) can be neglected.

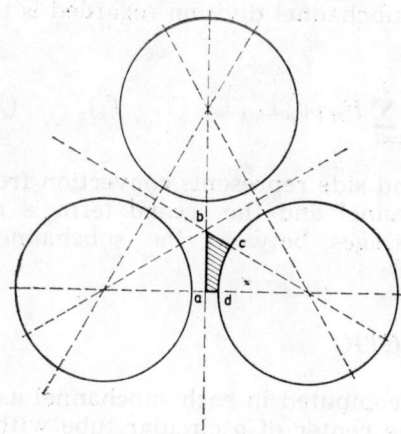

Fig. 11. Section of symmetry for triangular pattern; fuel rod radius = = 0.325 cm; P/D = 1.15; scale = 10×1; section analysed: a b c d.

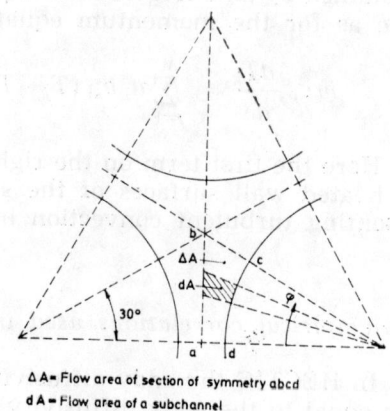

ΔA = Flow area of section of symmetry abcd
dA = Flow area of a subchannel

Fig. 12. Subchannel division inside section of symmetry.

Fluid flow calculation

For a single subchannel the momentum balance takes the following form:

$$\tau_\varphi \, dx \, R \, d\varphi = -dp \, dA + \rho \, d \, xd \left(\varepsilon_{\varphi \, m} \cdot Y_0 \frac{du_{l \cdot m \cdot \varphi}}{dz} \right). \tag{31}$$

friction = − pressure drop + peripheral turbulent mixing. Equation 35 is further illustrated in Fig. 13.

For the whole analysed section $a\ b\ c\ d$, (see Fig. 12), the momentum balance can be written:

$$\tau_{av} \, dx \, R \, \frac{\pi}{6} = -dp \, \Delta A. \tag{32}$$

Combining Eqs. (31) and (32) and introducing the pitch to diameter ratio, p/D, and the hydraulic diameter of both the whole analysed section D_E and the subchannel Ye in question, we finally obtain the dimensionless equation:

$$\tau_\varphi = \tau_{av} \frac{Y_E}{D_E} + \frac{\rho}{\frac{p}{D} R d\varphi} d\left(\varepsilon_{\varphi \cdot m} \frac{Y_0}{R} \frac{du_{e \cdot m \cdot \varphi}}{d\varphi} \right). \tag{33}$$

In order to apply Eq. (33) semiempirical correlations have to be introduced for ε and τ, by for instance applying Eqs. (38) and (41) respectively.

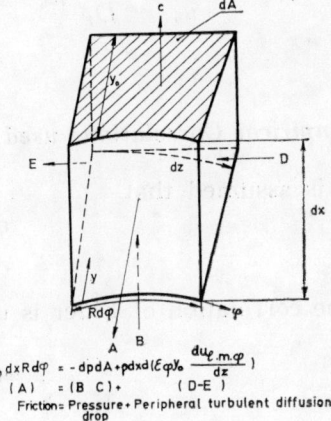

Fig. 13. Momentum balance for a volume element $dAdx$.

$\tau_\varphi dx R d\varphi = -dp dA + \rho dx d (\varepsilon_\varphi Y_0 \frac{d u_{l \cdot m \cdot \varphi}}{dz})$
(A)   = (B C) +   (D-E)
Friction = Pressure + Peripheral turbulent diffusion
           drop

Heat generation calculations

Figure 14 illustrates a subchannel with the different notations applied indicated. By means of this illustration the following energy balance can be established:

$$q_\varphi dx \, R d\varphi = u_{l \cdot m \cdot \varphi} C_p \rho \, dA \, dT - C_p \rho \, dxd \left(\varepsilon_\varphi Y_0 \frac{dT_{l \cdot m \cdot \varphi}}{dz} \right) \qquad (34)$$

Heat flow from   Energy increase   Peripheral
the wall   in the axial   turbulent
                  direction   diffusion

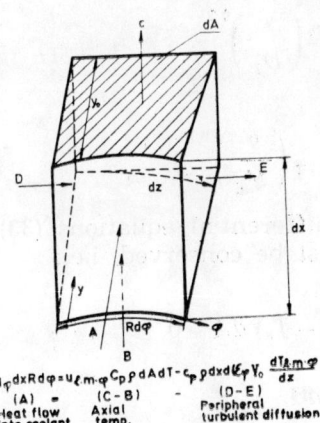

Fig. 14. Heat balance for a volume element $dAdx$.

$q_\varphi dx R d\varphi = u_{l \cdot m \cdot \varphi} C_p \rho \, dA dT - C_p \rho dxd (\varepsilon_\varphi Y_0 \frac{dT_{l \cdot m \cdot \varphi}}{dz})$
(A)   =   (C-B)   -   (D-E)
Heat flow   Axial   Peripheral
into coolant temp.   turbulent diffusion

For the whole analyzed section the energy balance can be written:

$$q A_H = \overline{m} \, C_p \, dT. \qquad (35)$$

Combining Eqs. (34) and (35) and introducing the same geometrical quantities as in the momentum equation, we obtain:

$$q_\varphi = \frac{u_{l \cdot m \cdot \varphi}}{u_b} \frac{Y_E}{D_E} q_{av} - C_p \underbrace{\frac{p}{\rho R}}_{D} \frac{d}{d\varphi} \left(\varepsilon_\varphi \frac{Y_0}{R} \frac{dTl \cdot m \cdot \varphi}{d\varphi} \right). \qquad (36)$$

Semiempirical Correlations used in Nijsings Theory

It is assumed that

$$\frac{\varepsilon_\varphi \cdot m}{\varepsilon_\varphi \cdot h} = 1.$$

The correlation of Elder is used for peripheral eddy diffusivity, i.e.:

$$\frac{\varepsilon}{\nu} = 0.0115 \, Re_b^{7/8}. \qquad (37)$$

The variation in the eddy diffusivity in peripheral direction is expressed as:

$$\frac{\varepsilon_\varphi}{\nu} = 0.0115 \left(\frac{u_{l \cdot m \cdot \varphi}}{u_b} \frac{Y_E}{D_E} Re \right)_b^{7/8} \qquad (38)$$

The Reynold number is based on the bulk properties in the theory of Nijsing.

The equation of Blassius is assumed to apply, i.e.:

$$\tau_u = 0.08 \frac{\rho u_b^2}{2} Re_b^{-0.25} \qquad (39)$$

The above equation can be rewritten as:

$$\tau_u = 0.04 \, \rho \, u_b^{1.75} \left(\frac{\nu}{D_E} \right)^{0.25} \qquad (40)$$

and for as subchannel we can write:

$$\tau_\varphi = 0.04 \, \rho \, U_{l \cdot m \cdot \varphi}^{1.75} \left(\frac{\nu}{Y_E} \right)^{0.25} \qquad (41)$$

The boundary conditions to the differential equations (33) and (36) are that continuity and symmetry must be conserved, i.e.:

$$\int_0^{\Delta A} u_{l \cdot m \cdot \varphi} \, (T_{l \cdot m \cdot \varphi} - T_b) \, dA = 0. \qquad (42)$$

$$at \quad \varphi = 0 \quad \frac{d\theta}{d\varphi} = 0 \qquad (43)$$

$$at \quad \varphi = \frac{\pi}{6} \quad \frac{d\theta}{d\varphi} = 0. \qquad (44)$$

Unfortunately Eq. (44) is not zero at $\varphi = \dfrac{\pi}{6}$ with the subchannel division chosen by Nijsing. This discrepancy is solved by introducing a modified hydraulic diameter. Unfortunately the whole theory is dependent on this modified diameter. This is illustrated in Fig. 15.

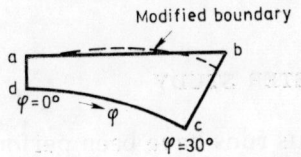

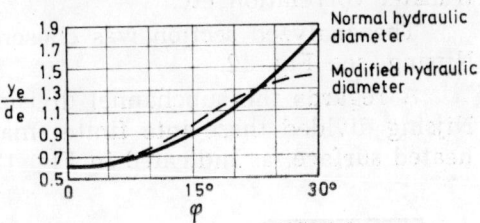

Fig. 15. Modified boundary as applied by Nijsing; P/D=1.10.

c) The Concept of Mixing Lengths

One of the major problems in turbulent flow is to calculate the mixing lengths correctly. Below, some approaches to this problem are shown:

a) HECTIC:

In HECTIC the mixing length is defined as the distance between the centroids between two neighbouring subchannels, measured normal to their common interphase.

b) NIJSING:

Nijsing defined his mixing length as the distance between the midpoints of two neighbouring subchannels measured along the boundary farthest away from the rod periphery.

3. Comparison of Different Mixing Models

INTRODUCTION

In order to develop an analytical model for subchannel analysis of fuel elements for the Swedish steam cooled fast reactor project, some initial studies of existing theories and models have been performed as outlied in the previous chapter.

A comparison between the two versions of the HECTIC programme and the analytical results of Nijsing were therefore performed. Some of the results of this study is described here.

This parameter study resulted in the discovery of some doubtuf aspects of the usually recommended method of arbitrary chosing the subchannles. The study also indicated that Nijsings treatment has certain assumptions which are questionable.

THE PARAMETER STUDY

Numerous runs have been performed with HECTIC II and III, varying the p/D ratio, heat input, mass flow rate, length of axial steps, heat transfer correlation etc.

The analyzed section was chosen identical to the one analyzed by Nijsing, see Fig. 12.

A regards the subchannel division within the section of symmetry, Nijsing divided these into finite small flow areas perpendicular to the heated surface, as indicated in Fig. 12.

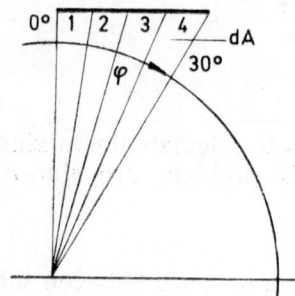

Fig. 16 Subchannel divisions as used in HECTIC.

$A(1) = 0.002300$ cm² $\quad$ $P(1) = 0.0425$ cm
$A(2) = 0.002615$ cm² $\quad$ $P(2) = 0.0425$ cm
$A(3) = 0.003325$ cm² $\quad$ $P(3) = 0.0425$ cm
$A(4) = 0.004505$ cm² $\quad$ $P(4) = 0.0425$ cm

$D_h(1) = 0.2165$ cm
$D_h(2) = 0.2461$ cm
$D_h(3) = 0.3129$ cm
$D_h(4) = 0.4240$ cm

$p/d = 1.151$. $D = 0.65$ cm

A similar subchannel division was applied in the HECTIC runs, by dividing the heated surface into 4 even surfaces in the manner as shown in Fig. 16. The surface heat flux was assumed constant over all the fuel element surfaces while the peripheral conduction in the canning was neglected. The latter assumption was adapted in order to make comparison with literature. However, it is stressed that the heat conducted peripherally by the canning is significant and will make the results presented in this report somewhat conservative. The results from the parameter study show the same tendencies when the p/D ratio is varied, as regards the distribution of heat transfer coefficients, surface-temperatures and coolant temperatures. Of this reason only the results from the runs with $p/D = 1.15$ are reported here. This is a value of particular interest for the steam cooled fast reactor project, being undertaken presently in Sweden. Of this reason the coolant medium has been chosen as steam for this parameter study.

RESULTS

a) Subchannel coolant temperatures

The variation in subchannel coolant temperatures resulting from HECTIC is shown in Figs. 17 and 18.

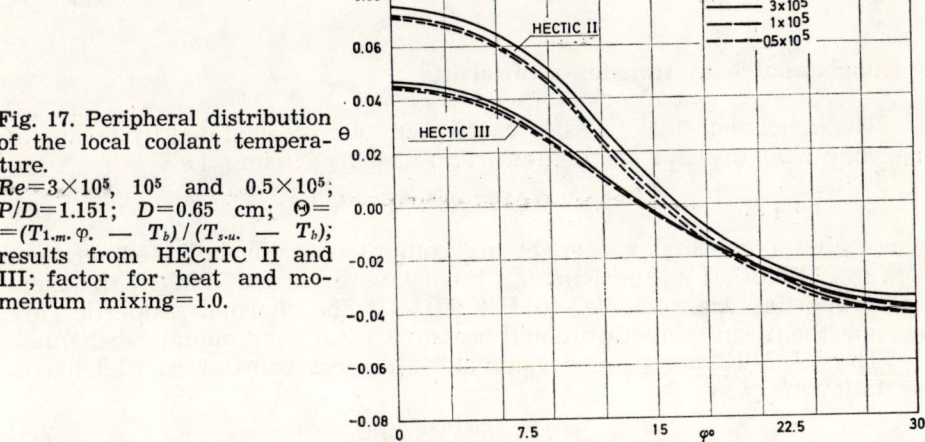

Fig. 17. Peripheral distribution of the local coolant temperature. $Re=3\times10^5$, 10^5 and 0.5×10^5; $P/D=1.151$; $D=0.65$ cm; $\Theta = (T_{1.m.} \cdot \varphi - T_b)/(T_{s.u.} - T_b)$; results from HECTIC II and III; factor for heat and momentum mixing=1.0.

Figure 17 shows the coolant temperature distribution for different Reynolds number when the mixing factors ε_m and ε_h are kept equal to 1.0, while Fig. 18 gives the same distribution for zero mixing (note the difference in the scale of the ordinate in these two figures).

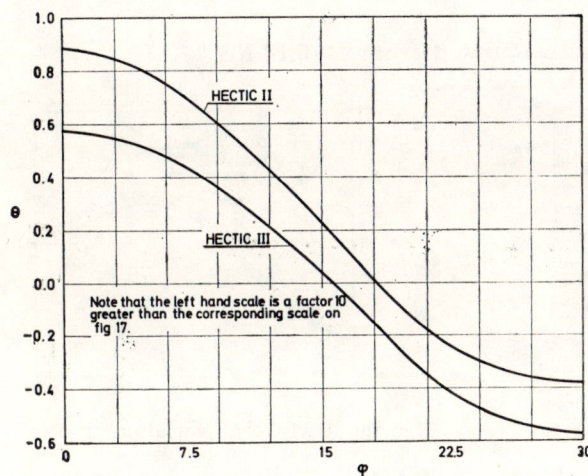

Fig. 18. Peripheral distribution of the local coolant temperature for $Re=3\times10^5$ and zero mixing — $P/D=1.151$, $D=0.65$ cm, $\Theta = (T_{1.m.} \cdot \varphi - T_b)/(T_{s.u.} - T_b)$; results from HECTIC II and III; factor for heat and momentum mixing=0.0.

The dimensionless local mean coolant temperature has been defined as:

$$\theta = \frac{T_{l \cdot m} \cdot \varphi - T_b}{T_{s \cdot u.} - T_b} . \qquad (45)$$

From Fig. 18 it can be seen that the peripheral variation in the coolant temperature is decreased by approximately a factor of 10 when mixing is introduced, and also that the variations in the peripheral direction are smaller in HECTIC III as compared to HECTIC II.

When comparing the dimensionless coolant temperature distribution around the periphery, it is seen (Fig. 17) that the Re is of minor importance in the range from $Re = 0.5 \times 10^5$ to $Re\ 3 \times 10^5$.

b) **Subchannel heat transfer coefficients**

The subchanel heat transfer coefficients were calculated by means of the same heat transfer correlation as applied by Nijsing, i.e.:

$$Nu = 0.0197\ Re^{0.2}\ Pr^{0.4}. \tag{46}$$

where all the thermal properties are computed at the bulk temperature both in HECTIC II and in Nijsing's theory, while in HECTIC III the thermal properties are variable. In HECTIC III the thermal properties are taken at the mean temperature and pressure in each individual subchannel.

The HECTIC programme calculates the heat transfer coefficients in the following way:

$$St = CH \cdot Re^{-EP} \cdot YST \cdot Pr^{-EH}, \tag{47}$$

where YST = adjusting factor.

Equations (50) and (51) can be expressed as:

$$St = \frac{Nu}{Re \cdot Pr} = 0.0197\ Re^{-0.18}\ Pr^{-0.6} \cdot YST. \tag{48}$$

The friction coefficient is calculated from $f = 0.18\ Re^{-0.2}$.

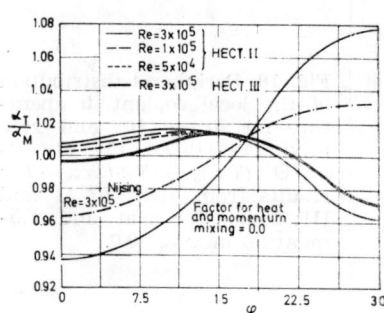
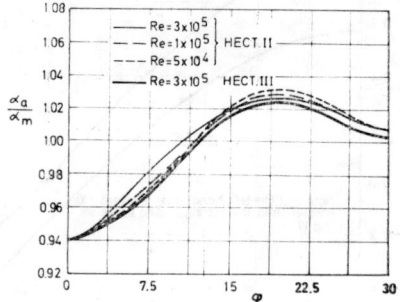

Fig. 19. Peripheral variations in the true heat transfer coefficient — $P/D = 1.15$, $R = 0.325$ cm; results from HECTIC II and III; factor for heat and momentum mixing = 1.0.

Fig. 20. Peripheral variation in the apparent heat transfer coefficient — $P/D = 1.15$, $R = 0.325$ cm; results from HECTIC; factor for heat and momentum fixing = 1.0.

In discussions of heat transfer in connection with subchannel analysis three different heat transfer coefficients are defined as outlined on page 5.

According to these definitions the subchannel heat transfer coefficients can be called true heat transfer coefficients.

In Figs. 19 and 20 the peripheral distribution of the true and apparent heat transfer coefficients are shown for Re numbers of 0.5×10^5, 1×10^5 and 3×10^5. These heat transfer coefficients have been made dimensionless by dividing them with their arithmetic mean value in each case. This is done in order to present dimensionless values in this report.

Figure 19, which shows the variation in the true heat transfer coefficient, shows that the highest heat transfer coefficients are obtained in subchannels 1 and 2 while subchannels 3 and 4 have lower coefficients. This is the opposite tendency of the results reported by Nijsing and is believed to be due to the difference in the way that the calculation of the hydraulic diameter has been performed by Nijsing and in HECTIC. This is dealt with in more detail later in this report. In the same figure (Fig. 19) the distribution of the true heat transfer coefficient has also been plotted when the momentum and heat mixing are set equal to zero. These latter values are in good agreement with theory and give a basis of comparison of the values obtained when mixing is introduced.

The peripheral variation in the apparent heat transfer coefficient is shown in Fig. 20. Here the tendency is the same as that obtained by Nijsing. However, since HECTIC calculates the heat transfer coefficient of each subchannel based on the subchannel local coolant temperature, the apparent heat transfer coefficients are direct results of the true heat transfer coefficients and does not have too much significance in this study.

However, knowledge of the distribution of the apparent heat transfer coefficient around a fuel rod would be of prime importance for the fuel element designer. Such a knowledge would enable the calculation of the temperature distribution around a fuel pin without having to go into subchannel divisions and thereby complicated theories.

Figures 19 and 20 also indicate the fact that the peripheral variation in the local heat transfer coefficient are nearly independent of the Reynold's number for fully developed turbulent flow. This same conclusion can be obtained by studying Nijsing's results.

c) **Variation of the mixing factors**

The HECTIC programme deals with the momentum and heat mixing in such a way that the magnitude of these mechanisms can be altered. This is done in the following way:

The total pressure drop in a subchannel is calculated by means of Eq. (25). In this equation it is introduced an artifical »mixing factor« by means of the constant F_M which is an input data to the programme. By altering this constant the influency of the turbulent mixing can be studied as follows:

Since the total pressure drop across any subchannel must equal the pressure drop across the other subchannels in each axial step, Eq. (25) is used to calculate the velocity in each subchannel and hence the mass flow distribution is obtained. The influency of the momentum mixing on the mass flow distribution can artificially be altered by the factor F_M in this equation. The total energy increase in a subchannel can be expressed by Eq. (30). In the same way, as explained above for the momentum mixing, the influency of heat mixing can artificially be altered by the factor F_H in Eq. (26).

Throughout this study it is referred to these two factor, F_M and F_H, when mentioning the »mixing factors«. Further it has been assumed throughout this study that the magnitude of the momentum mixing and heat mixing always are equal.

The influency of mixing on the coolant temperature distribution has been studied for $Re = 3 \times 10^5$ and mixing factors equal to 0/0, 1/1, 2/2 and 5/5. The results are presented in Fig. 21.

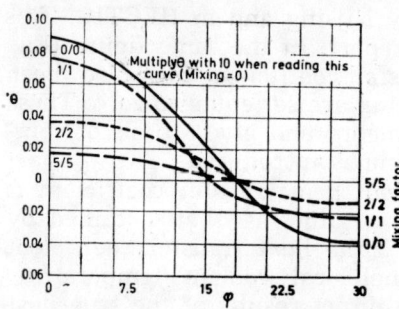

Fig. 21. Peripheral coolant temperature distribution vz., mixing factor — $P/D = 1.151$, $R = 0.325$ cm, $Re = 3 \times 10^5$, $\Theta = (T_{l.m} - T_b) / (T_{s.u.} - T_b)$.

As can be seen from this figure the coolant temperature distribution changes roughly by a factor of 10 when comparing 0 mixing and mixing equal to 1. Further it can be seen that dealing with mixing factors greater than 5 have very little influency on the temperature distribution for Re above 0.5×10^5 and this type of geometry (i.e. $p/d = 1.15$). (See also Figs. 17 and 23).

DISCUSSION

a) Subchannel division and hydraulic diameters

The subchannel division is supposed to be chosen in such a way that each subchannel can be treated as a separate passage where heat transfer and pressure drop correlations for circular pipes are applicable directly or in a modified form.

Subchannel divisions can be done in many different ways. Five basic types, applicable to rod bundles have been defined in chapter 1.

Here only the sectoral pattern is discussed since this is the method applied in HECTIC. The sectoral pattern is further divided into sectorial subdivisions as for instance shown in Fig. 12.

In order to obtain this kind of subchannels, symmetry is first used to reduce the fuel element configuration to the minimum symmetry sector. *a b c d*. Second, the symmetry sector is divided into parts as shown in Fig. 12. This latter subdivision is done arbitrarily as suggested in the Technical Discussion of the original description of HECTIC.

However, the parameter study described in this report indicated that this subdivision can affect the results considerably and does not seem to be arbitrary. An attempt is made to explain this in the following:

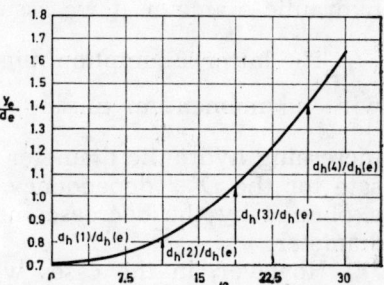

Fig. 22. Variation in the hydraulic diameter with periphery — $P/D=1.151$, $R=0.325$ cm.

In Fig. 22 the rations of the local hydraulic diameter to the hydraulic diameter of the complete symmetry section have been plotted as a function of peripheral position. From this figure it can be seen that the variation in the hydraulic diameter along the periphery is quite severe when having a symmetry section as in this case.

The hydraulic diameter obtained for channel 4 is about twice as large as the one obtained for channel 1 (see Fig. 22). However, if a different subchannel division had been chosen, say by keeping subchannel 1 and 2 as before but chosen subchannel 3 such that this channel included the area between the angles 15° and 30° (Fig. 22). Then the hydraulic diameter of this new section would only be about 1.65 times as big as the one obtained for section 1.

In performing the above mentioned calculations the hydraulic diameter has been computed, using the relationship:

$$D_E = \frac{4A}{P_{SCW}} = \frac{\pi[(\rho^2/\cos^2\varphi)-1]}{2p^2\sqrt{3}-\pi} \qquad (49)$$

where the extreme right hand side of the equation applies to the geometry shown in Fig. 12.

Now, how does the hydraulic diameter influence the heat transfer coefficient?

Assume that the heat transfer correlation is of the form:

$$Nu = k \cdot Re^a \cdot Pr^b \qquad (50)$$

$$\frac{\alpha D}{\lambda} = k \cdot \left(\frac{\rho VD}{\mu}\right)^a Pr^b \qquad (51)$$

$$\frac{w}{A} = \rho \cdot U. \qquad (52)$$

Assume further constant physical properties as is done in Nijsing's theory and in HECTIC II, i.e.

$$\alpha = \frac{K}{D_E}\left(\frac{w}{A}D_E\right)^a = KD_E^{(a-1)}\left(\frac{w}{A}\right)^a \qquad (53)$$

i.e. α is proportional to $D_E^{(a-1)}$. Since $(a-1)$ always is negative and generally in the range from -0.1 to -0.2 for most heat transfer correlations, we see that the heat transfer coefficient will decrease with increasing hydraulic diameter if we assume the $\dfrac{w}{A}$ value constant.

The latter assumption might seem somewhat artificial. In subchannels with no momentum mixing, the $\dfrac{w}{A}$ value will generally increase with increasing hydraulic diameter in such a way that it will merely compensate for the D^{a-1} dependency in the heat transfer correlation. This is explained by the decrease in pressure drop with increasing hydraulic diameter.

However, in the cases where mixing is included, the momentum exchange between subchannels try to flatten out the $\left(\dfrac{w}{A}\right)$ distribution peripherally, and the distribution of the $D_E^{(a-1)}$ value might become significant in comparison with the $\left(\dfrac{w}{A}\right)$ distribution. *This is the explanation why the true heat transfer coefficients obtained with HECTIC have the highest values in subchannel 1 and the lowest values in subchannel 4.* To illustrate this further the w/A distribution have been plotted in Fig. 23 for $Re = 3 \times 10^5$ both for zero and full momentum mixing ($F_M = 0$ and 1.0 respectively).

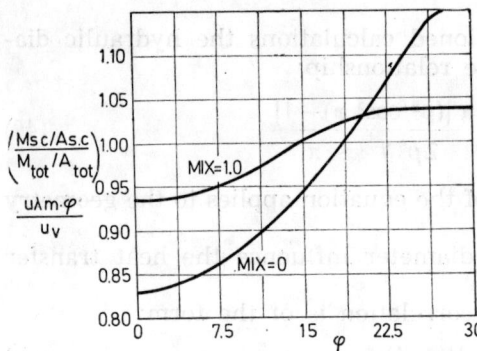

Fig. 23. Peripheral variation of mass per unit area for 0 mixing and for momentum mixing factor = 1 — $P/D = 1.151$, $Re = 3 \times 10^5$, $R = 0.325$ cm.

From the discussion given above it is seen that an arbitrary subchannel division can have serious effects on the heat transfer calculations.

A better subchannel division would have been the no-shear boundary. No-shear boundary is the surface along which the shear stress is zero. A no-shear boundary, however, is such a complicated function of flow, heat transfer and geometrical parameters that it is impracticable to apply such a subchannel division.

Now, the question that reamins unanswered is, how did Nijsing (1, 2 and 3) obtain his logical results with the same type of subchannel divisions? Nijsing goes to the step of making a slight alteration of the flow area as shown in Fig. 15. The total flow area has the same magnitude as the original one as can be seen from this figure. However, when plotting the hydraulic diameter, also shown in Fig. 15, the reslut is somewhat different. The peripheral variation in the hydraulic diameter resulting from the modified area is so much smaller than the »normal area hydraulic diameter« that this efect will result in a peripheral distribution of the true heat transfer coefficient which would give the correct tendency in the cases studied by means of HECTIC, i.e. α_T would increase with increasing Φ. Nijsing's justification for using such a modified boundary is the fact that the mathematical model requires that

$$\frac{d(y_e/d_e)}{d\phi} = 0 \quad \text{when} \quad \varphi = 30^\circ.$$

As has been mentioned earlier the distribution of the true heat transfer coefficient, should logically be so that α_T increases with increasing Φ, for the geometry considered here. Nijsing obtains this by means of the modified hydraulic diameter, while HECTIC has the opposite effect. — However, it is believed that the real justification for the »unlogical« distribution of the heat transfer coefficient in HECTIC is due to the arbitrarily chosen subchannels.

Therefore, it is stressed that in subchannel analysis the choice of subchannel divisions should not be believed to be completely arbitrary as this discussion, hopefully, has shown.

b) Dependence of physical properties

The subchannel analyses adapted in HECTIC II and also by most other existing theories applies constant physical properties for all the subchannels in each axial step. The physical properties are all calculated at the bulk temperature. This simplification is generally adapted since the mathematical analysis of heat and momentum mixing becomes lengthy and complicated when considering the temperature variation in the physical properties around a fuel pin. However, when considering such small spacing between the fuel pins as for instance is done in this study, the coolant temperature variation from subchannel to subchannel is important, especially when dealing with gaseous coolants as for instance steam. Further, if the calculations of the local heat transfer coefficient are made dependent upon the film temperatures, it can easily be seen that the temperature variations might vary considerably from surface to surface.

In order to get an opinion of the dependency of variable physical properties the resluts from HECTIC III is also shown in Figs. 17, 18, 19 and 20. From these figures it can be seen that a smaller peripheral variation in coolant temperatures is obtained when including these variations in the physical properties.

4. Conclusions

The following conclusions can be drawn from the study; when assuming normal heat — and momentum — mixing ($F_M = F_H = 1$).

1) The peripheral variation in the coolant temperature and in the true and apparent heat transfer coefficients is nearly independent of Reynold's number.

2) For Reynold's numbers greater than 0.5×10^5 and geometries with $p/d = 1.15$, mixing factors greater than 5 have no practical significance since the variations in temperatures and heat transfer coefficients around the fuel rods become negligible above this value.

3) The variation in the hydraulic diameter around a fuel pin might have an influence on the distribution of the local heat transfer coefficient, dependent upon how the sectorial subdivisions around the pin are chosen and the magnitude of the exponent of the Reynold's number in the correlation applied (i.e. from 0.8 to 0.9).

4) Conclusion № 3 involves the fact that an arbitrary subchannel division should not be performed. The subchannel division around a fuel pin should be carried out in such a way that extreme differences in the hydraulic diameters should be omitted.

5) The variation of physical properties with temperatures are important when calculating the distribution in the local heat transfer coefficients around a fuel pin in closely packed fuel bundels. This effect should therefore normally be included in a subchannel analysis.

6) The disagreements between the results obtained by means of the theory of Nijsing and the HECTIC programe are mainly due to the influency of the modified hydraulic diameter applied in Nijsing's theory.

7) The HECTIC programme is applicable for turbulent mixing in subchannel analysis in connection with reactor fuel elements. However, the results are dependent upon two important factors which are 1) the subchannel division and 2) the applicability of the eddy diffusivity correlation (i.e. the mixing factor chosen). Recent experimental investigations performed at AB Atomenergi indicate that the mixing factor for smooth rod bundles are dependent on geometrical parameters only.

NOMENCLATURE

A_k subchannel flow area of the k-th channel
A_{fsk} total frontal area of spacers in k-th passage
ΔA flow area of section $a\ b\ c\ d$ (see Fig. 12)
A_τ constant defined by Eq. (26)
A_H heated surface
C_{ds} spacer drag coefficient
C_{ik} mutual interphase distance between passages i and k divided by the distance between the nominal centroids of passages i and k normal to the mutual interphase
C_p specific heat of coolant

D	diameter of fuel rod
D_E	equivalent diameter
D_{EK}	hydraulic diameter of the k-th channel
F_M	artificial mixing factor for momentum
F_H	artificial mixing factor for heat
L_k	length of the k-th channel
M	number of surfaces which are wetted by passage k
N	number of passages with common borders with passage k
N_{AS}	number of axial steps
N_{SC}	number of subchannels
N_N	number of nodes of control volumes
P_{SCW}	wetted perimeter
P_{SCH}	heated perimeter
$Pr_{1.m\varphi}$	Prandtl's number in the subchannel
Pr_b	Prandtl's number based on bulk properties of coolant
R	radius of fuel rod
Re_b	Reynold's number based on bulk properties of coolant
$T_{1.m\varphi}$	local mean coolant temperature of the subchannel
$T_{s.\varphi}$	temperature of surface at angular position φ
$T_{s.u}$	average surface temperature
T_b	bulk coolant temperature
Y_0	distance between rod surface and line of symmetry taken perpendicular to the rod surface
Y_E	equivalent diameter of the subchannel in Nijsing's theory
a	constant defined by Eq. (28)
b_{kj}	wetted perimeter between the k-th passage and the j-th surface
f	friction coefficient
i, k	identification of channels
j	identification of surfaces
n	identification of axial steps
p	pressure
$\bar{p}$	average pressure
p'	fluctuation in the average pressure
p/D	gap perimeter (pitch/diameter)
q	heat flow
u	velocity in the x-direction
$\bar{u}$	average velocity in the x-direction
u'	fluctuation in the average velocity in the x-direction
$u_{1.m.\varphi}$	local mean velocity in the axial (x-)direction
v	velocity in the y-direction
$\bar{v}$	average velocity in the y-direction
v'	flucuation in the average velocity in the y-direction

w	velocity in the z-direction
$\bar{w}$	average velocity in the z-direction
w'	fluctuation in the average velocity in the z-direction
u_b	bulk coolant velocity
m'	coolant flow
a_φ	true heat transfer coefficient
a_u	mean heat transfer coefficient
a'_φ	apparent heat transfer coefficient
$\varepsilon_{r \cdot m}$	eddy diffusivity for momentum in the radial direction
$\varepsilon_{\varphi \cdot m}$	eddy diffusivity for momentum in the peripheral direction
$\varepsilon_{r \cdot h}$	eddy diffusivity for heat in the radial direction
$\varepsilon_{\varphi \cdot h}$	eddy diffusivity for heat in the peripheral direction
ε_{hik}	eddy diffusivity for heat between channels i and k
ε_{mik}	eddy diffusivity for momentum between channels i and k
τ_x	normal stress in the x-direction
τ_{xy}	shear stress acting on the x—y surface
τ_{xz}	shear stress acting on the x—z surface
μ	dynamic viscosity
∇	Laplacian operator
ρ	density of coolant
λ	thermal conductivity of coolant
ν	kinematic viscosity of coolant
φ	angle in radians

REFERENCES

1. N. Kattchee and W. C. Reynolds, HECTIC-II. An IBM 7090 Fortran computer program for heat transfer analysis of gas or liquid reactor passages, IDO—28595, 1962.
2. R. Nijsing, I. Gargantini and W. Eifler, Nucl. Eng. and Design, 4, 375—398 (1966).
3. C. B. Mayer, Risö report № 125, 1966.
4. A. C. Rapier and L. White, TRG 538 (w), 1963.
5. N. I. Buleev, AERE-Trans 957, 1963.
6. D. S. Rowe, BNWL—371 (PT. 1), 1967.
7. A. C. Rapier, TRG 1417 (w), 1967.

MEASUREMENTS CLOSE TO THE WALL IN A TURBULENT BOUNDARY LAYER

Z. ZARIĆ

*Boris Kidrič Institute of Nuclear Sciences,
University Beograd, Yugoslavia*

1. Introduction

It is the ultimate aim of all research to develop methods of prediction. In turbulent flows this is a difficult task.

Turlubence is a statistical phenomenon. All the quantities involved are therefore random variables characterized by probability distributions. The complete solution of the turbulent flow problem would then consist in the determination of the evolution in time of the probability distributions of the hydrodynamic, enthalpy, concentration and other fields, starting from a known set of the distribution functions at the initial moment, by employment of the corresponding fundamental equations. The task, such defined, is unfortunately too difficult at present.

Normal practice is therefore to use only the first and second moments of the distribution functions to characterize a random field, instead of the distribution function itself. By doing so the solution will be evidently incomplete if the distribution functions involved are not normal, which seems to be the case, at least in boundary layer flows. Even with this simplification we are still unable to close the system of equations without making some additional hypothesis concerning the flow structure. Most often the hypotheses are made concerning the relation between the Reynolds stresses and the mean motion, thus implying that the mean motion alone determines the flow. Experiments have shown that the ignorance of the influence of turbulent motion is possible only in simple flows. There is a possibility to provide more equations for the Reynolds stresses, making the solution dependent on some of the turbulence characteristics, but this involves the appearance of the additional unknowns in form of the higher order moments, with the net result of increasing the difference between the number of unknowns and the number of equations (19). Only reason for doing this is evidently the possibility of making more plausible hypothesis for the higher order moments.

Therefore, all the present methods of prediction have to be semi-empirical. However, there is an increasing need for even the semi-empirical methods of prediction, provided they are sufficiently reliable.

This can be seen from the example of the rough surfaces problem, which is very important in many applications. A great number of laboratories and individuals have for years been doing the research in this direction. The result is that whenever a specific type of roughness is employed an experiment has to be made. This is only one example, for modern technology demands the use of more and more complex flow configurations.

On this meeting a comprehensive and up-to-date review of the present methods of prediction of the turbulent boundary flows, based on various hypotheses, has been given. From the standpoint of the physical model employed these methods could be divided into two categories. In the first category are the methods based on hypothesis concerning the mean motion alone. Those of the second category employ in addition some existing information on the structure of turbulent motion.

It has been shown on this meeting and elsewhere that a surprisingly large number of boundary layer flows, including the effects of compressibility, heat and mass transfer, chemical reactions etc., could be predicted by the methods of the first category. However, it has been also shown that in some flow configurations this is not the case. That in some flows we are in need of the methods of the second category, with a corresponding need for more information concerning the turbulence structure.

Further it has been shown that even the methods of the first category are still lacking some information concerning the mean flow. As a good example data on the mean temperature distributions could be taken. Naturally, the situation is much worse in case of the information on the turbulence structure necessary for the second category methods. In both cases the really interesting region from the standpoint of transport processes, the viscous sublayer, has only recently yielded some of its incomplete details to the investigator.

Consequently the experiment still plays a very important role in the development of the predistion methods of the turbulent boundary layer flows: in providing necessary information, in checking the fitness of the method and even sometimes in inspiring new physical models. Only, the advanced stage of the development of prediction methods requires a correspondingly higher level of the experimental techniques in question. We are in need of experimental methods capable of producing information which is reliable, precise, local and dynamic in character. And all this in highly delicate conditions of the very wall vicinity, i.e. inside the viscous sublayer. By reliable it is meant as free as possible of any secondary disturbing effect, as calibration for instance in case of indirect methods. By precise and dynamic it is meant capable of providing precise data on dynamic characteristic in the important range of frequencies. And by local capable of doing point measurements.

It is the purpose of the present lecture to give a short critical review of existing measurement methods of relevant quantities which more or less satisfy the above requirements. It is restricted to the measurements in single phase, incompressible flows along the solid walls. As transport phenomena are most intense in the viscous sublayer and in its vicinity, we shall concentrate on methods enabling measurements very close to the wall. In the last section come possibilities of these methods will be illustrated by recent experimental results.

2. Methods of Measurement

2.1. WALL SHEAR STRESS

Determination of the local wall shear stress is of greatest importance since knowledge of its variation is required in all the prediction methods, including the simplest ones. It is therefore rather surprising to find out that, except in simplest flows, there is a general uncertainty on the reliability of the experimental data.

This comes out as a consequence of the fact that local wall shear stress is generally determined — not measured. The only employed direct technique, the floating element, is not really local in character, except for large, plane and smooth surfaces, because of its dimension. So the indirect, substandard methods, are extensively used. Of these, manometric methods and the methods based on heat or mass transfer analogies are most popular. All these, however, being substandard require careful calibration. And in order that the calibration could be done relatively easy, in simple or even laminar flows, these methods are all based on some hypothesis concerning the velocity, or velocity and temperature, or velocity and concentration, distributions in the inner wall layers.

Of the manometric methods, Preston and Stanton tubes have the widest use. Both are based on the universality of the velocity profile in the inner wall region. Of the two, the Stanton tube is less restrictive, requiring, due to its small dimensions, only the linearity of the profile in the wall vicinity. Even so Bradshaw [4] has shown that the calibration of the Stanton tube in laminar flows gives an error of about 30% when using it in turbulent flows. As possible explanations Bradshaw mentions the disturbancies caused by the tube edge or the dynamics of the viscous sublayer in turbulent flows. If this is so, it is doubtful whether calibration in simple flows, even turbulent, will give reliable results in the flows characterized by high intensity turbulence, for instance, which might have different turbulence structure in wall layers.

Methods based on heat or mass transfer analogies have the advantage of the negligible disturbance to the flow. So the thin hot film technique [3] and electrochemical methods [5, 18] have an extensive use lately. However the reliability of the calibration in simple flows in case of these methods is still more in question than in the case of manometric methods, as they are based not only on the hypothesis concerning the velocity profile but also on the hypothesis concerning temperature or concentration profiles.

It follows that indirect, substandard methods, could be employed with a fair degree of reliability only if the calibration is done in similar conditions. Only this requires another reliable method by which to calibrate. For these reasons lately, more and more authors turn back to the definition of the wall shear stress:

$$\tau_s = \mu_s \, (dU/dy)_s$$

and determine it from the velocity gradient at the wall [13, 15, 29, 33]. However this is not a simple task as the determination of the velocity gradient at the wall requires the entrance in the viscous sublayer by a micrcne dimension probe, and a corresponding precision in both velocity and distance measurement. Whether this method is qualified as direct or indirect depends on how the velocity measurement is done.

Regarding the dynamic characteristics of mentioned methods, manometric method could not be employed owing to its inherently poor sensitivity. On the other hand, the sensitivity of heat or mass transfer methods could be excellent. The micron thick heated films have a response time of the order of 1/100 microseconds, for instance. Very good dynamic response of these methods enable the determination of the statistical characteristics of the wall shear stress. Depending on the dynamic characteristics of the employed velocity measurement technique the method based on the determination of the velocity gradient on the wall could provide statistical characteristics of the wall shear stress too.

2.2. VELOCITY MEASUREMENT

Visualization techniques have been developed lately into valuable quantitative methods capable of the direct determination of velocities without causing any, or negligible, disturbancies to the flow. Some excellent results obtained with the stroboscopic technique using micron particles are presented on this meeting [8]. Very good results have also been reported lately using hydrogen bubble technique [24] as well as the flash photolysis technique [23]. To these laser-doppler technique, recently in development can also be added. So, a number of direct velocity measurement methods do exist.

Of the indirect methods, which have been used widely for years, we won't discuss the manometric method for it is not suitable for really point measurements close to the wall, and in addition has very poor dynamic characteristics. The hot-wire and hot-film techniques are most often used now-a-days for the velocity measurements near the wall, and for velocity fluctuations measurements. In liquid flows the electrochemical anemometry has been also used with success lately. However, this method does not differ in principle from the hot wire anemometry.

In what follows the relative merits and shortcommings of the hot wire technique, as opposed to the visualisation methods will be discussed to some extent, from the standpoint of: calibration and flow disturbance, wall and support effects, velocity direction determination, dynamic sensitivity and measurement statistics.

Hot wire method, being a substandard method, has an obvious disadvantage of requiring a careful calibration. The hot wire calibration is not a simple task, especially concerning the flow temperature depending effects, but it could be performed successfully with some care. To this disadvantage the signal nonlinearity effect has to be added, which however could be taken care of by electronic means or directly by use of the calibration curve. The influence of the velocity fluctuation components normal to the flow direction is another undesirable aspect of the hot wire anemometry. The signal from the wire corresponds to the intensity of the instantaneous velocity vector. This influence is still poorly understood.

The disturbance to the flow of a micron thick wire does not present a real problem. On the contrary, the effects of a near-by wall and the wire supports are far to be neglected and present serious difficulties for the measurements close to the wall. It is a known fact that near the wall heat transfer from the wire is sensibly increased due to the combined effect of the conduction and the modification by the wall presence of the heat convection from the wire. An example of the wall effect is given in Fig. 1 where the curve K_o represents wire response obtained by ap-

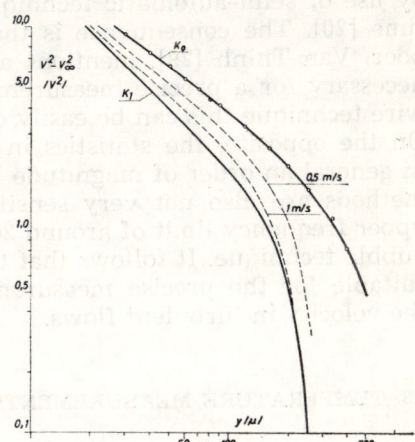

Fig. 1. Correction to hot-wire reading in the wall vicinity $K_0 — U=0$; K_1 — Wills correction (45).

proaching the wire to a steel wall in still air. The curve is very reproductible and can be successfully used as an efficient means for the precise determination of the distance from the wall, in conditions when optical methods are difficult to use. When the flow is present the corrections to the wall due to Wills [30] are often used. Following Coantic [6], curve K_1 in Fig. 1 is obtained for a given wire employing measured curve K_o and the Wills corrections. The corrections thus determined do not give satisfactory results for very low velocities close to the wall. In fact, it is not very plausible to suppose that the wire response for the zero velocity is given by the curve K_o and for velocities slightly larger than zero by the curve K_1. Fairly satisfactory results are obtained by empirical interpolation curves for low velocities, shown on the Fig. 1., Only this could not be very satisfactory since it is based on a great deal of empiricism. Thus the problem is still open.

Lately another undesirable effect is reported by Van Thinh [28]. It was determined that the inclination of the wire supports is influencing the signal from the wire, evidently caused by the conduction from the wire to the supports. Fortunately this effect could be taken care of by the appropriate positioning of the wire supports to the flow direction.

Another, perhaps most serious shortcoming of the hot wire technique is the insensitivity of the signal to the flow direction. Thus the determination of the mean and fluctuating velocity components normal to the flow direction, although in principle possible with X or slanted wires, is by no means local in character and is impossible in the wall vicinity.

Visualization methods, being direct, do not require any calibration. Moreover, no correction on the wall vicinity has been reported as necessary. And some of them are quite capable of the velocity component determination, i.e. of the determination of the instantaneous velocity vector. Being highly advantageous in the above respects the visualization methods are, however, not free of some shortcommings. Apart from being restricted to liquid flows, at least at present, the most serious disadvantages of these methods are connected with the difficulties in the application. Even by use of semi-automatic techniques, data elaboration takes considerable time [20]. The consequence is that the statistics is in general relatively poor. Van Thinh [29], mentions a 1000 points out of 2000 measurements necessary for a precise measurement of a mean point velocity. With hot wire technique this can be easily done, even an order of magnitude better. On the opposite, the statistics in present day visualization techniques is in general an order of magnitude lower. Besides, some of the visualization methods are also not very sensitive in the whole frequency range. An upper frequency limit of around 20 cps is reported in case of the hydrogen bubble technique. It follows that the visualization techniques are not very suitable for the precise measurement of a basically random variable as the velocity in turbulent flows.

2.3. TEMPERATURE MEASUREMENTS

Unlike hydrodynamic measurements, all temperature measurements are made by direct methods employing the same instruments which define the temperature in the International temperature scale. Of the limited experimental evidence in turbulent boundary layers the largest part has been provided by thermocouples. However, the dimensions of thermocouples and the necessary corrections in most applications make them unsuitable for the measurements close to the wall, and the poor dynamic response makes them unsuitable for dynamic measurements. Use of thin thermocouples has been reported but even with a 25 micron one, serious difficulties have been encountered regarding the necessary corrections due to the conduction to the supports.

The only other existing method which does not have shortcomings regarding dimensions and dynamic response is based on the resistance thermometer in form of a micron thick wire. After careful aging of the wire the reproductibility of this instrument is very good [26]. Besides, for relatively small temperature differences the wire response is sensibly linear. Sufficiently thin wire has a fair dynamic response up to approximately 1000 cps.

In spite of the obvious advantages, measurements employing this technique in boundary layer flows have been reported by but a few authors [10, 25, 32]. A discussion of this method is presented by Maye on this meeting, together with some preliminary results [16]. The method is not free of necessary corrections, the main being due to the conduction from wire to the supports. Necessary correction due to this effect could be considerable at high temperature gradients encountered close to the wall. As an example the magnitude of necessary corrections is given in

Fig. 2 for a one milimeter long, 5 micron thick tungsten wire, welded directly to 0,5 mm thick supports. Necessary corrections could be made smaller if instead of a naked wire, a »Wollaston« wire etched at the center is used [16]. However, unless the wire is very long the corrections are still necessary.

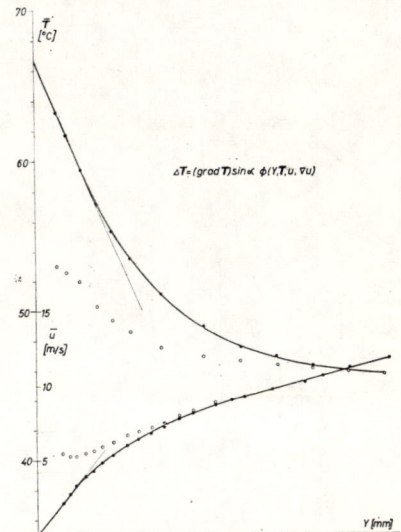

Fig. 2. Temperature and velocity corrections in the wall vicinity O —, uncorrected; — corrected.

The advantage of the method is also in the possibility of measuring instantaneous velocities and temperatures with the same probe. A technique has been proposed, based on the constant current anemometry, in which by supplying the wire with three different currents it is possible to determine simultaneously temperature, and velocity fluctuations and velocity-temperature correlations [11, 17]. In practice, due to a very low accuracy of the method measurements of much more then three currents were necessary to obtain fair results.

Another possibility for velocity and temperature measurement with the same probe is employed by the author and presented in Fig. 3 [32]. To eliminate lead resistance, 5 micron tungsten wire (1) is by four leads connected either to a high precision Mueller bridge (3) or to a constant-temperature anemometer (2). Mean temperature is measured by the bridge. For temperature fluctuation measurements the out-of-balance signal from the bridge is amplified by a low noise DC-amplifier (8a), with the R.m.s. value determined by a R.m.s. meter (4). Signals from the bridge or from the anemometer could also be registered on a tape recorder (9). Signals could be afterwards analysed by frequencies (filter-13, squarer-14, ADC converter-10a, scanner and timer-15), or using the data acquisition system (11) and a digital computer (16) probability distributions of signals could be determined. Unfortunately, simultaneous velocity and temperature measurement is not possible using this method.

Employing a wire thermometer sufficiently close to the wall, which could easily be done because of its dimensions, local surface temperature and wall heat flux could be determined from the slope of the temperature profile at the wall. This is an important advantage in flows with rapid

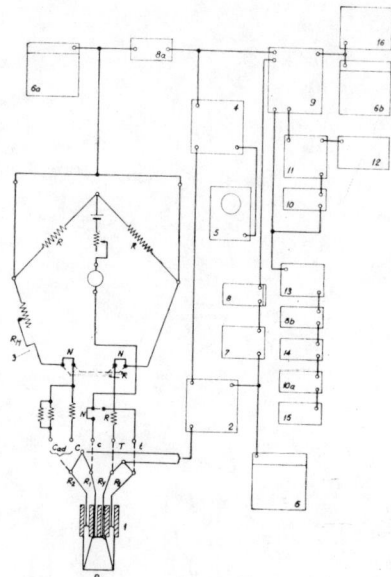

Fig. 3. Schema of the circuits for temperature and velocity measurement and statistical analysis.

temperature or heat flux variations along the surface. Local heat flux meters, employed in case of constant wall temperature conditions, are not very local in character due to its dimensions. Only recently heat transfer sensors employing thin hot film technique are used [1] having smaller dimensions and good dynamic characteristics.

3. Experimental Evidence on the Viscous Sublayer

Only five years ago, in a review article, Kestin [12] stated that the viscous sublayer is »normally inaccessible to direct measurement« so that »much of what is said about it is inevitably conjectural«. Regarding the hydrodynamics of the flow the situation has since considerably changed. Velocities and its various statistical characteristics have been measured well within the viscous sublayer, in some cases below $y^+ = 1$, with hot wire and visualization techniques.

A number of studies employing the hot-wire technique have been undertaken with a special aim of the investigation of the sublayer. Coantic [6] measured longitudinal velocity distributions as well as spectra, and space-time correlations in the air flow in a tube below $y^+ = 1$, correcting the values close to the wall. Van Tinh [29] is avoiding the corrections by working in thick layers along the glass walls. Bakewell [2] employed glycerine to provide a thick enough viscous sublayer. Coantic [6] detected a slight but definite influence of the Reynolds number on the constants of the logarithmic law and a much stronger influence on the turbulence intensity distributions. This has already been suggested by Fortier [7]. Measurements of the spectra and correlations of Coantic and Bakewell have contributed to the understanding of the nature of the sublayer.

By the flow visualisation techniques, employing hydrogen bubbles [13] and micron particles [9], velocities have also been measured below $y^+ = 1$. Khabahpasheva et al. [9] have also reported results of the intensities of fluctuations normal to the wall in this layer which is quite unique in all the sublayer studies.

Recent measurements by Popovich and Hummel [23] by the flash photolysis technique are illustrative of the statistical nature of the sublayer. They have found that the thickness of the sublayer, defined as the linear portion of the velocity profile, statistically changes following a nonnormal probability distribution very much skewed towards the greater thicknesses. The average thickness found was $y^+ = 6.17$ while the most probable was $y^+ = 4.3$. On the basis of somewhat poor statistics they have found no evidence of the nonlinearity below $y^+ = 1.6$, but the slope of this linear portion continuously changed. From the slope, statistical distributions of the wall shear stress were determined and found not far from the normal distributions with the average $\overline{\tau_s}$ being only 5% higher than the most probable, and the r.m.s. value being around 0.25 $\overline{\tau_s}$. Wall shear stresses as low as 0.1 $\overline{\tau_s}$ were detected.

Similar results have been obtained by the author in an air, zero pressure gradient flow, by measuring velocity probability distributions close to the wall by the previously described technique. In Fig. 4 the

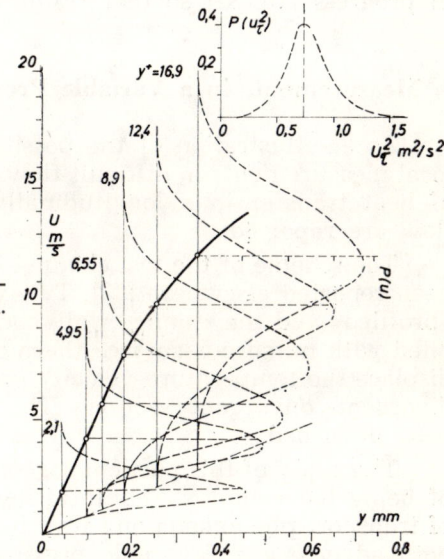

Fig. 4. Velocity probability density distributions at various distances from the wall.

velocity probability density distributions are plotted for various distances from the wall down to the $y^+ = 2.1$. The described corrections have been employed. It is seen that the distributions are non-normal. The mean velocity profile is also plotted and it is seen that the average and the most probable velocity differs systematically. By plotting a straight line through the values of the velocity at various distances having the same probability density, the probability density of the wall shear stress is obtained and shown in the corner of Fig. 4. The τ_s probability distribution is slightly skewed towards higher values as a consequence of the velocity probability distribution skewness, with the r.m.s. value being aroung 0.25 τ_s. It can also be seen that the average value of the sublayer thickness is around $y^+=6$, and the most probable around $y^+=4$. It is seen that the results are in very good agreement with the results obtained by flash photolysis technique.

These results are in further accordance with the measurements of the instantaneous shear stress obtained by hot film [3] and diffusion electrode [18] sensors. Mitchell [18], for instance, measured the probability distributions of τ_s from which the probability density distribution of the velocity in the sublayer was deduced and found slightly skewed towards higher velocities. Similar results have been reported for the mass transfer rate at the wall by van Shaw and Hanratty [27] and for the wall heat flux by Armistead and Keges [1]. Regarding wall heat flux such results could be obtainer by measurement of temperature probability density distribution using thin will resistance thermometer [32].

The rapid growt of the experimental evidence on the dynamics of the sublayer must provide a better understanding of the transport phenomena. This could be greatly helped by the similar temperature or concentration measurements in the sublayer. Unfortunately, such measurements are still lacking although some studies employing a thin wire thermometer are in progress [16, 32] so that results can be expected in the near future.

4. Measurements in a Variable Pressure Gradient Flow

As an illustration of the possibilities of the thin wire technique for local measurements in difficult flow configurations some results of a study of heat transfer in a longitudinally variable pressure gradient channel flow are reported.

The scheme of the test channel is shown in Fig. 5 and the details have been reported elsewhere [33]. Two wider channel walls, refered to as the »profiled« and the »plane« wall, could be heated electrically and are supplied with numerous surface thermocouples. A traversing mechanism can displace the temperature-velocity probe in any cross section along the last divergent-convergent section. The measurement technique has already been described.

The study of this particular geometry, having an obvious disadvantage of being too complex, is interesting from many aspects. In it the study of transport phenomena and wall turbulence can be made in conditions of both adverse and favorable pressure gradients as well as in regions of the transition from one to another. Similar studies have been reported [15] with the difference that the measurements were made only near the plane wall, and it is easy to suppose that the flow structure along the two walls differ. A previous study of heat transfer in similar flows [31] showed a significant intensification of the heat transfer comparable to that which can be obtained with rough surfaces. Only here the intensification in heat transfer rates is obtained on plane smooth surfaces easily accessible for the study of the transfer mechanism.

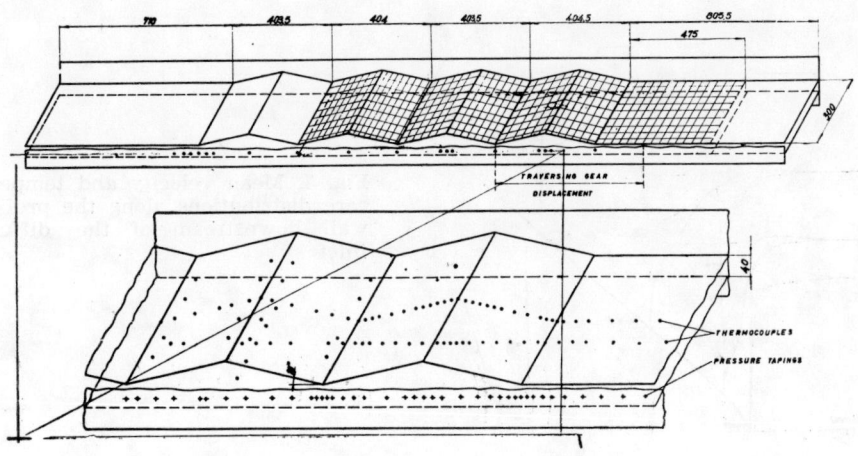

Fig. 5. Schema of the test channel.

The distributions of the local Stanton number determined from the surface temperature and the flow rate measurements, are shown in Fig. 6 for various Re numbers at the diffuser inlet. A steep increase in the local

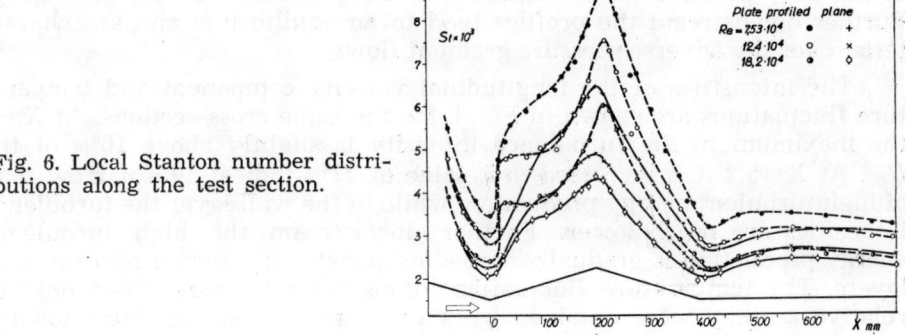

Fig. 6. Local Stanton number distributions along the test section.

St values can be noted downstream of the diffuser inlet and towards its outlet, as well as a significant difference of the St values for the profiled and the plane wall, in the same cross-section, reflecting different flow structures.

We will consider the region immediately downstream of the diffuser inlet along the profiled wall, where the increase of the St value is very steep. In Fig. 7 the mean velocity and temperature profiles in this region

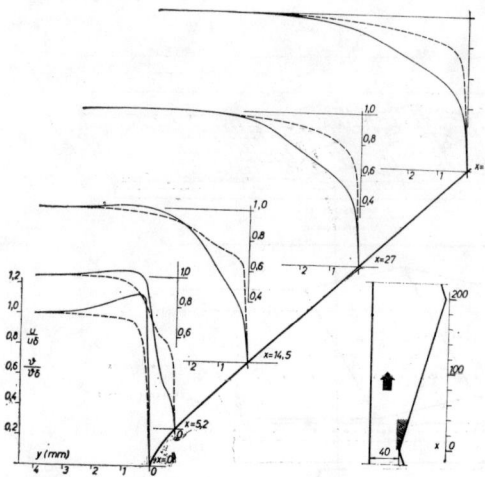

Fig. 7. Mean velocity and temperature distributions along the profiled wall downstream of the diffuser inlet.

are shown for the Reynolds number at the diffuser inlet of about 120 000. The velocities and the temperature differences are given in a ratio to the corresponding values at the edge of the boundary layer (U_δ, ϑ_δ) somewhat arbitrarily defined as the distance where the profiles flatten. The profiles at the inlet are typical of the laminarization of the wall layers, with a pronounced maximum in the velocity profile and a very steep gradient on the wall. Only 5 mm downstream the profiles completely change and exibit inflection points separating a wall layer from the rest of the flow, at the edge of which the velocity gradient is still very steep. Further downstream the profiles tend to an equilibrium and are characteristic for the adverse pressure gradient flows.

The intensities of the longitudinal velocity component and temperature fluctuations are shown in Fig. 8 for the same cross-sections. At $X=0$ the maximum in the turbulence intensity is slightly above 10% of the U_δ. At $X=5.2$ it is increased to a value of 27% indicating a narrow zone of high turbulent energy production, while in the wall layer the turbulence intensities are much lower. Further downstream the high turbulence energy production is gradually spreading penetrating slowly into the wall layers. The temperature fluctuation intensities can be regarded only as relative as they were obtained by a r.m.s. meter not sensitve to low frequencies.

The mean, as well as fluctuation intensity profiles for velocity and temperature very close to the wall are plotted in Fig. 9 for the cross-sections from $X=0$ to $X=5.2$. A gradual decrease in the velocity gradient on the wall is observed from $X=0$ to $X=1$, the inflection in the velocity profile being more and more pronounced. The profile at $X=2.2$ is characteristic in appearance to a reattachment profile although no indication of the separation has been detected. The mean temperature profiles are disturbed to a less extent being all the same unsimilar. The development

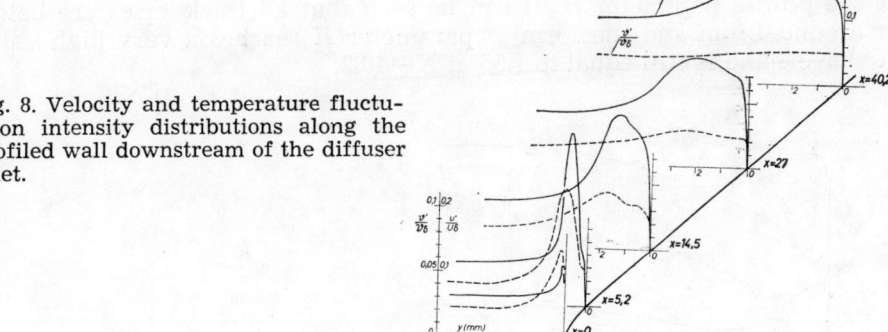

Fig. 8. Velocity and temperature fluctuation intensity distributions along the profiled wall downstream of the diffuser inlet.

of turbulence intensity profiles follows the development of mean velocity profiles. A growth of the turbulent energy is indicated by the value of $(u'/U_\delta)_{max} = 0.285$ at $X=2.2$, with $u'/U=0.57$ at this point. A gradual penetration of the high intensity turbulence in the wall layers is also observed. From the viewpoint of the measurement technique it is important to note the very rapid change of the flow characteristics along the surface. Very few methods of measurements on the wall or close to the wall are suitable in such conditions.

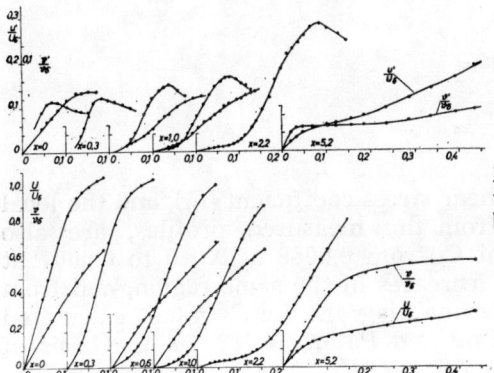

Fig. 9. Temperature and velocity mean value and fluctuation intensity distributions close to the wall.

From the slopes of the profiles close to the wall the wall shear stress as well as the surface temperature and the wall heat flux can be determined, and from the profiles various characteristic thicknesses and the Stanton number can be calculated. These are presented in Fig. 10. In the lowest part of Fig. 10 the pressure gradient distribution, determined from the static pressure distribution, and the maximum local temperature differences are plotted. The variation of the temperature differences, reflecting the variations of the surface temperatures, amounts up to 4.2/0.3 deg/mm, which once more illustrates the difficulties in local temperature measurements in such conditions. In the middle part of Fig. 10 various characteristic boundary layer thicknesses (δ^*, δ^{**}, δ^{**}_h) are plotted as well as the profile parameter H. It can be seen that all thicknesses are below or around 1 mm and the form of parameter H reaches a very high value of above 4 and is still equal to 1.55 at $X=40.2$.

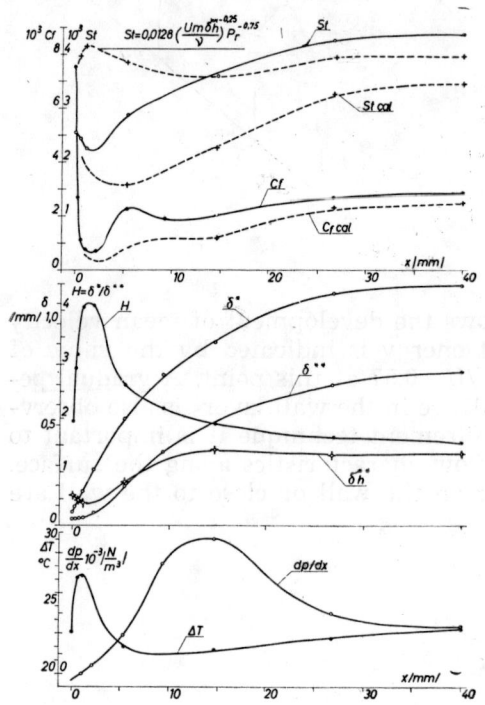

Fig. 10

The distributions of the local shear stress coefficient (C_f), and the local Stanton number (S_t), determined from the measured profiles, are also plotted in Fig. 10. A rapid drop of C_f from 0.0068 at $X=0$ to 0.0007 at $X=1$ is seen. The St value slowly decreases in the same region, and from $X=0$ it begins to increase. On the same diagram the C_f values calculated from the relations given by Spalding and Patankar [32] on the basis of the measured velocities in the inner layer (C_f cal), are also plotted. The agreement is within 10% at $X=40.2$ but is very poor around $X=5$ indicating unsuitability of the hypothesis concerning the velocity distribution the inner layer in this region. The St values calculated by the method suggested by Leont'ev [22], using the measured δ^{**}_h values, are also included in Fig. 10, as well as the values calculated from the relations given in [32], employing the measured temperatures and velocities in the inner region (St_{cal}). The agreement is in general poor, Spalding's prediction giving better results close to the diffuser inlet.

In Fig. 11 nondimensional velocity (U^+) and temperature (v^+) profiles at $X=0, 1, 5$ and 40 are plotted against the nondimensional distance y^+. The profiles are fairly unusual. For $X=5$ and 40 the curves obtained from the relation given in [32], by fittingt he constant E to the experimental data, are also plotted. The character of the relation is fairly satisfactory, although in a very limited range at $X=5.2$, except that the constant E varies very much. The non-similarity of the velocity and the temperature profiles at $X=1$ and 5.2 is very pronounced, which puts a serious doubt on the predictions based on the hypothesis of constant turbulent Prandtl number, and it can explain the disagreement between the calculated and the measured St values in Fig. 10.

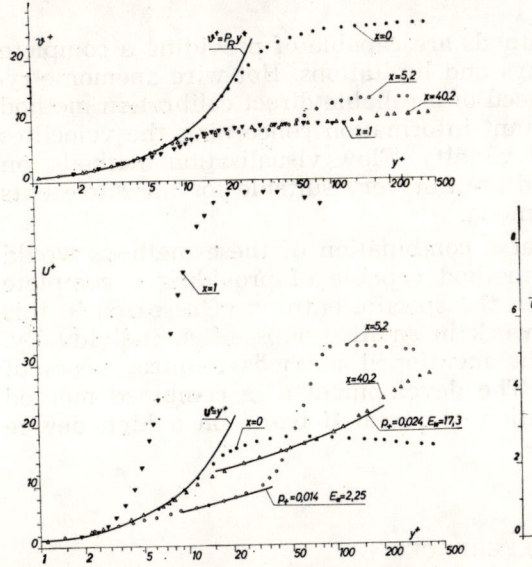

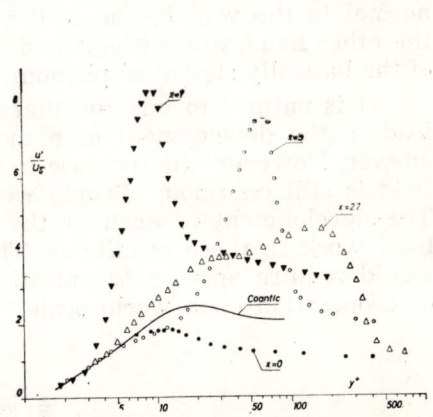

Fig. 11. Semilogarithmic plot of mean velocity and temperature distributions.

Fig. 12. Semilogarithmic plot of turbulence intensity distributions.

Nondimensional turbulence intensity (u'/u_τ) distributions at $X=0, 1, 5.2$ and 27 are plotted against y^+ in Fig. 12. The curve corresponding to the measurements of Coantic [6] in a straigt tube is also included. The profiles are very much non-universal, in favour of the hypothesis that the friction velocity alone does not govern the flow, and the influence of the flow further from the wall is very strong. This has also been stated by Khabahpasheva and others [9]. The profile at $X=0$ lies below the Coantic curve illustrating the effect of laminarization. The growth of the turbulent energy content is seen in profiles at $X=1$ and 5.2 and the penetration of the high intensity turbulence into the wall layers can be seen from the profile at $X=27$.

5. Conclusion

The development of the methods of prediction in turbulent flows is still strongly dependent on the experimental evidence in giving support to the various hypotheses and sometimes even in suggesting physical models. As the methods of predictions, and the physical models on which they are based, evolve they require more refined, more precise, and more delicate measurements.

A rapid development of the measuring techniques of the hydrodynamic quantities has been achieved in the past few years. Of these hot wire anemometry and various flow visualisation methods are able to produce some reliable results even in the very wall vicinity. This has led to a rapid growth of knowledge about the structure of the viscous sublayer which is of outermost importance to the understanding of the transport phenomena close to the wall.

However, none of these methods are capable of providing a complete picture, each having shortcomings and limitations. Hot wire anemometry, being an ndirect method, is in need of a reliable direct calibration method and is unable to provide important information concerning the velocities normal to the wall in the wall vicinity. Flow visualisation methods, on the other hand, are difficult and are not very suitable for measurements of the basically statistical phenomena.

It is natural to imagine that a combination of these methods would lead to the development of a method capable of providing a complete answer. However, the increase in the specific entropy of research in this field is still enormous. People work in small groups, often individually. The development of each of the mentioned methods requires years of hard work in these conditions. The development of a combined method would require another organisation of research based on a high degree of cooperation on a world scale.

REFERENCES

1. R. A. Armistead, J. J. Keyes Jr., J. Heat Transfer, 13 (1968).
2. H. P. Bakewell, Jr., Thesis, Pennsylvania State Univ., 1966.
3. B. J. Bellhouse, D. L. Schultz, J. Fluid Mech. 24, 379 (1966).
4. P. Bradshaw, M. A. Gregory, Aer Res. Con. RM 3202 (1959).
5. A. P. Burdukov, Int. Summer School, Herceg-Novi, Sept. 1968.
6. M. Coantic, Thesis, Univ. Aix-Marseille, 1966.
7. A. Fortier, La Houille Blanche, 3, 241 (1963).
8. E. M. Khabahpasheva, Int. Summer School, Herceg-Novi, Sept. 1968.
9. E. M. Khabahpasheva, B. V. Perepelica, E. S. Mihailova, V. V. Orlov, V. M. Karsten, G. I. Efimenko, Teplo- i Massoperenos, T. 1, Moskva, 1968.
10. D. S. Johnson, J. Appl. Mech. 24, 1 (1957).
11. D. S. Johnson, J. Appl. Mech. 26E, 3, 325 (1959).
12. J. Kestin, P. D. Richardson, Int. J. Heat Mass Transfer 6, 147 (1963).
13. S. J. Kline, W. C. Reynolds, F. A. Schraub, Rundstadler: J. Fluid Mech., 30, 741 (1967).
14. A. I. Leont'ev, Advances in Heat Transfer, Academic Press, V. 3, 33, 1966.

15. T. F. Makkarti, J. P. Hartnett, Teplo- i Massoperenos, T. 2, Minsk (1965).
16. J. P. Mayé, Int. Summer School, Herceg-Novi (1968).
17. J. Mathieu, Thesis, Univ. Grenoble (1959).
18. J. E. Mitchell, T. J. Hanratty, J. Fluid Mech. 26, 199 (1966).
19. A. S. Monin, A. M. Jaglom, Statisticheskaja Gidromekhanika, Pt. 1, »Nauka«, Moskva, (1965).
20. V. V. Orlov, J. PMTF, 4, 124 (1966).
21. S. V. Patankar, D. B. Spalding, Heat and Mass Transfer in Boundary Layers, Morgan-Grampian, (1967).
22. M. Petrović, A. Tošić, Inst. Boris Kidrič Rep. IBK—507 (1966).
23. A. T. Popovich, Hummel, AIChE J. 13, 5, 854 (1967).
24. F. A. Schraub, S. J. Kline, J. Henry, P. W. Runstadler, A. Littel, J. Basic Eng., 4, 429 (1965).
25. C. A. Sleicher, Jr., Trans. ASME, 693 (1958).
26. L. Taccoen, Thesis, Univ. Paris (1966).
27. P. Van Shaw, T. J. Hanratty, AIChE J. 10, 4, 475 (1964).
28. N. Van Th'nh, C.R. Acad. Sc. Paris, 264, 1150 (1967).
29. N. Van Thinh, Int. Summer School, Herceg-Novi (1968).
30. J. A. B. Wills, J. Fluid Mech. 12, 388 (1962).
31. Z. Zarić, Proc. 3rd Int. Conf. Atomic Energy, P/698 (1964).
32. Z. Zarić, Teplo-Massoperenos, T. 9, Minsk, 36 (1968).
33. Z. Zarić, JSME 1967 Symposium, Tokyo, 161 (1967).

STUDY OF TURBULENCE AT A WALL BY STROBOSCOPIC VIZUALIZATION METHOD

E. M. KHABAKHPASHEVA

Thermal Physics Institute, Siberian Division of the USSR Academy of Sciences, Novosibirsk, USSR

Extensive use of the stroboscopic visualization method takes place when studying turbulent flows at the Institute of Thermophysics of the Siberian Division of the USSR Academy of Sciences. Small light-scattering particles (of a few microns) are introduced in the flow and photographed in an impulse sidelight on a fixed-positioned film. The film tracks were heated with the help of a special semi-automatic device. The mean velocities and pulse characteristics were calculated with an electronic computer. The stroboscopic visualization method is used in the study of fluxes in the cases when there are difficulties in thermoanemometer measurements or in their interpretation.

The results of two investigations carried out in cooperation with B. V. Perepelitsa are reported in this communication.

1. Flow at a Wall Under Heat Transfer Conditions

The experiments were carried out in a rectangular channel of 20 x 40 mm cross-section and 2 m long with water and water-glycerine solutions. The upper and side walls of the channel were made of Plexiglass and the bottom of chrom-plated copper. In the first series of runs the bottom was heated with an electric heater and in the second — cooled by a counterflux of water.

Spherical particles of aluminium (2—10 µ) were introduced into the fluid to be studied. Up to 2000 tracks were measured and handled for each regime of the flow at the wall, the whole region being divided into 20 intervals so that 6—8 intervals fell within the viscous sublayer. From the velocity profiles obtained the velocity gradient in the viscous sublayer was determined and the shear stress on the wall calculated.

Figure 1 shows the experimental velocity profiles of water and water-gliceryne fluxes. The coordinates used are $W/v^* = f\left(\dfrac{y v^*}{\nu_\text{ж}}\right)$ where

$v^* = \sqrt{\tau_c/\rho}$ is the dynamic shear velocity and $\nu_ж$ viscosity determined at mean-calorimetric temperature. The values of Re and Pr ranged in the experiments from 10^4 to $4.5 \cdot 10^4$ and from 5 to 11, respectively, and viscosity within the channel cross-section ranged within the limits $0,5 \leqslant \nu_c/\nu_ж \leqslant 1,8$.

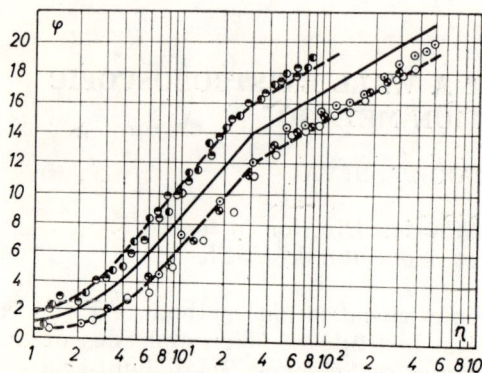

Fig. 1. Velocity profiles in dimensionless coordinates
a — heating
◐ — water, $Re=1500$, $\psi=0.6$
◑ — 20% water solution of glycerne $Re=13000$, $\psi=0.5$
b — cooling (water)
⊙ — $Re=45000$, $\psi=1.5$
◓ — $Re=36000$, $\psi=1.6$
○ — $Re=25000$, $\psi=1.8$
Solid line — universal velocity profile, dashed line-calculations by [2]:
a) at $Re=15000$, $\psi=0.6$;
b) at $Re=25000$, $\psi=1.8$.

The solid curve in Fig. 1 demonstrates the universal velocity profile (by three layer scheme [1]) in an isothermal flow. On heating the fluid, the velocity distribution curve in the universal coordinates passes higher than under the isothermal conditions. On cooling it passes lower. The dashed lines in Fig. 1 represent the results of calculations by the method described in [2] applied to the present experiment. The agreement between the experimental and calculated values is satisfactory.

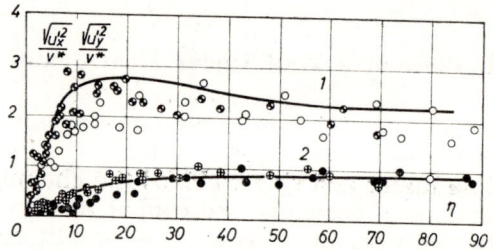

Fig. 2. Relative mean-square values of longitudinal (1) and transverse velocity pulsations.
◓ ⊕ — heating
○ ● — cooling
Solid line — Laufer's data [3].

Figure 2 shows the distributions of the mean-square values of the longitudinal and transverse velocity pulsations related to the shear velocity on the wall. The solid lines in the plot are the experimental results of the Laufer obtained for an air flux in a circular pipe [3] and our data for the water flow [4]. As seen from the plot, the mean-square velocity pulsation in the isothermal case and when heating are practically the same within the experimental range of viscosity. When cooling the fluid a certain decrease of both longitudinal and transverse velocity pulsations can be observed.

2. Turbulent Flux in Water with High Molecular Additives

As it has been shown by numerous investigations the small additions in water ($10^{-4} - 10^{-5}$) of some high molecular substances result in an appreciable decrease of friction in the turbulent fluxes. Velocity measurements with Pitot's tube in such fluxes have some disadvantages. First of all, they give no information about the velocity distribution in the vicinity of the wall. Moreover, the interpretation of such measurements is hindered by a possible occurrence of non-isotropic normal stresses in the flows of the solutions used. The use of a thermoanemometer has given no positive results either.

Studied were the mean and pulsation velocity distributions in water with a number of additives: polyacrylamide, polyethylen oxide, DNA and carboxymethylcellulose (CMC). The molecular weight of the first three substances is known to be greater than 10^6 and that of CMC — somewhat lower. Special experiments have shown that with polyethylen oxide and polyacrylamide concentrations in water ranging within 0.01—0.02% and $Re \approx 10^5$ the friction factor for an 8 mm pipe was 2—3 times lower as compared with the same value in pure water.

The mean velocity profiles for polyacrylamide and DNA additives are shown in Fig. 3 in the dimensionless coordinates. As seen from the

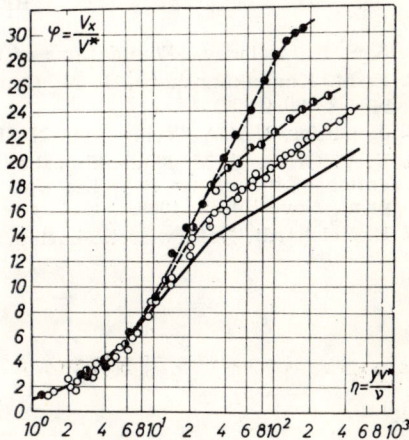

Fig. 3. Mean velocity profiles in dimensionless coordinates
○ ◐ — 0.015—0.025% water solution of DNA
● — 0.04% water solution of polyacrylamide.

plot, the points outside the viscous sublayer fall above the universal velocity profile characteristic of the usual fluids (solid line). The slopes of the profiles are greater in the buffer zone while in the turbulent core of the flux the slope of the logarithmic velocity profile is practically stable. A substantial increase in the buffer zone should be noted.

The obtained results show that strongest influence of the high-molecular additives on the flow occurs in the buffer zone where the viscous and turbulent shears are comparable. Polymer additives hinder the development of the turbulence and shifts the region of the developed turbulent flow to a larger distance off the wall.

Figure 4 shows the experimental mean-square values of longitudinal and transverse velocity pulsations as functions of the distance from the wall. It can be seen that the longitudinal velocity pulsations are a little larger in the solutions than in water (solid line) and the maximum, as it could be expected, is shifted to a larger distance off the wall. Simultaneously, (and what we think to be a most interesting and important result) the transverse velocity pulsations at the wall are reduced 2—3 fold by the addition of high-molecular polymers.

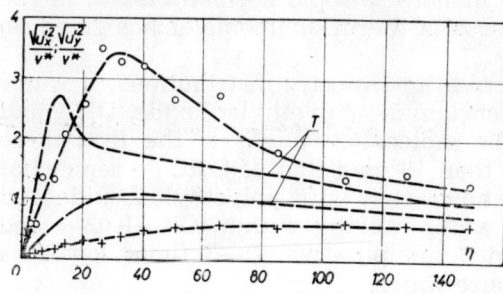

Fig. 4. Relative mean-square values of longitudinal and transverse velocity pulsation in 0.04% water solution of polyacrylamide.
I — Reichtrdt's data [5].

$$\bigcirc - \frac{\sqrt{\overline{u_x'^2}}}{v^*}; \quad + - \frac{\sqrt{\overline{u_y'^2}}}{v^*}.$$

REFERENCES

1. S. S. Kutateladze, »*Foundations of theory of heat transfer*«, Mashgiz, 1962.
2. E. M. Khabakhpasheva, J. M. Gruzdeva, Teplofizika vysokih temperatur, 4, 1, pp. 92—98 (1966).
3. J. Laufer, NACA Techn. Rep. N 1174, 1954.
4. V. V. Orlov, E. M. Khabakhpasheva, B. V. Perepelitsa, E. S. Michailova, Investigation of turbulence near the wall by the stroboscopic visualization method, Preprint, Novosibirsk, 1968.
5. H. Reichardt, Naturwissenschaften, 404 (1938).

HEAT TRANSFER FROM HOT WIRE AND FILM ANEMOMETER PROBES

M. R. DAVIS

Institute of Sound and Vibration Research, The University, Southampton, UK

1. Introduction

The steady calibration characteristics of hot wire and hot film probes are determined by the relationship between incident flow conditions and the surface heat transfer coefficient for the probes being used. Considerable attention has therefore been given to quantitative studies of heat transfer laws either empirically or theoretically. The case of a circular cylinder has been studied in detail because it has direct bearing on the behaviour of hot wire and cylindrical hot film probes which are very widely used. All types of heated anemometers will experience some loss of heat by thermal conduction into their mountings and electrical connection leads, depending upon the geometry of the probe. To interpret the measurements of overall power loss in terms of the absolute surface heat transfer rate it is necessary to solve for the distribution of temperature in the probe and its supports and to deduce the balance of energy flows from the probe.

2. The Surface Cooling of a Circular Cylinder

The theories which have been developed for the cooling of a circular cylinder can be divided into several groups dealing with continuum flow, natural convective flow, cooling by conduction through the fluid, slip flow and free molecular flow respectively. The results of this paper are confined to Mach numbers less than 0.5, and so compressibility effects will not be discussed. Almost none of the theories take the variation of fluid properties into account, and the heat transfer is derived in a linear form by neglecting the interaction of the thermal and flow processes.

For a circular cylinder losing heat into a continuum flow at right angles to the cylinder axis the law relating the Nusselt number ($N=hd/k$) to the Reynolds number ($R=Vd/\nu$ of the flow is of the form

$$N = A + B(R)^n \tag{1}$$

where A and B are constants. King (1914) derived this equation theoretically with a value of $n = \frac{1}{2}$. More recently Wood (1968) and Hieber and Gebhart (1968) have discussed the case where the Peclet number ($P = \sigma R$, where σ is the Prandtl number of the fluid) becomes small. Hieber and Gebhart give the expression

$$N = \frac{2}{Ln\left(\frac{8}{\gamma P}\right)} \left[1 - \frac{\lambda(\sigma)}{Ln\left(\frac{8}{\gamma P}\right) \cdot Ln\left(\frac{8}{\gamma R}\right)} - \frac{\nu(\sigma) - \frac{1}{2}\lambda(\sigma)}{Ln\left(\frac{8}{\gamma P}\right)\left[Ln\left(\frac{8}{\gamma R}\right)\right]^2} \right] \quad (2)$$

for the Nusselt number to $0\,[(1/LnR)^5]$. The constant γ is the ratio of specific heats of the gas, whilst λ and ν are functions of the Prandtl number, $\lambda(0.72) = 1.38$ and $\nu(0.72) = 0.4$. It can be seen that as the Reynolds and Peclet numbers tend to zero in this equation, the Nusselt number also tends to zero.

The limiting condition of R, $P = 0$ in a static gas is considered by Seneftleben (1953) and Mahony (1957). Mahony treated the cases of heat transfer from a sphere and from an infinite length circular cylinder with separate theoretical analyses. For the former the Nusselt number can be derived on the basis of conduction through a static fluid for very small Grashof number (G), the effect of buoyancy introducing a term of $0\,(\sqrt{G})$ in the Nusselt number. For the infinite cylinder the two dimensional radial conduction of heat does not provide a suitable model for the heat transfer process even at small Grashof number, and Mahony obtained solutions for this case by matching the temperature boundary condition between an inner conduction region and an outer convective region. The relationships which Mahony deduced were

$$N = -\frac{3}{Ln\,G} \quad \text{for long cylinders,} \quad L\sqrt{G} > 1 \quad (3a)$$

and $$N = \frac{1}{Ln\,2L} \quad \text{for short cylinders,} \quad L\sqrt{G} < 1 \quad (3b)$$

It appears that the limiting condition of heat transfer in a static gas represented by equations 3 should be considered as a lower bound for the Nusselt numbers at small Peclet numbers given by Eq. 2.

A simplified analysis of the rate of heat loss in slip flows was presented by Sauer and Drake (1956) based on a modified boundary condition due to heat transfer on a molecular basis close to the surface. This modified boundary condition was inserted into a simplified model for the outer flow. The results of Sauer and Drake are shown in Fig. 3 for different Reynolds numbers as a function of Knudsen number. This presentation easily permits the inclusion of a limiting case $R = 0$, although this is not included in Sauer and Drake's analysis.

Madden and Piret (1951) discussed the case of heat loss in static gases and used a theoretical model based on a molecular exchange for the first mean free path from the wire and conduction through the gas beyond that distance. The expression for the heat loss from a long cylinder then becomes

$$\frac{2}{N} = \frac{8\gamma}{(\gamma+1)\sigma}\left(\frac{\lambda}{d}\right) - Ln\left(1 + 2\left(\frac{\lambda}{d}\right)\right) + Ln\left(\frac{b}{d}\right) \tag{4}$$

where b is the diameter at which the gas temperature is T_0 (ambient temperature). In the limiting case of small Knudsen numbers the value of b would have to give a Nusselt number consistent with experimental measurements.

Theoretical predictions of the free molecular heat loss rate are well established for static or flowing gases. The analyses of Oppenheim (1953) and Schaaf (1964) derive the relationship between the Stanton number $\left[\left(\frac{\gamma}{\gamma+1}\right) \cdot \frac{N}{\alpha R \sigma}\right]$ and the Mach number, and show that the Stanton number tends to a constant for $M > 2$ where the mean gas flow predominates. The heat loss becomes insensitive to small Mach number flows under free molecular conditions. The only unknown factor which remains in predicting the heat transfer in free molecular flow is the thermal accommodation coefficient (α) at the surface.

3. Experimental Results

The experimental work which is reported here has been carried out using hot wire and cylindrical hot film probes operated in a constant resistance anemometer circuit. For the hot wire measurements solutions to the one dimensional conduction equation for the temperature distribution in the wire were found using numerical integration techniques. Similar detailed solutions have not been carried out for the cylindrical thin film probes, estimations of the Nusselt numbers being based on the film dimensions and average temperature. Consequently, strong emphasis should not be placed on the absolute values of the Nusselt numbers for the hot film probes (which appear rather low).

3.1. HEAT LOSS AT ATMOSPHERIC PRESSURE

Particular attention has been given to calibrations at extremely low speeds ($R < 0.1$) as there appears to be some uncertainty concerning the effect of buoyancy flows. Figure 1 shows a typical set of calibrations made using a metered flow through a pipe which could be rotated in a vertical plane. It is seen from these results that the calibrations approach the limit of zero flow quite smoothly, the zero speed reading being a smooth extrapolation of the measured data from the lowest experimental Reynolds number of 0.015. This result is apparently in conflict with some of the measurements of Collis and Williams (1959), who reported a minimum heat loss at a finite Reynolds number. The explanation for the absence of a significant buoyancy interaction in the results can be found in the theoretical work of Mahony (1957) who estimated that the heat transfer was negligibly influenced by small Grashof numbers for values of $L\sqrt{G}$ less than unity as already discussed. For the calibrations shown in Figure 1 this parameter has a value of about 0.8, whilst Collis and Williams used wires of much greater length to diameter ratio (5400, compared with 400 for the experiments reported here).

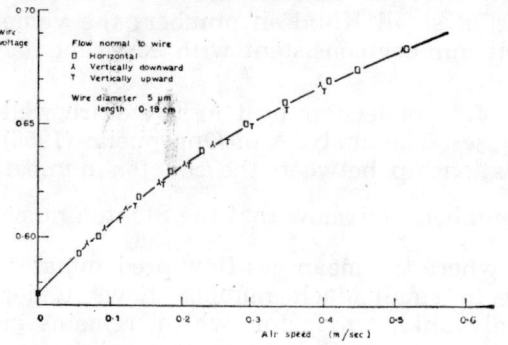

Fig. 1. Effect of gravity on low speed cooling.

The variation in the cooling relationships for a range of flow speeds is illustrated in Fig. 2, these results being taken from the calibrations of a number of 5 micron wires. For high speeds the index n tends to a constant value of 0.4 for the experiments made by the author, although other

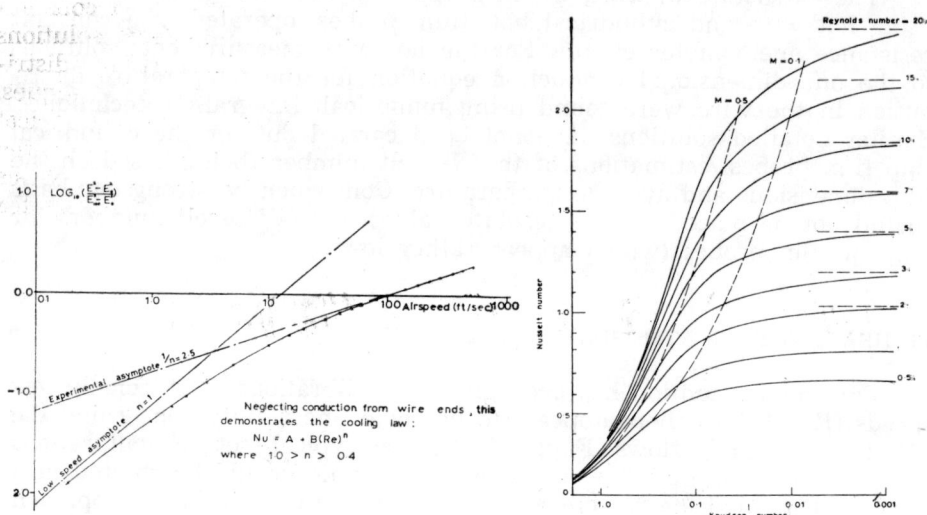

Fig. 2. Typical hot wire calibration at atmospheric pressure.

Fig. 3. Slip flow theory for a circular cylinder. (Sauer & Drake).

workers have reported values varying up to 0.5 for this index. At lower speeds the value of n tends to unity where the calibration takes on the approximate form of a straight line of constant slope (dE/dV).

3.2 HEAT LOSS IN VARIABLE DENSITY FLOWS

A series of experiments have been made using hot wire and cylindrical hot film probes mounted in the potential core of a one inch diameter nozzle at variable static pressure. Typical experimental calibrations are shown in Figs. 4 and 5 for a hot wire probe of 5.1 μm diameter and a hot film probe of 53.5 μm diameter. The thin film cylinder was manufactured by Thermo-Systems Inc., consisting of a painted platinum film on a quartz cylinder with gold painted connecting areas. The whole probe had been

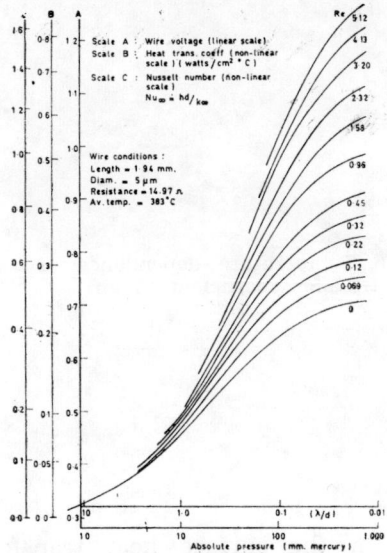

Fig. 4. Forced convection in air

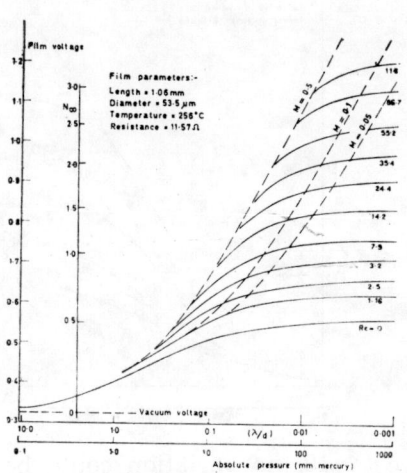

Fig. 5. Calibration of cylindrical thin film probe.

covered with a layer of protective quartz, deposited by an ionic sputtering technique. It can be seen how both the wire and film probes are influenced by variations in the Knudsen number. This effect is stronger for the hot wire probe, since the film probe is operated at approximately ten times smaller Knudsen numbers. Whilst the heat loss for the hot wire is still rising quite strongly with pressure at atmospheric pressure, the heat loss from the hot film probe has nearly reached a constant value (Figs. 4 and 5) at that condition.

The limiting case of zero incident flow has been included in the calibrations shown in Figs. 4 and 5, these results being found independent of the probe orientation. Also shown are the probe voltages with the probe in a low pressure vacuum sufficient to reduce the heat loss into the gas to a negligible quantity (.002 torr). Subsidiary scales on Figs. 4 and 5 show the values of the heat transfer coefficient and Nusselt number corresponding to the experimental measurements, the values for the hot film being only approximate.

A comparison between selected results from the measurements at different wire temperatures and Sauer and Drake's theoretical analyses is given in Fig. 6. It is clear that the value of the Nusselt number, based on free stream fluid conditions throughout this paper, becomes progressively more sensitive to the wire temperature as the Knudsen number is decreased. This behaviour corresponds to the transition from free molecular or slip flow towards continuum flow, and it can be deduced that the variation of the bulk thermal conductivity of the gas (air in this case) becomes more important in the vicinity of the wire as the flow approaches a fully continuum flow. In the limit of true free molecular flow,

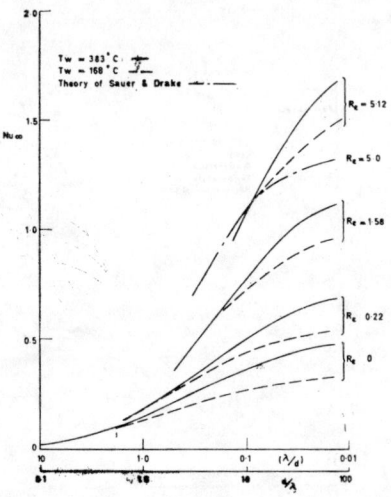

Fig. 6. Temperature dependance of heat transfer coefficient in air.

$Kn > 10$, no variation could be detected in the surface heat transfer coefficient due to changes in wire temperature. Measurements in free molecular flow were found to give a value for the surface accommodation coefficient of 0.86 for air flowing past a tungsten wire.

4. Concluding Discussion

It has been the purpose of this paper to show the range of flow conditions for which circular anemometer probes experience differing types of heat transfer mechanism. For the limit of continuum flow the laws of cooling are governed by the value of the Peclet number and a lower bound for the heat loss is determined by the conditions which in a static fluid. For increasing Knudsen numbers the heat transfer is progressively reduced until free molecular conditions are attained. In free molecular flow the heat loss is not strongly influenced by subsonic flows, and a corresponding reducing sensitivity to incident mass flux is found in the transition region from continuum flow. The non-linearity of the heat transfer with surface temperature is stronger in continuum flow where the bulk thermal conductivity of the fluid plays a more significant part in the cooling mechanism. In free molecular flow the heat transfer is linearly related to the surface temperature.

REFERENCES

1. L. V. Baldwin, J. C. Laurence, V. A. Sandborn, Heat Transfer from transverse and yawed cylinders in continuum, slip and free molecule flow. ASME Symp. on Unsteady Flow Measurements, 1962.
2. D. C. Collis and M. J. Williams, J. Fluid. Mech., **6**, p. 359 (1959).
3. R. J. Cybulski and L. V. Baldwin, NASA Memo. 4—27—59E, 1959.
4. P. O. A. L. Davis and M. J. Fisher, Proc. Roy. Soc., A. 280, p. 486 (1964).
5. M. R. Davis and P. O. A L Davies, Tech. Rep. 2, I.S,V.R., Southampton University, 1968.
6. J. W. Deardoff and G. E. Willis, J. Fluid. Mech., **28**, pt. 4 (1967).
7. C. A. Hieber and B. Gebhart, J. Fluid. Mech., **32**, pt. 1, p. 21 (1968).
8. L. V. King, Phil. Trans. Roy. Soc., A. 214, p. 373 (1914).
9. A. J. Madden and E. L. Piret, Heat transfer from wires to gases at sub-atmospheric pressures under natural convection conditions. I. Mech. E., General Discussion of Heat Transfer, p. 328, London, 1951.
10. J. J. Mahony, Proc. Roy. Soc., A. 238 (1957).
11. A. K. Oppenheim, J. Aero. Sci., p. 49 (January, 1959).
12. F. M. Sauer, and R. M. Drake, J. Aero. Sci., **20**, p. 175 (1953).
13. S. A. Schaaf, »Heat transfer in rarefied gases«, Chapt. 7 in Developments in Heat Transfer, M.I.T. Press, 1964.
14. H. Seneftleben, Z. Aug. Phys., **5**, p. 267 (1953).
15. W. W. Wood, J. Fluid Mech., **32**, pt. 1, p. 9 (1968).

MEASUREMENT OF THE VELOCITY NEAR A SMOOTH WALL AND THE DETERMINATION OF THE LOCAL WALL SHEAR STRESS IN A TURBULENT FLOW

NGUYEN VAN THINH

Laboratoire de Mécanique Expérimentale des Fluides,
Université de Paris, France

1. Introduction

The structure of turbulent flow near a smooth wall, particularly in the viscous sublayer, is of considerable importance to the understanding of the transport phenomena close to the wall. Recently, a few studies based on the flow visualisation have given some information about this structure but the lack of realiable results does not permit actually to draw up a general theory and more experimental research is still necessary.

This study is concerning with turbulent flow in a two-dimensionnal channel like in the classical investigation of Laufer [1], but a particular attention is brought to the wall region. In such flows, one of the various methods of determining the local mean wall shear stress is to take the velocity gradient at the wall and this necessitates precise measurements of this mean velocity.

Turbulence being a random phenomenon, it is necessary to treat it statistically, so we have tried a method of sampling the instantaneous values of the velocity, obtaining in this way its ensemble mean values. The present paper lists some results obtained by this method.

2. Measurements of the Velocity Near a Smooth Wall

The main studies of turbulent flow near a smooth wall have been made by means of the hot-wire but the results present a great amount of scatter. One of the effects modifying the hot-wire response is the influence of the inclination of the wire support with respect to the wall and we have pointed out this influence in an earlier note [2]. The non-linearity of the hot-wire response introduces also some errors when the fluctuations are important and it is necessary to use a linearizing circuit. In spite of these inconveniences, the hot-wire due to its small size, is the only element that enables to make measurements very near a smooth wall. It also permits a detailed study of the fine structure of the flow and, in the present case, it is particularly suitable for data processing.

3. Experimental Set-Up and Measuring System

The experimental set-up as well as the measuring system have been described in detail [3]. Figure 1 gives a schematic representation of this set-up. The fluid used is air which has been filtered. The sonic nozzle gives an absolutely constant mass flow. The Reynolds number R_* is about 1560, giving a viscous sublayer thick enough to permit measurements with a hot-wire in this region.

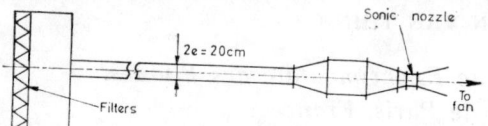

Fig. 1. General schematic of test apparatus.

Figure 2 represents the measuring system. The outputs from the hot-wire anemometer are fed into a linearizer, after having passed an isolation amplifier. Linearizing has been carefully made [3] and table 1 gives an example of calibration. During runs, the linearizer response was verified constantly with a direct-current voltage reference.

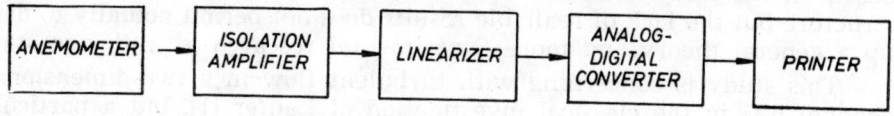

Fig. 2. Measuring system.

This response is converted in numerical form and registered by a printer. These printed values are then sampled in a complete random manner and fed into a small electronic desk computer.

The probes used have platinum-plated tungsten wires of 5 μ diameter, 1.8 mm length, giving a ratio $l/d = 360$. The wire supports were bended to be placed in a parallel direction to the wall.

4. Estimation of the Mean Values

In practice, the mean values are obtained by an analog integrating device and some difficulties may arise, especially if the signal is of low frequency. In some recent studies, digital averaging by means of a high speed digital computer has been used, but the analog data, after being digitized sequentially, are directly sent to the computer, the mean values obtained do not represent true ensemble means. To our knowledge, nobody has so far tried to make statistical estimation like the one performed in this study.

The different numerical values U_i of the velocity at a point constitute a sample of size N which served to give the mean velocity:

$$\overline{U} = \frac{1}{N} \sum_{i=1}^{N} U_i \qquad (1)$$

and the standard deviation:

$$u' = \left[\frac{1}{N} \sum_{i=1}^{N} (U_i - \overline{U})^2 \right]^{\frac{1}{2}} \qquad (2)$$

for different values of N.

For a given position, the values U_i introduced into the computer have been selected in a complete random manner in the record and one can obtain in this way a good estimation of the ensemble mean $E\ (U)$. The confidence interval is:

$$\overline{U} - k \leqslant E\{U\} < \overline{U} + k \qquad (3)$$

where k is proportionnal to $\dfrac{u'}{\sqrt{N-1}}$

As an example, for an intensity of turbulence of about 30% a value which is found in the viscous sublayer, the error is about 3% for $N=100$ with a chance of 9/10 and if it is assumed that the distribution is gaussian. Therefore, N must be sufficiently great. We have in practice taken $N=1000$ for $y_+ \leqslant 8$ to have accurate estimations in the linear region and $N=300$ elsewhere. Figures 3, 4, 5 give examples of the variation of the mean velocity $\overline{U}$ with N and for different values of $u'/\overline{U}$.

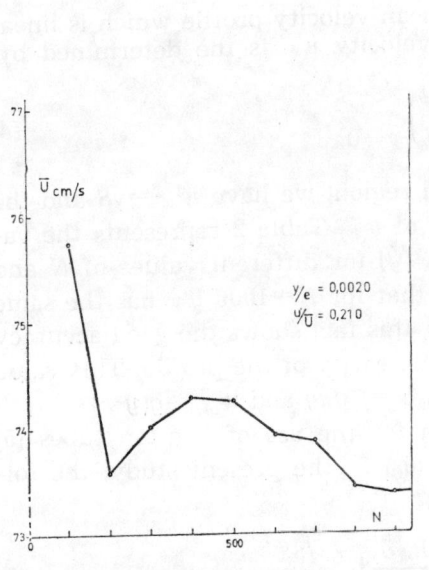

Fig. 3

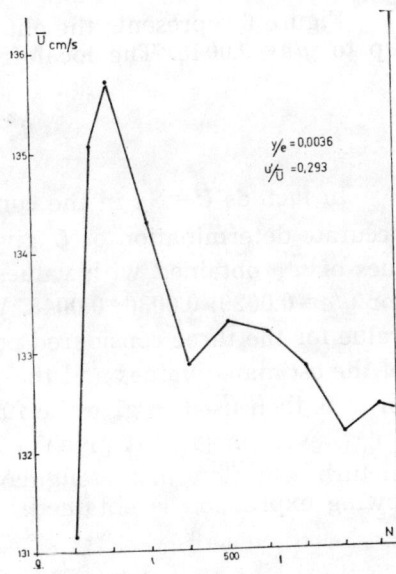

Fig. 4

We see that, to have some degree of the required accuracy, it is important to take the size N of the sample sufficiently high, particularly in the region of high turbulence intensity. Figure 4 is very demonstrative, it shows that, if one only takes about $N=100$ or 200, the error might be about 3%.

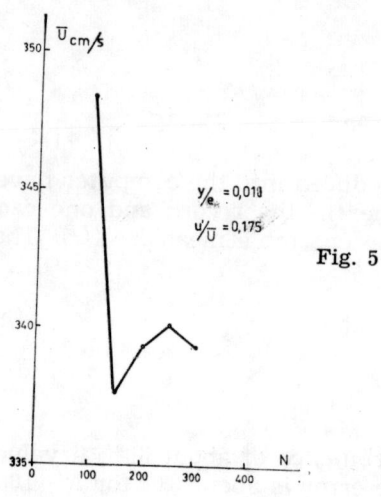

Fig. 5

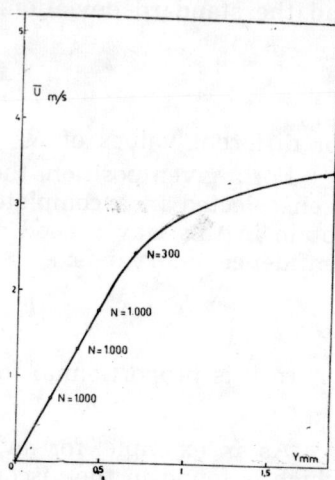

Fig. 6. Mean velocity profile.

5. Results

5.1. LOCAL WALL SHEAR STRESS

Figure 6 represents the obtained mean velocity profile which is linear up to $y/e=0.0048$. The local friction velocity u_* is the determined by:

$$u_*^2 = \nu \left(\frac{dU}{dy}\right)_{y=0} \tag{4}$$

In fact, as $\overline{U}=Sy$ in the considered region, we have $u_*^2 = \nu S$ and the accurate determination of $\overline{U}$ gives that of u_*. Table 2 represents the values of u_* obtained with values of $\overline{U}_N(y)$ for different values of N and for $y/e=0.0020; 0.0036; 0.0048$. We see that for $N=1000$ u_* has the same value for the three considered positions, this fact shows the good accuracy of the estimated values and the perfect linearity of the profile. This value of u_* is then used to give the functions $\overline{U}=f(y_+)$ and $u'_+ = g(y_+)$.

Stevenson [4] has given a formula for the use of Preston tubes [5] in turbulent flow in rectangular channels. In the present study, the following expression is obtained:

$$\frac{u_*}{u_m} = 0.1452 \left(\frac{u_m d_o}{2\nu}\right)^{-1/8} \tag{5}$$

which is valid for $355 \leqslant R_{do} = \dfrac{u_m d_o}{2\nu} < 1000$.

5.2. MEAN VELOCITY

Figure 7 gives the curve $\overline{U}_+ = f(y_+)$ which shows that $\overline{U}_+ = y_+$ up to $y_+ \neq 8$. This result seems to be different from the values which are generally accepted, but is in accordance with Kline's result [6].

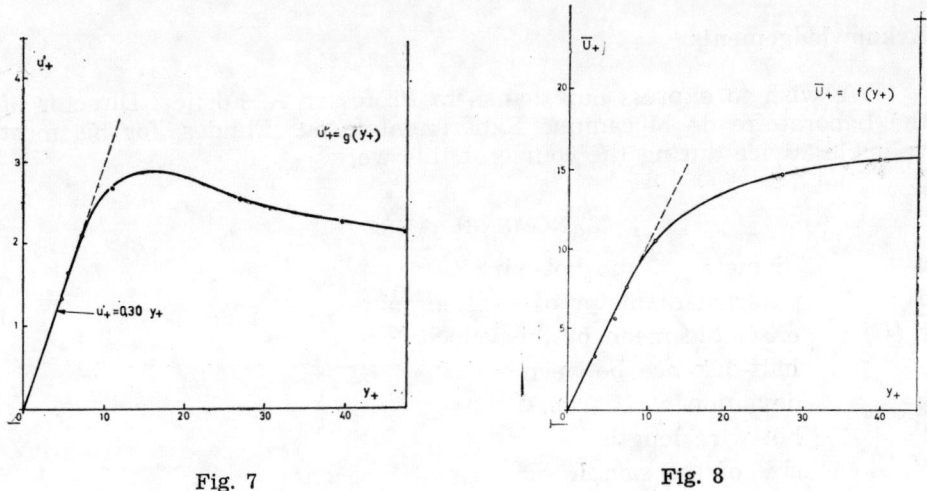

Fig. 7   Fig. 8

For turbulent flow between parallel planes, the law of the wall is written as: [7]

$$\overline{U}_+ = y_+ - \dfrac{1}{R_*} y_+^2 + \overline{a_5}\, y_+^5 \qquad (6)$$

and with $R_* \neq 1560$, we have found $\overline{a_5} \neq 0$.

5.3. FLUCTUATION

Figure 8 gives $u'_+ = g(y_+)$. We see that for $y_+ < 6$ we have practically:

$$u'_+ = 0.30\, y_+ \qquad (7)$$

result which is in good accordance with the limit value at the wall found by Mitchell [8] by an electrochemical method. The curve presents a maximum for $y_+ \neq 16$. The values are higher than those obtained in a previous study [2] by means of an analog R. M. S. voltmeter. The present results seem more reliable because in the range of low measured velocities, the frequencies of energetical fluctuations are very low and consequently the analog meters give values which are lower than the true values.

6. Conclusions

In turbulent flows, the different variables are random by nature and it is necessary to estimate the mean values by statistical methods. These methods are now possible by means of digital systems and this study illustrates an example. In the viscous sublayer where turbulence intensity is high, the sample size must be sufficiently great. The accurate determination of the mean velocity in this linear region of the profile gives at the same time the value of the local mean wall shear stress.

Acknowledgements

We wish to express our thanks to Professor A. Fortier, Director of the Laboratoire de Mécanique Expérimentale des Fluides, for his most valuable advice during the course of this work.

NOMENCLATURE

d	diameter of the hot-wire
d_0	external diameter of Preston tubes
$E\{U\}$	ensemble mean of the velocity
e	half-distance between walls
k	deviation of U from $E\{U\}$
l	hot-wire length
N	size of the sample
R_*	Reynolds number based on u and e
$\bar{U}$	estimated mean velocity
u'	standard deviation of U
$u'/\bar{U}$	turbulence intensity
u_m	measured velocity by the Preston tube
u_*	friction velocity
$\bar{U}_+$	$\bar{U}/u_*$
u'_+	u'/u_*
V_L	linearizer output
y	distance to the wall
y_+	yu_*/ν
ν	kinematic viscosity

REFERENCES

1. J. Laufer, NACA TN 2123, 1951.
2. N. Van Thinh, C.R. Ac. Sc., **264**, p. 1150—1152 (1967).
3. N. Van Thinh, Disa Information № 7, DISA : Herlev (Denmark).
4. M. Stevenson, Univ. of Maryland, T.N. BN-147, 1958.
5. J. H. Preston, J. Roy. Aero. Soc., **58**, p. 109 (1954).
6. S. J. Kline et al, J. Fl. Mech., **30**, 4, 741—773 (1967).
7. A. Fortier, La Houille Blanche, 3 (1963).
8. J. E. Mitchell et al, J. Fl. Mech., **26**, 1, p. 199—221 (1966).

TABLE I
Linearisation of the anemometer outputs

U cm/s	0	92,55	189,96	284,13	378,00
V_L volts	0	92,67	188,92	284,35	379,30

TABLE II
Values of u_* obtained for different values of N (cm/s)

y/e \ N	100	200	300	400	500	600	700	800	900	1000
0,0020	23.92	23.85	23.65	23.69	23.69	23.64	23.62	23.56	23.55	23.55
0.0036	23.43	23.83	23.71	23.58	23.62	23.61	23.58	23.52	23.55	23.54
0.0048	23.37	23.47	23.48	23.51	23.52	23.52	23.46	23.48	23.51	23.56

CALIBRATION OF A DISA HOT-WIRE ANEMOMETER AND MEASUREMENTS IN A CIRCULAR CHANNEL FOR CONFIRMATION OF THE CALIBRATION

B. KJELLSTRÖM and S. HEDBERG

AB Atomenergi, Sveden

1. Introduction

The objects of this investigation were
— to establish a suitable analytical expression which could be used for correlation of data from hot-wire calibration experiments,
— to run such experiments and use this expression for correlation of the results,
— and finally, in order to check the results of the calibration, to measure in a circular channel the shear stress and turbulence distributions and compare the results to the theoretical shear stress distribution and earlier data for the turbulence components.

This communication is a summary of a research report by the authors [19] issued by AB Atomenergi, Sweden, to which reference should be made for more detailed information and discussions.

2. Calibration Experiments

2.1. AN ANALYTICAL EXPRESSION FOR THE CALIBRATION CURVE

For evaluation of the results from the hot-wire measurements with a digital computer it is necessary to have an analytical expression for the calibration curve.

Such an expression can be obtained by combination of Collis' [3] empirical law for the heat transfer from the wire

$$Nu_{wi} \left(\frac{T_{wi} + T_a}{2 T_a} \right)^{0.17} = a + b \, Re_{wi}^c \tag{1}$$

with the direction sensitivity relation suggested by Hinze [6]:

$$u^2(\psi) = u^2 \left(\frac{\pi}{2} \right) \cdot (\sin^2 \psi + k^2 \cos^2 \psi) \tag{2}$$

where $u(\Psi)$ is the velocity to use in the Reynolds number in Eq. (1) if the wire is inclined at an angle Ψ to the flow direction.

Assuming now that the resistance of the wire is proportional to its temperature, that the ambient temperature is constant and that the resistance ratio R/R_a is kept constant, after insertion of the heat generation in the wire V^2/R and utilization of the definition of the Nusselt number Eqs. (1) and (2) can be combined to form

$$\frac{V^2 - V_0^2}{R(R-R_a)} = b\,(\rho u)^c\,(\sin^2\Psi + k^2\cos^2\Psi)^{c/2} \qquad (3)$$

In this relation the factor b, the exponent c and the direction sensitivity coefficient k^2 must be determined by calibration experiments.

V_0 is the voltage corresponding to zero velocity. This quantity may be determined in two ways, normally giving different values. One posibility is to measure the voltage over the wire when there is no net flow past it. This can be achieved in practice by measurements with the probe cover on. It is also possible to determine V_0 by extrapolation of results from calibration experiments in a certain range of ρu down to zero. This is in fact the method proposed in the work of Collis and Williams [3].

Both approaches were tried in this investigation. Since the heat transfer mechanism at the wire cannot be the same at very low and very high velocities, the first approach may be expected to give a variation of the exponent c with the velocity. This method was nevertheless preferred since the other approach, which may give a constant value of c within the range of particular interest, made it necessary to fit three constants to the results of the calibration experiments. It turned out that this caused great difficulties with the convergence of the calculations and this method was therefore abandoned.

2.2. THE EXPONENT IN COLLIS' LAW

In order to check the influence of the flow turbulence on the determination of c, measurements were made both in a low turbulence jet after a nozzle and in the turbulent flow in the center of the circular channel, later used for shear stress measurements. Except for the arrangement of the flow the techniques were similar in the two series of experiments. The hot-wire probe and a pitot tube were mounted in parallel on the probe support at a distance of 20 mm. Readings were first taken with the pitot tube. The support was then displaced 20 mm so that readings with the hot-wire probe could be made at the same point. Three repetitions were made. One stagnation pressure reading and, for the majority of the experiments, one voltage reading were made in each sequence. In one experiment 100 voltage readings were taken in each sequence in order to obtain better information about the accuracy of the measurements.

The results of these measurements are given in the tables after the text.

The exponent c was evaluated from the measurements by insertion of a measured value $(\rho u)_I$ and the associated value of the voltage, V_I in

Eq. (3). The same was done for the values $(\rho u)_{II}$ and V_{II} measured at a small interval in ρu from the first values. After logarithmation the two equations were subtracted, giving

$$c = \frac{\ln(V_1^2 - V_0^2) - \ln(V_{II}^2 - V_0^2)}{\ln(\rho u)_I - \ln(\rho u)_{II}} \qquad (4)$$

The value of c so calculated was assumed valid for the average conditions between the two measurements.

The complete results are given in the tables after the text.

Representative results for three sets of measurements are shown in Fig. 1. There is an apparent decrease of c with increasing values of ρu.

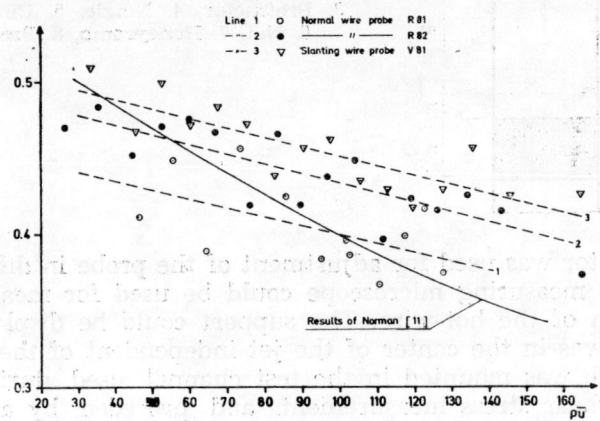

Fig. 1. Influence of $\rho \bar{u}$ on the exponent c in Collis' law

Similar observations have been made by Norman [10, 11], whose curve is also shown in Fig. 1 for comparison, and recently by Davis [4].

For each set of data a least square fit of a straight line

$$c = c_{av} + K[(\rho u) - (\rho u)_{av}] \qquad (5)$$

was made. It turned out that the differences in K for different sets of data were not significant, so that an average value could be used,

$$K = -0.0007782 \qquad (6)$$

with 95% confidence limits of ± 0.000122. This value of K seems to be valid independent of the air temperature in the range $320 > T > 280\ K$, a moderate flow turbulence (corresponding to channel flow) and the wire inclination in the range $54° > \Psi > 0$.

It can be observed that the variation found by Norman [10, 11] is much more pronounced. This disagreement is not considered very serious since the curve shown is based on only six observations.

2.3. THE DIRECTION SENSITIVITY COEFFICIENT

The direction sensitivity coefficient k^2 was measured by mounting the probe as shown in Fig. 2 in a jet after a nozzle.

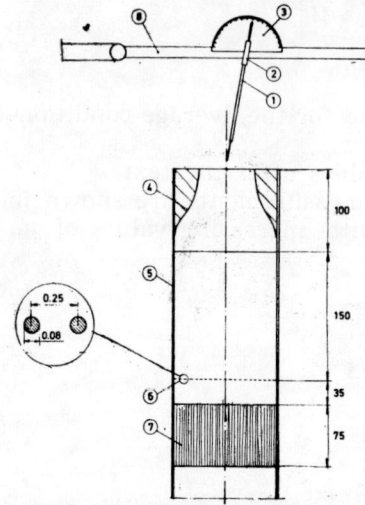

Fig. 2. Principal arrangement of the experiment for determination of the direction sensitivity coefficient.
1. Hot wire probe, 2. Revolving probe holder, 3. Protractor, 4. Nozzle, 5. Circular channel, 6. Net, 7. Honeycomb, 8. Probe support.

A protractor was used for adjustment of the probe in different inclinations and a measuring microscope could be used for measurement of the inclination of the hot-wire. The support could be displaced so that the hot wire was in the center of the jet independent of the inclination.

The nozzle was mounted in the test channel used during the turbulence and shear stress measurements and preceded by a honeycomb and a net. Measurements were made at different velocities. The results are given in Fig. 3.

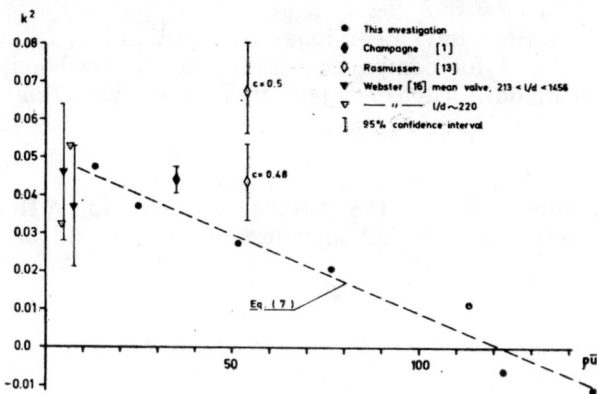

Fig. 3. Comparison of measurements of the direction sensitivity coefficient.

The results can be correlated by the equation

$$k^2 = 0.0505 - 0.000415 \, \rho u \qquad (7)$$

The evaluation of k^2 was made by fitting the constants B and $\dfrac{k^2}{1-k^2}$ in the equation

$$\sin^2 \psi - B (V^2 - V_0^2)^{2/c} + \frac{k^2}{1-k^2} = 0 \qquad (8)$$

to the experimental data obtained at the same velocity for different values of Ψ.

Equation (7) can easily be derived from Eq. (3) by utilization of well known trigonometric laws and the fact that ρu is constant.

For comparison, values of k^2 measured by earlier investigators have also been included in Fig. 3. It should also be added that Hinze [6] recommends values between 0.09 and 0.01, the lower values at high velocities. From the data of Rasmussen [13] two values of k^2, corresponding to different assumptions concerning the value of c (which was not given by Rasmussen), were calculated. This illustrates the large influence of the value for c used in the evaluation of k^2.

The agreement with earlier values of k^2 at low velocities and the somewhat vague recommendation of Hinze [6] is most satisfactory, which is a good support for the results of this investigation.

Measurements were only made for one hot-wire probe since the extensive investigation of Champagne [1] has shown that for different probes, under otherwise similar conditions, k^2 is mainly determined by the length-to-diameter ratio, and this was the same (around 220) for all probes used in the later experiments.

In this connection it should also be mentioned that many investigations, cf. for instance Patel [12], have neglected k^2. To study the effect of this the measurements in the circular channel described below were evaluated also with $k^2 = 0$.

3. Measurements in a Circular Channel

3.1. THEORETICAL CONSIDERATIONS

It is easy to show that for fully developed incompressible flow in a circular channel the shear stress distribution should be linear:

$$\frac{\tau}{\tau_w} = \frac{r}{R} \qquad (9)$$

In this experiment, however, measurements were made up to rather high velocities so that compressibility effects could not be neglected.

Strating from the Reynolds equations in the axial and radial directions for compressible flow in a circular channel

$$\bar{\rho}\,\bar{u}\,\frac{\partial \bar{u}}{\partial x} + \bar{\rho}\,\bar{r}\,\frac{\partial \bar{u}}{\partial r} = -\frac{\partial \bar{p}}{\partial x} + \frac{\partial}{\partial x}\left(\eta \frac{\partial \bar{u}}{\partial x} - \bar{\rho}\,\overline{u'^2}\right) +$$

$$+ \frac{1}{r}\frac{\partial}{\partial r}\left(\eta r \frac{\partial \bar{u}}{\partial r} - r\,\bar{\rho}\,\overline{u'r'}\right) + \frac{\eta}{3}\frac{\partial \bar{\theta}}{\partial x} \qquad (10)$$

and

$$\bar{\rho}\bar{u}\frac{\partial \bar{r}}{\partial r}+\bar{\rho}\bar{r}\frac{\partial \bar{r}}{\partial r}=-\frac{\partial \bar{p}}{\partial r}\frac{\partial}{\partial x}\left(\eta\frac{\partial \bar{r}}{\partial x}-\bar{\rho}\,\overline{u'r'}\right)+$$

$$+\frac{1}{r}\frac{\partial}{\partial r}\left(\eta r\frac{\partial \bar{r}}{\partial r}-r\bar{\rho}\,\overline{r'^{2}}\right)-\eta\frac{\bar{r}}{r^{2}}+\frac{\eta}{3}\frac{\partial \Theta}{\partial r} \qquad (11)$$

and assuming that after some distance from the inlet the axial variations of velocities and velocity correlations made dimensionless by division by the bulk velocity defined as

$$u_b=\frac{2}{R^2}\int_0^R \bar{u}\, r dr \qquad (12)$$

can be neglected, i.e. that

$$\frac{\partial}{\partial x}\frac{\bar{u}}{u_b}=\frac{\partial}{\partial x}\frac{\bar{r}}{u_b}=\frac{\partial}{\partial x}\frac{\overline{r'^{2}}}{u_b^2}=\frac{\partial}{\partial x}\frac{\overline{u'r'}}{u_b^2}=\frac{\partial}{\partial x}\frac{\overline{r'^{2}}}{u_b^2}=0 \qquad (13)$$

it is possible to show that the Reynolds equation in the axial direction, Eq. (10), can be written

$$\eta\frac{\partial \bar{u}}{\partial r}-\bar{\rho}\,\overline{u'r'}-\zeta=\frac{r}{2}\frac{dp_f}{dx} \qquad (14)$$

where the compressibility correction ζ can be calculated from

$$\zeta=\frac{1}{ru_b}\frac{du_b}{dx}\int_0^r\left[\bar{\rho}\left(\overline{u^2}+\overline{u'^2}\right)-\rho_b u_b^2\right]r dr \qquad (15)$$

and the friction pressure gradient is defined by

$$\frac{dp_f}{dx}=\frac{dp}{dx}+\rho_b u_b\frac{du_b}{dx} \qquad (16)$$

The assumption concerning the axial constancy of the dimensionless axial velocity $\bar{u}/u_b$ is supported by experiments of Deissler [5]. Local similarity considerations then justify the assumption concerning other velocity components and velocity correlations.

For detailed information on the considerations leading to Eqs. (14)—(16) reference should be made to the report by Kjellström and Hedberg [19].

Since the shear stress is

$$\tau_{xr}=\eta\frac{\partial \bar{u}}{dx}-\bar{\rho}\,\overline{u'r'} \qquad (17)$$

and the wall shear stress may be written as

$$\tau_W = \frac{R}{2}\frac{dp_f}{dx} \tag{18}$$

Equation (14) may also be written

$$\frac{\tau - \zeta}{\tau_W} = \frac{r}{R} \tag{19}$$

which can be compared to Eq. (9) which holds for incompressible flow.

With the exception of $\overline{u'r'}$ and $\overline{u'^2}$ (the latter is however of small importance), which are to be measured with the hot-wire anemometer, as will be described later, all quantities which come into Eq. (19) can be measured by standard methods with good accuracy. Eq. (19) therefore offers a valuable possibility of checking the accuracy of the hot-wire measurements.

3.2. THE EXPERIMENTAL ARRANGEMENT

The measurements were made at the outlet of a smooth circular channel shown in Fig. 4. A close-up view of the arrangements at the outlet of the channel is shown in Fig. 5. The diameter of the channel was 128.8 mm and the measurements were made at 61.1 diameters from the inlet, which should ensure fully developed flow.

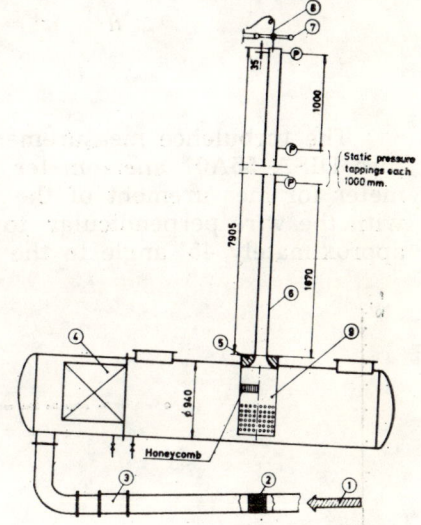

Fig. 4. Principal arrangement of the experiment.
1. Air from blower; 2. Flow rectifier; 3. Foster flow meter; 4. Cooler; 5. Wooden nozzle; 6. Test channel ø 128.8 mm; 7. Probe support; 8. Hot wire — or stagnation pressure probe; 9. Flow rectifier positioned before the inlet nozzle.

That the flow was fully developed is supported by the friction factor measurements over the last meter of the channel, the results of which (see Fig. 6) are in excellent agreement with the correlation of Nikuradse [9].

The mean velocity distributions, which were measured by a stagnation pressure tube, are also in good agreement with earlier results when plotted as the dimensionless velocity deficiency $\hat{u}^+ - u^+$, see Fig. 7 where a comparison has been made with the results of Nikuradse [9] and Coantic [2] and the correlation of Reichardt [15]

Fig. 5. View of the outlet of the test channel

$$\hat{u}^+ - u^+ = 2.5 \ln \frac{1+2\left(\frac{r}{R}\right)^2}{1-\left(\frac{r}{R}\right)^2} \qquad (20)$$

The turbulence measurements were made with a constant temperature DISA 55A01 anemometer together with a four-figure digital voltmeter for measurement of the mean voltage. Miniature probes 55A25, with the wire perpendicular to the flow, and 55A29 with the wire at approximately 45° angle to the flow, were employed.

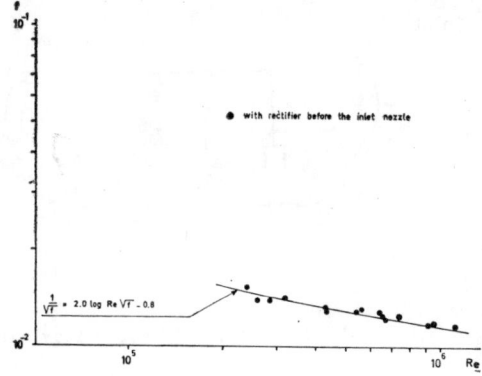

Fig. 6. Friction factors measured for $x/D = 57.2$

For each experiment a velocity traverse with a stagnation pressure probe was first made. This was followed by a traverse with the right angle hot-wire probe and two traverses with the slanting wire probe, the probe being rotated 180° around its own axis between the latter two traverses.

Before each traverse the angle between the wire and the flow direction was measured by means of a microscope, which can be seen in Fig. 5.

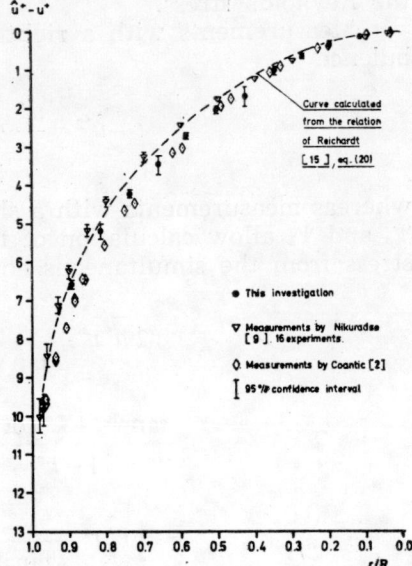

Fig. 7. Comparison of velocity distributions in circular channels.

3.3. EVALUATION OF THE HOT-WIRE MEASUREMENTS

Differentiation of Eq. (3), neglecting variations in the density in b, c and k^2, and division of the resulting equation by Eq. (3) give

$$\frac{2V\,dV}{V^2 - V_0^2} = c\left(\frac{du}{u} + \frac{\sin\psi\cos\psi - k^2\sin\psi\cos\psi}{\sin^2\psi + k^2\cos^2\psi}\,d\psi\right) \quad (21)$$

It can be shown that for small velocity fluctuations u', $\dot{r}'$ and $\dot{\varphi}'$ and small voltage fluctuations

$$du \sim u' \quad (22)$$

$$d\psi \sim \frac{\dot{r}'}{u} \quad (23)$$

$$dV \sim V' \quad (24)$$

Insertion of this in Eq. (21), squaring and taking the time mean value give after some rearrangement

$$\beta^2 \overline{V'} = \frac{c^2}{u^2}\left[\overline{u'}^2 + \left(\frac{2-k^2}{\tan\psi + k^2\cot\psi}\right)^2 \overline{\dot{r}'}^2 + 2\frac{1-k^2}{\tan\psi + k^2\cot\psi}\overline{u'\dot{r}'}\right] \quad (25)$$

where

$$\beta = \frac{2\overline{V}}{\overline{V^2} - \overline{V_0^2}} \qquad (26)$$

These equations allow calculation of the turbulence components and the Reynolds stress.

Measurements with a right angle probe $\Psi = 90°$ gives the axial turbulence

$$\overline{u'^2} = \frac{\overline{u}^2}{c^2} \cdot \left(\frac{2\overline{V}}{\overline{V^2} - \overline{V_0^2}}\right)^2 \overline{V'^2} \qquad (27)$$

whereas measurements with a slanting wire probe at two slanting angles Ψ_a and Ψ_b allow calculation of the radial turbulence and the Reynolds stress from the simultaneous equations

$$2\,\overline{u'r'} + \frac{1-k^2}{\tan\Psi_a + k^2 \cot\Psi_a} \cdot \overline{r'^2} =$$

$$\frac{\tan\Psi_a + k^2 \cot\Psi_a}{1-k^2}\left[\frac{\overline{u}^2}{c^2}\,\beta_a^2\,\overline{V_a'^2} - \overline{u'^2}\right] \qquad (28)$$

$$2\,\overline{u'r'} + \frac{2-k^2}{\tan\Psi_b + k^2 \cot\Psi_b}\,\overline{r'^2} =$$

$$\frac{\tan\Psi_b + k^2 \cot\Psi_b}{1-k^2}\left[\frac{\overline{u}^2}{c^2}\,\beta_b^2\,\overline{V_b'^2} - \overline{u'^2}\right] \qquad (29)$$

For each point of measurement the proper value of k^2 was claculated from Eq. (7) and the proper value of c from Eq. (5), c_{av} being determined by a least square fit to the simultaneous data for V and $\rho\overline{u}$ obtained during the traverse, and $(\rho\overline{u})_{av}$ calculated as the logarithmic mean value for the data from the traverse.

For comparison, evaluations were also made with $k^2 = 0$ and c constant equal to c_{av}.

3.4. MEASURED SHEAR STRESS AND TURBULENCE DISTRIBUTIONS

The measured shear stress and turbulence distributions are shown in Figs. 8—10.

The shear stress distribution is in excellent agreement with the theoretical distribution, Eq. (19). The agreement of the distributions of the axial and radial turbulence, with the results of earlier investigators, Coantic [2], Patel [12], Reichardt [15] and Laufer [8], is also very good.

It can therefore be concluded that the measurements in the circular channel confirm in a most satisfactory way the calibration measurements with the anemometer.

It remains only to check the reduction of accuracy which would have been obtained if a simpler method of evaluation had been used.

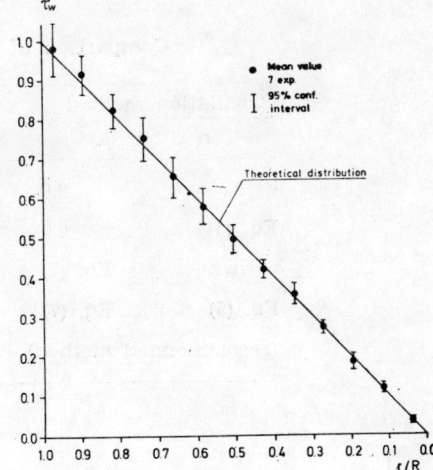

Fig. 8. Experimental and theoretical shear stress distributions

For comparison of the different evaluation methods, least square fits of the constants K_1 and K_2 in the equation

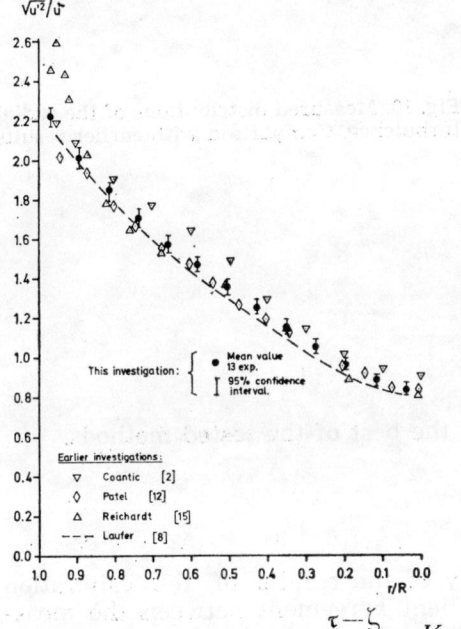

Fig. 9. Measured distributions of the axial turbulence. Comparison with earlier results

$$\frac{\tau-\zeta}{\frac{r}{R}\cdot \tau_W} = K_1 + K_2\left(\frac{r}{R}-0.5\right) \tag{30}$$

were made for the evaluated data.

For perfect agreement with the theory K_1 and K_2 should attain the values of unity and zero respectively.

The results of the comparison are shown in the table below.

Comparison of evaluation methods

Calculation method		K_1	K_2
c	k^2		
$c = c_{av}$	0.0	0.945	0.132
Eq. (5)	0.0	0.967	0.033
$c = c_{av}$	Eq. (7)	0.972	0.160
Eq. (5)	Eq. (7)	0.994	0.060
(recommended method)			

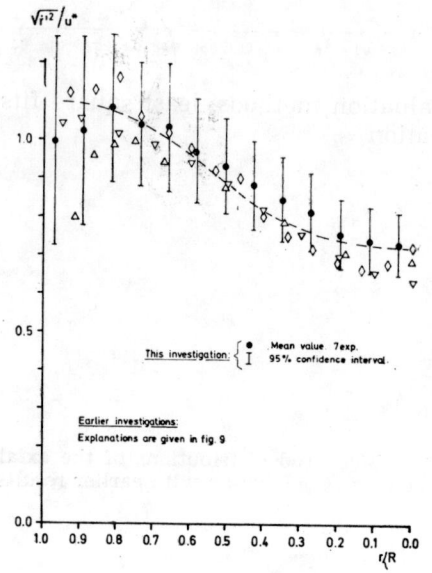

Fig. 10. Measured distributions of the radial turbulence. Comparison with earlier results

The fourth method is apparently the best of the tested methods.

4. Conclusions

A strong support for the validity of the results of the calibration experiments is obtained by the excellent agreement between the measured and theoretical shear stress distributions.

It can thus be concluded with a certain confidence that

— the characteristic equation derived, Eq. (3), is suitable for analytical representation of the calibration results.

- the voltage corresponding to zero velocity can be determined by measurements with no net flow past the wire. This implies variations with the velocity of the exponent in Collis' law and the direction sensitivity coefficient.
- these variations can with sufficient accuracy be expressed by linear relations, within the range studied in this investigation, ρu between 20 and 160 kg/sm².
- the direction coefficients of the straight lines representing the variation of the exponent in Collis law with ρu seem to be the same for all the probes used, independent of the fluid temperature.
- evaluation of the results with a constant value of c or neglecting the direction sensitivity coefficient gives less good agreement between measured and theoretical shear stresses.

NOMENCLATURE

A	flow area
a	constant
B	constant
b	constant
c	exponent in Collis' law
d	wire diameter
f	friction factor
K	constant
k	direction sensitivity coefficient
L	axial length in the test channel
l	wire length
$\dot{m}$	mass flow
p	pressure
R	radius of test channel
R	electric resistance
r	radius
$\dot{r}$	radial velocity
Re	Reynolds number
T	temperature
u	velocity, axial direction
u^*	friction velocity $u^* = \sqrt{\dfrac{\tau_w}{\rho}}$
u^+	dimensionless velocity $u^+ = \bar{u}/u^*$
u_b	bulk velocity
V	voltage
V_0	voltage at zero velocity

x	axial coordinate
β	defined by Eq. (26)
η	dynamic viscosity
ρ	density
τ	shear stress
Ψ	slanting angle of hot-wire
ς	correction term defined by Eq. (15)

SUBSCRIPTS

a, b	refers to the two hot-wire traverses made with V-probe
av	average
f	friction
tot	stagnation conditions
w	wall
wi	wire

SPECIAL SIGNS

$'$	fluctuating ex. u'
—	mean value ex. $\bar{u}$
$\wedge$	maximum ex. $\hat{u}$

Acknowledgements

This work was done at the Section for Heat Transfer and Component Experiments of AB Atomenergi, Studsvik.

The authors want to express their gratitude to Mr Jan E. Flinta, Head of the Laboratory, to Mr Bror Hedberg who prepared the test equipment and assisted during the measurements, to Mr Rolf Lindh who coded the computer program for evaluation of the data, to Messrs Andreas Brosos and Göran Karlsson who made some of the measurements, and to Mr Peter Lundborg who carried out many of the manual calculations.

REFERENCES

1. F. H. Champagne, Turbulence measurements with inclined hot-wires, Thesis, Washington Univ. Seattle, Wash., 1966.
2. M. Coantic, Contribution à l'étude théorique et expérimentale de l'écoulement turbulent dans un tube circulaire, Publ. Scientifiques et Techniques du Ministère de l'air, Paris, 1962.
3. D. C. Collis and M. J. Williams, J. Fluid. Mech., 6, 357—84 (1959).
4. M. R. Davis, Heat transfer from hot wire and hot film anemometer probes. Int. Summer School of Heat and Mass Transfer in Turbulent Boundary Layer, Herceg Novi, Sept. 9—21, 1968.
5. R. G. Deissler, NACA — TN — 2138, 1950.

6. J. O. Hinze, »Turbulence«, McGraw-Hill, New York, 1959.
7. B. J. Hoole and J. R. Calvert, J. Royal Aeronaut. Soc., **71**, pp. 511—13 (1967).
8. J. Laufer, NACA-TN-2954, 1953.
9. J. Nikuradse, VDI Forsch., Heft 356 (1932).
10. B. Norman, Hot-wire anemometer calibration at high subsonic speeds. DISA Information 1967:5, 5—19.
11. B. Norman, Division of Steam Engineering, Royal Institute of Technology Stockholm, Sweden. Personal communication, Dec. 1967.
12. R. P. Patel, Mech. Eng. Res. Lab., McGill University, Montreal TN 63—6, 1963.
13. C. G. Rasmussen, Dansk Industrisyndikat A/S Herlev Denmark. Personal communication, Dec. 13, 1965.
14. H. Reichardt, Die Naturwissenschaften, 26, 404—8 (1938).
15. H. Reichardt, ZAMM, 31, 208—19 (1951).
16. C. A. G. Webster, J. Fluid. Mech., 13, 307—12 (1962).
17. DISA Elektronik A/S Herlev Denmark, »DISA Constant temperature anemometer Instruction manual«, 1966. (Reg. № 9150 A 0213).
18. Ibid., »Instruction manual for DISA random signal indicator and correlator type 55A06«, 1962. (Reg. № 9150 A 5311).
19. B. Kjellström and S. Hedberg, Calibration experiments with a DISA hot-wire anemometer, AE-338, 1968.

Table 1. Results of experiments for determination of c (cont.)

Experiment № 680207
made in: Channel
$T_\infty = 282\ K$
$V_o = 5.45$ volt

with probe: R82
$P_{stat} = 0.998$ bar
$R/R_a = 1.8$

ρu kg/m² s	V volt	c	ρu kg/m² s
181.84	12.057		
148.25	11.700	0.3727	165.05
138.42	11.567	0.4286	143.34
129.70	11.443	0.4240	134.06
122.54	11.340	0.4155	126.12
115.39	11.230	0.4226	118.97
107.47	11.110	0.3966	111.43
100.61	10.987	0.4476	104.04
93.04	10.847	0.4368	96.83
85.77	10.710	0.4182	89.41
81.05	10.607	0.4647	83.41
70.97	10.397	0.4187	76.01
62.97	10.187	0.4666	66.97
56.85	10.013	0.4746	59.91
48.41	9.753	0.4071	52.63
44.69	9.507	0.4518	44.69
35.39	9.040	0.4843	35.39
26.56	8.720	0.4703	26.56

Table 1. Results of experiments for determination of c (cont.)

Experiment № 680208
made in: Channel
$T_\infty = 285\ K$
$V_o = 5.40$ volt

with probe: V81
$P_{stat} = 1.027$ bar
$R/R_a = 1.8$

ρu kg/m² s	V volt	c	ρu kg/m² s
178.81	11.000		
		0.4252	165.45
152.09	10.720		
		0.4236	145.98
139.87	10.580		
		0.4562	135.39
130.91	10.463		
		0.4278	127.25
123.59	10.370		
		0.4171	119.88
116.16	10.273		
		0.4286	112.46
108.76	10.170		
		0.4345	105.07
101.38	10.060		
		0.4619	97.48
93.58	9.930		
		0.4561	90.05
86.51	9.807		
		0.4376	82.67
78.83	9.670		
		0.4718	75.17
71.51	9.520		
		0.4830	67.73
63.95	9.350		
		0.4715	60.31
56.66	9.177		
		0.4990	52.63
48.59	8.953		
		0.4699	45.11
41.63	8.753		
		0.5094	33.09
24.55	8.093		

Table 1. Results of experiments for determination of c (cont.)

Experiment № 680405—1
made in: jet after nozzle with probe: V81
$T_\infty = 282\ K$ $P_{stat} = 0.996$ bar
$V_o = 5.48$ volt $R/R_a = 1.8$

ρu kg/m² s	V volt	c	ρu kg/m² s
197.18	11.210		
		0.4788	187.27
177.36	10.997		
		0.4645	164.40
151.44	10.700		
		0.4681	145.08
138.71	10.540		
		0.4729	134.24
129.78	10.420		
		0.4290	125.96
122.15	10.323		
		0.4884	118.32
114.49	10.207		
		0.4746	110.48
106.46	10.083		
		0.4765	102.32
98.19	9.947		
		0.4649	94.49
90.79	9.823		
		0.4967	88.15
85.50	9.723		
		0.4719	81.59
77.67	9.575		
		0.4872	74.19
70.71	9.430		
		0.5072	66.57
62.43	9.237		
		0.4893	59.01
55.58	9.070		
		0.5691	54.57
53.56	9.010		

Table 1. Results of experiments for determination of c (cont.)

Experiment № 680405—2
made in: jet after nozzle   with probe: V81
$T_\infty = 320$ K $P_{stat} = 0.997$ bar
$V_o = 5.49$ volt $R/R_a = 1.8$

ρu kg/m² s	V volt	c	ρu kg/m² s
53.28	9.233		
		0.4524	56.30
59.31	9.380		
		0.4211	63.45
67.59	9.553		
		0.4410	70.29
73.00	9.663		
		0.4286	76.34
79.68	9.787		
		0.4377	84.28
88.88	9.950		
		0.4627	92.15
95.42	10.065		
		0.4163	98.82
102.21	10.167		
		0.4143	105.82
109.43	10.270		
		0.3709	112.73
116.02	10.350		
		0.4256	119.77
123.51	10.450		
		0.3624	126.40
129.30	10.513		
		0.4065	135.95
142.60	10.667		
		0.3716	152.91
163.22	10.867		
		0.3628	172.78
182.35	11.033		

Table 1. Results of experiments for determination of c (cont.)

Experiment № 680417
made in: jet after nozzle   with probe: V81
$T_\infty = 281$ K $P_{stat} = 1.028$ bar
$V_o = 5.42$ volt $R/R_a = 1.8$

ρu kg/m² s	V volt	c	ρu kg/m² s
140.05	10.593		
		0.4786	136.39
132.73	10.493		
		0.4281	128.53
124.33	10.387		
		0.4758	120.71
117.10	10.280		
		0.4577	113.61
110.13	10.177		
		0.4868	106.09
102.06	10.047		
		0.4697	98.49
94.91	9.923		
		0.5035	91.06
87.20	9.777		
		0.4712	83.23
79.26	9.627		
		0.5766	75.61
71.96	9.447		
		0.4781	67.64
63.32	9.257		
		0.4964	60.01
56.71	9.093		
		0.5206	52.80
48.89	8.873		
		0.5190	44.94
40.99	8.627		
		0.5023	37.41
33.83	8.383		
		0.4502	29.97
26.11	8.111		

Table 1. Results of experiments for determination of c (cont.)

Experiment № 680423
made in: jet after nozzle   with probe: V81
$T_\infty = 282\ K$ $P_{stat} = 1.020$
$V_o = 5.429$ volt $R/R_a = 1.8$
Mean values of 100 readings.

ρu kg/m² s	V volt	c	ρu kg/m² s
139.164	10.6175		
		0.4390	132.74
126.307	10.4895		
		0.4243	122.08
117.847	10.3776		
		0.4595	113.91
109.965	10.2589		
		0.4545	105.68
101.397	10.1244		
		0.4548	97.77
94.138	10.0039		
		0.4635	90.96
87.781	9.8908		
		0.4037	84.29
80.792	9.7763		
		0.4747	76.37
71.944	9.5935		
		0.4736	67.78
63.608	9.4070		
		0.4744	59.88
56.152	9.2251		
		0.4910	52.41
48.667	9.0183		
		0.4984	44.89
41.114	8.7835		
		0.5257	37.26
33.399	8.4980		
		0.5462	30.91
28.428	8.2835		
		0.5734	24.45
20.462	7.8671		

Table 2. Results of experiment for study of the influence of the inclination of the wire on the exponent in Collis' law

ψ	u^1 m/s	V^1 volt	$\ln \rho u$	$\ln(V^2 - V_o^2)$	c
34°	72.49	9.680	4.5324	4.1580	0.4065
	61.37	9.458	4.3634	4.0893	
44°	72.49	9.998	4.5324	4.2514	0.4000
	61.37	9.766	4.3634	4.1838	
54°	72.49	10.240	4.5324	4.3188	0.4053
	61.37	9.994	4.3634	4.2503	

[1] Mean values of 5 measurements.
V_o=5.455 volt.

Table 3. Results of experiments for determination of the direction sensitivity coefficient

Experiment 680402—1					Experiment 680402—2			
	ψ^o	Measured values of $\overline{V}$ volt				ψ^o	Measured values of $\overline{V}$ volt		
		Reading 1	Reading 2	Mean value			Reading 1	Reading 2	Mean value
$\bar{u}=10.86$ m/s	35.0	7.41	7.40	7.405	$\bar{u}=20.69$ m/s	35.0	8.03	8.03	8.030
$\rho=1.21$ kg/m^3	37.0	7.44	7.44	7.440	$\rho=1.21$ kg/m^3	37.0	8.08	8.08	8.080
$\rho\bar{u}=13.132$ kg/m^2 s	39.0	7.49	7.48	7.485	$\rho\bar{u}=24.937$ kg/m^2 s	39.0	8.13	8.12	8.125
	41.0	7.52	7.52	7.520		41.0	8.17	8.17	8.170
$R/R_o=1.8$	43.0	7.55	7.56	7.555	$R/R_o=1.8$	43.0	8.21	8.21	8.210
	45.0	7.58	7.58	7.580		45.0	8.25	8.26	8.255
$V_o=5.45$ volt	47.0	7.61	7.61	7.610	$V_o=5.45$ volt	47.0	8.30	8.30	8.300
$c=0.5321$	49.0	7.64	7.64	7.640	$c=0.5226$	49.0	8.34	8.34	8.340
	51.0	7.67	7.67	7.670		51.0	8.37	8.37	8.370
	53.0	7.70	7.70	7.700		53.0	8.39	8.39	8.390
$k^2=0.0473$	55.0	7.71	7.71	7.710	$k^2=0.0371$	55.0	8.42	8.42	8.420

Table 3. Results of experiments for determination of the direction sensitivity coefficient (cont.)

	Experiment 680403—1					Experiment 680327—1			
		Measured values of $\bar{V}$ volt					Measured values of $\bar{V}$ volt		
	ψ^0	Reading 1	Reading 2	Mean value		ψ^0	Reading 1	Reading 2	Mean value
$\bar{u}=41.55$ m/s	35.0	8.84	8.84	8.840	$\bar{u}=59.59$ m/s	35.0	9.40	9.40	9.400
$\rho=1.24$ kg/m^3	37.0	8.90	8.90	8.900	$\rho=1.28$ kg/m^3	37.0	9.47	9.47	9.470
$\rho\bar{u}=51.574$ kg/m^2 s	39.0	8.96	8.96	8.960	$\rho\bar{u}=76.653$ kg/m^2 s	39.0	9.53	9.54	9.535
	41.0	9.01	9.02	9.015		41.0	9.60	9.60	9.600
$R/R_\omega=1.8$	43.0	9.07	9.07	9.070	$R/R_\omega=1.8$	43.0	9.65	9.65	9.650
	45.0	9.12	9.12	9.120		45.0	9.71	9.71	9.710
$V_o=5.44$ volt					$V_o=5.44$ volt	47.0	9.76	9.76	9.760
$c=0.5011$	49.0	9.21	9.22	9.215	$c=0.4809$	49.0	9.81	9.81	9.810
	51.0	9.25	9.25	9.250		51.0	9.86	9.86	9.860
	53.0	9.29	9.29	9.290		53.0	9.90	9.90	9.900
$k^2=0.0276$	55.0	9.33	9.33	9.330	$k^2=0.0204$	55.0	9.94	9.94	9.940

Table 3. Results of experiments for determination of the direction sensitivity coefficient (cont.)

Experiment 680327—2					Experiment 680404—1				
	ψ^0	Measured values of $\bar{V}$ volt				ψ^0	Measured values of $\bar{V}$ volt		
		Reading 1	Reading 2	Mean value			Reading 1	Reading 2	Mean value
$\bar{u}=87.47$ m/s	35.5	10.01	10.01	10.010	$\bar{u}=96.69$ m/s	35.0	10.10	10.10	10.100
$\rho=1.29$ kg/m³	37.5	10.08	10.08	10.080	$\rho=1.26$ kg/m³	37.0	10.18	10.18	10.130
$\rho\bar{u}=113.015$ kg/m²s	39.5	10.15	10.15	10.150	$\rho\bar{u}=122.198$ kg/m²s	39.0	10.25	10.25	10.250
	41.5	10.21	10.21	10.210		41.0	10.32	10.32	10.320
$R/R_o=1.8$	43.5	10.27	10.28	10.275	$R/R_o=1.8$	43.0	10.38	10.38	10.330
	45.5	10.33	10.34	10.335		45.0	10.44	10.44	10.440
$V_o=5.45$ volt	47.5	10.38	10.39	10.385	$V_o=5.45$ volt	47.0	10.50	10.50	10.500
	49.5	10.44	10.45	10.445		49.0	10.56	10.56	10.560
$c=0.4516$	51.5	10.49	10.50	10.495	$c=0.4442$	51.0	10.61	10.62	10.615
	53.5	10.54	10.54	10.540		53.0	10.66	10.66	10.660
$k^2=0.0113$	55.5	10.59	10.59	10.590	$k^2=-0.0063$	55.0	10.70	10.70	10.700

Table 3. Results of experiments for determination of the direction sensitivity coefficient (cont.)

Experiment 680327—3

	ψ^0	Measured values of $\bar{V}$ volt		
		Reading 1	Reading 2	Mean value
$\bar{u}=112.95$ m/s	35.0	10.40	10.40	10.400
$\rho=1.30$ kg/m³	37.0	10.48	10.48	10.480
$\rho\bar{u}=146.694$ kg/m² s	39.0	10.55	10.55	10.550
	41.0	10.62	10.62	10.620
$R/R_o=1.8$	43.0	10.69	10.69	10.690
	45.0	10.75	10.74	10.745
$V_o=5.44$ volt				
	47.0	10.80	10.80	10.800
$c=0.4244$	49.0	10.86	10.86	10.860
	51.0	10.92	10.92	10.920
	53.0	10.97	10.97	10.970
$k^2=-0.0111$				
	55.0	11.01	11.01	11.010

THE THEORY OF ELECTROCHEMICAL TECHNIQUE OF INVESTIGATING TURBULENCE AT A WALL

V. YE. NAKORYAKOV

Thermal Physics Institute, Siberian Division of the USSR Academy of Sciences, Novosibirsk, USSR

Electrochemical technique of friction measurement worked out by T. Hanratty and his colleagues [1] makes it possible to directly measure the shear stress and its fluctuation on the boundary of a solid body and an electrolyte flow.

This technique was used by the author of [1] to measure turbulence characteristics in the immediate vicinity of a pipe wall. The technique is based on measuring a limiting diffusion flow to an electrode mounted flush with the pipe wall. Under the conditions of diffusion the rate of reaction on the test electrode is determined by mass transfer intensity and the electrode works on a principle similar to that of a hot-wire anemometer with a constant surface temperature.

The present paper reports the results of a theoretical study of the test electrode functioning, carried out to specify the technique of measuring fluctuation of a shear stress at a wall.

1. The point of the technique lies in measuring the limiting diffusion current in an electrolytic cell made up of an electrolyte flowing in a pipe, a large electrode-anode and a measuring electrode-cathode mounted in the pipe wall flush with its surface.

The solution of $K_3Fe(CN)_6$ $K_4Fe(CN)_6$ in water with a solution of $NaOH$ serves as an electrolyte. This electrolyte composition prevents ion migrations under the action of the electric field and ensures the absence of precipitations on the electrodes during an experiment. Under the conditions of the limiting diffusion current the reaction rate on the test electrode is determined by the intensity of mass transfer between the electrode surface and the electrolyte flow. The mass flow is connected with the value of the limiting diffusion current by the ratio

$$q = I/AF$$

If the solution of the diffusion boundary layer equations yields the dependence $q = f(\tau)$, then the shear stress value is readily calculated. The solution of the diffusion boundary layer equations on the test electrode is facilitated by the fact that Prandtl's diffusion numbers, characterizing the rate of ion transfer, are rather large due to small values of electrolyte ion diffusion coefficients. This allows to use in calculations Levich's hypothesis [2] of the linear profile of velocity in the difusion boundary layer:

$$u = \frac{\tau}{\mu} y$$

2. Write the equations of the diffusion boundary layer and the boundary conditions in the form:

$$\frac{\partial C}{\partial t} + u \frac{\partial C}{\partial x} + v \frac{\partial C}{\partial y} = D \frac{\partial^2 C}{\partial y^2} \qquad (1)$$

$C=0$ at $y=0$, $C=C$ at $y=\infty$

Choose the following values as characteristic ones: the length of the plate L, the velocity $v=(\tau_0/\mu)L$, the time $t=\dfrac{L}{v}$. Then the diffusion boundary layer equation has the form:

$$\frac{\partial C}{\partial t} + u \frac{\partial C}{\partial x} + v \frac{\partial C}{\partial y} = \frac{1}{Pe} \frac{\partial^2 C}{\partial y^2} \qquad (2)$$

Here u, v, C, x, y are non-dimensional values.

$$u = \frac{\tau_0}{\mu} y + \sum_{n=1}^{N} \frac{\tau_{n-}}{\mu} + y \cos n\omega t + \frac{\tau_n}{\mu} - y \sin n\omega t \qquad (3)$$

assuming the average shear stress and its fluctuation ranges to be indepedent of the longitudinal coordinate. This assumption was used in [1]. It implies that in using this method stationary and fluctuation values of the shear stress are determined which are averaged along the length of the gauge. Strictly formula (3) is valid in case of average-stabilized flows and long-wave perturbances.

The non-dimensional expression for the longitudinal velocity component is written in the complex form of Fourier's series

$$u = y \left(1 + \sum_{n=-\infty}^{n=+\infty} v_n e^{in\omega t} \right), \quad n \neq 0$$

where v_{-n} is the conjugate complex number cv_n, and v_n is expressed as

$$v_n = \frac{\tau_{n+} - \tau_{n-}}{2}$$

Here $\tau_{n\pm}$ are normalized by the mean shear stress. All expressions to be used further will be considered non-dimensional.

The field of concentrations is also represented in the complex form of Fourier's series

$$C = C_0 + \sum_{n=-\infty}^{n=+\infty} C_n e^{in\omega t}, \quad n \neq 0 \qquad (3a)$$

where C_n is the complex amplitude.

Assuming c_n, v_n to be small values, we obtain the stationary diffusion boundary layer equation and the equation for the amplitude of concentration fluctuations

$$y \frac{\partial C_0}{\partial x} = \frac{1}{Pe} \frac{\partial^2 C_0}{\partial y_0} \qquad (4)$$

with the boundary conditions $C_n = 0$, at $y = 0$, $C_n = 0$ at $u = \infty$,

$$in\omega C_n + y \frac{\partial C_n}{\partial x} + y v_n \frac{\partial C_0}{\partial x} = \frac{1}{Pe} \frac{\partial^2 C_n}{\partial y^2} \qquad (5)$$

with the boundary conditions $C_n = 0$, at $y = 0$, $C_n = 0$ at $y = \infty$.

To solve Eq. (4) introduce a new indepedent variable $\eta = y \left[\dfrac{Pe}{x}\right]^{1/3}$ and obtain the equation

$$\frac{d^2 C_0}{d\eta^2} + \frac{1}{3} \eta^2 \frac{dC_0}{d\eta} = 0 \qquad (6)$$

with boundary conditions $C_o = 0$ at $\eta = 0$, $C_o = 1$ at $\eta = \infty$. The solution of Eq. (6) has the form

$$C_o = \frac{\int_0^{\eta} \exp(-\eta^{3/9}) d\eta}{\int_0^{\infty} \exp(-\eta^{3\,9}) d\eta} \qquad (7)$$

whence for the stationary mass transfer we obtain the expression (written in a dimensional form)

$$q = 0{,}542 \, C_\infty \left(\frac{\tau D^2}{\mu x}\right)^{1/3} \qquad (8)$$

Further, consider the most interesting case of small fluctuation frequencies, i.e. assume $n\omega < 1$. Then for the fluctuation amplitude of concentration it is possible to write the series

$$C_n = C_{n_0} + in\omega C_{n_1} + (in\omega)^2 C_{n_2} + \ldots$$

and to obtain from Eq. (5) the system of equations

$$y\frac{\partial C_{n_0}}{\partial x}+yv_n\frac{\partial C_0}{\partial x}=\frac{1}{Pe}\frac{\partial^2 C_{n_0}}{\partial y^2} \qquad (9)$$

$$C_{n_0}+y\frac{\partial C_{n_1}}{\partial x}=\frac{1}{Pe}\frac{\partial^2 C_{n_1}}{\partial y^2} \qquad (10)$$

$$C_{n,m-1}+y\frac{\partial C_{n,m}}{\partial x}=\frac{1}{Pe}\frac{\partial^2 C_{n,m}}{\partial y^2} \qquad (11)$$

with the boundary conditions $C_{n,m}=0$ at $y=0$, $C_{n,m}=0$ at $y=\infty$.

Equation (9), describing the quasi-steady approximation, can readily be shown to have the form

$$C_{n_0}=1/3\, v_n y\frac{\partial C_0}{\partial y}$$

Transform Eq. (10) by introducing new indepedent variables according to the correlations:

$$C_{n_1}=v_n Pe^{1/3} x^{2/3} F(\eta), \qquad \eta=y\left[\frac{Pe}{x}\right]^{1/3}$$

The following equation is obtained:

$$F''(\eta)+\frac{1}{3}\eta^2 F'(\eta)-\frac{2}{3}\eta F(\eta)=0{,}178\eta\exp\left[-\eta^{3/9}\right] \qquad (12a)$$

Equation (11) at $m=2$ is transformed into a similar form by substitutions:

$$\eta=y\left[\frac{Pe}{x}\right]^{1/3}, \qquad C_{n,2}=x^{4/3} F_1(\eta) Pe^{2/3} v_n$$

$$F_1''+1/3\,\eta^2 F_1'-4/3\,\eta F_1(\eta)=F(\eta)$$

Similar operations with Eq. (11) at $m=3$ yield the new indepedent variables $\eta=y\left[\frac{Pe}{x}\right]^{1/3}$, $C_{n_3}=x^2 F_2(\eta) Pe v_n$ and give the equation

$$Fe_2''+1/3\,\eta^2 F_2'-2\eta F_2=F_1 . \qquad (13)$$

By restricting the series of the fluctuation amplitude with respect to frequency to the first four terms, it is possible to write

$$C_n=1/3 v_n y\frac{\partial C_0}{\partial y}+in\omega\, x^{2/3} Pe^{1/3} F(\eta) v_n - \\ \omega^2 n^2 x^{4/3} F_1(\eta) Pe^{2/3} v_n - i\omega^3 n^3 x^2 F_2(\eta) Pe v_n + \ldots \qquad (14)$$

The functions $F(\eta)$, $F_1(\eta)$, $F_2(\eta)$ we determined by solving numerically Eqs. (12a), (12), (13), on the electronic computer EVM-20. The results are shown in Fig. 1

The computation was carried out by two parallel methods: the method of finite differences and that of Runge-Kutta. The derivates of the function F at $\eta=0$ have the values: $F'(0)=0{,}09903$, $F_1'(0)=0{,}0353$, $F_2'(0)=0{,}00974$.

Expressions (14) and (3a) represent the solution of the problem in consideration.

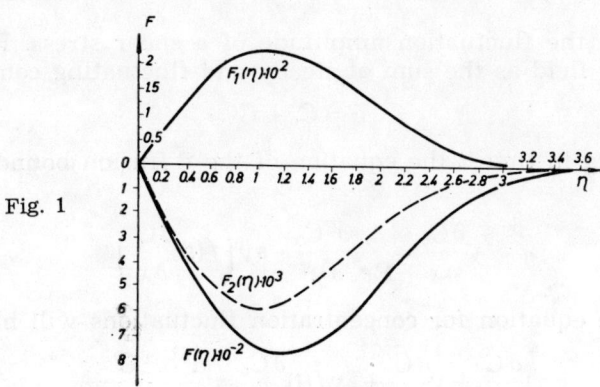

Fig. 1

The mass transfer to the cathode surface is written in the dimensional form of the series.

$$q = \frac{CD^{2/3}\tau_0^{1/3} 3^{-2/3}}{x^{1/3}\mu^{1/3}\Gamma(4/3)} + C_\infty \sum_{n=1}^{N}\left\{\left[\frac{\tau_n}{\tau_0} + \left(0{,}178\left[\frac{D^2\tau_0}{\mu x}\right] - \frac{\omega^2 n^2 \mu x}{\tau_0}F_1'(0)\right)\right] + \frac{\tau_n}{\tau_0} - \left[\frac{\omega n \mu^{1/3} x^{1/3}}{\tau_0^{1/3}}F'(0) - \frac{n^3\omega^3\mu^{-1/3}x^{5/3}}{\tau_0^{5/3}D^{1/3}}F_2^1(0)\right]\right\}\cos n\omega t + \left\{\frac{\tau_n}{\tau_0} - \left[0{,}178\cdot\left(\frac{\tau_0 D^2}{\mu x}\right)^{1/3} - \frac{\omega^2 n^2\mu x}{\tau_0}F_1'(0) - \frac{\tau_n}{\tau_0} + \left[\frac{\omega n \mu^{1/3} x^{1/3} D^{1/3}}{\tau_0^{1/3}}F_0' - \frac{\omega^3 n^3 \mu^{5/3} x^{5/3}}{\tau_0^{5/3}D^{1/3}}F_2'(0)\right]\right]\right\}\sin n\omega t.$$

Representing the experimental values of the diffusion current in the form

$$q = q_0 + \sum_{n=1}^{N} q_n + \cos n\omega t + q_n - \sin n\omega t$$

it is possible to calculate τ_0, τ^-, τ^+.

3. The reported calculation can be called »linear«, because the perturbations of friction and concentration were assumed to be infinitely small. Consider the fluctuations of a finite amplitude, restricting the analysis to monochromatic perturbations.

In this case we consider a field in the form

$$u = y + \varepsilon y \sum_{n=1}^{n=-1} 1/2\, e^{in\omega t} = y + \varepsilon y f(t), \quad (16)$$

Equation (16) is written in the non-dimensional form with $\varepsilon = \dfrac{\tau'}{\tau_0}$, where τ'— is the fluctuation amplitude of a shear stress. Represent the concentration field as the sum of steady and fluctuating component:

$$C = C_0 + C'$$

Averaging in time t the equation of the diffusion boundary layer (2) we obtain

$$y \frac{\partial C_0}{\partial x} = \frac{1}{Pe} \frac{\partial^2 C_0}{\partial y^2} - \varepsilon y \left(f(t) \frac{\partial C'}{\partial x} \right) \quad (17)$$

Then the equation for concentration fluctuations will have the form

$$\frac{\partial C'}{\partial t} + y \frac{\partial C'}{\partial x} + \varepsilon y f(t) \frac{\partial C_0}{\partial x} = \frac{1}{Pe} \frac{\partial^2 C'}{\partial y^2} + \\ + \varepsilon y \left(f(t) \frac{\partial C'}{\partial x} \right) - \varepsilon y f(t) \frac{\partial C'}{\partial x}. \quad (18)$$

The averaged term in Eq. (18) causes the deformation of mean concentration field while the difference between the averaged and the non-averaged terms in the right-hand side of Eq. (18) leads to the appearance of multiple harmonics in the fluctuation of the concentration field and of the diffusion current.

Assuming $\varepsilon \ll 1$, represent the solution of the system of Eqs. (17) and (18) as the series

$$C_0 = C_{01} + \varepsilon C_{02} + \varepsilon^2 C_{03} + \ldots \varepsilon^{n-1} C_{0n}$$
$$C' = C'_0 + \varepsilon C'_1 + \varepsilon^2 C'_2 + \ldots \varepsilon^n C'_n$$

By equating in Eqs. (17) and (18) the terms with the same powers the following system of differential equations is obtained

$$y \frac{\partial C_{01}}{\partial x} = \frac{1}{Pe} \frac{\partial^2 C_{01}}{\partial y^2} \quad (19)$$

$$y \frac{\partial C_{03}}{\partial x} = \frac{1}{Pe} \frac{\partial^2 C_{03}}{\partial y^2} - y \left(f(t) \frac{\partial C'_0}{\partial x} \right) \quad (20)$$

$$y \frac{\partial C_{0n}}{\partial x} = \frac{1}{Pe} \frac{\partial^2 C_{0n}}{\partial y^2} - y \left(f(t) \frac{\partial C'_{n-2}}{\partial x} \right) \quad (21)$$

$$\frac{\partial C'_1}{\partial x} + y \frac{\partial C'_1}{\partial x} + y f(t) \frac{\partial C_{01}}{\partial x} = \frac{1}{Pe} \frac{\partial^2 C'_1}{\partial y^2} \quad (22)$$

$$\frac{\partial C'_n}{\partial t}+y\frac{\partial C'_n}{\partial x}+yf(t)\frac{\partial C_{on}}{\partial x}=\frac{1}{Pe}\frac{\partial^2 C'_n}{\partial y^2}+$$
$$+y\left(\frac{\partial C'_{n-1}}{\partial x}f(t)\right)-y\frac{\partial C'_{n-1}}{\partial x}f(t) \tag{23}$$

The present paper is restricted to considering the equations for C_{03}, C_1', C_2', leaving aside the equations of the order ϵ^3 and higher.

Here it should be taken into account that $C_{02}=0$, $C_0'=0$.

Denoting the complex amplitude by C_n we have

$$C_1'=\sum_{n=-1}^{n=+1}C_n e^{in\omega t}, \quad n\neq 0.$$

Then the equation for the fluctuations of concentration takes the form

$$in\omega C_n+y\frac{\partial C_n}{\partial x}+\frac{1}{2}y\frac{\partial C_{01}}{\partial x}=\frac{1}{Pe}\frac{\partial^2 C_n}{\partial y^2}.$$

The solution of this equation at the zero boundary conditions can obviously be represented by series (14) at $v_n=1/2$.

By restricting the solution in the first approximation to two terms of the series, the expression for C'_n is written in the form

$$C_1'=\sum_{n=-1}^{n=1}\left[1/6\eta\frac{\partial C_0}{\partial \eta}+1/2\, in\omega x^{2/3}Pe^{1/3}F(\eta)\right]e^{in\omega t} \tag{24}$$

The solution of Eq. (19) is given by expression (7). Equation (20) for C_{03} is reduced to an ordinary differential equation by introducing a new indepedent variable

$$\eta=y\left[\frac{Pe}{x}\right]^{1/3}$$

The equation obtained has the form

$$\frac{d^2 C_{03}}{dy}+1/3\eta^2\frac{dC_{03}}{d\eta}+1/3\eta\frac{\partial}{\partial \eta}(f(t)\cdot C_1')=0 \tag{25}$$

By using expressions (16, (24)) the time-averaged term is calculated:

$$(f(t)\, C_1')=\frac{1}{6}C_0\frac{dC_0}{d\eta}$$

Then the deformation of the average initial profile of concentrations is expressed by the solution of the equation

$$\frac{d^2 C_{03}}{d\eta^2}+\frac{1}{3}\eta^2\frac{dC_{03}}{d\eta}+\frac{1}{18}\eta^2\frac{\partial C_{01}}{d\eta}+\frac{1}{18}\eta^3\frac{dC_{01}}{d\eta^2}=0 \tag{26}$$

which satisfies the boundary conditions $C_{03}=0$ at $\eta=0$, $\eta=\infty$.

This solution has the form:

$$C_{03} = \frac{3^{3/5}\Upsilon(1/3\eta^{3/9}) - 6\eta\exp(-\eta^{3/9}) - \eta 4\exp(-\eta^{3/9})}{210} - \frac{\Upsilon(1/3,\eta^{3/9})}{18\,\Gamma(1/3)}$$

The additional mass transfer due to velocity fluctuations is given by the formula (written in the non-dimensional form)

$$q \cong 0,03\, C_\infty \left[\frac{\tau_o D^2}{\mu x}\right]^{1/3} \left(\frac{\tau'}{\tau_o}\right)^2 \tag{27}$$

The comparison of (8) with (27) shows the contribution of the »fluctuation« mass transfer to the convection mass transfer to be negligible, hence the calculation of τ_o can be made in accordance with formula (8). All this, of course, is possible only if the shear stress fluctuation amplitude is small compared to the average shear stress. Further, consider Eq. (22) and study the possibility of occcurrence for subharmonic concentration fluctuations. The calculation of the difference of the averaged terms in the right-hand side of the equation yiedlds

$$y\left[f(t)\frac{\partial C'_i}{\partial x}\right] - yf(t)\frac{dC'_1}{\partial x} = -\frac{1}{2}y\frac{\partial}{\partial x}\sum_{n=-1}^{n=1} C_n l^{2\,in\omega t} \;;$$

The equation (22) is written as

$$\frac{\partial C'_2}{\partial t} + y\frac{\partial C'_2}{\partial x} = \frac{1}{Pe}\frac{\partial^2 C'_2}{\partial y^2} - \frac{1}{2}y\frac{\partial}{\partial x}\sum_{n=-1}^{n=1} \overline{C}_i l^{2in\omega t},$$

where C_n is given as series (14) at $v_n = 1/2$ the solution in the form we are looking for

$$C_2^1 = \sum_{n=-1}^{n=1} W n \, l^{2\,in\omega t} \;;$$

$$W_n = W_{no} + in\omega W_{n1} + (in\omega)^2 W_{n2} + \; . \; . \; . \; . \; . \; . \; ,$$

assuming $n\omega < 1$. Then for the quasi-steady »subharmonic« amplitude the following equation is obtained

$$y\frac{\partial W_{no}}{\partial x} = \frac{1}{Pe}\frac{\partial^2 W_{no}}{\partial y^2} - \frac{1}{2}y\frac{\partial}{\partial x}\left(\frac{1}{6}\eta\frac{\partial C_{o1}}{\partial \eta}\right);$$

$$\eta = y\left[\frac{Pe}{x}\right]^{1/3}$$

By passing from from the independent variables x, y to the variable η the ordinary differential equation is obtained:

$$\frac{d^2 W_{no}}{d\eta_2} + \frac{1}{3}\eta^2\frac{dW_{no}}{d\eta} + \frac{1}{36}\eta^2\frac{dC_o}{d\eta} + \frac{1}{36}\eta^3\frac{d^2 C_o}{d\eta^2} = 0$$

with the boundary conditions $W_n=0$, $\eta=0$. As the solution of this equation is given by that of Eq. (26) ($W_{no}=1/2 C_{03}$) we have

$$q'_2 \cong 0{,}03 C_\infty \left(\frac{\tau'}{\tau_o}\right)^2 \left[\frac{\tau_o D^2}{\mu x}\right]^{1/3} \cos 2\omega i \qquad (28)$$

The result obtained shows the necessity of taking into account the subharmonic fluctuations of the diffusion current in analyzing the results obtained on the basis of the linear theory. In the experiments made by J. E. Mitchell, L. P. Reiss and T. I. Hanratty [1] the ratio of the squared average shear stress had the value $(\tau'/\tau_o)^2 \approx 0{,}1024$ that corresponds to $\tau'/\tau_o = 0{,}32$. In this case the quasi-steady velocity fluctuations generate the subharmonic fluctuations of mass transfer, whose frequency equal the doubled frequency of quasi-steady fluctuations, and the relative amplitude has the value

$$\frac{q'_2}{q_{2,\,kbcm}} = 0{,}168 \left(\frac{\tau_1^2}{\tau_o \tau_2}\right),$$

where τ_1 — is the shear stress fluctuation amplitude of the first order harmonic,

τ_o — is the mean shear stress

τ_2 — is the second order harmonic amplitude

$q_{2\,kbcm}$ — is the mass flux fluctuation due to the second order fluctuations

q_2' — are false subharmonic fluctuations of mass transfer.

From the cited formula it is seen that in the experiments of T. Hanratty and I. Mitchell the subharmonic fluctuations of mass transfer are comparable in value with real doubled harmonics of mass transfer, which introduces an essential error into all spectral characteristics reported by them.

NOMENCLATURE

q	mass flux to the control electrode
I	diffusion current
A	Faraday's number
μ	viscosity
D	diffusion coefficient
C	concentration
Pe	Peklet's number
x, y	coordinates
ω	frequency
Γ, γ	complete and incomplete gamma functions
τ_o	mean shear stress
τ'	fluctuation amplitude of shear stress
C_∞	concentration in the volume of electrolyte

REFERENCES

1. I. Mitchell, T. Hanratty, Fluid Mech., **26**, Part 1, 199—221 (1966).
2. V. G. Levich, *Fisiko-khimicheskaya gidrodinamika*, Ed. by USSR Academy of Sciences, Moscov, 1952.

A METHOD OF TEMPERATURE MEASUREMENT CLOSE TO A WALL, IN A TURBULENT BOUNDARY LAYER

J—P. MAYE

Laboratoire de Dynamique des Fluides, Université de Poitiers, France

1. Introduction

In a turbulent boundary layer, measurement of mean temperature and temperature fluctuation, especially near a wall, requires the use of a thermometer of sufficient sensibility and above all of one which enables punctual measurements. Moreover, the thermometer used to measure fluctuations should have a very small thermal inertia and should be easily connected with anemometer circuits for correlation measurements between velocity components and temperature.

A resistance thermometer, the sensor of which is a very small platinium or tungsten wire (diameter: $d \approx 3\mu$; length: $l \approx 1$ mm) similar to those used in hot-wire anemometry (Fig. 1), seems to be the best instrument, at the present time, to satisfy previous requirements.

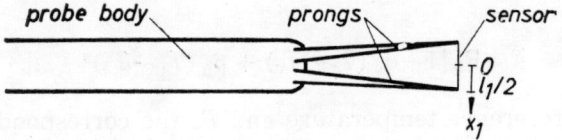

Fig. 1. The sensor

Such a thermometer has already been used in natural convection [1, 2, 3] or in other very particular cases such as mean temperature measurement in the wake of a hot wire set in a turbulent stream [4]; it has not yet been systematically used for the study of turbulent boundary layers in forced convection. In this last case available experimental results have generally been obtained with thermocouples (mean temperature) but then far from the wall, or by the constant current anemomenter technique [5] (fluctuations) wich uses three (or more) different and successive currents, but then with a low accuracy chiefly due to the necessity of several measurements and calculations for each result.

2. Principle of the Resistance Thermometer

The resistatnt wire is supplied with an electrical current of constant intensity I and so low that its temperature increase ΔT_1 due to Joule effect can be neglected in front of the temperature differences which are to be measured; for a given wire, ΔT_1 is also in relation with the fluid velocity (Fig. 2).

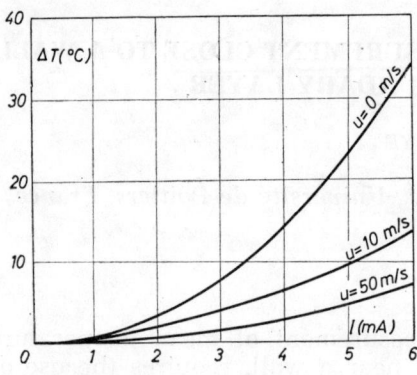

Fig. 2. Temperature increase ΔT_1 in relation to I and u for a platinum wire of 3μ diameter.

Under the precedent assumption it can be considered that:

$$T_1 = T + \frac{u^2}{2c_p},\tag{1}$$

where T and u are respectively local fluid temperature and normal velocity component on the wire. Then it is possible to determine T if T_1 and u are measured; now there is a single relation between T_1 and R, electrical resistance of the wire (under the assumption that the whole wire temperature is T_1):

$$R = R_0 \left[1 + \alpha_0 (T_1 - T_0) + \beta_0 (T_1 - T_0)^2 + \ldots \right],\tag{2}$$

where T_o is a reference temperature and R_o the correspondent wire resistance. For some metals (platinum, tungsten, ...) and under the assumption that $T_1 - T_o$ is not too high, it can be considered that:

$$T_1 - T_0 = K (R - R_0)\tag{3}$$

(platinum, $T_o = 0°C : \alpha_o \approx 4 \cdot 10^{-3} \Omega/°C$, $\beta_o \approx -6 \cdot 10^{-7} \Omega/(°C)^2$).
K is a wire characteristic $K = \dfrac{1}{\alpha_0 R_0}$ which can be determined by calibration. From (3) we deduct for a turbulent stream:

$$\overline{T_1} - T_0 = K (\overline{R} - R_0)\tag{4}$$

if T_o is not fluctuating (for instance: $T_o = T_\infty$).

Now, if we consider the voltage at the ends of the wire:

$$\overline{T_1} - T_0 = k(\overline{E} - E_0) \tag{5}$$

$$T'_{1\,rms} = k E'_{rms}, \tag{6}$$

where k is a constant $\left(k = \dfrac{K}{I}\right)$.

Equations (4) or (5) can be used to determine mean temperature T_1, either measurements are performed with a Wheatstone bridge or with a direct voltmeter, while (6) can be used to determine temperature fluctuations $T_1'_{rms}$ from readings on an integrator RMS voltmeter connected to a low noise amplifier. Most hot wire anemometers are now fitted to be used as resistance thermometers; meanwhile, in this case, the voltage is measured between two opposite tops of the Wheatstone bridge.

Yet, to determine T'_{rms} from $T_1'_{rms}$ it's necessary for the thermal inertia of the sensitive wire to be low enough. This inertia is in relation, chiefly, with the diameter of the sensor; a 3μ diameter wire, for instance, is sensitive up to about 1 KHZ and for higher frequencies its response will lag the actual change in temperature due to its own thermal inertia. To compensate for this the amplifier connected to the sensor must have a non-linear characteristic with frequency, as in constant current anemometers [6]. However, in a turbulent boundary layer, higher frequency fluctuations are also lower amplitude fluctuations and so their contribution to T'_{rms} is not very important; but for measurements of correlations between T' and the velocity components, their contribution can be more important.

It's also useful and easy to record voltage measurements with an automatic printer system, each mean value of a turbulent stream having to be deducted statistically from an important pattern of experimental points.

3. Error Due to Thermal Conduction Between the Sensor and Its Supports

The error due to radiance between hot walls and sensor is negligible for most applications, even very close to a wall, because of the small diameter of the wire [7]. The only error caused by the probe is that due to thermal conduction between the wire and its two supports. This conduction effect results, at first, of the greater diameter of the supports and then of their necessary inclination α relating to the stream isothermal surfaces, close to a wall (Fig. 3).

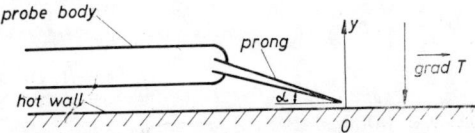

Fig. 3. Inclination of the prongs relating to the wall

Therefore, the supports are placed in a temperature gradient which causes a cooling of wire ends by thermal conduction; then, measured temperature T_m is lower than real temperature T_r (Fig. 4). Experiments have shown that this error $\delta T = T_r - T_m$ can reach several percents of the difference $T_r - T_\infty$ in a boundary layer, according to the shape of the supports and the proximity of the hot wall. We are now going to connect this error δT to some sensor parameters and then look for possible improvements.

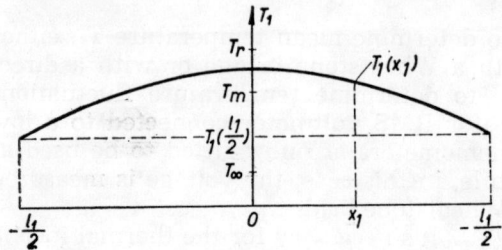

Fig. 4. Temperature distribution along the wire

Under the following assumptions: $T_1 = f(x_1)$, $T_1(x_1) = T_1(-x_1)$, wire on an isothermal line, a first expression for δT is:

$$\delta T = -\frac{2}{l_1} \int_0^{l_1} \theta_1(x_1)\, dx_1, \qquad (7)$$

where

$$\theta_1(x_1) = T_1(x_1) - T_r$$

If now we consider a wire element $d\tau_1$ (Fig. 5) we have:

$$q_1 + q_2 + q_3 = 0 \qquad (8)$$

Fig. 5. Heat flux on a wire element

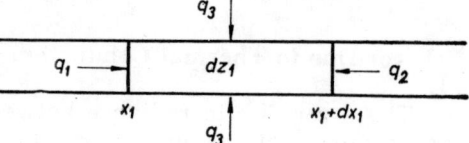

with:

$$q_1 = -\lambda_1 S_1 \left(\frac{dT_1}{dx_1}\right)_{x_1} \quad \text{thermal flux of convection}$$

$$q_2 = -\lambda_1 S_1 \left(\frac{dT_1}{dx_1}\right)_{x_1 + dx_1} \qquad (9)$$

$$q_3 = -h_1 p_1 \theta_1(x_1)\, dx_1 \quad \text{thermal flux of convection}$$

if we pose,

$$\omega_1 = 2\left(\frac{h_1}{\lambda_1 d_1}\right)^{1/2} \quad (= \text{constant}) \qquad (10)$$

from (8) and (9) we deduct:

$$\frac{d^2 \theta_1}{d x_1^2} - \omega_1^2 \theta_1 = 0 \quad \text{for} \quad 0 \leq |x_1| \leq \frac{l_1}{2} \tag{11}$$

Boundary conditions are:

a) $\left(\dfrac{d \theta_1}{d x_1}\right)_{x_1=0} = 0$

b) $\theta_1 \dfrac{l_1}{2}$ determined by the supports.

Then the solution of (11) is:

$$\theta_1(x_1) = \theta_1\left(\frac{l_1}{2}\right) \frac{\operatorname{ch} \omega_1 x_1}{\operatorname{ch} \dfrac{\omega_1 l_1}{2}} \tag{12}$$

and from (7) we deduct:

$$\delta T = -\frac{2}{l_1 \omega_1} \operatorname{th} \frac{\omega_1 l_1}{2} \theta_1\left(\frac{l_1}{2}\right) \tag{13}$$

This expression can be simplified in most cases.
Indeed ω_1 can be computed from:

$$\omega_1 = 2\left(\frac{\lambda_0}{\lambda_1}\right)^{1/2} Nu^{1/2} d_1^{-1} \tag{14}$$

where λ_0 is the thermal conductivity of the fluid and Nu is the Nusselt number of the wire, an empirical expression of which is, in relation to the Reynolds number Re of the wire, for instance that of King:

$$Nu = 0{,}32 + 0{,}56\, Re^{1/2} \tag{15}$$

For an air stream velocity of $1 m/s$ and a 3μ diameter, 1 mm length wire, it is found:

$$\omega_1 \approx 9\, mm^{-1}, \quad \operatorname{th} \frac{\omega_1 l_1}{2} \approx 0{,}999$$

Then, for most measurements in forced convection, a good expression for δT is:

$$\delta T = -\frac{2}{l_1 \omega_1} \theta_1\left(\frac{l_1}{2}\right) \tag{16}$$

where l_1 and ω_1 are characteristics of the sensor, whereas $\Theta_1\left(\dfrac{l_1}{2}\right)$ can be considered as a characteristic of the support, for given dynamic and thermal conditions of the stream; indeed, the support mass is necessary greater than the wire mass $\left(\dfrac{d_1}{d_2} \sim \dfrac{1}{100},\ l_1 \ll l_2\right)$ and so it is for its thermal capacity.

Equation (16) shows that δT is proportional to $\theta_1\left(\dfrac{l_1}{2}\right)$ and from (14) and (15), nearly proportional to d_1. The great difficulty is to determine $\theta_1\left(\dfrac{l_1}{2}\right)$ for each measurement. A sensor, the supports of which are special thermocouples, is going to be made in our laboratory. This probe will permit to measure simultaneously T_m with the resistance thermometer and $T_1\left(\dfrac{l_1}{2}\right)$ with the thermocouples and then to deduct:

$$T_r = T_m + \delta T = T_m - \dfrac{2}{l_1 \omega_1}\left[T_1\left(\dfrac{l_1}{2}\right) - T_r\right]$$

$$T_r = T_m \dfrac{1 - \dfrac{2}{l_1 \omega_1}\dfrac{T\left(\dfrac{l_1}{2}\right)}{T_m}}{1 - \dfrac{2}{l_1 \omega_1}} \qquad (17)$$

Therefore the corrective factor is:

$$\xi = \dfrac{1 - \dfrac{2}{l_1 \omega_1}\dfrac{T_1\left(\dfrac{l_1}{2}\right)}{T_m}}{1 - \dfrac{2}{l_1 \omega_1}} > 1 \qquad (18)$$

From (16) we see also that δT is proportional to $\dfrac{1}{l}$; but we are limited in the choice of l_1 by the mechanical resistance of the wire or by the characteristics of turbulence (for instance for measurement of fluctuations).

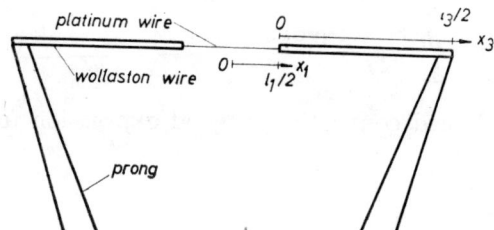

Fig. 6. The improved sensor

However we are going to see that for given supports and sensible wire (d_1, l_1) it is possible to improve the probe by using Wollaston wire of small external diameter $(d_3 \approx 30\mu)$, the silver sheath of which is scraped on l_1 length (Fig. 6).

The principle of such an improvement is that the two Wollaston parts attenuate the cooling at the ends of the platinum wire without notable increase of the electrical resistance of the sensor, if judicious sizes are chosen for the different elements. This improvement can be demonstrated from theorical study similar to the precedent one. Now (7) becomes:

$$\delta T = -2 \frac{\dfrac{R_1}{l_1}\displaystyle\int_0^{\frac{l_1}{2}} \theta_1(x_1)\, dx + \dfrac{R_3}{l_3}\displaystyle\int_0^{\frac{l_3}{2}} \theta_3(x_3)\, dx_3}{R_1 + R_3} \qquad (19)$$

because α_o (silver) = α_o (platinum).

Likewise (16) becomes:

$$\delta T = -2\, \theta_3 \left(\frac{l_3}{2}\right) \frac{\dfrac{R_1}{l_1 \omega_1} + \dfrac{R_3}{l_3 \omega_3}\left[sh\dfrac{\omega_3 l_3}{2} + k\dfrac{\omega_1}{\omega_3}\left(ch\dfrac{\omega_3 l_3}{2} - 1 \right)\right]}{(R_1 + R_3)\left(ch\dfrac{\omega_3 l_3}{2} + k\dfrac{\omega_1}{\omega_3} sh\dfrac{\omega_3 l_3}{2}\right)} \qquad (20)$$

where

$$k = \left[1 + \frac{\lambda_3}{\lambda_1}\left(\frac{d_3^2}{d_1^2} - 1\right)\right]^{-1}$$

If now we suppose, as for $\theta_1\left(\dfrac{l_1}{2}\right)$, that $\theta_3\left(\dfrac{l_3}{1}\right)$ is only dependant on the supports, that is to say:

$$\left[\theta_1\left(\frac{l_1}{2}\right)\right]_{(1)} = \left[\theta_3\left(\frac{l_3}{2}\right)\right]_{(2)}$$

we deduct from (16) and (20):

$$\frac{(\delta T)_{(2)}}{(\delta T)_{(1)}} = \frac{R_1 + R_3 \dfrac{l_1 \omega_1}{l_3 \omega_3}\left[sh\dfrac{\omega_3 l_3}{2} + k\dfrac{\omega_1}{\omega_3}\left(ch\dfrac{\omega_3 l_3}{2} - 1\right)\right]}{(R_1 + R_3)\left(ch\dfrac{\omega_3 l_3}{2} + k\dfrac{\omega_1}{\omega_3} sh\dfrac{\omega_3 l_3}{2}\right)} \qquad (21)$$

This quotient between δT (2) (probe with Wollaston wire) and δT (1) (probe with platinum wire only), for the same supports and sensible wire (d_1, l_1) and for the same dynamic and thermal stream conditions, has been computed for $u = 1$ m/s on the sensor and for several platinum wires in relation to Wollaston wire characteristics; for usual sensors (d_1 : 1µ to 5µ;

l_1 : 0.5 mm to 3 mm) it is found that the precedent quotient is practically independant of the sensible wire (l_1, d_1). The results are given on the following board (Fig. 7).

l_3 mm	1			5			10		
d_3 μ	10	30	50	10	30	50	10	30	50
$\frac{\delta T_{(2)}}{\delta T_{(1)}}$ %	80	96	98	16	51	72	5	16	36

Fig. 7. Board showing the improvement due to the Wollaston wire

This board shows the improvement due to the Wollaston wire; it is seen also that $\frac{\delta T_{(2)}}{\delta T_{(1)}}$ decreases with d_3 and l_3^{-1}.

However, it is always necassary to have: $R_3 \ll R_1$ (for instance $\frac{R_3}{R_1} \leq 10^{-2}$) and this condition limits the choice for d_3 and l_3 apart from problems of mechanical resistance of the wire.

Concluding Remarks About This Survey

The sensibility of the probe is proportional to $l_1 d_1^{-2}$ and its accuracy nearly proportional to $l_1 d_1^{-1}$. Then it's better to take d_1 minimum and l_1 maximum. The choice of these extreme values is conditioned by:
— the mechanical resistance of the wire
— the characteristics of the stream and the nature of measurements (mean temperature or fluctuations).

When l_{max} is imposed the method of Wollaston wire is then very interesting because it permits to increase the accuracy of the probe (precedent board). In particular, when it is desired to measure simultaneously $\bar{T}$ and T' with the same sensor, it's better to use such a probe.

4. Applications

The possibilities of such a resistance thermometer are illustrated by Figs. 8 and 9 which represent respectively a mean temperature profile and a profile of fluctuations measured in the turbulent boundary layer of a heated flat plate without pressure gradient.

Our purpose is not, here, to study these profiles; this will be done in a next and more complete paper. Meanwhile we can make some remarks about them. In particular we see that several experimental points are set in the viscous sublayer; (the first measurement corresponds to about $y_+ = 1$). Figure 9. also demonstrates that T'_{rms} is a linear function of y_+ close to the wall and then a good method to determine the distance y_+ from the wall with much accuracy is to extrapolate the curve $T'_{rms}(y_+)$ to $T'_{rms} = 0$. Furthermore, if y_+ is known with great accuracy it's possible by an extrapolation of the mean temperature profile to $y_+ = 0$ to deter-

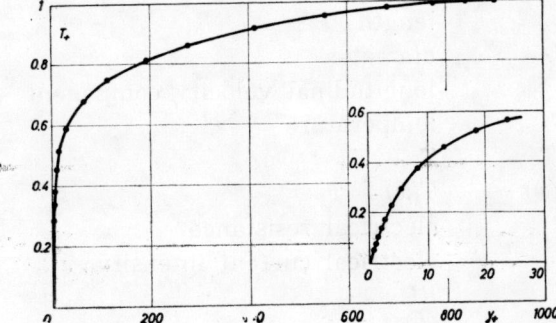

Fig. 8. Mean temperature profile ($Re_x = 10^6$, $T_w - T_\infty = 20°C$)

mine T_w and y_w, two important grandeurs for the study of the thermal boundary layer, a direct measurement of which is generally difficult; so, this technique uses the same instrument (the resistance thermometer) for measurements in the boundary layer and at the wall, which is a real advantage.

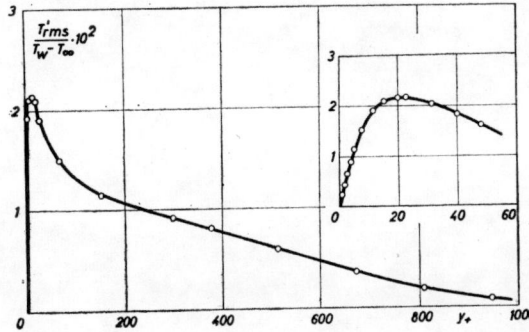

Fig. 9. Profile of fluctuations ($Re_x = 10^6$, $T_w - T_\infty = 20°C$)

5. Conclusion

This type of thermometer seems actually the best for temperature measurements in laminar or turbulent gaseous boundary layers, in free or forced convection. It can be used very close to a wall ($y \geq 0.02$ mm) without any interaction between the sensor and the wall. Its calibration curve is linear. The error due to thermal conduction between the wire and its supports, close to a heated wall, can be corected if the temperature of wire ends is known, and greatly diminished by use of fine supports and Wollaston wire. At last, the same probe and apparatus can be used to measure both temperature and velocity characteristics in the boundary layer and at the wall.

NOMENCLATURE

x	longitudinal co-ordinate
y	distance normal to the wall
l	length
d	diameter
u	longitudinal velocity component
T	temperature
δT	$T_r - T_m$
$\theta(x)$	$T(x) - T_r$
R	electrical resistance
I	electrical current intensity
E	RI
ρ	electrical resistivity
α_o, β_o	temperature coefficients
q	heat flux
λ	thermal conductivity
ν	kinematic viscosity
τ	shear stress

Dimensionles parameter

Re	Reynolds number
Nu	Nusselt number
T_+	$\dfrac{T_w - T}{T_w - T_\infty}$
y_+	$\dfrac{y \cdot \sqrt{\dfrac{\tau_w}{\rho}}}{\nu}$

Subscripts

r	real
m	measured
w	wall
∞	in free stream
1	platinum wire
2	support
3	Wollaston wire

Superscripts

—	arithmetic time average
′	fluctuating component

REFERENCES

1. A. A. Townsend, J. Fluid Mech., 5 (1959).
2. D.-J. Tritton, J. Fluid Mech., 16 (1962).
3. J. W. Deardorff and G. E. Wills, J. Fluid Mech., 28 (1966).
4. L. Taccoen, Etude de la diffusion turbulente de la chaleur dans un tube lisse, Thesis, Paris.
5. D. J. Johnson, J. Appl. Mech., 26 (1959).
6. J. Gaviglio, P.S.T., 385 (1962).
7. J. Gosse, Etude de la convection par les fils aux faibles nombres de Reynolds, Thesis P.S.T., № 22.

HEAT AND MASS TRANSFER IN A BOILING BOUNDARY LAYER

M. A. STYRIKOVICH

Institute for High Temperatures, USSR Akademy of Sciences, Moscow, USSR

The investigation of heat and mass transfer occurring in a boiling boundary layer is of great importance in practice and science. At moderate heat fluxes the boiling heat transfer coefficients are sufficiently high and practically in most cases affect but negligibly upon the total heat transfer coefficient through the heating surface involved.

At high heat fluxes the main trouble is to provide faultless operation of the heating surface; this being the case, for low relative pressures (P/P_{cr}) and small steam void fractions of the flow, a zone of reliable operation is usually nucleate boiling only. As far as we know, the boundary of this zone depends on a large variety of quantities and in so far as extensive experimental data may be differently interpreted, depending upon different physical models of the nucleate boiling burnout.

Therefore, alongside with accumulation of the data concerning the resultant dependence $q_{cr} = f(P/P_{cr}, w, i_m$ and other quantities), it is of great importance that direct measurements of these, or other parameters of the boiling boundary layer be taken.

Moreover, the dependences characterizing the process occurring within the zone of the imparied heat transfer so-called dry-out are likewise complex dependences, this concerns both the boundary of the zone involved and the value of the heat transfer coefficient within this zone.

Accumulated recently has been a great amount of data on such boiling boundary layer parameters as time-averaged distributions of steam void fractions with respect to the layer depth and fluctuations of the steam void fractions; distribution of the steam and water fractions within close proximity to the heating surface (the appearance of steam concentrations), with respect to temperature fluctuations of the heating surface with time, etc. These data are of particular concern to appreciate the correctness or wrongness of these or other physical models of the nucleate boiling process and the burnout.

Of great importance to characterize the heat transfer process are likewise the data concerning mass transfer within the boiling boundary layer and in particular its main integral characteristic, i.e. the average evaporation degree of the fluid in close proximity to the heating surface.

This value which is in fact a ratio of concentration of any substance (practically soluble in its liquid phase only and featuring no remarkable surface activity) in the fluid being in contact with the wall involved, and an average-consumable concentration of this substance in the fluid flowing through the channel, numerically cahracterizes the circulation ratio of the fluid flowing through the boundary layer and is of great importance in any physical model of the boiling boundary layer and the nucleate burnout.

Apart from the above said, the evaporation degree is of a particular concern for the deposits formations on the heating surfaces and it is necessary to set a permissible concentration of various impurities in water depending upon the requirements to prevent any deposition on the heating surfaces. It is of common knowledge that depositions may in a number of cases substantially effect upon the total heat transfer coefficient through the wall involved. Even more the depositions may efect the temperature of the metal on the steam-water side.

It is known that at high heat fluxes peculiar to a plurality of fields of up-to-date engineering including forced equipment employed in modern power engineering, only rather small-thickness depositions of the order of tens and even of hundredths of a millimeter are capable of increasing the temperature of the heating surface above the safe limits. Alongside with the abovesaid a number of water impurities specific to conventional and nucleate thermal power stations, such as the products of corrosion of structural materials, are so low soluble in the high-temperature water that even up-to-date highly efficient methods of water purification fail to prevent (under the acceptable conditions from the economical point of view) sedimentation of a solid phase. It is worth mentioning as an example that according to our last investigations the solubility of cobaltic oxides in the water at 280°C is somewhat lower than 1 μkg/kg, i.e. by 3 orders lower than that of the normal total content of impurities in the water of the boiling reactors.

However, the only knowledge of a true (equilibrium) solubility of this or other compound in the boiling water provides no possibility to determine permissible concentrations capable of preventing any deposition.

It is a well known fact that in the evaporation process of the fluid within the boiling boundary layer concentrations of impurities soluble in water, may be increased (under the conditions where the mass transfer with a flow core is very low) up to great values. Practically these values are confined by the limiting ratio between the concentrations in the water and the steam which is in equilibrium with the abovementioned ratio, i.e. by the fact that the impurities in the process of evaporation pass into vapour due to their solubility in the latter. This limiting evaporation degree depends upon both the properties of the given compound and the pressure involved and may be of the order of 10^2 to 10^9 and above at a pressure, for example, of 70 atm. abs. for various compounds. As we reach the critical point where by definition $K=1$, the value of maximum evaporation is decreased. However, only at very high pressures which are close to the rated ones, the abovementioned confinement for the most compounds being of interest in power engineering may be practically noteable. Here we should mention that second confinement of the evaporation degree associated with the increase in the boiling temperature observable as the concentrations in the highly forced apparatus facilitates, is of no practical effect. Therefore, in most cases only the mass transfer between the boundary layer near the surface and the flow core is capable of facilitating the concentration within the wall involved. In cases of strongly suppressed mass transfer, for example, when the wall is wetted with drops which are completely dried on this wall, deposition of the substance in question on the wall be even facilitated despite the fact that the concentration of the substance in the flow will be several orders lower than the saturation level. It is therefore clear, how important is the knowledge of the laws of mass transfer between the flow core and the boundary layer from the point of possible prevention of any deposition on the heating surfaces.

In a number of cases the mass transfer is likewise an important factor to prevent the corrosion or limit the rate of it on the heating surface involved. The importance of this fact is made more evident when referring to steam generators operating on organic fuels where deep evaporation occurred within the walls was associated with numerous cases of corrosion of the shield tubes. Recently a number of American electric companies have assigned more than 1,000,000 dollars to conduct investigations concerned on the semi-production installation. The investigators have been proved that with the absence of porous depositions on the wall, alkaline corrosion of the carbone steel is unobservable even in the zone of loads and steam void fractions which are close to the nucleate burnout. However, it only evidenced that the evaporation degree is lower than a dozen of hundreds, since the corrosion of low-carbon steel increases sharply with very high concentrations of the alkali in the solution (hundreds and thousands higher than flow water concentrations).

A more difficult case is encountered with the stainless austenitic steel corrosion where with the presence of oxygen it is sufficient that the fluid contacting the wall contain only 100 mg/kg of chlorion or hydroxylion to give rise to a very dangerous corrosion cracking of the metal involved. Therefore, with a view of ratesetting the permissible content of this fluid in the water of boiling reactors, or in the water of the second circuit of the pressurised water reactors, it is required that one should appreciate the evaporation degree more exactly. From this it is clear the practical importance of the integral indices of the mass transfer between the boiling boundary layer and flow core, i.e. the evaporation degree or circulation ratio through the boiling layer, for different duties of the boiling process.

However, of no lesser importance are these quantities for appreciation of a diversity of the physical models of the heat transfer occurring with the nucleate boiling and the burnout, that have been suggested recently.

Indeed, any physical model has been appreciated up till now mostly from the point of its correspondence to the accumulated experimental data concerning integral heat transfer characteristics, i.e. the dependence of the heat transfer coefficient upon a variety of determining factors. Under these conditions the availability of any additional information pertaining to the process involved may be of great concern. In particular, the knowledge of the evaporation degree and, consequently, the amount of water inflowing to the wall, provides the possibility to reject in most cases the models, as obviously being in no conformity with the experimental data, that correlated well with the heat transfer data only. How one can determine the evaporation degree within the sublayer? The question resolves itself into the measurement of the concentration of this or other impurities, i.e. the indicator in the fluid contacting the wall, since the measurement of the average concentration within the flow core actually pressents no difficulties.

It is the most simple thing that both the average value with respect to time and the value featuring the surface concentration of any salt in the fluid contacting the wall, may be fixed at the moment the solution within the wall reaches the saturation level, since at this moment (at least for a large number of salts with a negative temperature solubility coefficient) a continuous growth of depositions on the heating surface and the corresponding wall temperature average consumable salt concentration is increased gradually it is easily to fix the value of C beginning with which the T_w is continuously increased.

In this case the evaporation degree

$$Z = \frac{C_W}{C_{f.c.}} = \frac{C_{sat}}{C_{f.c.}}$$

where C_{sat} is taken with the temperature and pressure within the heating surface. In those cases where the solubility of the given salt substantially changes within the temperature range from T_w to $T_{f.c.}$ a certain indefinity in selecting the rated value of C_{sat}. Indeed, the temperature of the fluid contacting the wall involved varies continuously and is not the same at different points on the surface. Moreover, there is a local short-time inecreas in concentration within the zone of the bubble growth and especially near its root, which fact may lead to the formation of a thin ringlet of depositions. However, upon departure of the bubble, a nonevaporated fluid rushes to the wall the temperature of this fluid being somewhat above the temperature observed within the flow core, but is below that of the wall, and if this fluid is in fact an unsaturated solution, its contacting with the wall results in solubility of the depositions. As have been shown by special experiments, small depositions are likely to occur in the process of boiling of the unsaturated solution due to the presence on the surface involved of the microportions whose geometry sophisticates their contacting with the inflowing fluid, but these depositions tend to stabilize rapidly.

On the other hand, continuous growth of depositions is observable in the experiments of highly intensive mass transfer only in the cases where C reaches the value satisfying the saturation at the effective temperature $t_{eff.}$ which is in turn below the temperature T_w, but is above the temperature T_f.

In most cases due to the selection of a salt featuring low temperature factor of solubility the indefinity in selecting the $t_{eff.}$ may be confined by the permissible changes of the C_{sat}^{rat}, however in those cases where the ratio

$$C_{sat}^{T_w} \Big/ C_{sat}^{T_f}$$

is found to be considerably different from unity, usually at the experiments with substantial undercooling

$$T = T_w - T_f$$

the value of C_{sat}^{rat} is better to be set experimentally, by changing the parameters of the process toward deliberate decrease in the degree of evaporation. In this case C_f^{cr} corresponding to the onset of the continuous growth of the depositions increases gradually until it becomes practically constant which fact means that within the accuracy limits of the experiment the degree of evaporation is close to unity. At higher pressures and developed boiling where the temperature head is rather low the abovementioned indefinity is of no concern.

It is reasonable to select as an indicator a salt with a negative and possibly low temperature factor of solubility.

Apart from this, the salt should feature a sufficient solubility to provide for a rapid growth of depositions from the saturated solution; otherwise, it will result in substantial increase of the duration of the experiment required to achieve the temperature rise of the wall which deliberately exceeds its fluctuations. It is, of course, possible to determine the average consumable concentration which gives birth to deposition, by comparing the analitically determined concentrations at both the tube inlet and the outlet, however, at a small value of the deposition involved, the accuracy of the method may be insufficient. Therefore, it is advantageous that the solubility be not worse than tens mg/kg, which usually makes it possible to find out the temperature rise of the wall in 1 to 2 hours. Alongside with this, the solubility should be of a sufficiently low value, so that the properties of the saturated solution be practically identical to those of a pure water. This holds true of with a sufficiently high extent for the salts with $C_{sat}\left\{1 + 10\dfrac{gr}{kg}\right\}$ and even above, in so far as the properties of the outmost liquid phase are concerned. However, some cases may be encountered where even at a considerably lower solubility the presence of the salt greatly changes the conditions within the bounddary of the steam-water phases, and consequently, the entire hydrodynamics of a two-phase flow. As it is known from the experiments dealing with the steam bubbling through the water, when the latter contains even small amounts of ferrous oxides in colloid form, the salt concentration above certain limits sharply changes the properties of the bubble envelopes. This results in a substantial change in the structure of the steamwater mixture toward a decrease of an average size of the bubbles, the rate of their buoyancy, the height of the layer of the froth formed on the water surface and etc. The aforementioned change takes place only within a definite, for each particular salt, range of concentrations from

C'_{cr} to C''_{cr}. Within the range from pure water to C'_{cr} no influence of the concentration involved is observed, while above C''_{cr}, this influence is observed but negligibly.

The value of C'_{cr} for all the salts decreases with the pressure increase. It amounts a few grams per kg at 1 atm abs. and only hundreds and even tens of mg/kg at 140 atm abs. This fact restricts the selection of salts, but at the same time provides the possibility, as it will be better understood hereinbelow, to elucidate the mechanism of analogy between the heat transfer during the boiling process and during bubbling.

Now let me conclude general and methodical problems and to say a few words about the results that have already been obtained.

First investigations with resort to this method were conducted by me and by M. I. Reznikov many years ago, but they permitted only to appreciate the fact that for the case of free convection up to the heat fluxes of the order of 0.5 q_{cr} and p 100 atm abs., deep evaporation (tens and hundreds times as has supposed by many authors after Hall) is unobservable.

A series of investigations covering already the forced motion have been conducted recently at both the Mass Transfer Division of the Institute for High Temperatures, at the laboratory headed by E. I. Nevstruyeva, and at the Chair of Steam Generators of the Moscow Power Engineering Institute, headed by me

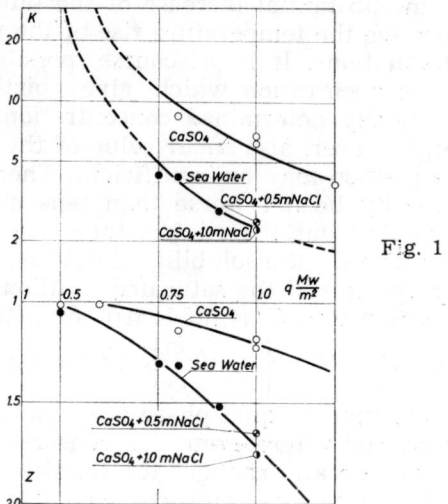

Fig. 1

The mass transfer investigations of pool boiling showed that at both low and moderately high pressures, the mass transfer tending to decrease with the increase of the heat flux, remains sufficiently intense up to the entire burnout.

As it is seen from Fig. 1, the evaporation degree Z at the atmospheric pressure up to the heat fluxes of the order of 0.6 q_{cr} is below 1.1 and even at (0.9 to 0.95) q_{cr}. does not exceed 1.33. Accordingly, the circulation ratio of the fluid through the boiling boundary layer $K=\dfrac{G_W}{G_V}$; amounts to more than 10 with $q=0.6\,q_{cr}$ and about 4 with $q \cong 0.9$ to $0.95\,q_{cr}$. Thus, even in close proximity to the burnout only 1/4 of the water inflowing to the wall is evaporated, whereas 3/4 of the water is withdrawn with the steam.

However, in the case of boiling of the solutions of increased concentration the picture differed markedly; as can be seen from graph Fig. 1, when passing from boiling of the monosolution of calcium sulphate, concentration of the saturation is of the order of 1 gr/kg, i.e. lower than critical into the solution of $CaSO_4$ in 0.1 and 1 m $NaCl$ solution the circulation ratios under the same environmental conditions sharply drop, from 2 to 5 times. A special investigation showed that the characteristic feature of the dependence of the circulation ratio of the total salt fraction is the same as for the bubble conditions. This is clearly seen from the graph Fig. 2 which represents variations of the circulation ratio at the load of 1.0 Mw/m² and 1 atm abs.

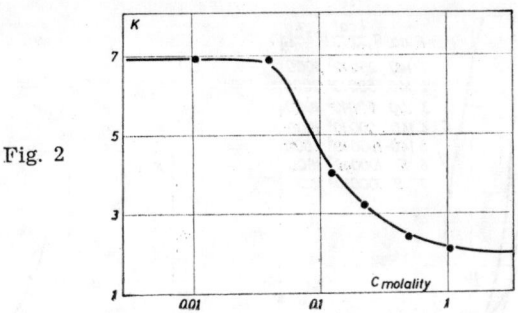

Fig. 2

Systematical investigations under the conditions of forced motion in tubes have been conducting only in 1—2 last years and resulted, as for the heat transfer in the process of boiling, in a very variated data which present a joint effect of a diversity of factors involved. Systematizing of these data requires long-time elaborate work, but even the data obtained are undoubtedly of great interest.

What are the determined dependences of the heat transfer under forced movement of a two-phase flow in tubes (at the experiments $d=4.6$ to 6.0 mm)? First of all, at low pressures, subcooled boiling or moderate steam qualities, the circulations ratios far from the burnout are fairly high, i.e. $K=30-50$. Approaching the burnout, the circulation ratios are first decreased rapidly and then slowly, thereby amounting in close proximity to the burnout to 2 to 5. The effect of particular parameters in question, especially at very low pressures of the order of 1 to 1.2 atm abs., is rather different, in all cases K tends to decrease toward the direction of approaching the burnout, though the rate of reduction is different (Fig. 3).

At high pressures (the herein-described experiments were conducted at 140 atm abs.), the circulation ratios are considerably smaller, but even in this case at moderate heat fluxes 0.27 to 0.58 Mw/m², far from the burnout they may reach $K=40$, but then, however, they are rapidly reduced (Fig. 3). It is worth noting that at high heat fluxes $q=1.2$ Mw/m², K remains relatively small, even within the zone of deep subcoolings, where as it is usually considered the convective heat transfer is a dominant factor, whereas the efect of evaporation is of a negligible concern. As approaching the burnout, the heat transfer weakens, however, even with $X=0.9$ X_{cr} it remains at the level $K > 2$. It is of a particular concern that this fact is observable within the zone which is usually considered as the burnout region of the second kind (dryout), (for example, with $q = 0.27$ to 0.58 Mw/m² and $x_{cr.}=0.25$ to 0.26). As far as we know, most of the »dryout« models arise from the pattern of drying the film slightly fed from the flow core and practically evaporating. This does not hold true of with $K > 2$ when from each kg of the fluid inflowing to the wall, only a smaller portion of half of the amount of the fluid is evaporated while the larger portion breaks contact with the wall and escapes to the flow core. Therefore, it is an urgent matter to subject to refinement the variations of the mass transfer directly in close proximity to the burnout.

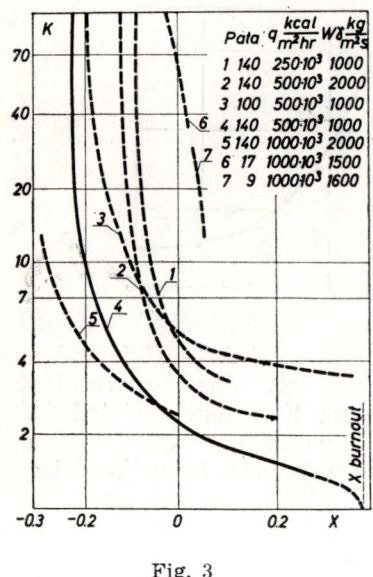

Fig. 3

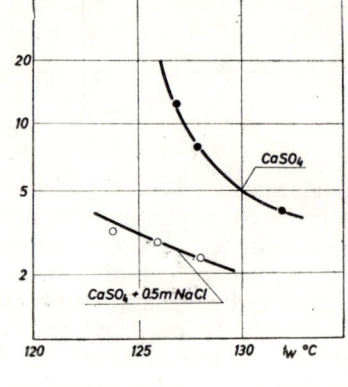

Fig. 4

Finally, as it is clearly seen from graph Fig. 4, the mass transfer occurring under the conditions of forced movement as well as under the conditions of natural convection is diminished sharply with high salt fractions. This fact presents certain interest from the point of hydrodynamics of a boiling sublayer and the analogy with bubbling requires further consideration.

On the whole, it is conceivable that the data already obtained make it possible to considerably refine a plurality of the concepts concerning the mechanisms of heat and mass transfer occurrable within two-phase flows, and in a number of cases, to raise a question to revise some approaches. We may hope that systematical investigations of heat transfer with resort to a salt — deposition method conducted presently at the Institute for High Temperatures and at the Moscow Power Engineering Institute in combination with other experimental approaches which E. I. Nevstrueva is going to speek on and taking into due account analytical methods provide the possibility to substantially enlarge our knowledge about the mechanisms of heat and mass transfer occurrable in two-phase flows.

NOMENCLATURE

q	heat flux
p	pressure
wj	mass flow rate
Z	the evaporation degree
C	the salt concentration
t	temperature
K	the circulation ratio
G	flow rate
d	the inside diameter of tube
x	steam quality
i	enthalpy

SUBSCRIPTS

cr	critical
sat	saturation
w	wall
f	fluid
$f.c.$	flow core
m	mixture

INVESTIGATIONS OF TWO-PHASE BOUNDARY LAYER CHARACTERISTICS BY VARIOUS METHODS

E. I. NEVSTRUYEVA

Institute for High Temperatures, USSR Academy of Sciences, Moscow, USSR

Investigations of a two-phase boundary layer even in the absence of phase transitions and under adiabatic conditions, present considerable difficulties. Moreover, when investigating a thermodynamically unequilibrium boiling layer featured by simultaneous processes of evaporation and condensation, extra particular difficulties are enconutered.

A general analytical description of the processes proceeding in an unequalibrium boiling layer, and especially the obtaining of exact analytical solutions, is yet a very difficult problem to be accomplished even for rather simpler particular cases. Therefore, in the present status of the field under discussion, to investigate the processes occuring in a boiling boundary layer, experimental methods are mostly resorted to.

One of the most promising and widely spread methods of investigating a boiling layer is the utilization of the similarity between the processes of boiling and bubbling. Such a method has already presented the possibility to obtain a plurality of worth-noting results in various institutes and establishments all over the world and especially at the Institute of Thermal Physics of the Siberian Division of the USSR Academy of Sciences.

However, it is worth mentioning that even in the case of utilization of the similarity between boiling and bubbling, it is necessary to take into consideration the key difference between the two of the processes described above. The cardinal difference is probably concerned with the origination of steam bubbles accompanying the process of boiling and the bubbles of gas blown through the porous surface involved. The formation of steam bubbles accompanying the process of boiling that evolve on the heating surface takes place in the form of a burst, the velocity of dislocation of the interface amounting to a few tens of metres per second. In this case some amount of fluid is ejected into the flow core at a high velocity and with a great kinetic energy. In bubbling the velocity of dislocation of the interface of the formed gas bubble and the kinetic energy of the fluid being ejected are supposed to be lower. Another difference between the process of boiling and bubbling is dealt with the fact that in the process of boiling the increased or diminished size of the bubbles involved is to the most extent due to evaporation or condensation, while in the process of bubbling the increased or diminished size of the bubbles is neglectable and may occur due to the pressure variation only.

To provide more exact similarity between boiling and bubbling, it might have been reasonable to blow in gas through a porous surface not at a constant flow rate or pressure but at pulsating pressure or flow rate whose frequency should be the same as the rate of evaporation. A certain interest may likewise be found in blowing in a fluid instead of gas which is supposed to give an opportunity to appreciate the fact of the process in question being influenced by the outburst of the fluid from the wall-attachment layer into the flow core.

All the problems involved are to a great extent open to discussion. However, it is worth noting that the utilization of the similarity between the process of boiling and bubbling may appear to be rather promising. The example of a certain new aspect of such similarity may be found in a qualitative analogy between frothing of boiling water in the drum and a sharply decreased intensity of mass transfer through the wall-attachment layer when the salt content of the solution involved tends to be above a certatin critical value. The existence of the above-described analogy which was communicated in the lecture by M. A. Styrkovich was for the first time revealed to exist at the Laboratory of two-phase systems when investigating mass-transfer in solutions of different concentrations.

All the investigations conducted at the Laboratory of two-phase systems of the Institute for High Temperatures, are based upon the assumption that, heat- and mass-transfer on one hand, and hydrodynamics and heat- and mass-transfer on the other hand, are indissolubly connected. Joint complex investigations of hydrodynamics and mechanism of heat- and mass-transfer, give an opportunity to a comprehensive approach to analysing the process proceeding in a two-phase unequilibrium boundary layer.

The methods dealing with the investigation of originating of boiling, can be subdivided into a number of principally different ones:

I. Methods of investigating microcharacteristics of boiling such as 1) bubble departure dimensions; 2) evaporation rates; 3) bubble growth rates; 4) number of effective evaporating centres; 5) ratio between delay time and bubble life time; and some other.

Such investigations are conducted mostly by the methods of visual observation or rapid film-shooting. Either of the above methods provides the possibility to obtain a qualitative and sometimes quantitative picture of the process in question. However, the resort to the film-shooting requires subsequent troublesome statistical processing of a large amount of frames and fails to give even a trustful qualitative picture when employed within the burnout conditions.

Promising results are obtained due to film-shooting process accomplished through a transparent heating surface, thereby enabling the surface wetted with the fluid within the burnout conditions to be obtained. However, even such data require subsequent thorough statistical processing of frames shot.

II. The methods of investigating averaged characteristics of a two-phase boiling layer. Fallen within the above mentioned methods are investigations dealing with time averaged local 1) flow or surface temperature, 2) velocities, 3) steam void fractions and 4) the maximum local steam void fractions. Pertinent to the all above-mentioned methods are investigations of circulation ratios by the salt methods.

III. The methods of investigating instantaneous characteristics of a two-phase wall-attachment layer such as: 1) temperatures of flow or surface; 2) steam void fractions; 3) pressures effective in the flow or nearby the wall, etc.

All these investigations are effective to find out some peculiarities inherent in mechanism of the processes proceeding within the wall-attachment layer and to make use of them to obtain more innocent ideas about mechanism of boiling or burnout.

So the investigations of steam void fraction distributions nearby the heating surface that have been conducted a few years ago at the Laboratory of two-phase systems, make it possible to appreciate the following things: 1) average statistical depth of a bubble-evolving layer, that is, the boiling layer; 2) to bring out two areas with respect to the depth of the boiling layer, that are featured by different manifestations of the effect of various factors such as fluid temperature and velocity; 3) to bring out the effect of various parameters such as pressure, temperature and velocity of the flow, as well as heat flux, upon both the nature of steam void fraction distribution and the depth of a boiling layer. Characteristic dependences of local steam void fraction at a certain distance from the heating surface upon the flow temperature are represented in Fig. 1, while represented in Fig. 2 are steam void fraction distribution nearby the heating surface at various parameters of the flow and heat fluxes.

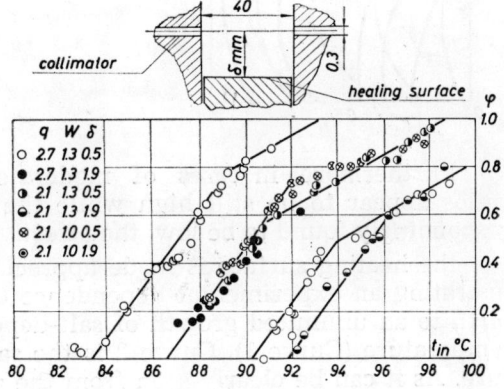

Fig. 1. A steam void fractions as a function of flow temperature at different distances from the heating surface.

The analyses of the results of these investigations and those of the investigation (1) concerning the study of temperature pulsation nearby the wall-attachment layer, hawe made it possible at one time to prove rather clearly a thermodynamical unequilibrium of a boiling layer and in particular to prove the fact that steam bubbles are free to penetrate throughout the depth of subcooled liquid thus condensing comparatively slowly.

Insofar as the utilization of the salt method in investigating the intensity of mass-transfer through a boundary layer is concerned, it has been spoken much in the lecture by M. A. Styrikovich. However, it should be noted that the use of the salt method makes it possible to appreciate not only circulation ratios (or degrees of evaporation of a wall-attachment layer), but also to determine certain other characteristics of a wall-attachment unequilibrium layer that are not to be appreciated by any other methods applied heretofore. Thus, the knowledge of circulation ratio makes it appreciable the amounts of heat consumed respectively to heat up the fluid to a certain temperature and to evaporate part of the fluid involved.

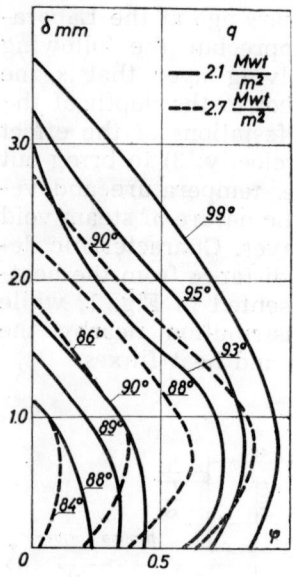

Fig. 2. A steam void fractions distributions near the heating surface.

Furthermore, in cases of moderate heat fluxes when circulation ratios appear to be still high while the degrees of evaporation are correspondingly found to be low, the effective temperature of the fluid hearby the heating surface is made appreciable. Fig. 3 represents a chart illustrating an experimental dependence of the flow concentrations giving birth to an unlimited growth of salt-deposition, versus the environmental temperature (Curve 1). Curve 2 is the salt solubility versus the temperature. As it can be clearly seen from the above chart, the effective temperature of the layer at which the salt-deposition process originates and, consequently, saturation is achieved, may actually be substantially lower than the temperature of the heating surface.

The investigations of steam void fraction effective within the burnout area have made it possible to reveal the fact that the burnout is not liable to originate at a certain constant value of steam void fraction as it has been supposed early by a number of investigators, but tends to change in a great variety of ranges (from 0.2 to 0.95) depending upon the parameters of the flow involved.

The concurrently conducted investigations of both the circulation ratios and distributions of steam void fractions have made it ascertainable that the velocities of counter-current flows of both steam and liquid phases outflowing from and inflowing to the heating surface, respectively, are not fixed ones but vary within wide ranges. A relative velocity of the phases tends to considerably vary as well. An assumption has been put forward on the basis on the investigations conducted, that a predominant factor for the burnout is supposed to be the kinetic energy of the fluid inflowing to the surface being cooled and overcoming the hydraulic resistance of the boiling layer.

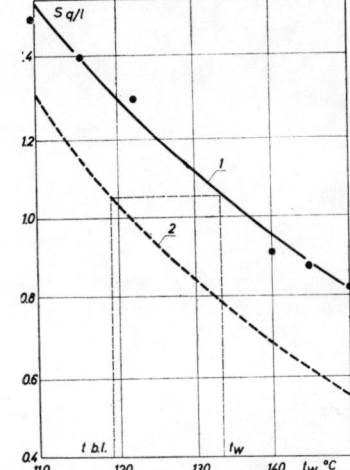

Fig. 3. A flow concentrations giving birth to an unlimited growth of saltdepositions as a function of the heating surface temperature.

From this viewpoint a reknown interest would be offered by a complex investigation of distributions of instantaneous values of steam void fraction and pressure pulsations in both the flow core and nearby the wall, which would allow of appreciating the variations of hydraulic resistance of the boundary layer with time. Of promising interest would likewise be a complex utilization of both the method of Y-ray attenuation and the salt method in studying the mechanism of frothing of the boundary boiling layer when transiting the critical salt-content.

In conclusion it should be noted that when investigating the processes proceeding in two-phase boiling layer all the methods are to be resorted to that may give an opportunity to appreciate both the mechanism and characteristics of the processes proceeding in boiling layer in all the aspects concerned [1—6].

REFERENCES

1. J. Clark Jiji, J. Heat Transfer, Ser. C, 3 (1964).
2. E. I. Nevstrueva, Kh. M. Gousales, Teploenergetica, 9 (1960).
3. M. A. Styrikovich, E. I. Nevstrueva, A. S. Mekhdi, Heat and Mass Transfer, 3, pp. 42 (1965).
4. G. G. Trestcher, Teploenergetica, 5 (1957).
5. I. G. Malenkov, Prikladnaja mechanika i technicheskaja phizika, 6 (1963).
6. M. G. Hubbard, A. E. Dukler, Proc. Heat Transfer and Fluid Mechanics Institute, pp. 100—121, 1966.

INVESTIGATION OF POOL BOILING HEAT TRANSFER OF WATER AND FREON-113 IN A WIDE RANGE OF TEMPERATURE HEADS

S. A. KOVALEV, V. M. ZHUKOV, Ya. A. KUZMA-KICHTA,
V. P. OGORODNIKOV

Institute for High Temperatures, USSR, Academy of Sciences, Moscow, USSR

1. Introduction

At present due to developlent of a number of branches of novel engineering, the problem of cooling down the surfaces adapted to operate under nonstationary regimes, by virtue of boiling of a heat carrier, is of great concern. To carry out heat calculations of such surfaces it is necessary to know heat transfer regularities within all of the regimes of boiling. If the regions of nucleate and film boiling are considered sufficiently well, the transition boiling due to its unstability remains to be seen. The available data concerning the effect of various factors upon heat transfer are rather contradictory and provide no possibility to ascertain heat transfer regularities within this region of the boiling process. Considered but poorly is the question associated with the effect of a low thermal conductivity coating upon heat transfer in transition and film boiling.

It is a primary object of this work to investigate experimentally boiling heat transfer of water and Freon-113 within a wide range of temperature heads under various surface conditions.

2. Experimental Installation

For investigation on boiling heat transfer of water and Freon-113 use is made of condensing vapour to heat up the test section involved. Such method makes it possible to maintain the test section temperature constant irrespective of the boiling regime, and, thus make the data obtainable in a wide range of temperature heads.

The heating steam is generated in a high-pressure steam generator by means of an electric heater and is condensed on the lower test section surface (Fig. 1). Selected as a test section is a horizontally arranged 36.6 mm dia. and 25 mm high copper disk which upper surface serves as a heating surface for the liquid. At a distance of 3.5 and 6.5 mm from the heating test section surface is provided with two 1 mm dia. radial borings wherein are embedded chromel-alumel thermocouples enabling the heating surface temperature to be measured. To compensate for heat losses from the boiler to the environmental medium, provision is made for auxiliary heaters.

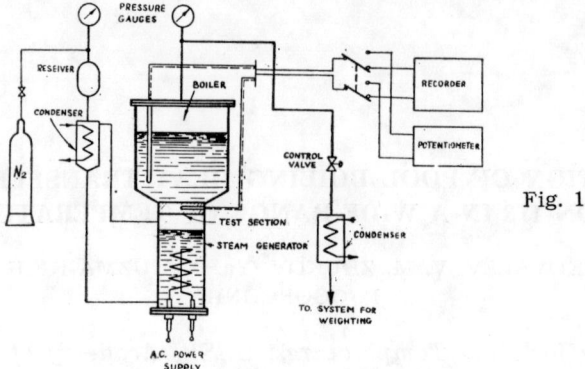

Fig. 1

The steam formed in the process of liquid boiling on the surface of the test section is condensed in the condenser cooled down by water. The pressure control in the boiler is accomplished through the use of a control valve provided between the boiler and the condenser. The heat flux is measured by weighting the amount of condensate of the liquid under consideration.

3. Results of Investigation

Boiling heat transfer of water. The investigation of boiling heat transfer of water at atmospheric pressure within a wide range of temperature heads (from 7 to 200°C) is attributed to the nucleate and transition

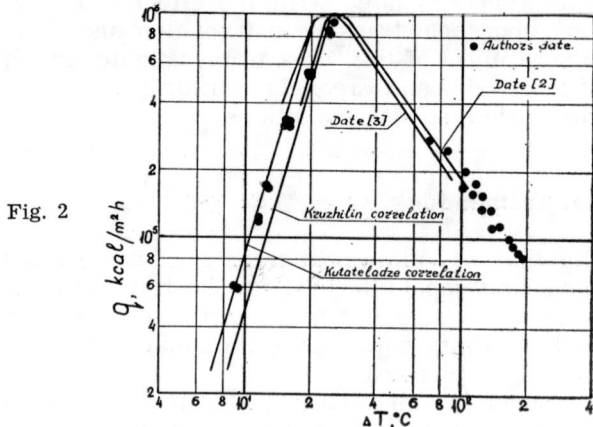

Fig. 2

boiling. The obtained transfer data in a transition boiling (Fig. 2) cover the ranges of heat flux variations from $80 \cdot 10^3$ to $280 \cdot 10^3$ Kcal/m²hr and temperature heads from 70 to 200°C. Within the range of temperature heads from 25°C to 70°C we failed to carry out the measurements of heat transfer, since within this range of temperature heads the boiling regime was unstable. The instability of the boiling regime was characterized by pulsations of wall temperature that tend to increase with the increase of heat flux, and by spontaneous transition to the nucleate boiling. As is shown in [11], the instability of the boiling regime was due to deterioration of heat equilibrium of the heating surface.

Figure 2 likewise represents comparing of our investigations results with the data obtained by other authors. It is seen from this figure that our data concerning the nucleate regime of water boiling are in good agreement with well known calculation dependences by Kruzhilin and Kutateladze. Within the investigated region of the transition boiling our data satisfactory agree with the data [2], as well as with the data [3].

Boiling heat transfer of Freon-113. For investigation use was made of Freon-113 corresponding to BTY-EY-120-56 (USSR Specifications). Low saturation temperature of Freon-113 at a normal pressure $(t_s = 47.6°C)$ and high temperature of the heating steam $(t_s \simeq 300°C)$ made it possible to carry out investigation within a wide range of temperature heads from 8 to 250°C which covers the nucleate, transition and film boiling.

The results of boiling heat transfer investigation of Freon-113 for pressures 1, 2 and 3 ata. are represented in Fig. 3. The data obtained

Fig. 3
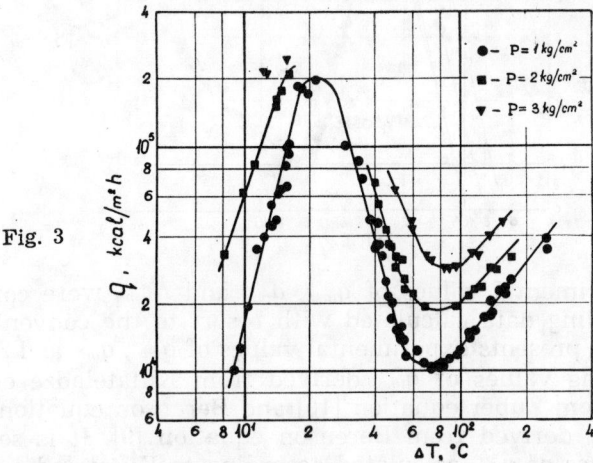

provide the possibility to determine the values of critical heat fluxes and their corresponding temperature heads. The increase of pressure results in improved heat transfer in all the boiling regimes (with $\Delta t = $ const) and increased values of q_{max} and q_{min}. So with $P=1$ ata q_{min} amounts to 10^4 Kcal/ /m²hr, since with $P=3$ ata q_{min} equals $2.9 \cdot 10^4$ Kcal/m²hr.

As with water boiling, in the transition regime of boiling of Freon-113 was observed an unstable heat transfer region. It is seen from Fig. 3 that within the regions close to the points of q_{max} and q_{min} a smooth variation of the boiling curve is observed. Thus, within the entire transition boiling, the experimental data cannot be described by a straight line connecting the points corresponding to critical heat fluxes. If we exclude from consideration the regions close to q_{max} and q_{min}, where variation of the dependence $q=f(\Delta t)$ is of a more complicated character, the experimental data may be sufficiently well approximated by straight lines and we may select empirical dependences as $q=c\Delta t^n$.

Comparison of the results of boiling heat transfer investigation of Freon-113 at $P=1$ ata. with the data obtained by other authors is represented in Fig. 4. It is seen from the figure that in so far as the nucleate boiling is concerned, there is satisfactory agreement with the data [5], which were obtained on the surface featuring the roughness approximately corresponding to that used by us ($R_z=6.3$ μk). The fact that the results of investigation [4], as well as the data [8] lie somewhat below may be explained by different means of treating the surfaces involved [1]. In so for as the transition boiling is concerned, there is a good agreement with the data [8] and as to the film boiling, there is the same agreement with the data [6, 7].

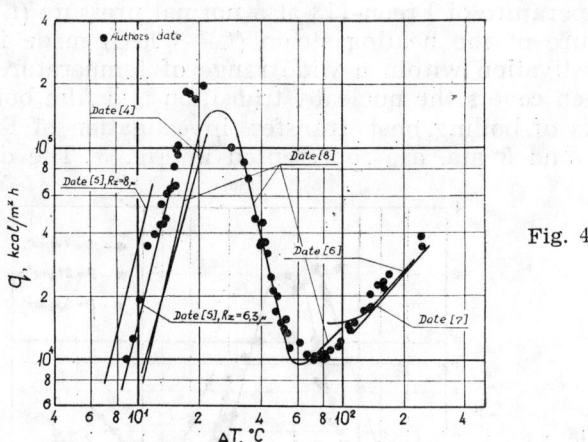

Fig. 4

The experimental values of q_{max}, q_{min} and Δt_{min} were compared with the corresponding data calculated with resort to the conventional equations. The table presents experimental values of q_{max}, q_{min} and Δt_{min} together with calculating values of q_{max} derived from Kutateladze equation [12], q_{min} derived from Zuber equation [10] and Berenson equation [9], and the values of Δt_{min} derived from Berenson equation [9]. It is seen from the table, the values of q_{max} calculated according to Kutateladze equation are in good agreement with the experimental data. As to the values of q_{min} Zuber and Berenson equations provide satisfactory agreement with experimental data. The departure of the value Δt_{min} calculated according to Berenson equation from experimental data for $P=1$ ata amounts to 12%, for $P=2$ ata $\sim$ 31% and for $P=3$ ata $\sim$ 54%.

The effect of low thermal conductivity coatings upon boiling heat transfer of Freon-113.

The present work concerns the investigation of the effect of low thermal conductivity coatings applied to the heating surface upon boiling heat transfer of Freon-113.

Utilized as such a coating was a phenol glue, grade BC-10T, since this glue meets the following requirements: it is heat resistant (its heat resistivity amounts $\sim 400°$ to $500°C$), it is resistant against Freon-113 and features thermal conductivity equal to 0.25 Kcal/mhr °C which is 1000 times lower as compared with copper.

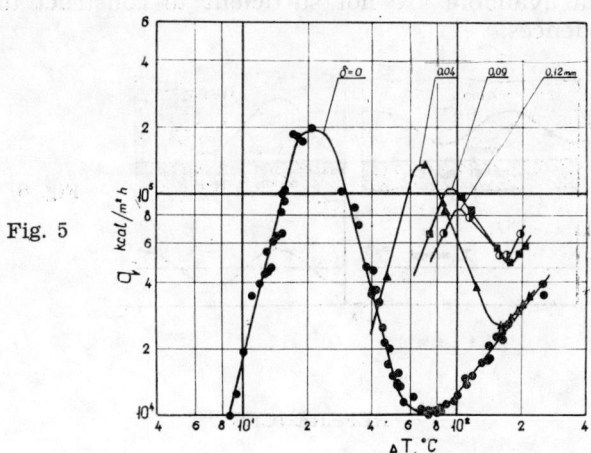

Fig. 5

Represented in Fig. 5 are the results of measuring the dependence $q = f(\Delta t)$ with boiling of Freon-113 on the pure surface and on the surface coated with a film of BC-IOT glue. The heating surface temperature was calculated with taking account of coating thermal resistivity. It is clearly seen, that with the increase of the coating thickness (0.04; 0.09; 0.12 mm) the shape of the boiling curve varies sharply.

The analysis of the data represented in Fig. 5 permits to conclude that application of a low thermal conductivity coating causes a somewhat decrease of q_{max} which is probably associated with variation of chemical nature of the surface in question. A considerable increase of q_{min} with application of a coating featuring a comparatively pure surface (for the thickness of 0.04 mm approximately as high as two times, and for the thickness of 0.12 mm, approximately as high as five times) may be explained as follows.

With film boiling the interface fluid-vapour is undergone to continuous fluctuations and features a wawe-shaped form, as is seen in Fig. 6. At low heat fluxes the fluid may approach particular points of the surface and may even get in contact with it. Drawing together of the fluid and copper surface is of no influence upon the copper surface temperature due to a rather high thermal conductivity of copper. On the contrary, contacting of the fluid with the heating surface which features a low thermal conductivity results in a sharp local decrease of the temperature of this surface [13] which tends to drop with the increase of the coating thickness involved.

A local decrease of the surface temperature causes a decrease of the amount of the generated steam and, consequently, results in a decreased steam film thickness. On one hand, this leads to increased heat transfer, and on the other hand, to ceasing of film boiling with rather high the surface average temperatures and heat fluxes.

At present we may note that the increase of q_{min} and the film boiling heat transfer coefficient in the case of a thin-film low thermal conductivity surface coating is undoubtedly of practical interest. However, the experimental data available are not sufficient to construct universal calculation dependences.

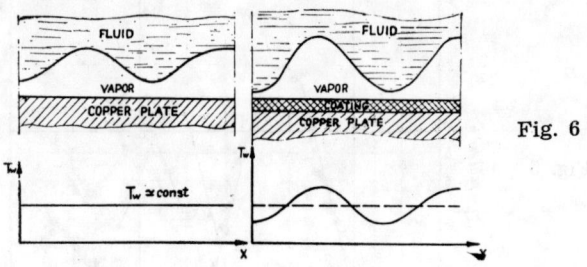

Fig. 6

REFERENCES

1. P. I. Berenson, Int. J. Heat Mass Transfer, 5, 5 (1962).
2. B. S. Ishigai, T. Kuno, Bull. of JSME, 9, 34 (1966).
3. R. H. Braunlich, Thesis, Mass. Inst. Tech., 1941.
4. G. V. Ratiani, D. I. Avaliani, Kholod. Tekh., 3 (1965).
5. G. N. Dan'lova, V. K. Belsky, Kholod. Tekh., 4 (1965).
6. D. E. Kautsky, I. W. Westwater, Int. J. Heat Mass Transfer, 10, 2 (1967).
7. M. L. Pomerantz, J. Heat Transfer, ser. C, 86, 2 (1964).
8. K. W. Haley, I. W. Westwater, Proc. Third Int. Heat Transfer Conf., vol. 3, 1966.
9. P. I. Berenson, J. Heat Transfer, 83, 3 (1961).
10. N. Zuber, Trans. ASME, 80, p. 711 (1958).
11. S. A. Kovalev, Int. J. Heat Mass Transfer, 11, 2 (1968).

TABLE

| P, ata | kcal/m²h | | kcal/m²h | | | | °C | |
| | Experiment | Equation Kutateladze | Experiment | Equations | | | Experiment | Equation Berenson |
				Berenson		Zuber		
1	195000	195000	10000	9160	13650	10330	76	67
2	240000	248000	19000	16750	24500	18550	80	105
3	270000	268500	29000	20150	32000	24250	90	138

CRITICAL HEAT FLUXES IN BOILING LIQUID METALS

G. I. BOBROVICH

Thermal Physics Institute Siberian Division of the USSR Academy of Sciences, Novosibirsk, USSR

Publications on this subject are very scarce due to great technological difficulties in studying heat transfer in boiling metals at high temperatures.

Only in recent years have there appeared in this country and elsewhere first publications on determining critical heat fluxes in boiling potassium, sodium, caesium and rubidium [1—5] under the conditions of natural convection.

Very few contributions have been made on the experimental determination of critical heat fluxes in the forced motion of alkali metals. These contributions are not discussed here.

The conventional technique of studying the boiling mechanism of non-metallic fluids is inapplicable to boiling metals and many characteristics of their boiling process remain unknown without visual observation.

To study the mechanism of determining critical heat flows in boiling alkali metals an experimental set was mounted at our Institute which permitted high-speed X-ray photography.

The first experimental data ever obtained on the mechanism of potassium boiling under the conditions of free convection are reported in [6].

1. The Experimental Set

A schematic diagram of the experimental set is represented in Fig. 1 where the principal units of the set are shown. Some of them are common to all sets of this type and require no comments. There seems to be a necessity to describe only the boiler and the text section. The boiler was made of a 6 mm thick sheet stainless steel in the form of a parallelepiped. For the experiments two boilers of this type were fabricated with the following dimensions:

1) $280 \times 180 \times 100$ mm and 2) $280 \times 180 \times 80$ mm

In the front and back facets of the boiler hollows were milled out in the shape of rectangular window to leave the wall 1.5 mm thick. These hollows were made to diminish the absorption of X-rays by the boiler walls.

Three thermocouple bushes were welded into the upper part of the boiler at a distance of 10 mm from the upper formative line of the test section. In the lower part of the boiler there were two thermocouple bushes, one of which was welded at 10 mm from the lower and the other at the level with the upper formative line of the test section. Two thermocouple bushes were welded into the vapour space of the boiler to control the vapour temperature.

The boiler was provided with an external heater. The power was fed to the boiler from a transformer permitting a smooth adjustment of the voltage in the range from 0 to 250 volts.

The heat-transfer surface of the test-sections was made of the stainless steel. The external diameter of the sections varied from 9 to 14 mm. The length of the heat transfer part of all the test sections amounted to 80 mm.

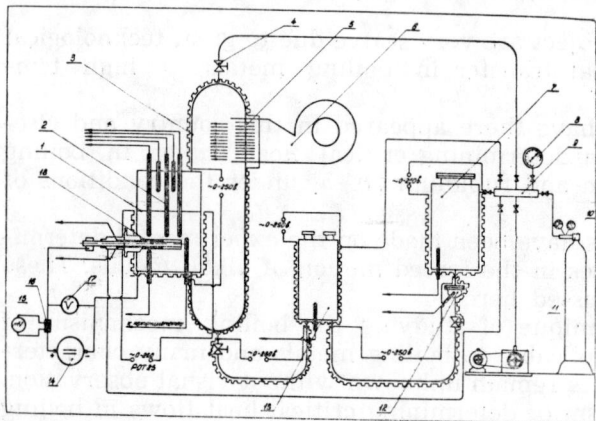

Fig. 1. Schematic diagram of experimental set.
1. Boiler. 2. Bushes for thermocouples. 3. Vapour discharge pipe socket. 4. Bellows valve. 5. Condenser. 6. Ventilator. 7. Charging tank. 8. Monovacuumeter. 9. Gasvacuum collector. 10. High-purity helium cylinder. 11. Forvacuum pump. 12. Filter. 13. Intake tank. 14. Direct current generator. 15. Millivoltmeter. 16. Shunt. 17. Voltmeter. 18. Test section.

The test sections were arranged horizontally in the boiler.

A ube ground out of tantalum or molybdenum served as a heating element. The heating element wall was 0.25 mm thick. A special compensator insured unrestricted thermal expansion of the heating element. The clearance between the internal diameter of the test section and the external diameter of the heating element was 0.5 mm and was filled with boron nitride which served as both an isolating and heat-conducting medium. The heating element worked in the medium of high-purity helium. Such a design of the test section allows to obtain heat flux higher than 4 Mwt/m².

The temperature of the test section surface was measured by chromel-alumel thermocouples mounted at the top and at the bottom of the middle part of its heating surface. The diameter of the electrode was 0.15 mm. The hot joint of the thermocouples was located at a distance of 0.5 mm from the external surface of the test section.

The accuracy of the test section surface treatment corresponded to the class 6 of GOST-2789-59. The test section was heated electrically from a direct current generator with independent excitation which allowed to smoothly regulate the heat transfer over the test section.

2. High speed X-ray Filming of the Boiling Process of Liquid Metals

The principle of high-speed X-ray photography is represented in Fig. 2. A boiling alkali metal is X-rayed and the obtained picture on an amplifying X-ray screen is transported by means of a light-powerful objective onto the photocathode of electron-optical transducer (EOT). From the anode of the EOT the boiling process of alkali metals was either photographed or observed visually. The shooting was done with a high-speed camera »Pentacet-35«. The camera is provided with a frame time marker.

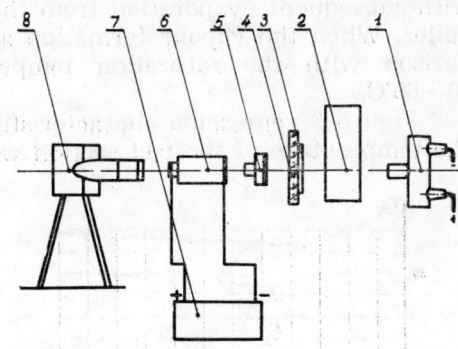

Fig. 2. Principle of X-ray high-speed film shooting.
1. X-ray tube. 2. Boiler. 3. Amplifying screen. 4. Lead glass. 5. Objective. 6. Electron-optical transformer. 7. High-voltage supply suorce. 8. High-speed camera.

3. A Study of the Critical Heat Fluxes in Caesium Boiling

The critical heat fluxes in the boiling of liquid metals were studied on caesium as the latter can be most conveniently investigated in a wide range of pressures (from vacuum to 3—4 bars) owing to its lower saturation temperature in comparison with lithium, sodium and potassium. The metal used was of high purity, contained in glass ampules in accordance with RETU-117-59. The set was filled with caesium by placing the glass ampules into the charging tank which was then vacuumized. Then the tank was filled with high-purity helium and under an excessive pressure the ampules were broken by a special device and the melted caesium was pressed through a system of filters into the inlet tank. The final purification of caesium was effected by pressing it through a filter made of a porous pipe of stainless steel with a diameter of few microns. The purified caesium was passed from the inlet tank into the boiler. The level of caesium in the boiler was controlled by X-raying.

Before the experiment the vapour space of the boiler was vacuumized with the external heater switched on, and the caesium was heated up to 300°C with the bellows valve opened to remove the residues of helium dissolved in caesium. The further increase of temperature was effected with the bellows valve closed. All the experiments aimed at determining the critical heat fluxes in the boiling caesium were carried out at the saturation temperature, in 10 mm diameter test section. The section surface was treated to reach the class 6 of cleaniliness according to GOST--2789-59. The first series of experiments was carried out in this section in the pressure variation range from 0.04 to 1.7 bars. In this pressure interval the boiling process proved unsteady for all heat fluxes up to critical ones. As is known, the boiling process of sodium [7, 8] and potassium [6] is unsteady on sufficiently smooth heating surfaces.

Visual observations of the caesium boiling process through EOT in the first series of experiments have revealed that in close-to-critical heat fluxes a large vapour bubble is formed over all the heating surface from many smaller ones joined into one. After this the bubble tears itself off the surfaces, the vapour formation on the test section surface ceases and the heat transfer occurs through natural convection and heat conduction with subsequent evaporation from the free surface of the caesium in the boiler. When the vapour formation stops, the wall superheating in comparison with the saturation temperature sharply increases reaching 50—60°C.

Figure 3 represents characteristic records of the difference between the temperature of the test section wall and that of saturation.

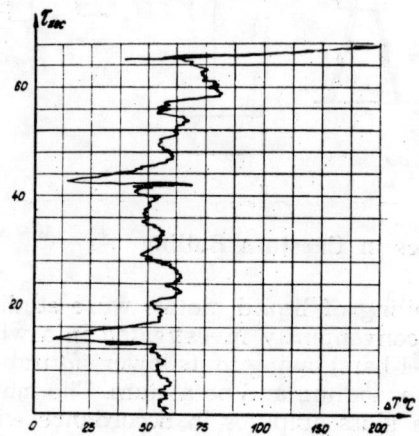

Fig. 3. Records of temperature differences »Wall-l'quid«. Unsteady boiling (first series of experiments) $P=0.15$ bars.

The critical heat flux occurs at a slight increase in the heat flux after unsteady boiling begins. The crisis in the boiling occurs after the large bubble breaks off, i.e. during heat transfer by natural convection rather than at the moment of the formation of the large bubble. However, the very moment when the crisis begins cannot be observed visually. The crisis was registered by a sharp jump in superheating the test section surface (by hundreds of degrees) with a subsequent break in the electric power supply to the test section. In the diagram representing the records the difference between the temperature of the test section wall and that of the caesium saturation, the beginning of the crisis is marked by a sharp drop of the test section wall temperature followed by its sharp rise.

The results of the first series of experiments aimed at determining the critical heat fluxes in the unstable boiling of caesium are represented in Fig. 4 (curve 1). The graph reveals no essential dependence of the critical heat flux on the pressure but still a certain tendency to the decrease in the critical heat flux is observed as the pressure increases. It should be noted that at a pressure above 2 bars caesium boiled steadily and the heat flux increased two-fold in comparison with the critical heat flux in unstable boiling. The crisis in the unsteady boiling caesium occurred at superheating the test section wall in the interval of 55—75°C. Thus the first series of experiments have shown that crisis to occur in the unsteadily boiling of caesium immediately after the natural convection regime sets in. An analogous phenomena was also observed when ethyl alcohol was boiling in vacuum on sufficiently smooth heating surface [9]. In this case, crisis immediately followed the natural convection regime, but the values of the critical heat flux were much lower than those at the crisis with the preceding nucleate boiling on the heating surface (nucleate boiling of organic liquids in vacuum was obtained by creating artificial centers of bubble formation on the heating surface).

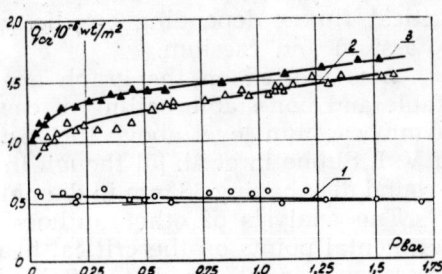

Fig. 4. Dependence of q_{kp} on P for caesium.
1. Unsteady boiling
2. Steady boiling ($h=30$ mm).
3. Steady boiling ($h=60$ mm).

The second series of experiments aimed at determining critical heat fluxes in boiling caesium was carried out in the same test section as the first one, but to secure stable boiling artificial vaporization centers were created in the test section (two on the upper component and one on the bottom one). The geometry of an artificial centre of vaporization is shown in Fig. 5. In the presence of these centers the boiling of caesium in the test section was stable in the whole range of pressures studied.

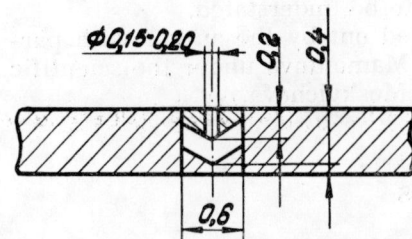

Fig. 5. Geometrical dimensions of artificial centres.

The experimental values of critical heat fluxes depending on pressure for the second series are plotted in Fig. 4 (curve 2). As follows from the graph, the critical heat fluxes at stable caesium boiling increased two- or three-fold in comparison with unstable boiling.

This series of experiments was also made to study the influence of the caesium level above the test section. Curve 2 in Fig. 4 corresponds to the values of the critical heat fluxes at the level of 30 mm above the upper component of the test section. As a result of visual observation, the dimensions of the bubble departure diameters were found to average about 40 mm. Therefore the level of caesium above the section was increased up to 60 mm after that the values of the critical heat fluxes became somewhat higher (curve 3 in Fig. 4).

Figure 6 represents the results of the experimental determination of

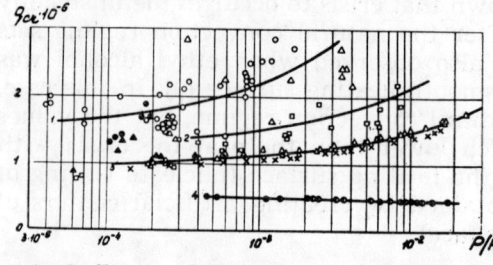

Fig. 6. Dependence of q_{kp} on P/P_{kp}
1. sodium experiments [1, 2. 4]
2. potassium R. F. Balzhiser experiments from [10]
3. caesium, experiments [5]
x. caesium, our experiments (stable boiling)
4. caesium, our experiments (unstable boiling).

critical fluxes depending on the pressure reported in [10] for sodium, potassium and caesium.

Also plotted on the graph are our experimental points during the stable and non-stable boiling of caesium (curves 1 and 3, Fig. 4). At the 60 mm caesium level above the test section, these data agree with those of V. I. Subbotin et al. [5] though in the works [5] caesium was boiling on a round disk having 38 mm in diameter.

The analysis of other authors' data suggests that some of the experimental points on the critical heat fluxes at the boiling of sodium and potassium were obtained at unstable boiling.

It should be noted in the conclusion that, as seen from observations of the stable boiling up to the crisis the latter is natural in boiling of metallic and non-metalic fluids, being analogous.

However, due to the high heat conductivity of alkali metals, a great part of the heat even under the near-to-crisis regime is transferred by means of conduction, natural convection, turbulization of the boiling liquid by emergent bubbles. Therefore, the values of the critical heat fluxes calculated after the formulas obtained on the basis of the hydrodynamic theory of boiling crisis turn out to be understated.

Investigation reported here was carried out by the author with participation of B. P. Avksentyuk and N. N. Mamontova under the scientific direction of S. S. Kutateladze and V. N. Moskvicheva.

REFERENCES

1. R. C. Noyes, J. Heat Transfer, ASME Ser. C, **85**, 2 (1963).
2. V. I. Subbotin et al., Pap. 328 3rd Int. Conf. on Peaceful Uses of At. En. (Geneva, 1964).
3. B. F. Gaswell and R. E. Balzhiser, Chem. Eng. Progr. Symp. Series Heat Transfer, Los Angeles, **62**, 64, 1966.
4. R. Noyes, H. Lurie, Third Int. Heat Transfer Conf., pap. 160, vol. 5, Chicago, 1968.
5. V. I. Subbotin et al., Pap. French-Soviet Sympos., Grenoble (France), Dec., 1966.
6. G. I. Bobrovich, B. P. Avksentyuk, N. N. Mamontova. ISME Semi-International Symposium, Tokyo, 1967.
7. Marto, W. R. Rohsenow, Teploperedacha, Trudy amerikanskogo obshchestva inzhenerov-mekhanikov, 38—50, **2**, 1966.
8. Marto, W. R. Rohsenow, Teploperedacha, Trudy amerikanskogo obshchestva inzhenerov-mekhanikov, 51—59, **2**, 1966.

ELECTROCHEMICAL TECHNIQUE OF FLOW STUDIES

A. P. BURDUKOV

Thermal Physics Institute, Siberian Division of the USSR Academy of Sciences, Novosibirsk, USSR

Of late more and more publications appear dealing with the use of the electrochemical technique of studying fluid motions. The technique consists in measuring the values of a limiting diffusion current in an electrochemical cell formed by nickel or platinum electrodes and an equimolar solution of $K_3Fe(CN)_6$ and $K_4Fe(CN)_6$, in the presence of a large excess of NaOH with 0.005—0.1 N — concentration of ferri- and ferrocyanide and $2N$ concentration of an alkali. The following reactions occur at the surface of the electrodes:

$$\text{cathode} - Fe(CN)_6^{-3} + e \rightleftarrows Fe(CN)_6^{-4}$$

$$\text{anode} - Fe(CN)_6^{-4} \rightleftarrows Fe(CN)_6^{-3} + e$$

The cathode is generally used as a polarized electrode, i.e. the surface of the anode must be much larger. The value of the limiting diffusion current is determined by the rate of ion feeding the cathode surface. Thus, the electrochemical technique is analogous to the thermoanemometric one with a constant temperature of the wire as the ion concentration at the surface of the sensitive element is constant (and equal to zero). As the potential for the reduction of oxygen from the solution is of the same order as in the basic reaction the solution must be thoroughly freed from oxygen by passing nitrogen periodically through the fluid. All elements of the setting are made of stainless steel, plexiglass or vinilite to exclude concurrent reactions.

The current flowing through the cathode depends on the coefficient of mass transfer to the electrode surface, i.e. on the velocity field in the vicinity of its surface.

As the Schmidt numbers for such a solution are rather high (approximately 2400), the diffusion layer is much thinner than the hydrodynamic one and sufficiently small longitudinal dimensions of the gauge in most cases of practical interest its thinness is within (the range of) the viscous sub-layer, i.e. the electrochemical technique yields measurement of local friction at the wall.

Thus, the electrochemical technique is very helpful in studying local velocities of the flow and local friction at a wall.

The velocity and friction measurement by the electrochemical technique is schematically represented in Fig. 1. The figure also shows the volt-ampere characteristics of the process.

The potential from the storage battery 8 is fed to the cathode 1 and the earthed anode 2 through the 30-ohm divider 3.

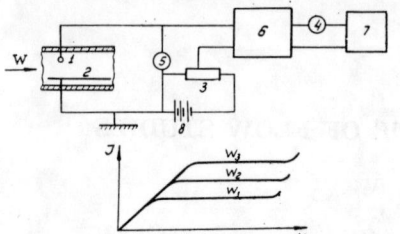

Fig. 1. Measurement scheme and volt-ampere characteristics.

The semi-conductor balanced direct-current amplifier 6 has a low (about 100 ohm) input resistance for decreasing the tension fluctuations on the cathode at fluctuations of mass transfer coefficients and, hence, of the current in the electrochemical cell.

1. Measurement of Shear Stress at a Wall in Gas-Liquid Flows

Flows of gas-liquid mixtures occur in many domains of chemical technology, in thermoenergetics etc. One of the main parameters characterizing gas-liquid systems is a fall of pressure in a channel. In case of one-phase flows this parameter is determined simply while in two-phase ones, due to the occurence of phase sliding, the knowledge of discharge contents of the components does not yield the values of real phase velocities, i.e. real gas contents. The experimental technique reported in publications [3—4] for determining gas contents are not sufficiently reliable.

Out of the theoretical calculation models the simplest is the homogeneous one presupposing the identity of phase velocities in the channel. The pressure losses are calculated in this case according to the familiar correlation for a one-phase flow where the viscosity is taken as that of the liquid phase or the averaged value of the phase viscosities.

In the American reference literature wide use is made of the Lockart and Martinelly correlation [5] establishing a connection between the pressure losses on friction in two-phase and one-phase flows.

To calculate the pressure drop and the distribution of density and velocity in a two-phase flow Levi [6] used a technique based on Prandtl mixture length theory for a one-phase flow. The density and velocity distribution and the pressure drop are determined by regarding the two-phase system as a medium in which the turbulent transfer of momentum and density is identical. The comparison of this technique of calculation with experimental data does not give the same answer as to its efficiency. Besides, the experiments themselves do not yield the same data on measuring friction in two-phase flows.

It should be noted that no techniques have been so far developed for complex condition of flows such as, say, an unstable two-phase flow.

The electrochemical technique is rather promising for the study of gas-liquid flows. Using the assumption of the either dependence of the longitudinal velocity component on the transversal coordinate within the range of the laminar sub-layer it is possible to obtain a dependence between the mass flow to the probe and the friction at the wall [7].

The experiments in measuring the friction in the flowing two-phase mixture were conducted on a horizontal and a vertical pipe. The scheme of one of the set (the horizontal channel) is represented in Fig. 2.

Both sets consist of closed circulation contours including the centrifugal pump 2 made of stainless steel the separator 1, the flowmeters 4, 11 and a stabilization section.

To prevent the oxidation of the electrolyte by the air oxygen due to the great closeness of the decomposition potentials of oxygen and ferrocyanide, nitrogen was used as a light phase. The temperature of the liquid was kept constant at 25°C with the help of the refrigerator 5. The temperature of the two-phase flow at the input and at the output of the experimental section was measured by copper-constant thermocouples with the accuracy of 0.1°C.

The experimental section 3 is designed as an 11 mm diameter 100 mm long tube made of plexiglass. Mounted along the tube are nine electrochemical probes 6 made as rings of nickel foil 0.2—0.5 mm thick. The gas is blown in through the porous channel walls of the experimental section.

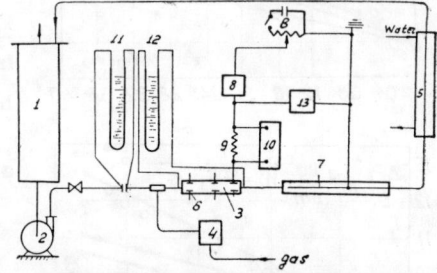

Fig. 2. Scheme of the set for measuring friction in two-phase flow

The quality of the probes was checked by taring in friction measured electrochemically and in pressure drop while the liquid flowed.

The anode 7 made of the same nickel foil was placed downstream. The measurement circuit has a 20 ohm resistor. The value of the regulated tension and the limiting current in the circuit was measured by the voltmeter 13 and the milliampermeter 8.

The current fluctuations were recorded on the loop oscillograph 10. The concentration of ferrocyanide ions in the solution was measured by volumetric chemical titration.

In the experiments with the vertical pipe-with a 15 m diameter similar ring-shaped probes were distributed along the channel height with 50—100 cm intervals. A vertical plexiglass pipe about 3 m long was used. The gas-liquid mixture was made in the lower part of the channel, in the mixer, at a distance of 150 calibres from the first probe.

The experimentation scheme is the same as that in the set with the horizontal channel.

The experiments on a horizontal pipe were conducted in the range of the liquid velocity variations 1.5—8 mps with the gas velocity 1.2—10 mps; on the vertical pipe, with the liquid velocity up to 1.5 mps.

The results obtained are presented in Fig. 3 as dependences of the friction pressure drop on the non-dimensional gas velocity W_o''/W_o' for the liquid velocities $W_o'=2.8$ and 8.06 mps.

The points correspond to the results of the experiments with electrochemical probes and of those with a drop of static pressure. The lines 1, 2, 3 correspond to the results of calculation in accordance with the Lockart-Martinelli [4] techniques, the homogeneous model and the CKTI recommendations.

As seen from the graphs, the divergences in $\Delta P_{mp}/L$ measured by the pressure drop and by electrochemical technique reach in some cases the values of about 20%. This can be explained by losses on acceleration occurring in an unstable two-phase flow.

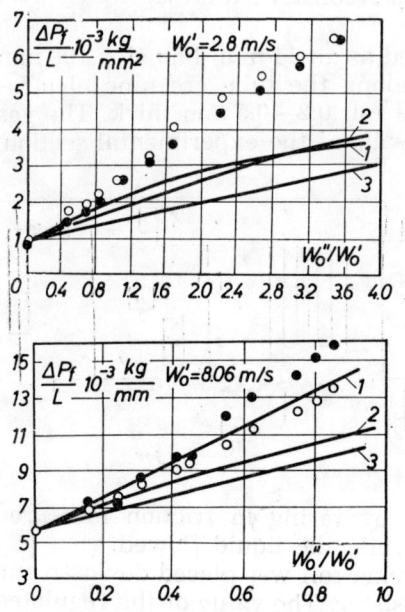

Fig. 3. Dependence of pressure loss due to friction $\dfrac{\Delta P_{mp}}{L} \cdot 10^{-4} \left[\dfrac{H}{MM^2}\right]$ on W_o''/W_o' for a horizontal pipe
○ — averaged values of τ measured electrochemicaly
● — result of experiments on measured statistical pressure drop.

The Lockart-Martinelli technique yields good agreement with experiments for the velocity $W_o'=8.06$ mps in the range of variations W_o''/W_o' up to 0.9. At lower velocities of the liquid (2.8 mps) this technique yields smaller values, almost twice as small as the experimental ones. The calculations on the homogeneous model and on the norms of CKTI yield even worse results.

The results of measurements on the vertical pipe were treated in a similar way.

Thus the electrochemical technique of shear-stress measurement in the flow of two-phase mixtures allows to measure the local average and fluctuating shear-stress at a wall.

2. Determination of Shear Stresses on a Vertical Wall in the Liquid Film Streaming Down

The study of the thin film flow regime is of both theoretical and practical interest. Films are widely used in various apparatuses: evaporators, rectifiers, refrigerators etc.

The flow of the film can be laminar and turbulent; besides depending on the Froude and Weber numbers at the surface of the film there can be gravitational and capillary waves.

Due to this diversity the film flow is characterized by several of Reynold's critical numbers which characterize, for example, the appearance of capillary, gravitational waves etc.

The following flow regimes are considered most frequently: smooth-laminar, wavy-laminar and turbulent. A flow regime is a complex function of physical properties of a liquid, flow velocity, inlet conditions, channel slope angle etc. Of great interest are the conditions under which the laminar films flow turns into turbulent.

There are many experimental data showing the region of the corresponding Reynolds numbers. Besides, the data obtained by different authors differ considerably and lie in the range $Re_{kp} = 60—800$.

Nearly all papers on wave formation on the surface of liquid films report the presence of a smooth region immediately after the film is formed. The measurement of shear stresses at the wall as the film flows is of considerable interest for studying its flow regime. A technique of »floating« elements is immediate measurements of the shearing force of the flow by a preliminarily tared mobile element of the wall. The main shortcomings of this technique — the difficulties in constructing a gauge of high quality and the inevitability of perturbances — exclude the application of this technique to studying the thin film flow.

At a balanced flow τ can be determined by measuring the mean thickness of the film δ:

$$\tau_w = \rho \cdot g \cdot \delta$$

where ρ — is density and g is gravity acceleration. The most familiar methods of measuring δ are the contact method; the method of light absorption measurement; the radioscopy method and the shadow method.

The shortcomings of these methods are the requirement of strict flow uniformity and the difficulties in analizing and taking into account the wave formation. The electrochemical technique of measuring shear stresses allows to measure the steady and fluctuating components of shear stress in the region of both steady and unsteady flows.

The experimental study of the film flow has been carried out on a vertical plexiglass 500 mm long and 50 mm diameter cylinder assembled from 50 mm long rings between which circular nickel 0.5 mm thin probes are mounted flush with the cylinder surface. After the assemblage the surface of the working area was carefully ground on a circular grinding machine. During the experiment the working area was hermetically covered with a plexiglass casing. The flow film was made through a distribution slot with a variable width of 0—3 mm. The shear stresses were determined in the range of Reynolds numbers from 30 to 110. In the graph presented in Fig. 4 the dependences of electrochemically measured shear stress on Reynold's numbers are compared with the formulas obtained by S. S. Kutateladze and M. A. Styrikovich for smooth turbulent films [8] and by H. Brauer [9]. The divergency from the formula of Kutateladze and Styrikovich at small Reynolds numbers can be attributed to the not purely turbulent character of the film flow.

As shown by P. L. Kapitsa [10], at $Re > 6$ on the surface of the films there occur waves with an amplitude amounting to half the average thickness of the film. This can account for the divergence from Bauer's formula obtained as a result of treating the experimental data on film thickness measurement by the contact method.

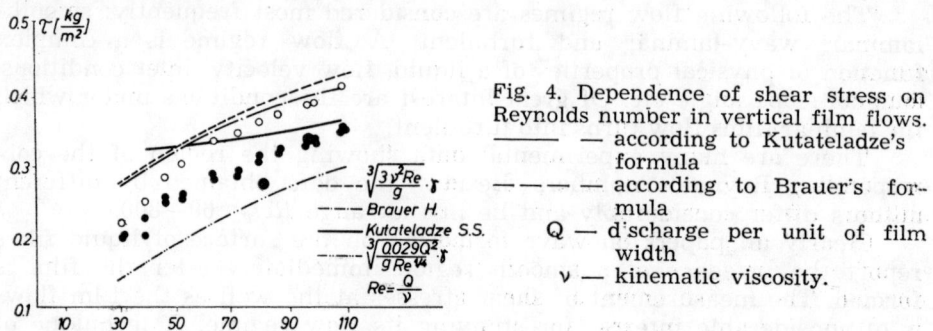

Fig. 4. Dependence of shear stress on Reynolds number in vertical film flows.
——— — according to Kutateladze's formula
– – – – — according to Brauer's formula
Q — discharge per unit of film width
ν — kinematic viscosity.

3. Application of an Electrochemical Anemometer to the Study of Gas-Liquid Flows

As is known, the study of liquid flows with the help of thermoanemometers meets with a number of difficulties. The scheme of an electrochemical anemometer is known for its simplicity, greatly weakened »wall« effect, absence of effects similar to the heat run-off along the current conductors of the anemometer and to the heat inertia of the wire etc.

Wire electrodes 50—100 microns in diameter with a nonisolated part 1—3 mm long were used as sensitive elements of the anemometers.

The test of the probe inertia was carried out with the help of a mechanical pulsator with simultaneous recording on a loop oscillograph of signals from the electrochemical anemometer and the indicator of shift at fluctuations. The test has shown the absence of inertia in the considered range of frequencies 0—100 cps.

The velocity field was studied in the vicinity of a separating bubble. The experiments were carried out in a plexiglass tank using a stroboscope to determine the separation frequency of bubbles generated by blowing nitrogen into the electrolyte through nozzles of different diameters.

The separating bubble diameter was determined by measuring the volume of the bubbles.

The signal from the thermoanemometer probe checking consecutively the volume of the liquid in the vicinity of the bubble was directed to the loop oscillograph. The analysis of the experiments reveals the presence of steady eddies. The velocity distribution picture for one of the experiments with the determined separating bubble diameter $d=5$ mm at the separation frequency of 22.5 cps is represented in Fig. 5.

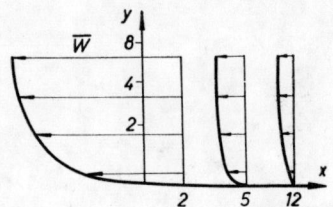

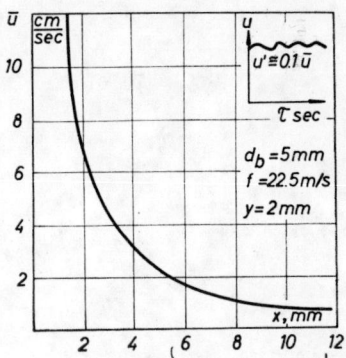

Fig. 5. Velocity distribution in the vicinity of a separating buble.

On the basis of the experiments reported here a model is being prepared for calculating the heat exchange process in boiling.

REFERENCES

1. E. Mitchell, I. Hanratty, J. Fluid Mech., **26**, part 1 (1966).
2. L. P. Peiss, T. J. Hanratty, AIChE J., 9, 159 (1963).
3. A. A. Armand, Izvestiya VTI, № 1, 1946.
4. H. S. Isbin, N. C. Sher, K. E. Eddy, AIChE J., (March 1957).
5. V. L. Streetor, *Handbook of Fluid Dynamics*, p. 17—3, 1961.
6. S. Levis, Teploperedacha, Seria C, **85**, 2 (1963).
7. S. S. Kutateladze, V. Y. Nakoryakov, A. P. Burdukov, V. A. Kuzmin, Trudy 3. Vsesoyuznoy konferentsii po teplomassoobmenu, Minsk, **2**, 367, 1968.
8. S. S. Kutateladze, M. A. Styrkovich, Gidravlika gazozhidrostnykh sistem, GEI, 1958.
9. H. Brauer, VDI-F, 457 (1956).
10. P. L. Kapitsa. Zh. Eksp. Teor. Fiz. **18**, 3 (1948).

Fig. 3. Filling pen. The heat visoon heavy permanents caused the pressure in the spaye addled in levels, by distribution theorie ion but, of the con ponents and the determined separating bubble diameter = 6 mm at the equation holdensyeg at 30 p.s is represented in Fig.

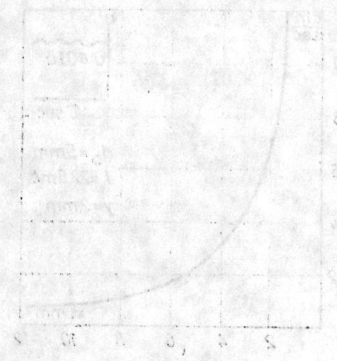

Fig. 4. Velocity distribution in the vicinity of a separating bubble.

On the basis of the experiments reported here a model is being prepared for calculating the heat exchange process in boiling.

REFERENCES

ON BOILING CRISIS AT THE CORE STREAM OF VAPOR-WATER MIXTURE

I. G. DRUKER

The Institute of Theoretical and Applied Mechanics, Siberian Department of the Academy of Sciences of the USSR, Novosibirsk, USSR

In this paper the model allowing the explanation of the negative effect of increasing the velocity and vapor content on the value of heat load, at which there exists a considerable deterioration of heat transfer to vapor-water mixture flowing in ducts is considered. The above phenomenon occurs in a wide range of pressure, velocities and vapor contents. The model is schematic but, as seen from the treatment of the experimental data, it gives an applicable approximation.

When the mixture moves in the duct, the vapor bubble, growing on the wall, is effected by the hydrodynamic force arising from high velocity gradients near the wall. The difference of total pressures and velocities on different streamlines leads to the existence of the resulting force directed to the center of the duct. The accurate calculation of this force under conditions of viscous flow round the bubble is a difficult problem.

Making use of the dimensional theory and calculations for inviscid flow round the bubble, the ratio of force affecting the bubble can be obtained.

$$F = K_0 \frac{\pi}{12} \omega^2 D^4, \qquad (1)$$

where K_0 must be on order of 0.2.

It is interesting to have presentation of the value of the indicated force. For this purpose we calculate its ratio to Archimedes' force at the same pressure. We shall obtain

$$\frac{F}{F_A} = \frac{K_0 \, \omega^2 \, D}{2g \, (\rho' - \rho')} \qquad (2)$$

From the force balance affecting the single bubble, the ratio for the separated diameter, which is similar to Laplace ratio can be obtained

$$D_0 = \sqrt[3]{\frac{12 \, \theta^2}{K_0} \cdot \frac{\sigma}{\omega^2 \rho'}} \qquad (3)$$

The numerical estimations described below, show that the bubble grows, separates and moves mostly within the viscous sublayer; in contrast to boiling under conditions of natural convection, the hydrodynamic force related to the gradient decreases when the bubble moves as it is proportional to the relative velocity of the motion of a bubble in water.

In the dimensionless variables the equations of the motion of a bubble after separating may be approximately written in the form

$$\frac{18\mu}{\left(\frac{1}{2}\rho' + \rho''\right) D_0^2 \omega} (\bar{y} - \bar{u}) = \frac{d\bar{u}}{d\bar{t}},$$

$$\frac{\pi}{6} k(\bar{y})(\bar{y} - \bar{u}) - \frac{3\pi\mu}{\rho' D_0^2 \omega} \bar{v} = \left(\frac{1}{2} + \frac{\rho''}{\rho'}\right) \frac{\pi}{6} \frac{d\bar{v}}{d\bar{t}}. \tag{4}$$

From this we obtain two criteria of similarity

$$\frac{\rho' D_0^2 \omega}{\mu} = \overline{Re}, \qquad \frac{1}{2} + \frac{\rho''}{\rho'} = N \tag{5}$$

The former is some fictitious Reynolds number related to the flow over the bubble, the latter is the weak function of pressure.

From the analysis of the system (4) it follows that the distance which the bubble passes in the transverse direction, is 2—3 diameters of the bubble. Thus, if the ratio D_0/δ_A (0.3—0.5) is less than some critical one, the bubble cannot go out of the sublayer.

Now it is worth to make some estimations. With the zero vapor content flow we have

$$\frac{D_0}{\delta_A} = 0.32 \sqrt[3]{\frac{\theta^2}{K_0}} \frac{\sigma^{1/3} (a \rho')^{1/24}}{w^{7/24} \mu^{3/8}} \tag{6}$$

Consider the particular case: $P = 100$ ata, $2a = 0.01$ m, $\gamma' w_0 = 1500$ kg/m² per hour. Then assuming $\theta = 6°$, $K_0 = 0.1 \div 0.4$, we shall obtain from (2) that F/F_A has the order of some tens. The latter is indirectly confirmed in the experiments with using horizontal tubes where in practice no differences of appearing a crisis of boiling on upper and below generatrices are observed. This would not occur if Archimedes' force did not play an essential role.

We can get from (6) that $D_0/\delta_A = 0.6 \div 1.0$. The increasing of vapor content, as it will be shown below, leads to the decreasing of D_0/δ: at $x = 0.2$, for instance, $D_0/\delta = 0.40 \div 0.65$.

Thus the estimations show that bubbles must be removed from the viscous sublayer, however, the worse D_0/δ_A, the nearer it to critical, i.e. the higher the velocity and vapor content of mixture.

Putting the values w and D_0 into Re and ignoring the numerical coefficient, we shall obtain

$$\overline{Re} \sim \left(\frac{D_0}{\delta_A}\right)^2 \tag{7}$$

that is, the region geometry, in which the bubble motion takes place, is described by the known criterion.

From the above we can assume that the regime of the deterioration of heat transfer has another nature in comparison with the boiling crisis under conditions of natural convection. As the force separating the bubble from the wall is much times Archimedes' force, the deterioration of heat transfer cannot be explained by the conditions on the very wall. The main moment here is not the bubble separation but its further rejection hindered because of the decreasing of the rejecting force to zero as the bubble is increased by the water flow. The full velocity of vapor rejection from the wall can be assumed to be proportional to some specific velocity of a buble, for instance, the velocity on the viscous sublayer boundary.

Assume that

$$\overline{Q}^* = \frac{Q^*}{\gamma'' r \omega D_0} = c \overline{v} = f(\overline{Re}, N) \tag{8}$$

or

$$\overline{Q}^* = \frac{Q^* (a\rho')^{\frac{1}{12}} \mu^{\frac{1}{4}}}{\gamma'' r \sigma^{\frac{1}{3}}, w^{\frac{7}{1}}} = f\left(\frac{\sigma^{\frac{2}{3}} (a\rho')^{\frac{1}{12}}}{w^{\frac{7}{12}}, \mu^{\frac{3}{4}}}, \frac{1}{2} + \frac{\rho''}{\rho'} \right) \tag{9}$$

With the heat load close to critical the motion of individual bubbles cannot be treated especially through the undisturbed wall layer. Therefore even the accurate calculation of the motion of a single bubble does not give the possibility to determine theoretically a critical load. However, at the same time it is evident, that the bubble velocity is essential. When it increases, the bubbles will start to accumulate in the wall layer with great loads and this will ultimately lead to the change of a flow regime and deterioration of heat transfer.

The treatment of experimental data showed that the criteria governed not only the bubble motion but, to a great extent, the approach of the crisis of heat transfer.

The relation (9) can be generalized in the case of non-zero vapor content, using friction data with the stream of vapor-water mixture.

For this purpose in (9) instead of w, we should put the expression

$$w = w_0 \left[\frac{1}{1-x} \left(\frac{\rho'}{\rho''} x + 1 - x \right) \right]^{\frac{n}{2}}, \tag{10}$$

where

$$n = \frac{0{,}8 \varphi + 1{,}08 - 2 \varphi_0}{0{,}9 - \varphi_0} \tag{11}$$

Making use of (10) the treatment of experimental data of some authors was carried out. The results are presented in Fig. 1. The complex

$\dfrac{\overline{Re}}{N}$ instead of two criteria $\overline{Re}$ and N turns out to be used and this does not follow from the system (4), but in practice, it is more convenient. The data for the rectangular duct give qualitatively the same dependence as for circular tubes, the values of loads being something less.

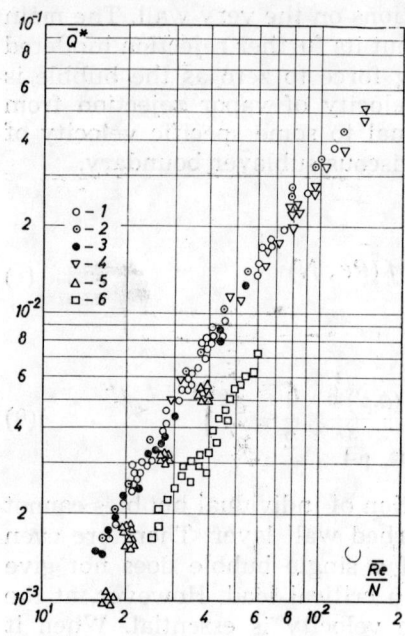

1 — [1] $p=100$ ata; 2 — [1] $p=60$ ata; 3 — [1] $o=140$ ata; 4 — [2] $p=50, 100$ ata; 5 — [3] $p=100, 150$ ata; 6 — [4] $p=70$ ata; (rectangular tube).

NOMENCLATURE

a	radius of the pipe
c	factor of proportionality
D, D_o	diameter and separated diameter of a bubble.
F	hydrodynamic force
F_A	Archimedes' force
g	acceleration of a free drop
K_o	coefficient of hydrodynamic force
N	criterion
P	pressure
Q^*	critical heat load
t	time
u	velocity along the axis of the duct
v	velocity across the duct
w	velocity of vapor-water mixture
w_o	weight velocity of vapor-water mixture

x	weight vapor content
y	transverse coordinate
γ''	specific weight of vapor
δ_Λ	thickness of viscous sublayer
μ	dynamic viscosity
φ	volume vapor content
ρ'	water density
ρ''	vapor density
σ	surface tension
Θ	boundary angle
ω	velocity gradient

$\bar{t} = t\omega$, $\bar{u} = u/\omega D_o$, $\bar{v} = v/\omega D_o$, $\bar{y} = y/D_o$.

REFERENCES

1. V. I. Subbotin, O. L. Peskov, B. A. Zenkevich, N. D. Sergeev, The article in the collection »Teploperedachai gidravlika v dvukh faznikh sredakh« Gosenergoizdat, 1963.
2. N. S. Alferov, R. A. Ribin, The article in the collection »Konvektivnaya teploperedacha v dvukhfaznom i odnofaznom potoke«, Gosenergoizdat, 1964.
3. V. N. Smolin, V. K. Poljakov, V. I. Esikov, »Atom. Energiya« 16, 5 (1964).
4. Tippets, Trans. ASME, Ser, C. 86, 1 (1964).
5. G. Lamb, Gidrodinamika, ONTI, 1947.
6. I. G. Druker, PMTF, 2 (1966).
7. I. G. Druker, Teploenergetika, 4 (1967).

POST BURNOUT HEAT TRANSFER TO MIST FLOW

WAREN M. ROHSENOW

Massachusetts Institute of Technology, Cambridge, USA,

and

EUGENII FEDOROVICH

Polytechnical Institute, Leningrad, and Massachusetts Institute of Technology, 1967—1968

1. Introduction

The design of nuclear reactors has in the past been limited in general to operation in the pre-dryout range where the heating surface remains wet and either nucleate boiling or evaporation at the liquid-vapor interface takes place. Heat flux and quality are kept below the so-called critical condition. In recent years some consideration has been given to designing reactors to operate through the critical condition into the post-critical (or post-burnout) region where the wall is dry and perhaps a mist flow of liquid droplets and vapor exists.

The once-through steam generators have long been designed to operate into this region. New application in nuclear power plants, nuclear rockets, and Rankine cycles for space systems have renewed our interest in this process. Further, these applications are visualized as employing a wide variety of fluids, heat fluxes and pressures. In addition to water, interest focusses on cryogenic fluids such as hydrogen, nitrogen and on hydrocarbons such as methane and propane, and on liquid metals such as mercury and potassium.

Many research results for post-burnout (post-critical) region have been interpreted in terms of a heat transfer coefficient based on wall temperature minus the saturation temperature, ignoring the presence of significant non-equilibrium which appears as highly superheated vapor coexisting with liquid droplets. This approach does not permit prediction of actual vapor temperature, actual quality and heat transfer coefficient.

At times designers have neglected the presence of any liquid droplets after the critical point and based their calculations on heat transfer to pure vapor which is overly conservative.

It is the purpose of this review to survey and compare some of the more important works in this area, to emphasize the importance of the presence of the non-equilibrium and to suggest an appropriate method for predicting heat transfer in this region. The significance of the non-equilibrium is exemplified by the observation [1] that liquid droplets can exist in superheated nitrogen vapor when the heat added is more than three times that necessary to produce dry saturated vapor (>300% quality).

2. Nature of Experimental Results

At lower heat fluxes the critical point (change to dry wall) may occur at higher qualities, while at high heat fluxes the entire heated wall may be dry and large chunks of liquid may be present in the core. In either case the flow down the passage usually becomes at very low quality a mist or dispersed flow of droplets in the vapor and the wall is dry.

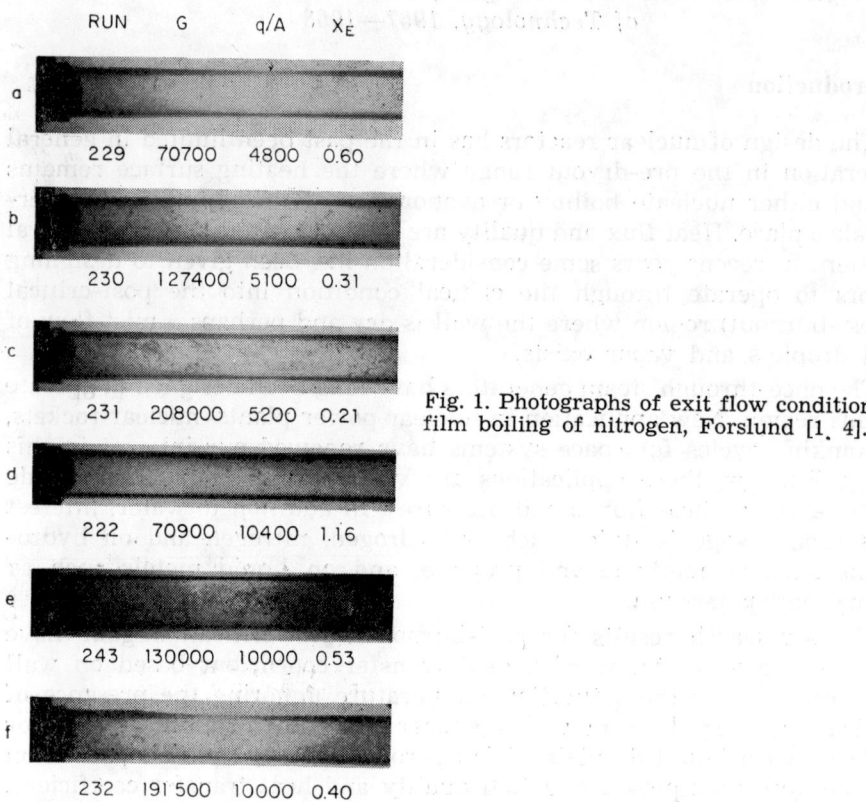

Fig. 1. Photographs of exit flow conditions for film boiling of nitrogen, Forslund [1, 4].

Photographs (Fig. 1) of this post burnout flow for nitrogen in 0.323" diameter tube show the mist flow region. It is noted that a significant number of droplets are present at equilibrium qualities, X_E, greater than unity, where:

$$X_E \equiv \frac{\int_0^L (q/A)_W \pi D \, dL - W \Delta H_{s.c.}}{W H_{fg}} \qquad (1)$$

Typical wall temperature profiles are shown in Fig. 2 when a particular flow rate is established and the heat flux is raised to the magnitude shown. These data were taken recently in the M.I.T. Heat Transfer

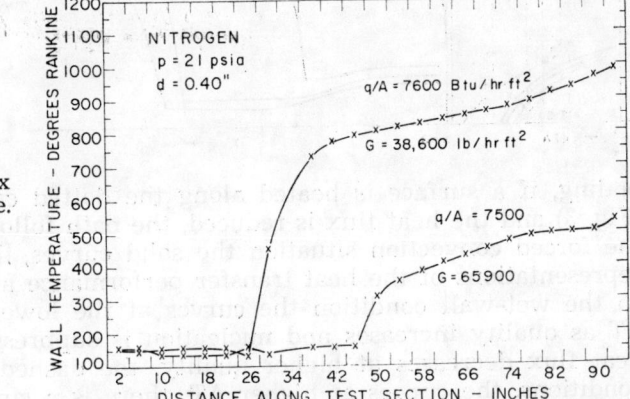

Fig. 2a. Effect of mass flux on wall temperature profile.

Laboratory for nitrogen flowing in 0.400″ tube. The critical point moves to higher quality when the flow rate, G is increased at constant q/A (Fig. 2a) and when q/A is decreased at constant flow rate (Fig. 2b). In the post-critical region the wall temperature rises significantly.

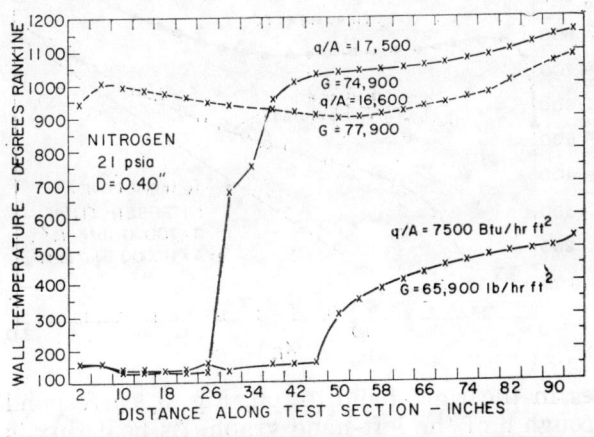

Fig. 2b. Effect of heat flux on wall temperature profile.

In some heat flux range it is possible to obtain the dotted curve of Fig. 2b by first raising the heat flux to a very high magnitude and reducing the heat flux to the test value. In other words either the dotted curves or the solid curves of Fig. 2b may be obtained depending on the history of heating. At a sufficiently low heat flux it is not possible to obtain the dotted curve, but only the solid curve. This suggests the existence of a kind of Leidenfrost point for forced convection boiling similar to but different in magnitude from the pool-boiling Leidenfrost point. This is a hysteresis effect similar to that experienced in pool boiling. In pool

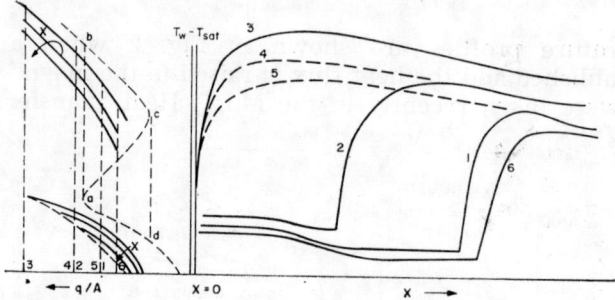

Fig. 3. Hysteresis effect in pool and forced convection boiling.

boiling, if a surface is heated along the dotted curve from a to b to c (Fig. 3) and the heat flux is reduced, the path followed is c to d to a. For the forced convection situation the solid curves, Fig. 3, are schematic representations of the heat transfer performance at a given mass flux G. In the wet-wall condition the curves at the lower ΔT move to smaller ΔT as quality increases and nucleation is suppressed. Also the critical heat flux decreases at higher quality, the dashed line. In the dry-wall conditions, the curves at higher ΔT, there is a similar shifting as vapor quality changes.

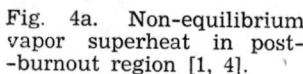

Fig. 4a. Non-equilibrium vapor superheat in post--burnout region [1, 4].

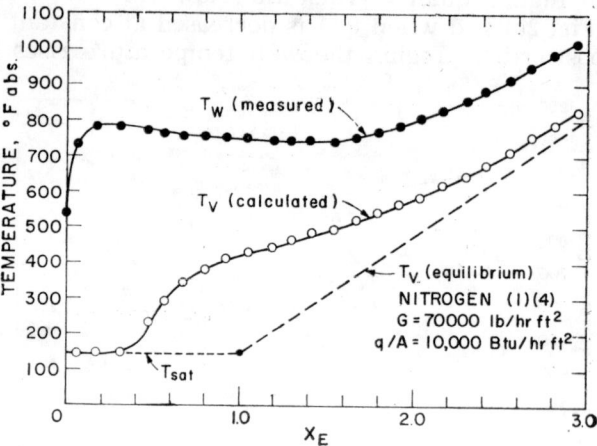

The performance curves in the right-hand graph, Fig. 3, correspond to the heat flux levels 1 through 6 of the left-hand graph. As heat flux is raised in Fig. 3 to above the $(q/A)_{crit}$ for the tube inlet condition, 3, the entire tube wall is dry and the operating points are all at the very high ΔT. As q/A is decreased to 4 and 5 the operating points remain at high ΔT throughout the tube. When (q/A) is decreased to some kind of Leidenfrost effect for forced convection the curve shifts to 6 and reestablishes normal operation.

This forced convection Leidenfrost effect needs further study. Simon et al. [2] attempted to establish a criterion for it but their suggestion involves many questionable assumptions and has not been verified by data. Further work on this phenomenon is needed and is being pursued at the M.I.T. Heat Transfer Laboratory.

Fig. 4b. Non-equilibrium vapor quality calculated from measured wall temperatures [1, 4].

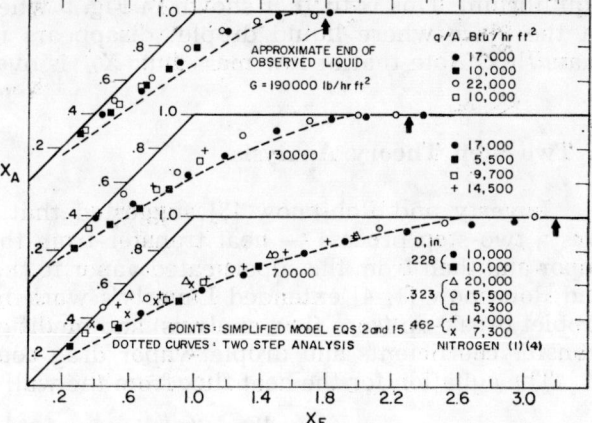

To obtain a clearer explanation of the shape of the wall temperature profile refer to Fig. 4a which was obtained [1, 4] from measurements and an analysis, to be discussed later, for the vapor temperature. Significant

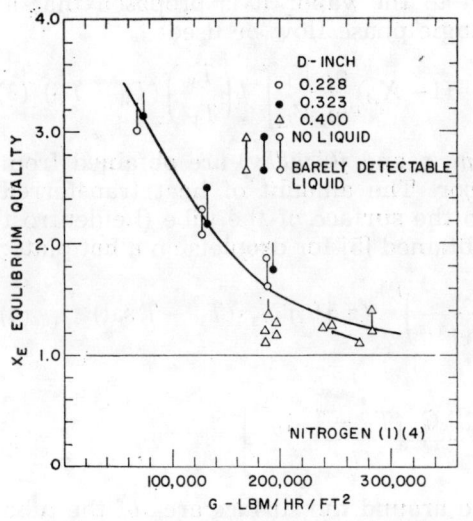

Fig. 5. Effect of flow rate on non-equilibrium [1, 4].

vapor superheat and non-equilibrium are shown to exist beyond the critical point and beyond the $X_E = 1.0$. The shape of the wall temperature profile is influenced both by the increasing vapor velocity and vapor superheat along the tube. The increasing vapor velocity tends to decrease the wall temperature and the increasing vapor temperature tends to increase the wall temperature thus often producing a minimum in the wall temperature curve. An alternate way of viewing the non-equilibrium is shown in Fig. 4b where the actual quality X_A, is plotted as a function of equilibrium quality X_E. The solid line represents complete equilibrium. It is also observed that increasing the mass flux decreases the extent of non-equilibrium. The mass flux has the most significant effect on non-equilibrium. This is further shown in Fig. 5 where the equilibrium quality at the place where liquid droplet disappears is shown as a function of mass flux. Note that at low mass flux X_E is over 3 for this data.

3. Two-Step Theory Analysis

Laverty and Rohsenow [3] suggested that the heat transfer process was a two-step process — heat transfer from the wall to the superheated vapor and then from the superheated vapor to the liquid droplets. Forslund and Rohsenow [1, 4] extended Laverty's work by including the effects of droplet break-up (core flow analysis) and modified the wall-to-droplet heat transfer coefficients and droplet-vapor drag coefficients.

The equation for the heat flux from the wall is:

$$(q/A)_W = (q/A)_{W,V} + (q/A)_{W,l} \qquad (2)$$

The analysis will presume a one-dimensional model with uniform droplet size, vapor velocity and droplet velocity at any axial position along the tube.

For heat transfer from the wall to the vapor it is proposed that a modified form of this equation for single phase flow be used:

$$(q/A)_{W,V} = A \cdot \left(\frac{k_V}{D}\right)\left(\frac{G_T D}{\mu}\right)^m Pr^n \left[X_A + (1-X_A)\frac{\rho_V V_V}{\rho_l V_l}\right]^m f\left(\frac{T_W}{T_V}\right)(T_W - T_V) \quad (3)$$

where the empirical coefficients A, m, n and $f(T_W/T_V)$ are obtained from single phase correlation for the vapor. The amount of heat transferred from the wall to the droplets next to the surface of the tube (Leidenfrost effect) is a modified form of results obtained [5] for droplets on a hot plate:

$$(q/A)_{W,l} = k_1 \left[\frac{k^3 H_{fg}^* g \rho_V \rho_l}{(T_W - T_{sat}) \mu (\pi \delta^3/6)^{1/3}}\right]^{1/4} (\pi \delta^2/4) N_2 \cdot (T_W - T\,\text{sat}) \qquad (4)$$

where:

$$H_{fg}^* = H_{fg}\left[1 + \frac{7}{20} \cdot \frac{C_L(T_W - T_{sat})}{H_{fg}}\right]^{-3}$$

The number of droplets dispersed around the surface area of the tube N_2 may be estimated by:

$$N_2\left(\frac{\text{drops}}{\text{ft}^2}\right) = k_2 \left(N_3 \cdot \frac{\text{drops}}{\text{ft}^3}\right)^{2/3} \qquad (5)$$

where N_3 is the known volume density of droplets existing in the tube.

$$N_3 = G_T (1-X_A)/\rho_l V_l \cdot \frac{\pi \delta^3}{6} \qquad (6)$$

and k_2 is the correction factor to take into account the type of packing in the region of influence of the tube wall. The magnitude of the product $k_1 k_2$ will be determined from data for the forced flow droplet region.

The heat transfer rate between the vapor and the droplets is calculated as follows: (1), (4).

The total heat flux to the droplets is the sum of the heat transfer from vapor to droplets and from the wall to the droplets adjacent to the wall. The heat transfer from the wall to the droplets next to the wall is assumed to be distributed equally among all of the droplets since there exists a continual exchange of droplets between the wall region and the core. Therefore, the heat flux per unit area of total droplet surface is written as follows:

$$(q/A)_{\delta,\text{total}} = h_{\delta,c}(T_V - T_{\text{sat}}) + \left(\frac{q}{A}\right)_{w,l} \frac{\pi D}{\pi \delta^2 N_s \frac{\pi D^2}{4}}, \qquad (7)$$

where

$$h_{\delta,c} = \left(2 + 0.55 \left(\frac{(V_V - V_l)\delta}{\nu_f}\right)^{1/2} Pr_f^{1/3}\right) \qquad (8)$$

for a single sphere in a vapor.

The change in actual quality along the tube is:

$$\frac{dX_A}{dl} = \frac{(1-X_o)}{\delta_0^3} 3\delta^2 \frac{d\delta}{dl}, \qquad (9)$$

where:

$$\frac{d\delta}{dl} = \frac{1}{V_l}\frac{d\delta}{dt} = \frac{2(q/A)_{\delta,\text{total}}}{V_l H_{fg} \rho_l \delta}$$

Here X_o and δ_o are known magnitudes at a given position along the tube.

From an energy balance the change in vapor temperature is:

$$X_A C_{pv}\frac{dT_V}{dl} = H_{fg}\frac{dX_e}{dl} - \left\{H_{fg} + C_{pv}(T_V - T_{\text{sat}})\right\}\frac{dX_A}{dl} \qquad (10)$$

In the accelerating vapor flow the droplet acceleration in terms of a drag coefficient is:

$$a = \frac{dV_l}{dt} = V_l\frac{dV_l}{dl} = \frac{3 C_D \rho_V (V_V - V_l)^2}{4 \rho_l \delta} \pm g \qquad (11)$$

where $+g$ is for downward and $-g$ for upward flow.

The drag coefficient C_D may be defined from experimental data. One such set of data for evaporating droplets is given by Ingebo [7] as follows:

$$C_D = \frac{27}{Re_\delta^{0.84}} \qquad (12)$$

Forslund [1, 4] in analyzing nitrogen film boiling data suggested a more complex set of relations to evaluate C_D.

From studies of oil sprays into air [3] and single droplets of various liquids in air streams it is observed that droplets break up at a particular Weber number defined as:

$$We_{cr} = \frac{\rho_V (V_V - V_l)^2 \delta}{\sigma} \tag{13}$$

The magnitude of We_{cr} is between 7 and 15.

The relation between V_v and V_l can be found from the continuity equation:

$$V_V = \frac{G_T X_A}{\rho_V} \left(1 - \frac{G_T (1 - X_A)}{V_l}\right)^{-1} \tag{14}$$

4. Calculation procedure for two-step model

The heat flux $(q/A)_v$, mass flux G_T, pressure p and liquid and vapor properties are given initially for fixed tube geometry (D and L).

As a result of calculations, wall temperature T_v, actual quality X, vapor temperature T_v must be predicted as a function of position along the tube, l.

The calculation is started at an equilibrium condition in the post burnout flow $(X_E = X_A)(T_V = T_{at})$. The location of this point can be obtained from the condition $X_E > X_{E\,Burnout}$, which can be found from burnout correlations for this particular type of flow and liquid.

At this starting point assume $dV_l/dt = 0$ and $We = We_{cr}$, then from (11) and (13) solve simultaneously for $(V_v - V_l)_0$ and δ_0 with C, determined from Eq. (12). V_V and V_l can be deteermined with the use of Eq. (14). Then T_W can be calculated from Eq. (2) and tssuming a specific value of produce $k_1 k_2$. For nitrogen data Forslund [1, 4] found $k_1 k_2 = 0.2$ gives good agreement with data.

The next step of calculations is to determine conditions at Δl, a small distance down the tube. From Eqs. (9, 10, 11) calculate

$$d\delta/dl, \quad \frac{dX_A}{dl}, \frac{dT_u}{dl}, \frac{dV_l}{dl}$$

from Eq. 2 using $(q/A)_{w,l}$ and Eqs. (7, 8) and from Eq. (1):

$$\frac{dX_E}{dl} = \frac{(q/A)_w \pi D}{W H_{fg}}$$

From these derivatives obtain the new magnitudes of δ, X, T_V, V_l and X_E at the location Δl from the starting point.

Then repeat the calculation for successive position along the tube. Wherever the Weber number reaches the critical value break up each droplet in two.

5. Comparison with Experimental Results

Comparison of analytical and experimental results (two-step model) can be made by comparing the calculated tube wall temperature with the measured. This comparison for the Forslund's experimental data for nitrogen is shown in Fig. 6. These calculated results used $k_1 k_2 = 0.2$ and $We_{cr} = 7.5$. Also shown are curves for $k_1 k_2 = 0$ which means the Leidenfrost

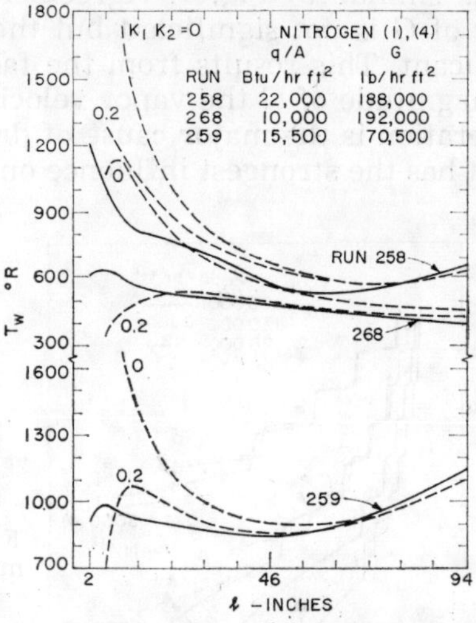

Fig. 6. Comparison of predicted wall temperature profiles with measured temperatures for nitrogen [1, 4].

effect is neglected. Also in Eq. (3) the following constants were used: $A = 0.035$, $m = 0.743$, $n = 0.4$, $f(T_W/T_V) = 1$ and properties were evaluated at the T_V.

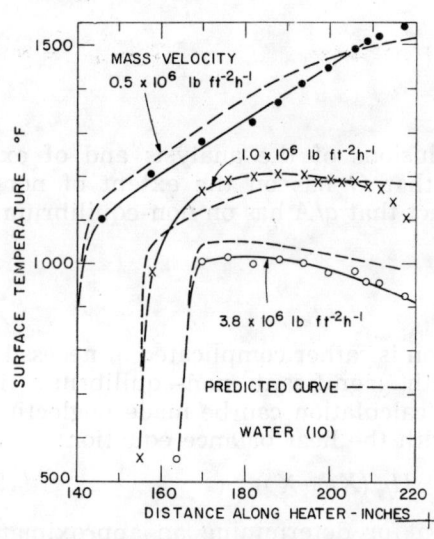

Fig. 7. Comparison predicted wall temperature profiles with measured temperature for water [10].

A similar, though slightly modified, procedure was used by Bennett et al [10] for treatment of water film boiling data (Fig. 7) taking $k_1 k_2 = 0$.

Also in Eq. (3) $A=0.0133$, $m=0.84$, $n=0.33$, $f(T_W \, l/v)=1$ and properties were avaluated at $T_f = \frac{1}{2}(T_W + T_V)$. In both cases the agreement of the theory with experiment is good.

During the course of the calculations for nitrogen [1, 4] droplet size is obtained along the length. These results are shown in Fig. 8 along with the data points of the droplet sizes obtained from observation of the photographs similar to Fig. 1. Agreement is surprisingly good. Note that the effect of G is not significant but the effect of q/A on droplet size is quite significant. This results from the fact that for the same q/A regardless of the magnitude of G the vapor velocity and acceleration is the same. Vapor acceleration is the major cause of droplet evaporation and break-up therefore it has the strongest influence on droplet size.

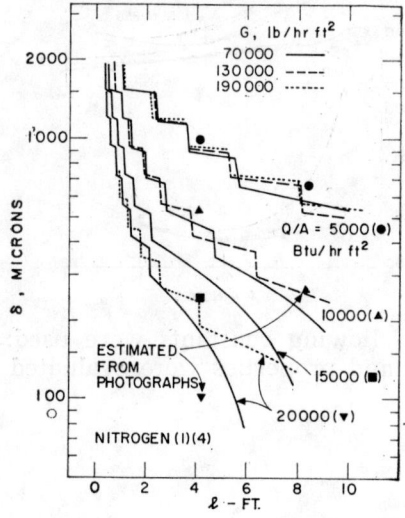

Fig. 8. Droplet diameter: Effect of heat and mass flux [1, 4].

One of the most important conclusions of the analysis and of experimental data is the strong effect that G has on the extent of non-equilibrium, Fig. 4b, and the small effect that q/A has on non-equilibrium.

6. Simplified Two-Step Model

The analysis in the preceding section is rather complicated of necessity because the film boiling process with significant non-equilibrium is complicated. A simplified, less correct calculation can be made neglecting $(q/A)_{W,l}$ in Eq. (2) and using Eq. (3) with the heat balance equation:

$$C_{pv} X_A (T_V - T_{sat}) = H_{fg} (X_E - X_A) \tag{15}$$

This provides a simple calculation for determining an approximate magnitude of T_V and X_A from measured values of T_W and (q/A). The data points in Fig. 4b were obtained by this method and show reasonably good agreement with the more complicated two-step model including the core flow analysis. The discrepancies are greater at lower qualities and high mass flux.

7. Homogeneous Model

A simplified homogeneous model was proposed by Hsu et al [11] where all of the properties were taken as the weighted average of vapor and liquid properties, weighted on the basis of void fraction. There is assumed a core at saturation temperature and a region near the wall with superheat vapor. The analogy between turbulent heat-and-momentum transfer is used and axial slip between vapor and liquid is neglected. Further the radial distribution of void fraction is assumed to be the same as the radial distribution of velocity. The calculation leads to a calculated heat transfer coefficient which, when compared with experimental data, is as much as 50% different from the measured values in some experiments. Errors thus large and smaller may be more simply obtained by using Eq. (3) with T_V taken as T_{sat}.

8. Empirical Correlations

Many researchers have attempted to correlate forced convection film boiling heat transfer data by variations of the following kind of relation:

$$\frac{Nu_{expt}}{Nu_{calc}} = f(X_{tt}) \text{ or } f(x) \qquad (16)$$

where X_{tt} is the Martinelli two-phase parameter and x is the local quality. Here Nu_{expt} is based on $(T_W - T_{sat})$ and Nu_{calc} is a reference Nusselt number calculated from a modified single phase forced convection equation of the form:

$$Nu_{calc} = A\, Re^m\, Pr^n \qquad (17)$$

where A, m and n are constants and Re is the Reynolds number based on an average velocity of the mixture. The properties in Eq. (17) are evaluated either at saturation temperature or at an average of saturation and wall temperatures. Sometimes Eq. (17) is multiplied by $(\mu_{sat}/\mu_w)^{0.14}$.

Some researchers multiply the right side of Eq. (17) by very complex functions of a velocity of other qualities, the function relation being determined empirically.

An excellent survey of these various proposed correlations is presented by Giarratano and Smith [12] who discuss the works of others, [13, 14 and 15], as well as their own. The various proposed correlations are compared [12] with three different sets of data for hydrogen.

Glickstein and Whitesides [16] presented experimental data for methane, propane and butane -1 and attempted to correlate this data by an equation of the form of (16) and (17) with the right side of Eq. (16) replaced by:

$$f\left(\frac{q''(v_f - v_b)}{V h_{fg}}\right)$$

Williamson et al [17] presented experimental data for hydrogen and nitrogen heated by water and attempted to correlate the data by a modified single phase equation such as Eq. (17).

Chi [18] presented data for hydrogen primarily in the slug flow region where he observed significant flow and temperature oscillations.

Bishop et al [19] present data for water at high pressure, both sub- and super- critical. In the pressure range 2420—3120 psia, $G=0.5$ to 2.5×10^6 1 b/hrft^2, $X=0.10$ to 1.0 and $q''=0.1$ to 0.65×10^6 Btu/hrft^2 the data for large L/D were correlated within 17% by:

$$\left(\frac{hD}{k}\right)_W = 0.098 \left(\frac{D V_b \rho_W}{\mu_W}\right)^{0.80} Pr_W^{0.83} \left(\frac{\rho_V}{\rho_l}\right)^{0.50} \quad (18)$$

There is the possibility that at these very high pressures the non-equilibrium effect of superheated vapor is not very large, hence an equation such as Eq. (18) correlated data in this range rather well.

Parker and Grosh [20] recognized the possibility of the existence of non-equilibrium in the form of superheated vapor and presented a limited amount of data for water. No correlation was suggested.

Miropolski [21] presented data in the range of pressure 580—3190 psia, $G=0.88—1.03 \times 10^6$ 1 b/hrft^2, $q''=0.074—0.33 \times 10^6$ Btu/hrft^2, and correlated the data within 25% by:

$$\frac{hD}{k_V} = 0.023 \left(\frac{GD}{\mu_V}\right)^{0.8} Pr_W^{0.8} \psi \theta \quad (19)$$

where

$$\psi \equiv \left[x + \frac{\rho_V}{\rho_l}(1-x)\right]^{0.8}$$

$$\theta \equiv 1 + 0.1 \left(\frac{\rho_l}{\rho_V} - 1\right)^{0.4} (1-x)^{0.4}$$

Predictions from Eqs. (18) and (19) are in good agreement with each other at this high pressure region for water.

Swenson et al [22] obtained data for water at 3000 psia which is correlated well by equation (18).

Polomik et al [23] obtained water data for pressures of 800 to 1400 psia in annular geometries which was correlated by a modified single phase equation.

The common feature of all of these proposed correlations is that the non-equilibrium vapor superheat is neglected and the heat transfer coefficient is based on $h = q''/(T_{wall} - T_{sat})$.

None of the proposed correlation are good predictors over a wide range of conditions and fluids.

For steam-water mixtures at very high pressures simple correlations such as Eqs. (18) and (19) appear to agree reasonably well with data. This type of equation is not valid for water at lower pressure and for cryogenic fluids in general. Equations of the form of Eq. (16) can be made to agree with data over a limited range but in general are valid to within 50 to 100% as shown in reference [12]. Further, none of these empirical correlations can predict the amount of non-equilibrium superheat and actual quality.

9. Measurements of Actual Quality and Vapor Temperature

Several attempts have been made to measure the superheat vapor temperature directly by using thermocouple probes protected from the liquid droplets. Some of these probes require a slight amount of suction flow across the thermocouple. Deflectors of various forms are intended to keep the droplet off the thermocouple. In general these measurements have not been successful since very small droplets get carried into the probe and strike the thermocouple. Also liquid is sometimes carried into the probe by surface tension. Forslund [4] tried three different forms of probes for nitrogen and Mueller [24] used a double tube probe for steam-water mixture. Our conclusion is that although some amount of superheat was measured, the actual superheat was much greater than the measured values.

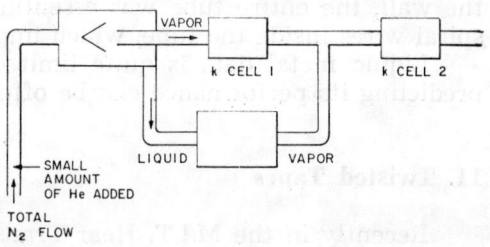

Fig. 9. Apparatus for helium tracer technique for quality measurement [1, 4].

A somewhat unique measuring technique for measuring actual quality was suggested by Peter Griffith and the results reported by Forslund [1, 4]. The apparatus shown schematically in Fig. 9, involves adding a trace amount of helium to the total nitrogen flow. At the exit the liquid is separated and a conductivity cell (k cell 1) measures the percent helium in this exit vapor. The separated liquid is evaporated and mixed with this vapor; k-cell 2 measures the percent helium in the total flow. The ratio of the two readings is the actual quality (Xa) at the exit of the test section. Then T is calculated from equation (15). Some of the measured points are plotted in Fig. 10 showing very good agreement between measured results and results predicted from the two-step model theory. These results in Fig. 10 are for average actual quality. Forslund [4] also traversed the tube radially and found a significant radial variation in quality with lower quality (more liquid) being in the central position of the core.

Other techniques such as gamma-ray or X-ray attenuation produce good results at low quality but are not very successful at high vapor qualities which is an important region of interest in this problem.

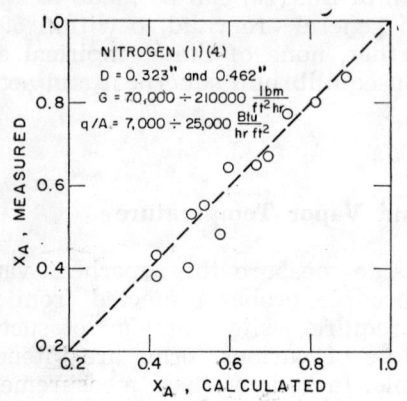

Fig. 10. Comparison of measured and calculated actual quality for nitrogen [1, 4].

10. Liquid Metals

Peterson [25] and Brooks [26] report some data in the post burn-out region for potassium heated by sodium. It was deduced that significant vapor superheat existed with liquid droplets present. Some data was presented by Koestler [27] for mercury. Because mercury does not wet the wall, the entire tube was essentially in film boiling. They also used spiral wires inside the tube, which improved the heat transfer.

Liquid metal data is quite limited. No general conclusion regarding predicting its performance can be offerred at this time.

11. Twisted Tapes

Recently in the M.I.T. Hear Transfer Laboratory the data in Fig. 11 was obtained for nitrogen. The upper curve was obtained in a tube without

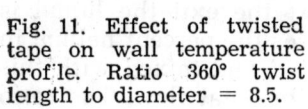

Fig. 11. Effect of twisted tape on wall temperature profile. Ratio 360° twist length to diameter = 8.5.

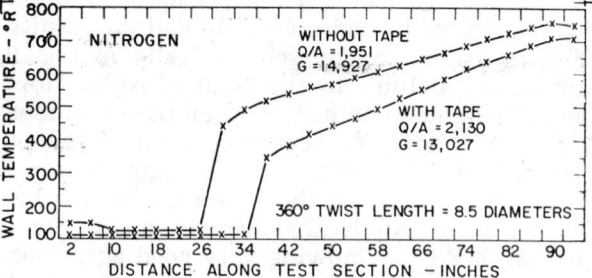

a twisted tape and the lower curve for the same tube with a twisted tape. The tape tends to keep the liquid on the wall at higher quality and delay the dry-out point. Also the wall temperatures are lower or the heat transfer coefficients are higher for the same (or slightly higher) heat flux and the same flow rate.

Twisted tapes, of many different geometries, have been used in an attempt to raise the critical heat flux but the post-burnout region with twisted tapes has not been studied extensively.

12. Conclusions

1. There is significant non-equilibrium present in forced convection, post-burnout film boiling in the form of high vapor superheat with liquid droplets present in the flow. Actual quality may be very much less than equilibrium quality. The amount of this non-equilibrium is strongly influenced by the mass flux and less by the heat flux and of course is influenced by the fluid properties.

2. The two-step model [1, 4] presented here is recommended for use in predicting performance in the post-burnout region because it takes into account this non-equilibrium.

3. Homogeneous models for correlation which are based on $(T_{wall} - T_{sat})$ and ignore the presence of non-equilibrium are not recommended except for water at very high pressures where the existence of non-equilibrium appears to be less. There are large discrepancies between such homogeneous prediction equations and data for cryogenic fluids.

No correlations exist for liquid metals.

4. An hysteresis effect exists in forced convection film boiling similar to the hysteresis effect in pool boiling. As heat flux is increased at constant mass flux the critical quality decreases monotonically until the entire heated tube is in a dry-wall condition. On subsequently reducing the heat flux the entire tube remains in this dry wall condition until at a lower heat flux the vapor film collapses in the erly part of the tube and the previous wall temperature pattern is re-established.

5. Significant difficulties were found [4, 24] in direct measurement of two important characteristics of non-equilibrium two-phase flow-superheated vapor temperature (T_V) and actual quality (X_A) in addition to traditional measurement of the heat transfer coefficient.

Progress in measurement of X_A has been made [4] but there are no reliable methods for direct measurement of T_V.

6. Artificial methods for keeping good contact between the liquid and wall, such as internal twisted tapes, are effective in increasing heat transfer coefficients, delaying the onset of the dry-wall condition, and decreasing the extent of non-equilibrium.

ACKNOWLEDGEMENT

The data presented in Figs. 2a, 2b and 11 were obtained by Scott Hynek, Ware Fuller and E. Fedorovich as part of a current experimental program being carried out in the M.I.T. Heat Transfer Laboratory.

NOMENCLATURE

A	inside surface area of tube
C_D	drag coefficient
C_p	specific heat
D	tube diameter
G	mass velocity ρV
g	gravitational acceleration
h	heat transfer coefficient.
H	enthalpy
H_{fg}	heat of vaporization
k	conductivity
L	tube length
l	position along the tube
N	numerical concentration of droplets
Nu	Nusselt number hD/k or $h\delta/k$
P	pressure
Pr	Prandtl number $\mu C_p/k$
q/A, q''	heat flux
Re	Reynolds number $\rho VD/\mu$ or $\rho V\delta/\mu$
T	temperature
t	time
V	velocity
v	specific volume
W	mass flow rate, $1bm/hr$
X	flowing quality of vapor
X_A	actual quality
X_E	equilibrium quality
X_{tt}	Martinelli parameter
δ	droplet diameter
μ	viscosity
σ	surface tension
ρ	density
ν	kinematic viscosity

SUBSCRIPTS

A	actual
b	bulk
E	equilibrium
f	film $T_f = (T_W + T_b)/2$
l	liquid
sat	saturation
T	total
v	vapor
w	wall

REFERENCES

1. R. P. Forslund and W. M. Rohsneow, 10th Nat. Heat Transfer Conf., Philadelphia, Aug. 1968, A.S.M.E. paper 68-H-164.
2. F. F. Simon, S. S. Papell and R. J. Simoneau, N.A.S.A. TN D-4307, February 1963.
3. W. F. Laverty and W. M. Rohsenow, A.S.M.E. J. Heat Transfer, 89c, pp. 90—98 (Feb. 1967).
4. R. P. Forslund and W. M. Rohsenow, M.I.T. Report № 75312—44 Nov. 1966.
5. K. J. Baumeister, T. D. Hamill and G. J. Schoessow, US-AIChE № 120, 3rd Int. Heat Transfer Conf., August 7—12, 1966.
6. Tsubouchi and Sato, »Heat Transfer between Single Particles and Fluids in Relative Forced Convection«, Chem. Eng. Prog. Symp. Series, Vol. 55, 1960.
7. R. D. Ingebo, NACA TN 3762, Sept. 1956.
8. R. D. Ingebo, NACA TN 3265, Oct. 1954.
9. N. Isshiki, Report № 35, Transportation Technical Research Institute, Tokyo, Japan.
10. A. W. Bennett, G. F. Wewitt, H. A. Kearsey and R. K. F. Keeys, A.E.R.E. — R 5373, Harwell, England, 1967.
11. Y. Y. Hsu, G. R. Cowgll, R. C. Hendricks N.A.S.A. TN D—4149, Dec. 1967.
12. P. J. Giarratano and R. V. Smith Paper № H—1, Cryogenics Engineering Conference, Rice University, Houston, Texas, Aug 1965.
13. R. C. Hendricks, R. W. Graham, Y. Y. Hsu and R. Friedman N.A.S.A. TN D—765, 1961.
14. H. H. Ellerbrock, J. N. B. Linvingood and D. M. Straight, N.A.S.A. SP—20, 1962.
15. U. H. Von Glahn, N.A.S.A. TN D—2294, 1964.
16. M. R. Glickstein and R. H. Whitesides Jr., A.S.M.E. paper 67—HT—7, National Heat Transfer Conference, Seattle, Washington, August 1967
17. K. D. Williamson Jr., J. R. Bartlit and R. S. Thurston, AIChE Preprint № 38C, AIChE Annual Meeting, New York, Nov. 1967.
18. J. W. H. Chi, A.S.M.E. Paper 65—WA/HT—32, Annual Meeting, Chicago, Illinois, Nov. 1965.
19. A. A. Bishop, R. P. Sandberg, L. S. Tong, »Forced Convection Heat Transfer to Water after the Critical Heat Flux at High Sub Critical Pressures: W CAP—2056 Part V, Westinghouse Electric Corp, Pittsburg, Pa.
20. J. D. Parker and R. J. Grosh, A.S.M.E. Paper 62—HT—47 National Heat Transfer Conference, Houston, August 1962.
21. Z. L. Miropolski, Teploenergetika, 5 (1963).
22. H. S. Swenson, J. R. Carver and G. Syoeke, A.S.M.E. Paper 61—WA—201, Nov. 1961.
23. E. E. Polomik, S. Levy and S. G. Sawoihka, GEAP—3703, General Electric Co., San Jose, California.
24. R. E. Mueller, GEAP—5423 General Electric Co., San Jose, California, April 1967.
25. J. R. Peterson, Contractor Report N.A.S.A. CR—842, August 1967.
26. R. D. Brooks, »Alkali Metals Boiling and Condensing Investigations« 2nd Annual Meeting, Liquid Metal Technology, General Electric Co., May 1962.

RELATIONSHIP BETWEEN THE STRUCTURE OF THE TWO-PHASE LAYER NEAR A HEATING WALL AND THE MECHANISMS OF THERMAL EXCHANGE

R. SEMERIA

Center of Nuclear Studies, Heat Transfer Section, Grenoble, France

1. Benefits of Thermo-dynamic Studies of the Two-Phase Structure Near a Heating Wall

The essential problems from the point of view of industrial applications always concern the evaluation of the heat or mass exchange coefficient, of the void fraction, or of the pressure drop for the two-phase flows.

The temptation is always very great to search for an equivalent single-phase flow leading to the same results (for example, the same exchange coefficient). In particular with small concentrations of vapor, it is logical to consider the vapor as a small disturbance of the liquid flow: on a horizontal plate, vapor bubbles improve the preexisting natural convection; this is taken into account by modifying the dilatation coefficient of the liquid without adding new mechanisms. [1]

As soon as the local volumetric concentration exceeds a few per cent, the preceding method loses all physical justification, while perhaps remaining a convenient means of correlation. It then becomes necessary to analyze the two-phase structure in more detail, and two methods may be utilized:

— The descriptive analysis of elementary mechanisms, especially applicable in organized periodic structures (film boiling, columns of isolated bubbles);

— The statistical analysis of the parameters of the structure associated with a formulation of general flow equations.

These two methods are complementary: the first leads to the elaboration of structural »models«, sometimes sufficient to correlate a certain number of experimental results; the second can utilize these models in their field of validity for the choice of necessary hypotheses or physical correlations.

2. Analysis of a Two-Phase Layer on a Wall

A model of the phenomenon is being sought that is close enough to reality to allow the deduction of all its characteristics (distribution of phases, velocities and temperatures in space and time) based on a limited number of parameters (for example, the average temperature of the wall, and the number of nuclei).

Only the case of boiling on a heating wall will be considered.
The experimental methods utilized are mainly:
— photography and high speed movies
— measurement of the void fraction
— measurement of the wall and fluid temperatures.

The results obtained allow a good qualitative description of the boiling. Quantitative visual studies are possible especially with small bubble populations (low heat flux densities, thin wires) [2, 3, 4] and lead to models that are more and more satisfactory [5, 6, 7].

2.1. SMALL BUBBLE DENSITY

One can then concentrate the study on the functioning of a single site, or even the behavior of a single bubble.

Our conception of the events that take place will be briefly reviewed:

a) — Birth of a bubble

This is the activation of a site for which the equilibrium has been broken by the overpressure of vapor which results from the heating of the wall and the liquid film. The essential and measurable characteristic parameter of a site is the local wall temperature at the moment of its activation.

The models of activation proposed up to now assume, in essence, at the moment of activation (Fig. 1a):

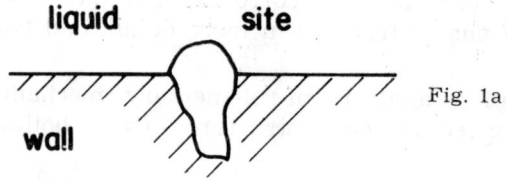

Fig. 1a

— a permanent hydrodynamic field
— a bubble emerging from the site [8, 9, 10].

These models have been verified in forced convection with high sub-cooling and at high velocities, conditions for which the wall super-heating for activation is high. The criteria for the beginning of boiling are based upon macroscopic detection, which are by order of decreasing sensibility:

acoustic detection, visible bubbles, perturbation of the single-phase heat transfer coefficient, perturbation of the single-phase pressure drop [11].

In truth, it is often necessary to distinguish between the first activation from the periodic activations which give some tens of bubbles per second; in the second case, the field is no longer permanent and the temperature of activation of the site is often much lower than that of the first activation.

It has also been observed that some sites, before functioning normally, manifest themselves by the periodic ejection (200 to 5000 cps) of hot liquid perpendicular to the wall, with the possible formation of very small bubbles (0.1 mm at 1 atm), of very short duration (50 microseconds) (Fig. 1b).

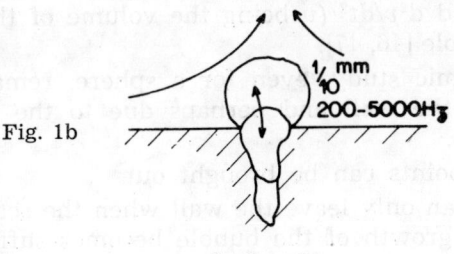

Fig. 1b

vibrating site

In conclusion, the initial behavior of the nucleus is still imperfectly known, although this is important for very rapid transitient heating and for boiling of mixtures and of liquid metals [12, 13]. Predictions of populations of site are still insufficient and phenomena like tribonucleation [14] cannot always be neglected.

b) — *Bubble growt on the wall*

This problem has often been studied theoretically and recent articles have provided further improvements [15]. The principal difficulties that remain for the case that is of interest here are:

— The bubble is not spherical during its growth and no theory at the present time allows its profile to be calculated: this would require calculations for a free surface moving in a real fluid with solid-liquid-gas contact, a situation where the physics of the phenomena of vaporization are poorly known.

The vapor pressure inside the bubble must itself be calculated as a function of the thermal and hydrodynamic field around the bubble.

— The thermal and hydrodynamic field is very complex, whether it be in a pool or in convection. Strioscopy has provided a good picture of the thermal field and indirectly, of the fluid motion [16, 17]. It is now thought that during the initial growth of the bubble on a flat wall, the thermal film trapped under the bubble provides the major part of the vapor, especially in sub-cooled liquid: on a thin wire, the role of this microlayer diminishes with respect to that of the entrained film.

A study of the film adhering to the wall and of the temperature in the neighborhood of the site would permit a better appreciation of the role of the microlayer. Professor Gregorig has rightfully recalled that the thermal field within the wall must be accounted for [18].

c) — *Departure of the bubble — Duration of its presence and contact at a site.*

It is now granted that all the forces, static and dynamic, acting on the bubble must be considered in order to analyze the conditions of its detachment from the wall. However this implies a determination of the derivatives dv/dt and d^2v/dt^2 (v being the volume of the bubble) and of the shape of the bubble [46, 47].

The hydrodynamic study, even for a sphere, remains complex due to the presence of the wall and perhaps due to the influences of the neighboring bubbles.

The following points can be brought out:

— the bubble can only leave the wall when the reaction of the fluid associated with the growth of the bubble becomes sufficiently small (or even negative): a minimum residence time is thus calculated the bubble being not perforce attached but only pushed. Because of this, it can slide on the wall.

— The contact angle of the bubble in a dynamic regime is not necessarily that of equilibrium: the microlayer, if it does not vaporize, will flow slowly (case of high pressure), and the force of adhesion to the wall will be very small compared to the dynamic forces (during the growth) and the force of gravity (at the end of growth).

— Between the instant when the conditions of release are satisfied and the instant when the bubble, if it is actually attached to the wall, leaves it, there occurs a period of deformation preceding the rupture.

— The rupture is followed by a very rapid movement of the rim, the bubble vibrates intensely, while the foot disappears in about 1/10 000 seconds. The details of the rupture are still poorly known.

d) — *The time during which a bubble remains on its site*

Is not necessarily the same as that of its presence on the wall, but they are very close in the case of a horizontal flat wall without convection. It has been ascertained experimentally that this time fluctuates very much.

e) — *The time of contact*

This is the time during which no visible bubble exists on the site being studied. It is therefore supposed that the neighboring temperature is lower than that of activation. This time period quickly becomes small

(Fig. 2) as soon as the average wall temperature surpasses that of activation by a few degrees. It fluctuates even more than the preceding one for a wall with an imposed heat flux.

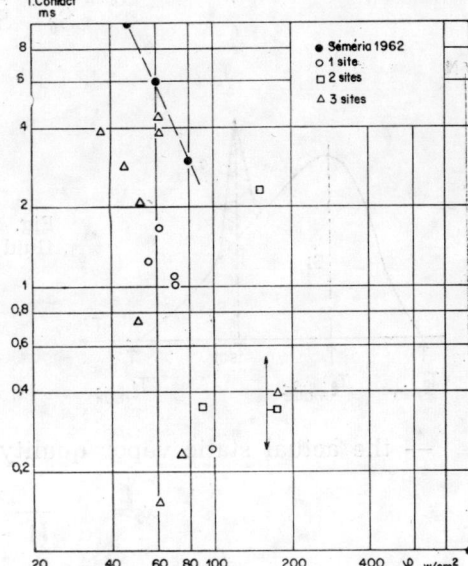

Fig. 2. Time of contact versus the heat flux.

f) — *Local volumetric concentration of vapor and local instantaneous liquid temperature*

Boiling upon a surface is a two-phase structure that can be studied with techniques already utilized in conduits (resistance probes, etc.) This system is distinguished by its high level of local thermal disequilibrium which can be observed with a local instantaneous measurement of the phase temperature at the point under consideration.

The ideal local, instantaneous probe should indicate at each instant:

a) the liquid phase l, or vapor v, from which comes the vapor concentration α, by integration.

b) the temperature T which will be either that of liquid T_l, or that of vapor T_v, in correlation with the preceding signal.

c) the vector velocity $\vec{V}$ which will be either $\vec{V}_l$ or $\vec{V}_v$ in correlation with the preceding results.

Studies of the liquid with microthermocouples are presently being developed; but its analysis and exploitation are still insufficient [19, 20, 21, 22]. The microthermocouple technique allows one to obtain the signal $T(t)$ with a response time lower than 200 microseconds; it is also more often possible to deduce a volumetric concentration of vapor [23].

For a signal $T(t)$, the histogram of which has the form given in Fig. 3, one defines:

— the average temperatures of each phase $\overline{T}_g$ and $\overline{T}_l$ and the concentration of vapor.

$$\alpha = \frac{S_0}{S_0 + S_1 + S_2}$$

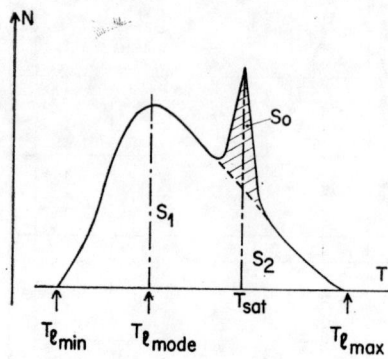

Fig. 3. Histogram of the local instantaneous fluid temperature.

— the actual static vapor quality:

$$x_r = \frac{\alpha}{\alpha + (1-\alpha)\rho_l/\rho_v}$$

— a measure of liquid superheat

$$\tau_{sat} = C\frac{\overline{T}_{12}-T_{sat}}{\alpha}\alpha_2 \text{ with } \alpha_2 = \frac{S_2}{S_2+S_1}$$

and a static quality at equilibrium:

$$x_{eq} = x_r - (1-x_r)\frac{(\overline{T}_{sat}-\overline{T}_1)C}{\alpha}$$

It is also possible to make an analysis in time allowing one to obtain a good indication of the granulometry of the two-phase mixture.

In pool boiling, we have obtained at this time the following results for local boiling with a small amount of sub-cooling:

— on a site, in the thermal layer (distance to the wall inferior to 0.2 mm), the liquid flow-back takes place in a saturated or super-heated liquid and not in sub-cooled liquid.

— the bubble drags a small amount of super-heated liquid behind it.

— the site and its column of bubbles at as a thermal sink as shown in the plot of the most probable isotherms around a natural site ($Y=0$) functioning on the upper crest of a horizontal tube of stainless steel, 1 mm in diameter and 0.12 mm thick, which is indirectly heated (Pyrotenax wire) (Fig. 4).

The minimum and maximum temperature plots (Figs. 5 and 6) show the great amplitude of the temperature fluctuations of the fluid in the neighborhood of the site.

The experiments were pursued in local boiling with forced convection and the local measurements led to, for example, the vapor distribution in the section of the channel (Figs. 13, 14, 15).

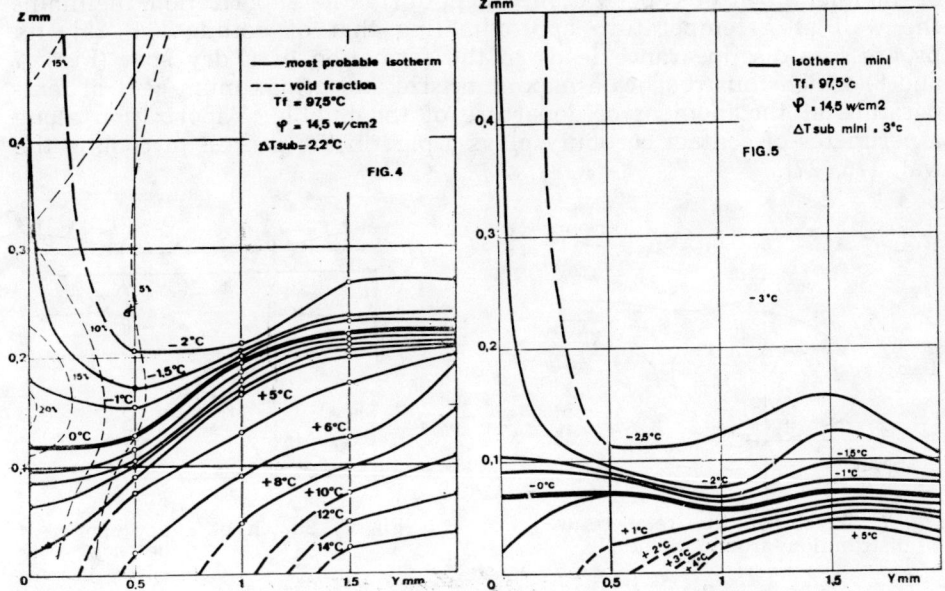

Fig. 4   Fig. 5

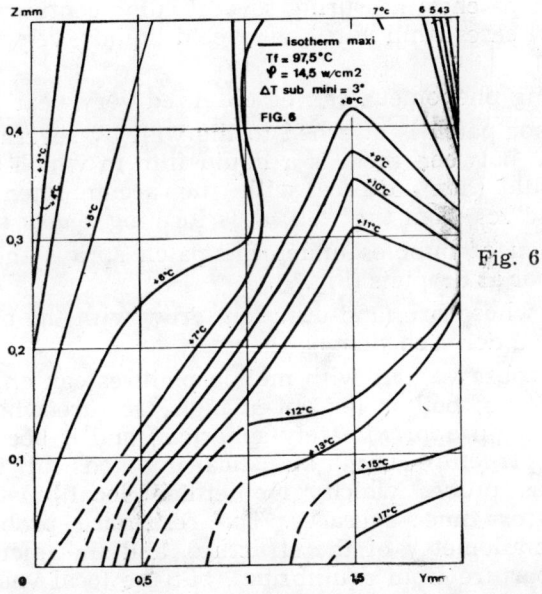

Fig. 6

g) — The wall temperature

With low heat flux densities, current studies have confirmed the role of the microlayer of vaporization which, during its vaporization, maintains the wall at a temperature approximating that of boiling [24, 25]. Its progressive disappearance leads to the formation of a dry spot (Figs. 7 and 8) which thus reaches a maximum size and a maximum central temperature at the moment of departure of the bubble. Vaporization along a perimeter of contact certainly plays a part in the process of cooling the wall [26, 27].

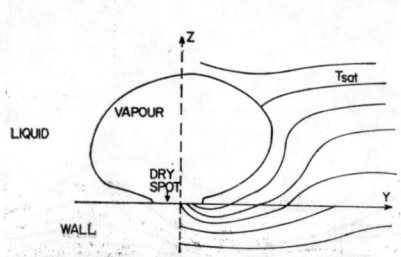

Fig. 7. Sketch of the temperature distribution around a bubble.

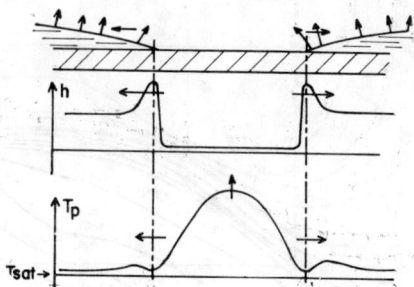

Fig. 8. Sketch of a dry spot.

2.2. HIGHER BUBBLE DENSITIES

The granulometry of the vapor is now much coarser because the bubbles coalesce on or near the wall; at atmospheric pressure, it is possible to observe vapor patches measuring several cubic centimeters, except for some exceptional cases (thin wires, very cold liquids, very high velocities, polluted liquids).

The following phenomena can be observed very near the wall:

— each vapor patch is attached to the wall by several stems in perpetual evolution; between them is a liquid film in which bubbles can be born, which would then coalesce with the accumulated vapor before detaching themselves from the wall, thus forming a new stem.

The explosion of bubbles under the patch sprays the liquid of the film into the vapor as droplets (Fig. 9).

A dry spot, whose size (and duration) grow with the concentration of vapor, is formed under each stem on the wall.

In this case, observations with motion pictures can only yield a qualitative description of boiling (except at high pressure where the smaller vapor patches remain approximately spherical) and it becomes necessary to handle this structure with statistical methods of twophase flow by using various probes which give either local instantaneous valeus, or spatial or time averages. The resistance probes permit the study of the granulometry of the structure, and the microthermocouple measures its departure from equilibrium, but the local velocities are still inaccessible. Only the indirect determination of the rate of recirculation in the neighborhood of the wall gives an indication of the resupplying of the layer on the wall in the cold liquid [28].

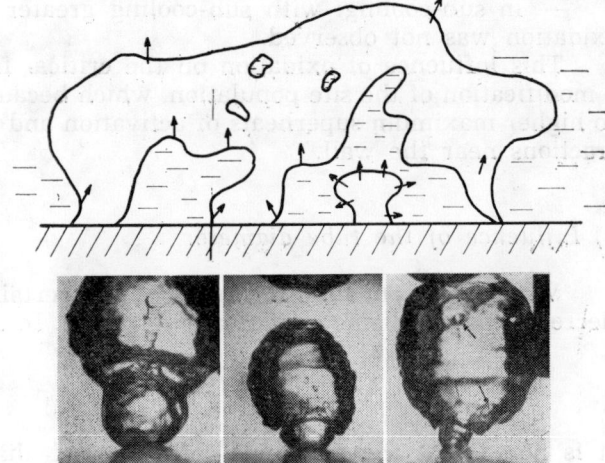

Fig. 9. Two phase structure at high heat flux.

3. Critical Dryout

The hydrodynamic theories which predict the vapor blanketing of the wall provide a maximum value of the heat flux density which correlates a major portion of the experimental results.

However, many authors have noted some significant divergences with the preceding theories. We shall set forth below some results obtained with horizontal stainless steel 18—8 tubes of different diameters and thicknesses, which were immersed in a vessel filled with water at various pressures and temperatures of liquid.* The reddening of the tubes was considered as the criterion which gives the value of maximum power.

The conclusions drawn from 980 measurements and 72 destroyed tubes are as follows:

a) Influence of the tube history in the liquid

— at boiling temperature: the first value or the first two values of the critical flux are often lower by 10 to 20% than the values later obtained on the same tube. This result can be attributed to the oxidation of the tube at the moment of reddening; in fact, if the new tube is reddened in the air before immersing it in the water, a stable value of the critical flux is obtained with the first reddening.

---

* Experiments carried out in the Heat Transfer Section with the collaboration of Messrs Dahan, Loiseleur and Gouzy.

The influence of the surface history has been mentioned often [29, 30] and still remains difficult to predict. This result is important for industrial applications because safety considerations based on the critical value must be derived from the first value.

— in sub-cooling: with sub-cooling greater than 10°C, the effect of oxidation was not observed.

This influence of oxidation on the critical flux can be explained by a modification of the site population, which because of this oxidation leads to higher maximum superheats of activation and thus leads to higher void fractions near the wall.

b) *Influence of the tube diameter*

With a constant tube thickness, it is ascertained that the critical flux decreases as the diameter increases (Fig. 10). In setting:

$$K \frac{\varphi_{critical}}{\alpha \sqrt{\rho_v} \sqrt{\sigma_g (\rho_l - \rho_v)}}$$

it is noted that the combined influence of diameter and pressure is complex, which was foreseable since the granulometry of the vapor in the neighborhood of the critical flux is of the same order as the dimensions of the tube. Likewise, in local boiling it is ascertained that the coefficient C of Kutateladze's formula:

$$\varphi \text{crit}_{sub.} = \varphi \text{rc}_{sat.} \left[1 + C \left(\frac{\rho_l}{\rho_v} \right)^{0,8} \frac{C_p \Delta T_{sub}}{\alpha} \right]$$

increases with the diameter (Fig. 11). Similar results have been obtained with tubes [31] and wires [31, 32].

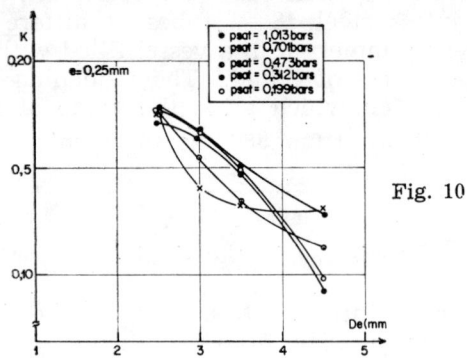

Fig. 10

c) *Influence of the thickness of the tube*

With a constant diameter it is ascertained that the thickness is not a parameter *(K* and *C* remaining constant) when it is greater than 0.25 mm (Fig. 12). The critical flux decreases significantly for lower values, which affirms that in this case, the thermal characteristics of the wall directly influence the critical phenomenon.

It must then be considered that boiling on a heating wall creates on the surface a certain distribution of dry spots which fluctuate in time and space [27, 33, 34, 35]; their rewetting remains possible as long as they themselves have not reached certain critical characteristics (dimensions, lifetime, maximum temperature) for which the equivalent heat flux can be either greater — as is most frequently the case — or lower than the hydrodynamic critical flux (in the case of small thicknesses and of thin wires with low site populations).

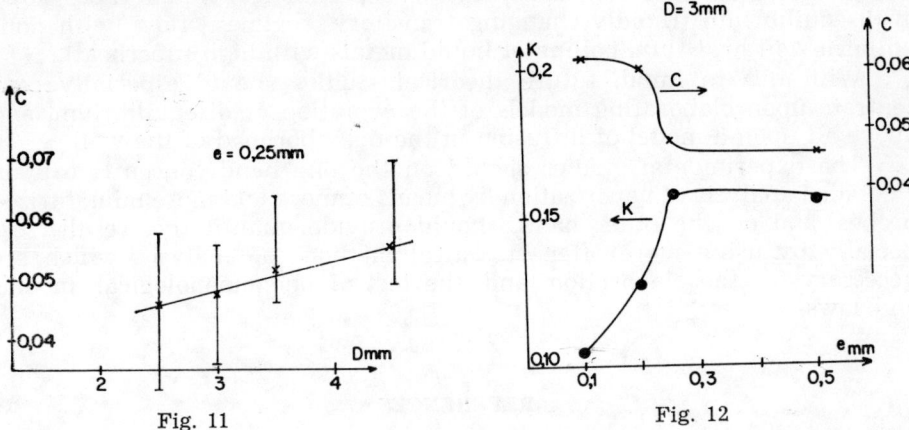

Fig. 11   Fig. 12

The experimental verification of such a concept is only possible with the usage of microthermometers, to measure the wall temperature, which affect only slightly the thermal characteristics and nucleation; the present techniques let us hope for such a delicate (0.1 mm and less) measurement in the near future [36, 37, 38]. The problem of the rewetting of the dry spots seems more accessible than that of their formation.

— In 1965 we proposed a simplified model of the dry spots: the calefaction spot [39], whose disappearance was possible only if the flow of heat towards the wet parts permitted the maintenance of stability or the reabsorbtion of the dry spot. A similar idea was explored by Kovalev [40]. The limiting case was considered, where the dry part was in calefaction, which means that the surface temperature exceeded the Leidenfrost temperature. Simon, Parel and Simoneau [41] have attempted to use this analysis for predicting the minimum heat flux of boiling nitrogen circulating in an electrically heated tube. In this analysis, in spite of the great number of simplifying hypotheses, one more difficulty is encountered: the heat exchange coefficient on a strongly non isothermal wall, especially near the dry boundary, fluctuates greatly and can attain very high values compared to the usually measured values which are spatial and temporal averages. Boiling on non-isothermal surfaces (Vapotron Process) [42, 43, 44] can take advantage of this effect.

— The formation of dry spots under the vapor patches [45] follows from a very complete analysis of the behavior of the liquid film (feeding, pulverization by nucleation, evaporation, spontaneous rupture, flow ...), thus permitting the influence of the wall properties upon the drying of the surface to be taken into account.

4. Conclusions

Although the measurements and models at the present time satisfy the most immediate demands of the designer, it must be noted that all extrapolations, the basic test of the hypotheses about the physics of the phenomenon, are generally disappointing, especially for liquid-vapor flows in disequilibrium (rapidly-changing transitory regimes, flow with non-uniform wall heat flux, boiling of liquid metals with high superheat).

With this in mind, future theoretical studies should especially concentrate upon elaborating models of the evolution of disequilibrium, and above all upon a model of diffusion in the neighborhood of the wall.

The experimental studies should on the one hand concentrate upon a detailed analysis of vaporization by means of more and more miniaturized probes, and on the other hand, should provide quantitative results, especially by using more often statistical methods of analysis, which are necessary for the elaboration and the test of phenomenological models and laws.

REFERENCES

1. N. Zuber, Int. J. Heat Mass Transfer, **6**, 1, 53—78 (1963).
2. C. Y. Han and P. Griffith, Int. J. Heat Mass Transfer, **8**, 905 (1965).
3. C. J. Rallis and H. H. Jawurek, Int. J. Heat Mass Transfer, **7**, 1051—68 (1964).
4. M. Cumo, RT/ING (67) 15, Roma 1967.
5. E. Ruckenstein, Int J. Heat Mass Transfer, **9**, 229 (1966).
6. S. J. D. Van Stralen, Int. J. Heat Mass Transfer, **9**, 995—1046 (1966) and **10**, 1469—1498, 1905—1907 (1967).
7. R. W. Graham and R. C. Hendricks, NASA. TN.D. 3943, 1967.
8. Y. Y. Hsu, J. of Heat Transfer, Paper № 61.WA.177 (1961).
9. A. E. Bergles and W. M. Rohsenow, J. of Heat Transfer **86C**, 3, 365 (1964).
10. E. J. Davis and G. H. Anderson, AIChE J. **12**, 4, 774 (1966).
11. M. Behar, M. Courtaud, R. Ricque and R. Semeria, Third Int. Heat Transfer Conf., Paper 113, Chicago, 1966.
12. P. J. Marto and W. M. Rohsenow, J. of Heat Transfer, 183—204, May (1966).
13. J. G. Collier and P. G. Kosky, AERE.R.5436 1967.
14. A. T. J. Hayward, Brit. J. Appl. Phys., **18**, 641 (1967).
15. W. J. Bornhorst and G. N. Hatsopoulos, J. of Applied Phys., 840—853, December (1967).
16. M. Behar and R. Semeria, C. R. Acad. Sci., Paris, **257**, 2801 (1963).
17. K. Torikai, Bull. JSME **10**, 41, 817 (1967).
18. R. Gregorig, Chemie Ing. Technik **39**, 1, 13—20 (1967).
19. G. G. Treschov, Teploenergetika **4**, 44 (1957). English translation CTS № 465.
20. I. D. R. Grant and T. D. Patten, Symp. Boiling Heat Transfer, Manchester Sept. 1965, Inst. of Mech. Eng. 180 part 3C.
21. B. D. Marcus and D. Dropkin, J. of Heat Transfer, 333 (1965).

22. R. W. Bobst and C. P. Colver, AIChE Preprint 30, 9th National Heat Transfer Conf., SEATTLE 1967.
23. R. Semer:a and J. C. Flamand, Rapport TT № 81, Service des Transferts Thermiques 1967.
24. N. Madsen, Symp. Boiling Heat Transfer, Paper 14, Manchester Sept. 1965 — Inst. of Mech. Eng.
25. C. E. Bonnet, Emacke and R. Morin, EUR 1622f EURATOM 1964.
26. L. A. Skinner, Physics of Fluids **10**, 3, 502 (1967).
27. L. A. Hale. A.S.M.E. Preprint 67—HT—68, Seatle, 1967.
28. E. I. Nevstrueva and A. S. Mekhdi, Teplofiz. Vis. Temp., **809**, 5 (1964).
29. S. Faggiani et all, Ric. Sci. Série 2—A, 3, 5, 591; № 7, 981; № 7, 1007 — 1963.
30. L. S. Sterman and Y. J. Vilemas. Int. J. Heat Mass Transfer, **11**, 347 (1968).
31. S. S. Kutateladze, N. V. Valoukina, I. I. Gogonin, Inzh. F'z. Zh., **12**, 5 (1967).
32. J. H. Lienhard and K. Watanabe, J. of Heat Transfer, p. 94, Feb. (1966).
33. D. B. Kirby and J. W. Westwater, Chemical Engng. Progress Symposium series **61**, № 57, 238, 1965.
34. M. Carne and D. H. Charlesworth, Chem. Eng. Progress Symp. Series **62**, № 64, 24.
35. F. Tachibana, M. Akiyana, H. Kawamura, J. Nucl. Sci. and Technology **4**, 3, 121 (1967).
36. J. Gouault, C. R. Acad. Sci., Paris, Groupe 5, **261**, 23, 5007 (1965).
37. Y. Bard and E. F. Leonard, Int. J. Heat Mass Transfer, **10**, 1727 (1967).
38. V. Kmoniček, Int. J. Heat Mass Transfer. **9**, 199 (1966).
39. R. Semeria and B. Martinet, Symp. Boiling Heat Transfer, Paper № 3, Manchester 1965, Inst. of Mech. Eng.
40. S. A. Kovalev, Int. J. Heat Mass Transfer, **9**, 1219 (1966).
41. F. F. Simon, S. S. Papell and P. S'moneau, NASA—TN—D 4307, 1968.
42. C. A. Beurtheret, 7e Journées de l'Hydraulique, Paris, Société Hydrotechnique de France p. 118, 1962.
43. Application of the Vapotron Process to Boiling Water Nuclear Reactor. — Rapport EUR 3384f 1968.
44. F. S. Lai and Y. Y. Hsu, AIChE J. **13**, 4, 817 (1967).
45. Y. Katto and S. Yokoya, Int. J. Heat Mass Transfer, **11**, 993 (1968).
46. F. Jansen, Chemical Engng. Sci., **22**, 1779 (1967).
47. L. G. Hamburger, Int. J. Heat Mass Transfer, **8**, 1369 (1965).

THE GENERAL EQUATIONS OF TWO-PHASE SYSTEMS APPLIED TO FLASHING FLOWS

J. M. DELHAYE, A. FIORE, Ph. VERNIER, B. BROISE

Center of Nuclear Studies, Heat Transfer Section, Grenoble, France

0. Introduction

The general equations of two-phase flows established in [1] and [2] are of such a complex character that, in order to apply them to particular situations, it is necessary to consider very special cases where, certain effects being negligible, various hypotheses may be made to simplify the equations.

The experimental study of G. Barois and J. Huyghe [3] concerning the adiabatic flashing of water flowing upwards, is particularly well suited for examination by the theoretical equations because of certain characteristics of their equipment, which include the approximate unidimensional character of the flow as well as the absence of frictionnal pressure loss due to the dimensions of the test channel cross-section (100 mm × 100 mm).

Water at a temperature T_{LE} travels upwards through the vertical test-channel, which has a uniform cross-section of area A and is made with adiabatic walls. For this flow, there is an appropriate piezometric line $p(z)$ and a corresponding saturation curve $T_{sat}(p)$ (Fig. 1). Some micro-

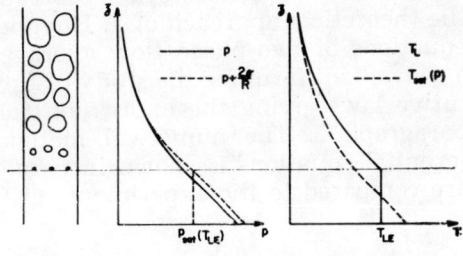

Fig. 1. Development of a flow with autovaporization.

bubbles of radius R_o already exist in the water. We suppose that a microbubble is in reality a cavity void of air whose radius R_o is constant and that all the microbubbles have the same radius R_o. When the pressure $\left(P + \dfrac{2\sigma}{R_0}\right)$ inside a microbubble reaches the saturation pressure $p_{sat}(T_{LE})$ corresponding to the inlet water temperature T_{LE}, vaporization begins.

The problem consists of describing the evolution of the phenomenon in the channel down-stream from the front of the beginning of vaporization, and in particular of determining:

01 — The »desequilibria« Θ and Θ^* between water and vapor defined by:

$$\Theta = T_L - T_{sat}\left(p + \frac{2\sigma}{R}\right)$$

$$\Theta^* = T_L - T_{sat}(p),$$

where p is the static pressure at the cross-section of the channel which is deduced from the piezometric line.

02 — The vapor fraction α in the cross-section defined by

$$\alpha = \frac{A_G}{A}$$

where A_G is the area of the section through which vapor flows. We shall also call α the void fraction in order to conform to current usage.

03 — the vapor quality x: the ratio of vapor mass flow rate to the total mass flow rate at a cross-section of the channel.

04 — the piezometric line $p(z)$

Two methods are possible for approaching this subject:

a) a study based on the growth of a bubble controlled by heat exchange by conduction [4].

b) a global study of the flow in which the heat exchange is accomplished by convection [5].

Analyses of the first type are mathematically complex and not very realistic because they do not take into account the interactions of the bubbles. Our research work is proceeding along the second path and follows the theoretical approach of J. Huyghe et al [6]. In paragraph 1 the general equations of two-phase flow averaged in space and in time are put into a convenient form for the study of flows with autovaporization. The constitutive laws giving the expression for the source of vapor are studied in paragraph 2. The numerical method used to solve the system of differential equations is shown in paragraph 3. Finally, the theorical results are compared to the experiments of G. Barois, in paragraph 4.

1. Simplification of the General Equations

The basic equations will be those presented by one of the authors in reference [7] from which we shall take the nomenclature and the terminology.

The flow will always be assumed to be steady state.

1.1. EQUATION OF CONSERVATION OF MASS FOR THE VAPOR

The equation of continuity for the G-phase with instantaneous variables, averaged across the section for gas flow, is written:

$$\frac{d}{dz}(R_G <\rho_G w_G>) + \frac{1}{A}\oint_{J(z,t)} \rho_G(\vec{V}_G - \vec{V}_i)\vec{n}_G \frac{dl}{\vec{n}_G \cdot \vec{n}_{GJ}} = 0$$

In this equation R_G is the void fraction in a section, ρ_i the density of the vapor, A the area of the cross-section of the channel, J the trace of the interface in the plane of the section, dl a curved element of J, $\vec{n}_G$ the normal unit vector on a point of the interface directed from the vapor towards the liquid, $\vec{n}_{GJ}$ the normal unit vector on a point of J located in the plane of the section and directed from the vapor towards the liquid, $\vec{V}_G$ the velocity at a point in the vapor phase, w_G the component of $\vec{V}_G$ along the Oz axis of the conduit, $V_i \vec{n}_G$ the velocity of displacement of a point on the interface and the $<>$ averaging operator in the gas flow section. The lower indices G and L refer to the vapor and the liquid.

In setting:

$$m_y = \frac{1}{A}\oint_{J(z,t)} \rho_G(\vec{V}_G - \vec{V}_i)\vec{n}_G \frac{dl}{\vec{n}_G \cdot \vec{n}_{GJ}}$$

and in averaging statistically, we get:

$$\frac{d}{dz}\overline{R_G <\rho_G w_G>} + \overline{m_J} = 0$$

where $\overline{}$ is the statistical averaging operator which is equivalent to the time average if we accept the ergodic hypothesis [8].

By definition of the vapor quality, x, we have:

$$x\,G = \overline{R_G <\rho_G w_G>}$$

where G is the mass flow rate per unit cross-section area of the conduit.

The equation of continuity for the G — phase can finally be put in the following form:

$$\frac{d}{dz}(x\,G) + \overline{m_J} = 0 \qquad (1)$$

1.2. TWO-PHASE EQUATION OF CONSERVATION OF MASS

The two-phase equation of continuity averaged across the whole section is written:

$$\frac{d}{dz}(R_G <\rho_G w_G> + R_L <\rho_L w_L>) = 0$$

By averaging statistically, we get:

$$\frac{d}{dz}(\overline{R_G <\rho_G w_G>} + \overline{R_L <\rho_L w_L>}) = 0$$

or:

$$\frac{dG}{dz} = 0 \qquad (2)$$

The latter relationship is obvious. It indicates that the total mass flow rate is conserved.

1.3. TWO-PHASE MOMENTUM EQUATION

The instantaneous momentum equations for a phase, averaged across a section, are written:

$$\frac{d}{dz}(R_K <\rho_K w_K \vec{V}_K>) - R_K<\rho_K \vec{F}> - \frac{d}{dz}(R_K<\overline{\overline{\&}}_K \cdot \vec{n}>) +$$

$$+\frac{1}{A}\oint_J [\rho_K \vec{V}_K \times (\vec{V}_K - \vec{V}_i)\vec{n}_K - \overline{\overline{\&}}_K \cdot \vec{n}_K] \frac{dl}{\vec{n}_K \cdot \vec{n}_{KJ}} -$$

$$(K = G \text{ or } L) \quad -\frac{1}{A}\oint_{b_k} \overline{\overline{\&}}_K \cdot \vec{n}_K \frac{dl}{\vec{n}_K \cdot \vec{n}_{Kb}} = 0$$

where $\vec{F}$ is the exterior force per unit mass, $\overline{\overline{\&}}$ the stress tensor, $\vec{n}$ the unit vector on the axis Oz, b_k the intersection of the conduit wall and the plane of the section located in phase K ($K = G$ or L).

The preceding equations written for $K=G$ and $K=L$ are added together, while taking into account the interfacial condition relative to the momentum shown in [7] and in assuming the surface tension σ to be constant, to give:

$$\frac{d}{dz}(R_G<\rho_G w_G \vec{V}_G>) + R_L<\rho_L w_L \vec{V}_L>) - (R_G<\rho_G \vec{F}> + R_L<\rho_L \vec{F}>)$$

$$-\frac{d}{dz}(R_G<\overline{\overline{\&}}_G \cdot \vec{n}> + R_L<\overline{\overline{\&}}_L \cdot \vec{n}>) + \frac{1}{A}\oint_J \sigma\left(\frac{1}{R_1 \vec{n}_1 \cdot \vec{n}_G} + \frac{1}{R_2 \vec{n}_2 \cdot \vec{n}_G}\right)\vec{n}_G$$

$$\frac{dl}{\vec{n}_G \cdot \vec{n}_{GJ}} - \frac{1}{A}\oint_{b_G} \overline{\overline{\&}}_G \cdot \vec{n}_G \frac{dl}{\vec{n}_G \cdot \vec{n}_{Gb}} - \frac{1}{A}\oint_{b_L} \overline{\overline{\&}}_L \cdot \vec{n}_L \frac{dl}{\vec{n}_L \cdot \vec{n}_{Lb}} = 0$$

where $\vec{u}_i$ is the unit vector of the radius of curvature, R_i, directed from the center of curvature towards the corresponding point on the surface.

Let us take the projection on Oz of the preceding equation and replace $\vec{F}$ by its value $-g\vec{n}$, where g is the acceleration of gravity:

$$\frac{d}{dz}\left(R_G <\rho_G w_G^2> + R_L <\rho_L w_L^2>\right) + (R_G <\rho_G> + R_L <\rho_L>)g$$

$$-\frac{d}{dz}\left(R_G <\bar{\bar{\&}}_G \cdot \vec{n}>_{oz} + R_L <\bar{\bar{\&}}_L \cdot \vec{n}>_{oz}\right) - \frac{1}{A}\oint_J \sigma\left(\frac{1}{R_1\vec{u}_1\vec{n}_G} + \frac{1}{R_2\vec{u}_2\vec{n}_G}\right)\frac{(\vec{n}_G)_{oz}\,dl}{\vec{n}_G \cdot \vec{n}_{GJ}}$$

$$-\left[\oint_{b_G}(\bar{\bar{\&}}_G \cdot \vec{n}_G)_{oz}\frac{dl}{\vec{n}_G \cdot \vec{n}_{Gb}} + \oint_{b_L}(\bar{\bar{\&}}_L \cdot \vec{n}_L)_{oz}\frac{dl}{\vec{n}_L \cdot \vec{n}_{Lb}}\right] = 0$$

The stress tensor $\bar{\bar{\&}}_K$ is made up of a deviator $\bar{\bar{\tau}}_K$ and a term of reversible pressure:

$$\bar{\bar{\&}}_K = \bar{\bar{\tau}}_K - p_K \bar{\bar{U}} \quad (K = G \text{ or } L)$$

where $\vec{U}$ is the unit tensor.

Our first hypothesis consists of neglecting the viscous stress and its variation along Oz with respect to respectively, the reversible pressure and its variation along Oz:

$$<\bar{\bar{\tau}}_K \cdot \vec{n}>_{oz} \ll <p_K> \quad (K = G \text{ or } L)$$

$$\frac{d}{dz}<\bar{\bar{\tau}}_K \cdot \vec{n}>_{oz} \ll \frac{d}{dz}<p_K> \quad (K = G \text{ or } L)$$

Furthermore, we set:

$$\tau_0(z) = \frac{1}{\pi D}\left[\oint_{b_G}(\bar{\bar{\&}}_G \cdot \vec{n}_G)_{oz}\,dl + \oint_{b_L}(\bar{\bar{\&}}_L \cdot \vec{n}_L)_{oz}\,dl\right]$$

Since we have:

$$A = \frac{rD^2}{4}$$

$$\vec{n}_G \cdot \vec{n}_{Gb} = \vec{n}_L \cdot \vec{n}_{Lb} = 1$$

We deduce from this:

$$\frac{d}{dz}(R_G <\rho_G w_L^2> + R_L <\rho_L w_L^2>) + (R_G <\rho_G> + R_L <\rho_L>)g$$

$$-\frac{d}{dz}(R_G <p_G> + R_L <p_L>) + \frac{4\tau_0}{D} - s = 0$$

With:

$$s = -\frac{1}{A}\oint_J \sigma\left(\frac{1}{\vec{R_1}\vec{u_1}\cdot\vec{n_G}} + \frac{1}{\vec{R_2}\vec{u_2}\cdot\vec{n_G}}\right)(\vec{n_G})_{oz}\frac{dl}{\vec{n_G}\cdot\vec{n_{GJ}}}$$

In averaging statistically, we get:

$$\frac{d}{dz}(R_G\langle\rho_G w_G^2\rangle + R_L\langle\rho_L w_L^2\rangle) + (R_G\langle\rho_G\rangle + R_L\langle\rho_L\rangle)g$$

$$+\frac{d}{dz}(R_G\langle p_G\rangle + R_L\langle p_L\rangle) + \frac{4\tau_0}{D} + \bar{s} = 0 \tag{3}$$

We will now calculate the term due to the surface tension $\bar{s}$. In order to do this, s will be evaluated at a given instant. Surface tension is a physical property of the two-phase system which intervenes only when the vapor bubbles are small. With these conditions, they can assumed to be spherical with radius R. We then have (Fig. 2):

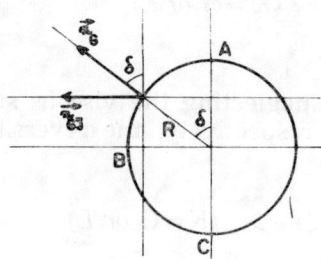

Fig. 2. Calculation of s.

a) from A to B:

$$(\vec{n_G})_{oz} = +\cos\delta$$

$$s = -\frac{1}{A}\cdot\frac{2\sigma}{R}\oint_J \cos\delta\frac{dl}{\cos\left(\frac{\pi}{2}-\delta\right)}$$

$$s = -n\cdot\frac{2\sigma}{R}\cdot\frac{\cos\delta}{\sin\delta}\cdot 2\pi R\sin\delta = -4n\pi\sigma\cos\delta$$

where n is the number of nucleii per unit area of cross-section.

b) from B to C:

$$(\vec{n_G})_{o\beta} = -\cos\delta$$

$$s = 4n\pi\sigma\cos\delta$$

We thus deduce immediately that $\bar{s} = 0$ because of the symmetry.

As in the case of the equations for conservation of mass we shall introduce into Eq. (3) the vapor quality defined by:

$$xG = \overline{R_G < \rho_G w_G >}$$

$$(1-x)G = \overline{R_L < \rho_L w_L >}$$

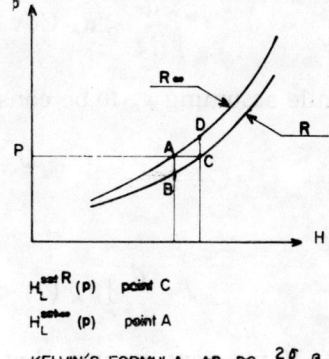

Fig. 3. Enthalpy of saturation in the presence of a curved interface.

Now Eq. (3) has the terms:

$$\overline{R_G < \rho_G w_G^2 >} \text{ and } \overline{R_L < \rho_L w_L^2 >}$$

Certain hypotheses must therefore be made in order to write these terms as functions of x and G. We can use either the variables averaged across a section of type R and $< >$, or the local variables of type α and $-$, since we know from [1, 2, 7 and 8], that we have the following relationships:

$$\left. \begin{array}{l} \overline{R_K} = \mathrel{\triangleleft} \alpha_K \mathrel{\triangleright} \\[4pt] \overline{R_K < f_K >} = \mathrel{\triangleleft} \alpha_K \bar{f_K} \mathrel{\triangleright} \end{array} \right\} \text{ with } K = G \text{ or } L \quad \begin{array}{l}(4)\\[4pt](5)\end{array}$$

in which α_K is the time fraction during which phase K is present, f_K is any function for the K-phase, and $\mathrel{\triangleleft} \mathrel{\triangleright}$ is the averaging operator in the entire section, A. We have chosen the latter method because the quantities that arise are susceptible to more obvious physical interpretations than those arising in the first process.

$$\frac{d}{dz} \overline{R_K < \rho_K w_K^2 >}$$

1.3.1. We first evaluate the term of type

By virtue of (5), we have:

$$J = \frac{d}{dz} \overline{R_K < \rho_K w_K^2 >} = \frac{d}{dz} \mathrel{\triangleleft} \alpha_K \bar{\rho_K} \bar{w_K^2} \mathrel{\triangleright}$$

In choosing the usual definition for the term of covariance of the average of a product, we have:

$$J = \frac{d}{dz}\left(\langle \alpha_K \overline{\rho_K} \cdot \overline{w_K^2} \rangle + \langle \alpha_K \operatorname{Cov} \overline{\rho_K w_K^2} \rangle \right)$$

We shall make the following hypothesis:

$$\frac{d}{dz} \langle \alpha_K \operatorname{Cov} \overline{\rho_K w_K^2} \rangle \ll \frac{d}{dz} \langle \alpha_K \overline{\rho_K}\, \overline{w_K^2} \rangle$$

while assuming $\overline{\rho_K}$ to be constant in section A, this allows us to write:

$$J = \frac{d}{dz}\left(\overline{\rho_K} \langle \alpha_K \overline{w_K^2} \rangle \right)$$

or:

$$J = \frac{d}{dz}\left[\overline{\rho_K} \left(\langle \alpha_K \rangle \langle \overline{w_K} \rangle^2 + \operatorname{Cov} \langle \alpha_K \overline{w_K^2} \rangle \right) \right]$$

If we now assume that:

$$\frac{d}{dz} \operatorname{Cov} \langle \alpha_K \overline{w_K^2} \rangle \ll \frac{d}{dz}\left(\langle \alpha_K \rangle \langle \overline{w_K} \rangle^2 \right)$$

and:

$$\operatorname{Cov} \langle \alpha_K \overline{w_K^2} \rangle \ll \langle \alpha_K \rangle \langle \overline{w_K} \rangle^2$$

then the quantity J becomes:

$$J = \frac{d}{dz}\left(\overline{\rho_K} \langle \alpha_K \rangle \langle \overline{w_K} \rangle^2 \right)$$

Now, if x_K is defined by:

$$x_K = \begin{cases} x & \text{if } K = G \\ 1 - x & \text{if } K = L \end{cases}$$

We have:

$$x_K\, G = \overline{R_K \langle \rho_K w_K \rangle} = \langle \alpha_K \overline{\rho_K w_K} \rangle$$

But in assuming:

$$\operatorname{Cov} \overline{\rho_K w_K} \ll \overline{\rho_K}\, \overline{w_K}$$

It follows that:

$$x_K\, G = \langle \alpha_K \overline{\rho_K}\, \overline{w_K} \rangle$$

Now, since we have made the hypothesis that ρ_K is constant in A, we obtain:

$$x_K\, G = \overline{\rho_K} \langle \alpha_K \overline{w_K} \rangle$$

Finally, if:

$$\text{Cov} \prec \alpha_K \overline{w}_K \succ \ll \prec \alpha_K \succ \prec \overline{w}_K \succ$$

The value of $x_K G$ is given by:

$$x_K G = \overline{\rho}_K \prec \alpha_K \succ \prec \overline{w}_K \succ$$

We then conclude that the expression J becomes:

$$J = \frac{d}{dz}\left(\frac{x_K^2 \, G^2}{\overline{\rho}_K \prec \alpha_K \succ}\right)$$

or because of (4):

$$J = \frac{d}{dz}\left(\frac{x_K^2 \, G^2}{\overline{R}_K \cdot \overline{\rho}_K}\right)$$

1.3.2. *The term $R_K \langle \rho_{.K} \rangle$ of Eq. (3) can also be written as:*

$$\prec \alpha_K \, \rho_K \succ = \prec \alpha_K \succ \overline{\rho}_K = \overline{R}_K \cdot \overline{\rho}_K$$

1.3.3. *Finally, we calculate the term* $\dfrac{d}{dz}(\overline{R_G \langle p_G \rangle} + \overline{R_L \langle p_L \rangle})$

we set:

$$\Pi = \frac{d}{dz}(\overline{R_G \langle p_G \rangle} + \overline{R_L \langle p_L \rangle})$$

from (5) we have:

$$\Pi = \frac{d}{dz}(\prec \alpha_G \overline{p}_G \succ + \prec \alpha_L \overline{p}_L \succ)$$

We assume that each phase is isobaric.

Consequently:

$$\Pi = \frac{d}{dz}(\prec \alpha_G \succ \overline{p}_G + \prec \alpha_L \succ \overline{p}_L)$$

Or:

$$\Pi = \frac{d}{dz}(\overline{R}_G \cdot \overline{p}_G + \overline{R}_L \cdot \overline{p}_L)$$

We choose as the unknown the pressure $\overline{p}_L$, which is measured by static pressure holes situated along the test channel. The pressures $\overline{p}_G$ and $\overline{p}_L$ are related by the interfacial condition relative to the momentum as shown in [7]. This condition is written for a constant surface tension and for spherical bubbles of radius R:

$$\sum_{K=G,L} [\rho_K \vec{V}_K \times (\vec{V}_K - \vec{V}_L) - \overline{\overline{\mathscr{E}}}_K] \vec{n}_K - \frac{2\sigma}{R} \vec{n}_G = 0$$

The forces due to phase change and to viscosity have a negligible influence compared to those due to surface tension. Consequently, we shall write:

$$\overline{p}_G - \overline{p}_L = \frac{2\sigma}{R}$$

The radius $R(z)$ is defined as being the radius of the bubble when its center is in the plane of the section. The pressure in the bubble is assumed to be constant when the bubble crosses this plane. The equation for Π then becomes:

$$\Pi = \frac{d}{dz}\left[\overline{R}_G\left(\overline{p}_L + \frac{2\sigma}{R}\right) + \overline{R}_L \overline{p}_L\right]$$

$$\Pi = \frac{d}{dz}\left(\overline{p}_L + \overline{R}_G \frac{2\sigma}{R}\right)$$

we set:

$$P^* = \overline{p}_L + \overline{R}_G \frac{2\sigma}{R}$$

and it follows that:

$$\Pi = \frac{dP^*}{dz}$$

Finally, Eq. (3) becomes:

$$G^2 \frac{d}{dz}\left[\frac{x^2}{\overline{R}_G \cdot \overline{\rho}_G} + \frac{(1-x)^2}{\overline{R}_L \cdot \overline{\rho}_L}\right] + (\overline{R}_G \overline{\rho}_G + \overline{R}_L \overline{\rho}_L) g + \frac{dP^*}{dz} + \frac{4\tau_0}{D} = 0 \qquad (6)$$

1.4. TWO-PHASE ENTHALPY EQUATION

The two-phase enthalpy equation, in which we have neglected the terms of:

— volumetric compression
— dissipation

— longitudinal conduction,

is put in the following form [7]:

$$\sum_{K=G,L} \left\{ \frac{d}{dz}(R_K < \rho_K H_K w_K >) + \frac{1}{A} \oint_J [H_K \rho_K(\vec{V}_K - \vec{V}_i) + \vec{J}_K] \vec{n}_K \frac{dl}{\vec{n}_K \cdot \vec{n}_{KJ}} \right.$$

$$\left. + \frac{1}{A} \oint_{b_K} \bar{\bar{\tau}}_K \vec{n}_K \frac{dl}{\vec{n}_K \cdot \vec{n}_{Kb}} \right\} = 0$$

where H_K and $\vec{J}_K$ are the K-phasic specific enthalpy and the K-phasic heat flux.

The interfacial condition relative to the enthalpy is written [7] for the case of constant surface tension and spherical bubbles of radius R:

$$\sum_{K=G,L} \left[\rho_K \left(\frac{1}{2} \vec{V}_K^2 + H_K \right)(\vec{V}_K - \vec{V}_i) + p_K \vec{V}_i - \bar{\bar{\tau}}_K \cdot \vec{V}_K + \vec{J}_K \right] \vec{n}_K$$

$$+ \sigma \, \text{div} \, \vec{V}_i - \sigma \vec{V}_i \cdot \frac{2\sigma}{R} \vec{n}_G = 0$$

We can neglect:

— the kinetic energy per unit of mass

— the kinetic energy per unit of mass $\frac{1}{2} \vec{V}_K^2$ with respect to the specific enthalpy H_K

— the terms $p_K \vec{V}_i$ and $\bar{\bar{\tau}}_K \vec{V}_K$ with respect to the heat flux $\vec{J}_K$.

Since on the other hand surface tension has a negligible role in so far as the energy is concerned, there remains:

$$\sum_{K=G,L} [\rho_K H_K(\vec{V}_K - \vec{V}_i) + \vec{J}_K] \cdot \vec{n}_K = 0$$

The two-phase enthalpy equation is therefore reduced to the following equation:

$$\sum_{K=G,L} \left\{ \frac{d}{dz}(R_K < \rho_K H_K w_K >) + \frac{1}{A} \oint_{b_K} \vec{J}_K \cdot \vec{n}_K \frac{dl}{\vec{n}_K \cdot \vec{n}_{Kb}} \right\} = 0$$

We put:

$$\Phi(z) = \frac{1}{\pi D} \left(\oint_{b_G} \vec{J}_G \vec{n}_G \, dl + \oint_{b_L} \vec{J}_L \cdot \vec{n}_L \, dl \right)$$

Since:

$$A = \frac{\pi D^2}{4}$$

and:
$$\vec{n}_G \cdot \vec{n}_{Gb} = \vec{n}_L \cdot \vec{n}_{Lb} = 1$$

we deduce:
$$\frac{d}{dz}(R_G \langle \rho_G H_G w_G \rangle + R_L \langle \rho_L H_L w_L \rangle) - \frac{4\Phi}{D} = 0$$

In averaging statistically, it follows that:
$$\frac{d}{dz}(\overline{R_G \langle \rho_G H_G w_G \rangle} + \overline{R_L \langle \rho_L H_L w_L \rangle}) - \frac{4\overline{\Phi}}{D} = 0$$

We set:
$$E = \frac{d}{dz}(\overline{R_K \langle \rho_K H_K w_K \rangle})$$

From Eq. (5) we have:
$$E = \frac{d}{dz}\overline{\langle \alpha_K \rho_K H_K w_K \rangle}$$

$$E = \frac{d}{dz}\langle \alpha_K \,\overline{\rho_K}\,\overline{H_K}\,\overline{w_K} + \alpha_K \,\mathrm{Cov}\,\overline{\rho_K H_K w_K}\rangle$$

Af before, we shall make a hypothesis about the covariant term. It is assumed that:
$$\frac{d}{dz}\langle \alpha_K \,\mathrm{Cov}\,\overline{\rho_K H_K w_K}\rangle \ll \frac{d}{dz}\langle \alpha_K \overline{\rho_K}\,\overline{H_K}\,\overline{w_K}\rangle$$

It has already been assumed $\overline{\rho_K}$ and $\overline{p_K}$ are constant in A; it therefore follows that $\overline{H}_K$ is also constant in A. Consequently:
$$E = \frac{d}{dz}(\langle \alpha_K \overline{w_K}\rangle \overline{\rho_K}\, \overline{H_K})$$

Then, all that is required is to add the following condition to the hypothesis already made about
$$\mathrm{Cov}\,\langle \alpha_K \overline{w_K}\rangle:$$

$$\frac{d}{dz}\mathrm{Cov}\,\langle \alpha_K \overline{w_K}\rangle \ll \frac{d}{dz}\langle \alpha_K \langle \rangle w_K\rangle$$

in order to obtain (while using the expression for $x_K G$ previously established):
$$E = \frac{d}{dz}(x_K G \overline{H}_K)$$

and for the complete equation:

$$G\frac{d}{dz}[x\overline{H}_G+(1-x)\overline{H}_L]-\frac{4\overline{\Phi}}{D}=0 \tag{7}$$

1.5. DERIVED SYSTEM OF EQUATIONS

In a flow with autovaporization without friction, we have:
$$\overline{\Phi}\equiv 0$$
$$\overline{\tau}_o\equiv 0$$

Equations (1), (6) and (7) become:

$$\begin{cases} G\dfrac{dx}{dz}+\overline{m}_J=0 \\[2mm] G^2\dfrac{d}{dz}\left[\dfrac{x^2}{\overline{R}_G\cdot\overline{\rho}_G}+\dfrac{(1-x)^2}{\overline{R}_L\overline{\rho}_L}\right]+(\overline{R}_G\overline{\rho}_G+\overline{R}_L\overline{\rho}_L)g+\dfrac{dP^*}{dz}=0 \\[2mm] \dfrac{d}{dz}[x\overline{H}_G+(1-x)\overline{H}_L]=0 \end{cases}$$

Henceforth we shall put:

$$\overline{R}_G=\alpha \quad \overline{R}_L=1-\alpha$$
$$\overline{\rho}_G=\rho_G \quad \overline{\rho}_L=\rho_L$$
$$\overline{H}_G=H_G \quad \overline{H}_L=H_L$$

With these new notations, the system becomes:

$$\begin{cases} G\dfrac{dx}{dz}+\overline{m}_J=0 & (8) \\[2mm] G^2\dfrac{d}{dz}\left[\dfrac{x^2}{\alpha\rho_G}+\dfrac{(1-x)^2}{(1-\alpha)\rho_L}\right]+[\alpha\rho_G+(1-\alpha)\rho_L]g+\dfrac{dP}{dz}=0 & (9) \\[2mm] \dfrac{d}{dz}[xH_G+(1-x)H_L]=0 & (10) \end{cases}$$

2. Dependent Variables and Constitutive Laws

2.1. CHOICE OF DEPENDENT VARIABLES

The utilization of dependent variables x, p and Θ make the system linear with respect to the derivatives of these variables; they were chosen for this reason. In fact, x is a physical variable that is not measurable in the present status of two-phase flow metrology. On the other hand, the void fraction α can be measured and is deduced from x with the classic equation:

$$\alpha = \frac{x}{x+\beta\gamma(1-x)}, \qquad \beta = \frac{\rho_G}{\rho_L} \qquad (11)$$

where γ is the ration of the gas velocity to the liquid velocity, often called »slip«. This means that all the equations will require this variable γ which will become a system parameter. Putting the results as a function of the parameter γ, is a priori unsatisfactory because γ is not susceptible to any physical interpretation. It would be preferable but much more complicated to take w_G and w_L as dependent variables. In this case, it would naturally be necessary to increase the number of system equations while taking into account the phase equations. Such a study is currently being undertaken.

2.2. CONSTITUTIVE LAW OF VAPORIZATION

This involves evaluating the term:

$$\overline{m_J} = \frac{1}{A}\oint_{J(z,t)} \overline{\rho_G(\vec{V}_G-\vec{V}_i)\vec{n}_G \frac{dl}{\vec{n}_G\cdot\vec{n}_{GJ}}}$$

The interfacial condition approached for the energy is written [7]:

$$[\rho_G H_G(\vec{V}_G-\vec{V}_i)+\vec{J}_G]\vec{n}_G + [\rho_L H_L(\vec{V}_L-\vec{V}_i)+\vec{J}_L]\vec{n}_L = 0$$

For the mass, the following interfacial condition applies:

$$\rho_G(\vec{V}_G-\vec{V}_i)\vec{n}_G + \rho_L(\vec{V}_L-\vec{V}_i)\vec{n}_L = 0$$

In combining these two equations we obtain:

$$\rho_G(\vec{V}_G-\vec{V}_i)\vec{n}_G = \frac{1}{H_G-H_L}(\vec{J}_L-\vec{J}_G)\cdot\vec{n}_G$$

But, since:

$$H_G-H_L = Z$$

And if we put:

$$\vec{J}_L-\vec{J}_G = \vec{\varphi}_J$$

It follows that:

$$\overline{m_J} = \frac{1}{A}\oint_{J(z,t)} \overline{\frac{1}{Z}\vec{\varphi}_J\cdot\vec{n}_G\cdot\frac{dl}{\vec{n}_G\cdot\vec{n}_{GJ}}}$$

If we define an average flux φ_J by the relationship:

$$\varphi_J = -\frac{1}{J}\oint_J \vec{\varphi_J}\cdot\vec{n_G}\frac{dl}{\vec{n_G}\cdot\vec{n_{GJ}}}$$

we get:

$$\overline{m_J} = -\frac{1}{A\,Z}\overline{J\cdot\varphi_J}$$

By hypothesis, we shall assume the covariant term between J and φ_J to be negligible. Finally we get:

$$\overline{m_J} = -\frac{1}{A\,Z}\overline{J}\cdot\overline{\varphi_J}$$

Or in eliminating the ―――― to simplify the writing:

$$m_J = -\frac{1}{A\,Z}J\cdot\varphi_J$$

2.2.1. Evaluation of the interfacial perimeter per unit area

We shall assume that a cross-section of the channel contains N identical spherical bubbles with diameter $d(z)$. In supposing that the cross-section under consideration cuts all the bubbles along a diametric plane, we have:

$$J = N\pi\alpha$$

where:

$$d = 2\sqrt{\frac{\alpha}{n\pi}}$$

and n is the number of bubbles cut by the cross-section per unit area. We then deduce:

$$\frac{J}{A} = 2\sqrt{\pi n\alpha} \qquad (12)$$

It would be more exact to calculate the most probable diameter of the bubble sections cut by the plane of the cross-section. Under these conditions the calculation gives:

$$J = \frac{\pi^2}{4}Nd$$

$$d = \sqrt{\frac{6\alpha}{n\,\pi^{1/2}}}$$

$$\frac{J}{A} = \frac{1}{4}\pi^{7/4}\sqrt{6\,x\,\alpha}$$

In considering the very schematic nature of all hypotheses concerning the constitutive law of vaporization, we have kept the first values of J and of d.

2.2.2 Variation of the flux at the interface

Let h_J be the heat transfer coefficient at the interface. We have:

$$\varphi_J = h_J \cdot \Theta$$

When the vapor bubbles are spherical, the value of h_J that we propose is that value which applies to heat transfer between a sphere and the fluid that surrounds it. We therefore adopt the classic equation:

$$Nu_J = 2{,}0 + 0{,}6\, Re^{0{,}5} \cdot Pr^{2/3}$$

where the Nusselt numbers Nu_J and the Reynolds numbers Re are based upon the bubble diameters.

$$d = 2\sqrt{\frac{\alpha}{x\,\pi}}$$

The Prandtl number Pr is defined by:

$$Pr = \frac{\nu_L}{\alpha_L} = \frac{\nu_L\, \rho_L\, C_L}{\lambda_L}$$

where α_L is the thermal diffusivity, ν_L is the kinematic viscosity, and λ_L is the thermal conductivity of the liquid. Consequently:

$$h_J = \lambda_L \sqrt{\frac{n\pi}{\alpha}} + 0{,}6\, \lambda_L^{1/3}\, \nu_L^{1/6}\, \rho_L^{2/3}\, C_L^{2/3}(w_G - w_L)^{1/2} \left[\frac{1}{2}\left(\frac{n\pi}{\alpha}\right)^{1/2}\right]^{1/2}$$

Now, we have:

$$w_G - w_L = w_L(\gamma - 1)$$

and since:

$$w_L = G\left(\frac{1-x}{\rho_L} + \frac{x}{\gamma\, \rho_G}\right)$$

it follows that:

$$h_J = \lambda_L \sqrt{\frac{n\pi}{\alpha}} + 0{,}6\, \lambda_L^{1/3}\, \nu_L^{1/6}\, \rho_L^{2/3}\, C_L^{2/3} \left[\frac{G(\gamma-1)}{2}\left(\frac{1-x}{\rho_L} + \frac{x}{\gamma\, \rho_G}\right)\left(\frac{n\pi}{\alpha}\right)^{1/2}\right]^{1/2} \quad (13)$$

2.2.3. Expression for the vapor source term

Consequently, at the beginning of the autovaporization phenomenon we have:

$$m_J = -\frac{1}{Z} \cdot \frac{J}{A} \cdot h_J \cdot \theta \quad (14)$$

where $\dfrac{J}{A}$ and h_J are given by Eqs. (12) and (13). Photographs taken during test runs [3] show that independent bubbles only exist for a very short distance of a few centimeters. The flow pattern changes drastically and an extremely agitated regime follows the bubbly flow. This modification of the flow regime must therefore be included in the calculation and we have ascertained that the best solution is to choose an equation of the following form for the source term, m_J

$$m_J = -\frac{1}{Z} \cdot \frac{J}{A} \cdot h_J \cdot \theta \cdot \psi(\alpha) \cdot \alpha^{-1/2}$$

where $\dfrac{J}{A}$ and h_J are still given by Eqs. (12) and (13). The function $\psi(\alpha)$ is made up (Fig. 4) of a cubic for $\alpha \leqslant \alpha_1$ and of a parabola for $\alpha \geqslant \alpha_1$. It meets the following conditions:

$\psi(\alpha) = 0$ for $\alpha = 0$ and $\alpha = 1$

$\dfrac{d\psi}{d\alpha} = 0$ for $\alpha = 0$ and $\alpha = \alpha_1$ (value for which $\Psi(\alpha) = m$)

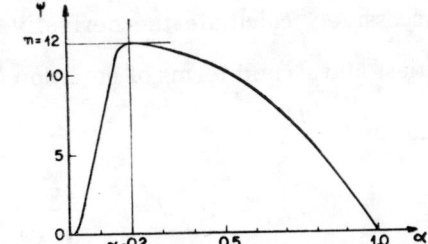

Fig. 4. Function $\psi(\alpha)$.

the cubic equation is written:

$$\psi(\alpha) = m \frac{\alpha^2}{\alpha_1^2} \left(3 - \frac{2\alpha}{\alpha_1}\right) \quad \text{with} \quad 0 \leqslant \alpha \leqslant \alpha_1 \tag{16}$$

The parabolic equation is written:

$$\psi(\alpha) = \frac{m}{(1-\alpha_1)^2} [\alpha(2\alpha_1 - \alpha) + 1 - 2\alpha_1] \text{ with } \alpha_1 \leqslant \alpha \leqslant 1 \tag{17}$$

the value of α_1 corresponds to a maximum value of the interfacial perimeter; thus it corresponds to the change of flow pattern described above. We have taken $\alpha_1 = 0.20$, a probable value of the critical void fraction at the experimental pressures (200 g/cm² absolute). At the present, we do not have a method for determining the value of m. It seems that for the experimental conditions of G. Barois and J. Huyghe [3] it is necessary to keep a value of m equal to 12. When the void fraction α is small, the two laws of variation of the interfacial perimeter have a slope at the origin equal to infinity for Eq. (14) and equal to zero for Eq. (15) in which the interfacial perimeter is:

$$J\psi(\alpha) \alpha - \frac{1}{2} \sim \psi(\alpha)$$

This is necessary, in addition, because in order to be thorough, the inertia of the liquid which opposes the growth of the vapor bubbles must be taken into account. The necessity of taking the derivative of $\psi(\alpha)$ to be zero at the point $\alpha=0$ is tied to the usage of the interfacial condition:

$$p_G = p_L + \frac{2\sigma}{R}$$

and to the hypothesis that the phases are isobaric. In fact, when we study the growth of a vapor bubble in an infinite medium, we derive:

$$pG = p_{L\infty} + \frac{2\sigma}{R} + \text{some inertia terms in } \frac{dR}{dt} \text{ and in } \frac{d^2R}{dt^2}$$

2.3. TRANSFORMATION OF THE MOMENTUM EQUATION

In this paragraph we shall introduce into the momentum equation (9) the derivatives of the dependent variables x, p and Θ. In order to do this we shall successively calculate the derivatives of $P = p + \frac{2\sigma}{R}$ and of P^*, as well as the first and second terms of equation (9).

2.3.1. Calculation of $\frac{dP}{dz}$

$$P = p + \frac{2\sigma}{R}$$

$$R = \frac{d}{2} = \frac{\alpha^{1/2}}{(\pi n)^{1/2}}$$

We deduce from the above:

$$\frac{dP}{dz} = \frac{dp}{dz} - \sigma (\pi n)^{1/2} \alpha^{-3/2} \frac{d\alpha}{dz}$$

Then α is related to x by Eq. (11). Consequently, by taking into account the variation of the vapor density with pressure, it follows that:

$$\frac{dP}{dz} = \frac{dp}{dz} - \sigma (\pi n)^{1/2} \frac{[x+\beta\gamma(1-x)]^{2/3}}{x^{3/2}} \frac{1}{[x+\beta\gamma(1-x)]^2}$$

$$\left\{ \frac{dx}{dz} [x+\beta\gamma(1-x)] - x \left[\frac{dx}{dz} + \frac{\gamma}{\rho_L} \left(\frac{d\rho_G}{dp}\right)_P \cdot \frac{dp}{dz} - \frac{\gamma x}{\rho_L} \left(\frac{d\rho_G}{dp}\right)_P \right. \right.$$

$$\left. \left. \cdot \frac{dP}{dz} - \beta\gamma \frac{dn}{dz} \right] \right\}$$

After simplification we get:

$$\frac{dP}{dz} = \left[1 - \sigma(\pi n)^{1/2} \frac{1-x}{x} \alpha^{1/2} \frac{\Upsilon}{\rho_L} \left(\frac{d\rho_G}{dp}\right)_P\right]^{-1} \cdot \frac{dp}{dz}$$

$$- \left[1 - \sigma(\pi n)^{1/2} \frac{1-x}{x} \alpha^{1/2} \frac{\Upsilon}{\rho_L} \left(\frac{d\rho_G}{dp}\right)_P\right]^{-1} \sigma(\pi n)^{1/2} \frac{\beta \Upsilon \alpha^{1/2}}{x^2} \frac{dx}{dz}$$

2.3.2. Calculation of $\dfrac{dP^*}{dz}$

$$P^* = p + \alpha \frac{2\sigma}{R}$$

with:

$$R = \frac{d}{2} = \frac{\alpha^{1/2}}{(\pi n)^{1/2}}$$

We then deduce:

$$\frac{dP^*}{dz} = \frac{dp}{dz} + \sigma(\pi n)^{1/2} \alpha^{-1/2} \frac{d\alpha}{dz}$$

By using Eq. (11), we get:

$$\frac{dP^*}{dz} = \frac{dp}{dz} + \sigma(\pi n)^{1/2} \frac{[x + \beta\Upsilon(1-x)]^{1/2}}{x^{1/2}} \cdot \frac{1}{[x + \beta\Upsilon(1-x)]^2}$$

$$\left\{\frac{dx}{dz}[x + \beta\Upsilon(1-x)] - x\left[\frac{dx}{dz} + \frac{\Upsilon}{\rho_L}\left(\frac{d\rho_G}{dp}\right) \cdot \frac{dP}{dz} - \frac{\Upsilon x}{\rho_L}\left(\frac{d\rho_G}{dp}\right)_P \frac{dP}{dz} - \beta\Upsilon \frac{dx}{dz}\right]\right\}$$

Upon replacing $\dfrac{dP}{dz}$ by its value in terms of the derivatives of p and of x obtained in the preceding paragraph, we derive, after simplification:

$$\frac{dP^*}{dz} = \left\{1 - \sigma(\pi n)^{1/2} \frac{1-x}{x} \alpha^{1/2} \frac{\Upsilon}{\rho_L}\left(\frac{d\rho_G}{dp}\right)_P \left[1 - \sigma(\pi n)^{1/2} \frac{1-x}{x} \alpha^{1/2} \frac{\Upsilon}{\rho_L}\left(\frac{d\rho_G}{dp}\right)_P\right]^{-1}\right\} \cdot$$

$$\cdot \frac{dp}{dz} + \sigma(\pi n)^{1/2} \beta\Upsilon \frac{\alpha^{3/2}}{x^2} \left[1 - \sigma(\pi n)^{1/2} \frac{1-x}{x} \alpha^{1/2} \frac{\Upsilon}{\rho_L}\left(\frac{d\rho_G}{dp}\right)_P\right]^{-1} \frac{dp}{dz}$$

2.3.3. Transformation of the term $\dfrac{d}{dz}\left[\dfrac{x^2}{\alpha\,\rho_G}+\dfrac{(1-x)^2}{(1-\alpha)\,\rho_L}\right]$

After having replaced α by its value given in Eq. (11) and $\dfrac{dP}{dz}$ by the term established in paragraph 2.2.2, we get:

$$\frac{d}{dz}\left[\frac{x^2}{\alpha\,\rho_G}+\frac{(1-x)^2}{(1-\alpha)\,\rho_L}\right]=$$

$$=\left\{\frac{1}{\rho_L}\left[2x(1-\Upsilon)\left(1-\frac{1}{\beta\,\Upsilon}\right)+\Upsilon-2+\frac{1}{\beta\,\Upsilon}\right]-\frac{x(1-\Upsilon)-1}{\rho_L^2\,\beta\,x}\left(\frac{d\rho_G}{dp}\right)_P\right.$$

$$\left[1-\sigma\,(\pi n)^{1/2}\,\frac{1-x}{x}\,\alpha^{1/2}\,\frac{\Upsilon}{\rho_L}\left(\frac{d\rho_G}{dp}\right)_P\cdot\sigma\,(\pi n)^{1/2}\,\alpha^{1/2}\right\}\cdot\frac{dx}{dz}+$$

$$+\frac{x[x(1-\Upsilon)-1]}{\rho_L^2\,\beta\,x}\left(\frac{d\rho_G}{dp}\right)_P\left[1-\sigma(\pi n)^{1/2}\,\frac{1-x}{x}\,\alpha^{1/2}\,\frac{\Upsilon}{\rho_L}\left(\frac{d\rho_G}{dp}\right)_P\right]^{-1}\frac{dp}{dz}$$

2.3.4. Transformation of the hydrostatic term:

$$\alpha\,\rho_G+(1-x)\,\rho_L=\rho_L\,\frac{\beta\,[x(1-\Upsilon)+\Upsilon]}{x+\beta\,\Upsilon\,(1-x)}$$

Consequently, the two-phase momentum Eq. (9) can be written:

$$\left\{\frac{G^2}{\rho_L}\left[2x(1-\Upsilon)\left(1-\frac{1}{\beta\,\Upsilon}\right)+\Upsilon-2+\frac{1}{\beta\,\Upsilon}\right]+\sigma\,(\pi n)^{1/2}\,\frac{\alpha^{3/2}\,\beta\,\Upsilon}{n^2}\right. \tag{18}$$

$$+\frac{1}{\rho_L}\left(\frac{d\rho_G}{dp}\right)_P\frac{\sigma\,(\pi n)^{1/2}\,\alpha^{1/2}}{x}\left[1-\sigma\,(\pi n)^{1/2}\,\frac{1-x}{x}\,\alpha^{1/2}\,\frac{\Upsilon}{\rho_L}\left(\frac{d\rho_G}{dp}\right)_P\right]^{-1}$$

$$+\left[\sigma\,(\pi n)^{1/2}\,\alpha^{1/2}\,\frac{\Upsilon^2}{x^2}(1-x)\,\beta-G^2\,\frac{x(1-\Upsilon)-1}{\rho_L\,\beta}\right]\bigg\}\frac{dx}{dz}$$

$$+\left\{1+\frac{1}{\rho_L}\left(\frac{d\rho_G}{dp}\right)_P\left[1-\sigma\,(\pi n)^{1/2}\,\frac{1-x}{x}\,\alpha^{1/2}\,\frac{\Upsilon}{\rho_L}\left(\frac{d\rho_G}{dp}\right)_P\right]^{-1}\right.\cdot$$

$$\cdot\left[G^2\,\frac{x[x(1-\Upsilon)-1]}{\rho_L\,\beta^2\,\Upsilon}-\sigma\,(\pi n)^{1/2}\,\frac{1-x}{x}\,\alpha^{3/2}\,\Upsilon\right]\bigg\}\frac{dp}{dz}=$$

$$=\rho_L\,g\,\frac{\beta\,[x(1-\Upsilon)+\Upsilon]}{x+\beta\,\Upsilon\,(1-x)}$$

2.4. TRANSFORMATION OF THE ENTHALPY EQUATION

It is assumed that the vapor is saturated. Under these conditions, its enthalpy is that of saturation taken at the pressure in the neighbourhood of an interface of radius R:

$$H_G = H_G^{\text{sat } R}(P) = H_L^{\text{sat } R}(P) + Z(P)$$

Kelvin's formula is then (Fig. 3):

$$H_L^{\text{sat } R}(P) = H_L^{\text{sat } \infty}(P) + \frac{2\sigma}{R} \beta \left(\frac{dH_L^{\text{sat } \infty}}{dp}\right)_P$$

where $H_L^{\text{sat } \infty}(P)$ is the enthalpy of saturated liquid at a pressure P in the neighbourhood of a plane interface. This function, $H_L^{\text{sat } \infty}$, is that which is given in the thermodynamic tables. Thus, we have for the vapor:

$$H_G = H_L^{\text{sat } \infty}(P) + \frac{2\sigma}{R}\left(\frac{dH_L^{\text{sat } \infty}}{dp}\right)_P + Z(P)$$

On the other hand, the liquid is superheated and we can take the following equation as a definition of the superheat, Θ:

$$H_L = H_L^{\text{sat } R}(P) + C_L \theta$$

where C_L is the specific heat of the liquid. Taking into account Kelvin's formula, this equation becomes:

$$H_L = H_L^{\text{sat } \infty}(P) + \frac{2\sigma}{R}\beta \left[\frac{dH_L^{\text{sat } \infty}}{dp}\right]_P + C_L \theta$$

The above expressions for the enthalpies H_G and H_L are introduced into the enthalpy equation (10). We then have:

$$\left[\left(\frac{dH_L^{\text{sat } \infty}}{dp}\right)_P + x\left(\frac{dS}{dp}\right)_P + \frac{2\sigma}{R}\left(\frac{dP_G}{dp}\right)_P\left(\frac{dH_L^{\text{sat } \infty}}{dp}\right)_P + \beta\left[\frac{dH_L^{\text{sat } \infty}}{dp}\right]_P\right.$$
$$\left. + \frac{2\sigma}{R}\beta\left[\frac{d^2 H_L^{\text{sat } \infty}}{dp^2}\right]_P\right]\frac{dP}{dz} + [Z(P) - C_L \theta]\frac{dx}{dz} + C_L(1-x)\frac{d\theta}{dz}$$
$$- \beta_l \frac{dH_L^{\text{sat } \infty}}{dp} \frac{dp}{dz} = 0$$

Now we have the following inequalities:

$$\frac{2\sigma}{R}\left(\frac{d\rho_G}{dp}\right)_P \left(\frac{dH_L^{\text{sat } \infty}}{dp}\right)_P \ll \left(\frac{dH_L^{\text{sat } \infty}}{dp}\right)_P$$

$$\beta \left(\frac{dH_L^{\text{sat } \infty}}{dp}\right)_P \ll \left(\frac{dH_L^{\text{sat } \infty}}{dp}\right)_P$$

$$\frac{2\sigma}{R}\beta \left(\frac{d^2 H_L^{\text{sat } \infty}}{dp}\right)_P \ll \left(\frac{dH_L^{\text{sat } \infty}}{dp}\right)_P$$

The validity of these inequalities depends on the pressure. We have numerically verified these inequalities for the experimental conditions of G. Barois and J. Huyghe [3]. Consequently, upon replacing $\dfrac{dP}{dz}$ by its value and in retaining the significant terms in the coefficient of $\dfrac{dp}{dz}$, it follows that:

$$\left\{\left[\frac{dH_L^{sat}}{dp}\right]_P + x\left[\frac{dZ}{dp}\right]_P\right\}\left[1 - \sigma\,(\pi n)^{1/2}\,\frac{1-x}{x}\,\alpha^{1/2}\,\frac{\Upsilon}{\rho_L}\left(\frac{d\rho_G}{dp}\right)_P\right]^{-1} \cdot \frac{dp}{dz} \quad (19)$$

$$+ \left\{Z(P) - C_L\theta - \left[\left(\frac{dH_L^{sat}}{dp}\right)_P + x\left(\frac{dZ}{dp}\right)_P\right] \cdot \left[1 - \sigma\,(\pi n)^{1/2}\,\frac{1-x}{x}\,\alpha^{1/2}\,\frac{\Upsilon}{\rho_L}\left(\frac{d\rho_G}{dp}\right)\right]^{-1}\right.$$

$$\left. \sigma\,(\pi n)^{1/2}\,\frac{\beta\,\Upsilon\,\alpha^{1/2}}{x^2}\right\}\frac{dx}{dz} + C_L\,(1-x)\,\frac{d\Theta}{dz} = 0$$

3. Numerical solution

3.1. SYSTEM OF EQUATIONS AND INITIAL CONDITIONS

3.1.1. *The system of equations to be solved* is composed of:

a) — The equation of conservation of mass (8) associated with the expression for the source term m_J (15), in which the interfacial exchange coefficient h_J is given by (13) and the function $\psi(\alpha)$ by (16) and (17).

b) — The two-phase momentum equation (18)

c) — The two-phase enthalpy equation (19).

Furthermore, the auxiliary unknown, α, is related to x by Eq. (11).

3.1.2. *The unknowns are:*

a) the quality, x

b) the pressure, p

c) the disequilibrium, Θ

3.1.3. *The data are:*

a) The physical properties of the liquid:
— The enthalpy at saturation, $H_L^{sat}(p)$, of the liquid
— The latent heat of vaporization, $Z(p)$, of the liquid
— The liquid density, ρ_L

- The thermal conductivity, λ_L, of the liquid
- The kinematic viscosity, ν_L, of the liquid
- The specific heat, C_L, of the liquid
- The vapor density, $\rho_G\ (p)$
- The surface tension, σ

b) The mass flow rates G, per unit area of the cross-section

c) The acceleration of gravity, g

d) The number, n, of vapor bubbles per unit area of the cross-section

e) The initial conditions for:
   - The pressure, p_o
   - The radius of nucleii, R_o

3.1.4. *The parameters of the calculation are:*

a) the slip γ

b) the value of m arising in the function $\Psi(\alpha)$

3.2. NUMERICAL METHOD UTILIZED

The preceding system has been integrated by using the Runge-Kutta methods with variable steps (variant Kutta-Merson) on an I.B.M. 360-40 computer.

4. Results

4.1. INFLUENCE OF VARIOUS PARAMETERS

A calculation has been made for the following conditions:
- water-steam flow
- $G = 10$ g cm²s⁻¹
- $n = 1$ cm⁻² (we shall return to this value later)
- $p = 200$ g cm⁻² for $z = 0$
- $R_o = 50$ μm

Experimental determination of the nucleii radius, R_o, was not made during the tests of G. Barois and J. Huyghe. However, we can estimate the order of magnitude of R_o from the results of O. Ahmed and F. G. Hammitt [9]. These authors obtained a histogram of the volume of the nucleii as a function of the amount of gas dissolved in the water at a pressure equal to 1.75 bar. This volume varied between 200 and 2200 μm³, which corresponds to a radius between 3,6 and 8,1 μm. However, we have used a nucleus radius of $R_o = 50$ μm. On the other hand, the value that we have chosen for n (1 nucleus per square centimeter) is very high. In fact, these values for R_o and n take into consideration the effects of back-mixing caused by large eddies due to the agitation of the flow; these eddies carrying with them many large nucleii.

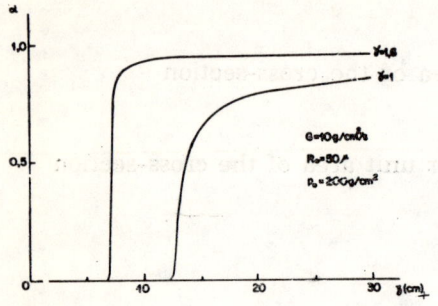

Fig. 5. Development of the void fraction.

The parametric analysis of m and γ shows that the value of γ has no perceptible influence upon the values of α and of p. On the other hand, the curves $\Theta^*(z)$ strongly depend on the slip. Figures 5, 6 and 7 give the

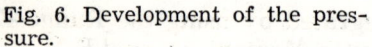

Fig. 6. Development of the pressure.

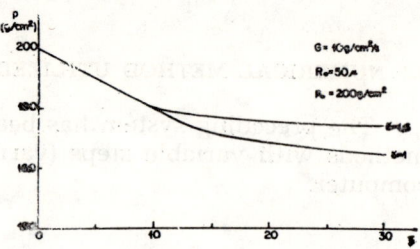

curves $\alpha(z)$, $p(z)$ and $\Theta^*(z)$ in terms of the slip parameter γ. We find again on these curves the separation of the flow into two very distinct regimes. The value of the slip, γ, is important only in the curve $\Theta^*(z)$ relative to the regime of agitated flow. It must be noted that a slip value of 1 gives

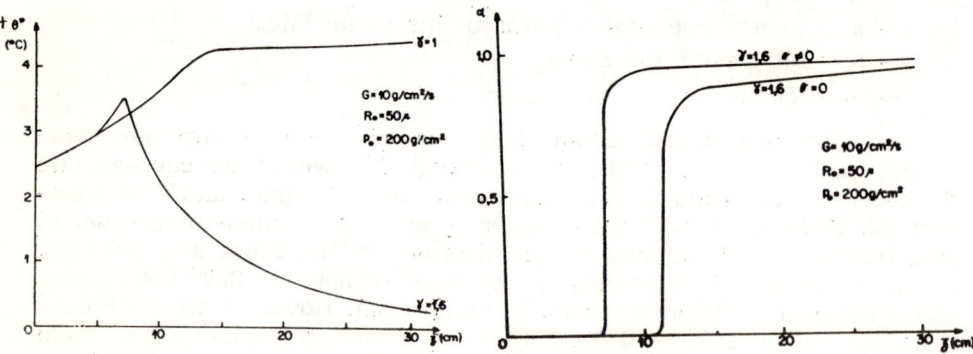

Fig. 7. Development of the superheat.

Fig. 8. Development of the void fraction.

very different results. This is explained by the absence of the term for exchange by convection in Eq. (13) giving the expression for h_J.

The influence of surface tension is very important at the beginning of the autovaporiaation phenomenon. Curves 8, 9 and 10 show the spreads obtained between the theoretical curve for $\gamma=1,6$ and a theoretical curve for the same value of slip but with $\sigma=0$.

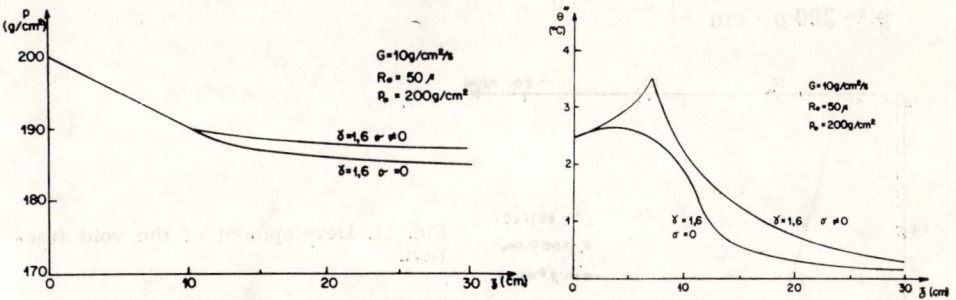

Fig. 9. Development of the pressure.   Fig. 10. Development of the superheat.

Order of magnitude calculations made with the momentum and enthalpy equations, (18) and (19), show that the equations reduce to very different forms at the beginning and at the end of the phenomenon. That is:

4.1.1. *At the beginning:*

Momentum:

$$\sigma(\pi h)^{1/2} \alpha^{3/2} \beta \frac{\gamma}{x^2} \frac{dx}{dz} + \frac{dp}{dz} = -\rho_L g$$

Enthalpy:

$$\left(\frac{dH_L^{sat}}{dp}\right)_P \frac{dp}{dz} - \left(\frac{dH_L^{sat}}{dp}\right)_P \sigma(\pi n)^{1/2} \beta \gamma \frac{\alpha^{1/2}}{x^2} \frac{dx}{dz} + C_L \frac{d\theta}{dz} = 0$$

4.1.2. *At the end:*

Momentum:

$$\frac{G^2}{\rho_L z \gamma} \frac{dx}{dz} + \frac{dp}{dz} = -\rho_L g \frac{\beta \gamma}{x}$$

Enthalpy:

$$\left(\frac{dH_L^{sat}}{dp}\right)_P \frac{dp}{dz} + Z\frac{dx}{dz} + C_L \frac{d\theta}{dz} = 0$$

4.2. COMPARISON WITH EXPERIMENTAL RESULTS

Runs were achieved by G. Barois and J. Huyghe in the following conditions:

$$\left.\begin{array}{l}G = 11{,}3 \; g \cdot cm^{-2} \cdot s^{-1} \\ T_{LE} = 62{,}14°C \\ p = 200 \; g \cdot cm^{-2}\end{array}\right\} \text{ at } z = 0$$

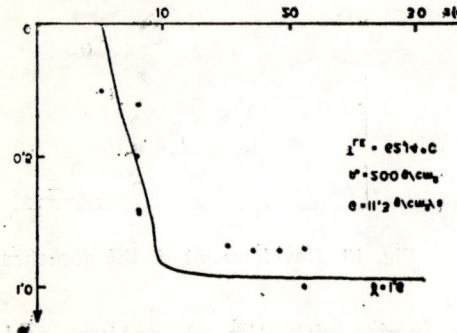

Fig. 11. Development of the void fraction.

The solution has been computed from these data for $n = 1 \; cm^{-2}$ and $\gamma = 1.6$. The experimental flow had an oscillatory pattern due to the loop and we have taken this phenomenon into account by averaging the theoretical results on a 2 cm amplitude sinusoidal oscillation. Figures (11), (12) and (13) show void fraction, pressure drop and superheat versus z with the experimental data of G. Barois and J. Huyghe.

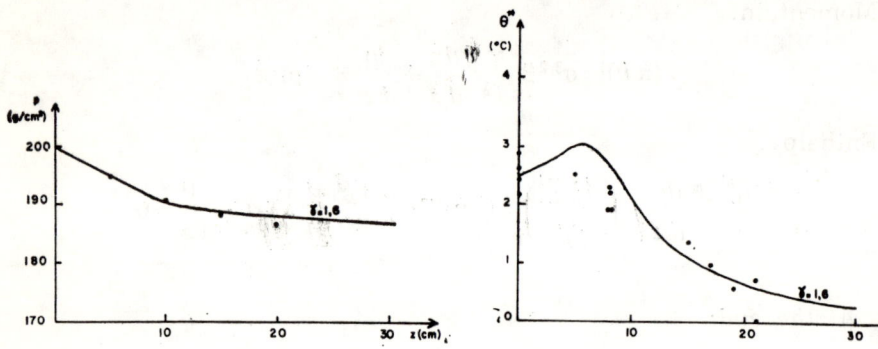

Fig. 12. Development of the pressure.   Fig. 13. Development of the superheat.

5. Conclusions

We have presented an example of a practical application of the general equations of two-phase flow. The adiabatic flow of a water-steam mixture in autovaporization was illustrated. The equations, which were solved numerically, were obtained by making hypotheses in accordance with the characteristics of the chosen flow. On the one hand, the restrictions with respect to the covariant terms were related to the unidimensional aspect; on the other hand, the restrictions concerning dissipation and the energy of compression are typical of flows with small variations of velocity and pressure. It was also necessary to make additional hypotheses in order to close the system, such as the constitutive law of vaporization. Therefore, all these equations do not represent a model postulated a priori, but an attempt to represent a real phenomena in terms of basic physical principles.

The two principal new ideas that come out of this study are, first, the constitutive law of vaporization which allows the liquid temperature to be considered as an unknown in the problem; then, the introduction of vaporization.

However, there still remain difficulties to be overcome among which can be mentioned: the size and distribution of the nucleii; and the taking into account of interfacial friction, which would have the advantage of eliminating the parameter of slip actually used of giving the role of dependent variable to the void fraction, α. It is also possible that the influence of the longitudinal conduction should be taken into consideration in view of the very agitated nature of the flow and the large slope of the temperature profile. Moreover, with respect to this consideration, we would utilise the assumptions made in chemical engineering concerning back mixing in direct contact exchangers.

REFERENCES

1. J. M. Delhaye, La Houille Blanche, 5, 559 (1967).
2. P. Vernier, J. M. Delhaye, Energie Primaire, 4, 1—2 (1968).
3. G. Barois, J. Huyghe, Etude expérimentale de l'autovaporisation d'un écoulement ascendant adiabatique d'eau dans un canal de section uniforme, Colloque EUROMECH № 7, Grenoble, 23—26 Avril 1968.
4. D. C. Leslie, A.E.E.W., R 505, 1966.
5. A. N. Dickson, R. S. Silver, Desalination, 2, 175—195 (1967).
6. J. Huyghe, G. Barois, A. Michel, H. Mondin, Desalination, 4, 209—219 (1968).
7. J. M. Delhaye, Equations fondamentales des écoulements diphasiques C.E.A. R 3429 (to be published).
8. J. M. Delhaye, C.R.A.S., Série A, t 267, 1968, p. 660.
9. O. Ahmed, F. G. Hammitt, The University of Michigan: College of Engineering, report 08 466—4—I; 07 738 — Février 1968.

MAXIMUM SUPERHEATING OF LIQUIDS IN FLASHING AND BOILING

M. STEFANOVIĆ

Boris Kidrič Institute of Nuclear Sciences, Vinča, Yugoslavia

1. Introduction

The boiling process has been studied very intensively in the course of the last two decades. By analyzing these studies we can state that the main problem in the investigation of boiling is the influence of the heated surface i.e. the process of the establishment of the bubble generation centers. Considering the influence of the heated surface and the process of establishing the bubble generation centers (nucleation) there is a lot of contradictory opinions. It is known little about the physical essence of this process.

Several years ago in The Heat Transfer Department of the Boris Kidrič Institute the researches of the heated surface influence on boiling started. For this purpose the boiling of a number of liquids on the mercury surface has been studied, which could be considered as nearly ideal smooth surface in a careful experimental process.

By these experiments the answer for the problem of the influence of the heated surface on boiling has not been found. One came to the useful conclusion that the roughness of the heated surface does not influence essentially boiling and that its influence was overestimated and in order to explain the process of the establishing bubbles generation centers is not to be searched in the geometry of the heating surface and in the gases absorbed in it, but in the processes of heat or energy transport from the heating surface to the boiling liquid and in the stability of the superheated liquid. On the basis of these conclusions one began the study of the maximum superheating of the liquids. The aim was to study teoretically and experimentally the maximum superheating of the liquids, and to determine on the basis of this in which way and how the system parameters and their changes influence the superheating of liquids, that can be achieved. Of special interest are the systems with heating surfaces on which the boiling process is carried out and they were treated particularly. This paper is a short synthesis of the whole methodology by which this problem is to be solved, i.e. it explains in a more detail what in the introduction has been said and proves it by experimental results.

2. Theoretical Approaches to the Study of the Maximum Superheating of liquids

It is often used the method by which is analised the liquid phase stability on the basis of general thermodynamical balance conditions in order to explain the achieved maximum superheatings of liquids.

The theory of the thermodynamical balance was developed by Gibbs, generalizing Lagrange's principle of the virtual displacement in mechanics. According to it, the unbalancing of the system is a result of virtual deviations of the internal parameters of the system from their balance values. These parameters are correlated as follows:

$$f_S(b_1, b_2, b_3) = 0 \qquad (s = 1, \ldots k \leqslant m) \qquad (1)$$

The changes of the internal parameters that enable these connections — virtual displacements satisfy the equations:

$$\sum_{i=1}^{m} \frac{\partial f_S}{\partial b_i} \delta b_i = 0 \qquad (2)$$

On the basis of these relations and general conditions of the system balance in different cases there are particular conditions of balance for the corresponding system.

It is not difficult to show, starting from the basic thermodynamic equations for nonstatic processes (in the case of the isolated system $TdS > dU + pdV$), that the general conditions of the system balance are being determined by extreme values of thermodynamic potentials. For the isolated system maximum of entropy is $\delta S = 0$, $\delta^2 S < 0$; for the system of the constant volume in the thermostat the minimum of free energy $\delta F = 0$; $\delta^2 F < 0$, and so on.

Searching particular conditions of balance for the one component single phase system, i.e. using the general conditions of the system balance and Eqs. (2), the conditions are obtained

$$C_A > 0; \quad \left(\frac{\partial A}{\partial a}\right)_T < 0 \qquad (3)$$

which are expressed considering the stability of the liquid phase as:

$$\left(\frac{\partial P}{\partial V}\right)_T = 0$$

In the case the boundary of the stability of the liquid phase balance condition observing some continual changes (without phenomenon of the vapor phase) is determined with the relation

$$\left(\frac{\partial P}{\partial V}\right)_T = 0 \qquad (5)$$

In order to use this condition it is necessary to know the equation of state.

If we analyze this method for the determination of the possible superheating of the liquid phase, it is necessary to conclude that by means of it the systems in which there are small fluctuations — balance systems are treated, though the state is being determined by means of the temperature and external system parameters »a_i«, i.e. by them the internal system parameters are determined »b_i«. In the case of greater fluctuations, the internal parameters are not any more the functions of the external parameters and temperature, but this unbalance state is to be defined by supplementary independent parameters (external force fields, adiabatic partitions) which gives the possibility that an unbalanced system should be considered as a balanced one.

Beside that, in the case in treating the balance system it is necessary to use the equation of the state in order to determine superheating of the liquid phase, but most of the state equations have a series of deficiencies that cause faults in use of this method.

A more complete approach to the examination of the maximal superheating of liquids is that based on the premises of the kinetic theory. Here the phenomenon of the steam bubble is considered as a possible fluctuation in the superheating metastable liquid and the probability of the phenomenon of the fluctuation. Therefore, the free energy necessary for the generation of the bubbles of a fixed diameter r is determined, taking into account the action of the surface forces

$$\Delta F = \frac{4\pi}{3} \left(\mu_2 - \mu_1 \frac{r^3}{v_2} + 4\pi r^2 \sigma \right) \tag{6}$$

and observing the probability of the bubble generation of the diameter r in the statistic balance system through the Gibbs' function of the distribution of the particles in the molecular forces field

$$N_g = C e^{-\frac{\Delta F}{KT}} \tag{7}$$

The kinetic theory, adopting this distribution, treats the system in the stable balance as a system in which the number of bubbles with g molecules, that are turned to the bubbles with $g+1$ molecule, is the same as the number of bubbles with $g+1$ molecule, that is turned to bubbles with g molecules, that is

$$N_g \cdot \alpha_g \cdot S_g = N_{g+1} \cdot \beta_{g+1} \cdot S_{g+1} \tag{8}$$

the system in the metastable balance is treated as a system in which

$$N_g \cdot \alpha_g \cdot S_g > N_{g+1} \cdot \beta_{g+1} \cdot S_{g+1} \tag{9}$$

i.e. there is a flow of bubbles with g molecules to the bubbles with $g+1$ molecule and when we express it through the function of the distribution of bubbles in a unbalanced state f_g:

$$I_g = f_g \cdot \alpha_g \cdot S_g = f_{g+1} \cdot \beta_{g+1} \cdot S_{g+1} \tag{10}$$

The kinetic theory considers the stationary process i.e. presumes that the bigger bubbles go out of the system and the system mass is being kept constant by the addition of the liquid phase.

If β_{g+1} is eliminated from Eqs. (8) and (10), we get

$$I_g = N_g S_g \alpha_g \left(\frac{f_g}{N_g} - \frac{f_{g+1}}{N_{g+1}} \right) \tag{11}$$

At the same time we can write that the change of the number of the bubbles of a fixed class is equal:

$$\frac{\partial f_g}{\partial t} = I_g - I_{g+1} \tag{12}$$

This is in fact the equation that Beker and Döring got.
Solving it as it is written with the final differences Döring got the relation that shows what is the probability that in the unit of the liquid content arises a bubble $(J=1)$:

$$J = N \left[\frac{6\sigma}{(2+P/P_S)\pi m} \right]^{1/2} e^{-\frac{\lambda}{KT} - \frac{16\pi\sigma^3}{3KT(P_S-P)^2}} \tag{13}$$

This relation is often used, but the papers of Zeldovich and Frenkel mean by all means a step forward in the solution of this problem. From the final differences they passed to the differential calculus where Eq. (11) gets the form:

$$I_g = -N(g) \cdot D(g) \frac{\partial}{\partial g} \left[\frac{f(g)}{N(g)} \right], \tag{14}$$

where $D(g) = \alpha_g S_g$ \hfill (15)

and Eq. (12) passes to:

$$\frac{\partial f_g}{\partial t} = \frac{\partial I(g)}{\partial g} \tag{16}$$

Solving these equations together with Eq. (7) they came to the differential equation, that is the basic equation of the boiling kinetics of Zeldovich and Frenkel:

$$\frac{\partial f_g}{\partial t} = \frac{\partial}{\partial g} \left(D_g \frac{\partial f_g}{\partial g} \right) + \frac{1}{KT} \frac{\partial}{\partial g} \left(f_g \frac{\partial \Delta F}{\partial g} \right) \tag{17}$$

Taking into account that D_g is slightly dependent on g compared with fg and that the process is stationary $\frac{\partial f_o}{\partial T} = 0$, i.e. $I_g = I = $ const. Zeldovich found a simple method of solving this equation. Further it is not difficult to get an equation similar to that of Döring:

$$I \cdot g_{KP} = J = 2 N r_{KR}^2 \frac{P_S}{K\pi} \sqrt{\frac{\mu_1 - \mu_2}{3m}} g_{kr} \cdot e^{-\frac{4\pi\sigma r_{kr}}{3KT}} \tag{18}$$

Analysing these premises it can be concluded: the relations (13) and (18) have been got when the free energy necessary for the generation of the critical bubble is incorporated into the relation (7), i.e. the bubble that is in equilibrium with the surrounding liquid. It is clear that smaller bubbles are not viable, i.e. that they arise and disappear soon due to the fluctuations, and those bigger than the critical one stay on and grow to the size at which they withdraw from the system. That means that the bubble is a spherical one and that the pressure in it is dependent only on the diameter and that there is no pressure fluctuation in the bubbles of the same size and the stability of these bubbles is defined as with the relation (6). Likewise, as Kagan exposed it, the viscous and inertial forces of the liquid were not taken into account, as well as the process of the heat transport in the surroundings of the bubble. Rusanov showed that principally into consideration may be taken also the bubbles smaller than the used one, and Kagan gave a process with which into consideration are taken also the forces that were neglected so far. However, these contributions have not found any practical application, except that by suitable choice of the bonudary and starting conditions. It was shown that the Kogan's relation in these condition is reduced to the Döring's relation that is generally used. By changing the boundary conditions that would reflect more really the actual systems, i.e. taking into account the convective movements in the system, the heat transfer processes in it and specially the process in the boundary layer on the heating surface, where from the liquid is boiling many problems of the boiling theory could be cleared up, and especially the dependence of superheating of the heating surface by the boiling heating flux.

Here only the comparisons of the existing analyses with the experimental results will be exposed and it will be tried to conceive in which way the above said should be done:

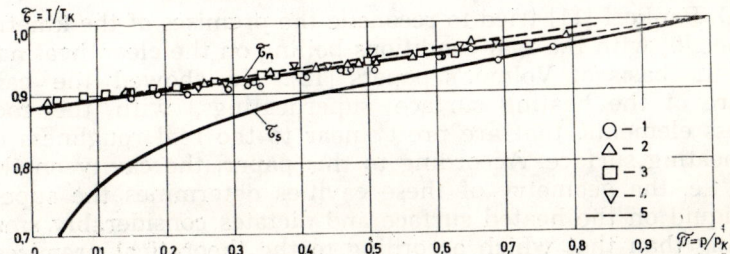

Fig. 1. Maximal superheating temperature of liquid VS generalized thermodynamical coordinates.

1. n-pentane; 2. n-hexane; 3. n-heptane; 4. ethyl-ether; 5. experimental results from Boris Kidrič Institute — n-heptane, n-hexane.

It is said that so far the Döring's relation mainly has been used for the comparison with the experimental results. Scripov [3] showed that superfeating obtained in this relation is in excellent agreement with the overheating obtained on the basis of the analysis of the liquid phase stability if the used state equation is good. On the Fig. 1 the saturation curve and the superheating curve obtained in the above exposed theoretical relations are given in thermodynamic coordinates. Experimental results (21) of liquid drops superheating in the column of the heated liquid and the results of the superheating researches in the bulk for N-heptan and

N-hexan (18) are shown on Fig. 1. It is to be noted that with these experimental results are in accordance also the results in papers [4] and [6] in which are also examined the superheating of liquid drops.

On the basis of this we can conclude that the results of the experiment are in good agreement with the theoretical superheating obtained on considerably idealized conditions. In the case of fluid drops there is the surface tension on the boundary of drops, the heat transport into the drop and also the drops are of very small dimensions, while in the case of the superheating of a bulk of liquid (18) there was also convective flow in the liquid with temperature graedients that should not be neglected. However, the neglection of deviations from idealized theoretical conditions cannot surpass certain limits, which are shown in the best way by boiling from the heating surfaces where it is impossible to realize such big superheatings. Where are these limits, which deviations are important and in what size, is to be solved in the future.

3. Liquid Superheating at Boiling on Heating Surfaces

The results of the examinations of liquid superheating at boiling on solid or liquid metal heating surfaces show considerably smaller superheatings than the superheatings obtained on the basis of these two theoretical approaches. Therefore R. B. Dean [10] concludes that W. Döring's kinetic theory has a very small practical application. The fact that the conditions at which the kinetic theory could be applied without any limitation are practically unrealizable.

S. G. Bankoff [11] tried to reconcile the premises of the kinetic theory of Fischer [9] with the real conditions boiling on the clean heating surface and on the basis of Volmer's papers [12]. He showed the calculation procedure of the heating surface superheatings, with the models of roughness elements, that are pretty near to the real roughness elements of the heating surface. According to this paper, the cavity on the rough surface, i.e. the geometry of these cavities determines the superheating of the liquid on the heated surface and dictates considerably smaller superheating than that which according to the theoretical premises can be obtained in liquid without any contact with the heating surface or on a smooth surface. On the basis of this work, the influence of the absorbed air on the decrease of the superheating with boiling seems to become clear.

Bankoff starts from Volmer's expression [12] for the reversible work of the steam phase spherical sector creation on a smooth solid surface in the case of the transition from the liquid phase into a steam one:

$$W = \left[\frac{4\pi r^2 (\mu_2 - \mu_1)}{3v^2} + 4\pi r^2 \sigma \right] \left(\frac{2 + 3\cos\theta - \cos^3\theta}{4} \right) \qquad (19)$$

For the bubble of a critical size this expression obtains the form:

$$W_{max} = \frac{16\pi\sigma^3}{3(P_S - P)^2} \cdot \frac{2 + 3\cos\theta - \cos^3\theta}{4} = \frac{16\pi\sigma^3}{3(P_S - P)^2} \cdot f(\theta) \qquad (20)$$

Using the already exposed assumptions of the kinetic energy theory and Clausius-Clapeyron's equation, Bankoff comes to the overpressure in the bubble respectively to the superheating of liquids up to which the nucleation does not start:

$$\Delta P = \frac{\rho_L}{\rho_L - \rho_G} \cdot \frac{16\sigma^3 \cdot f(Q)\pi}{3\,KT\ln(6N^{2/3}KT/h)} \qquad (21)$$

As the real conditions of wetting for most combinations of liquids and solid surfaces $\Theta < 90°$ and out of this expression great overpressures or superheatings are obtained so is the probability of bubbles generation on a solid smooth surface very small.

Bankoff further on the basis of these expressions mathematically analyzes the probability of the bubbles generation on different shapes of roughness on the heating surface and comes to the conclusion that according to the superheatings, which are experimentally stated, the nucleation centres can be only the slightly wetted conical cavities and cracks in heating surface. The openings of these cavities and cracks dictate the size of the critical bubble, as well as the superheating, at which the nucleation starts. It can be concluded that roughness dictates the superheating.

However the boiling at the mercury surface with relatively small superheatings [1] and at the electrochemically polished surfaces [13], on which by electronic microscopes roughness elements were not noticed that could be the centres of nucleation, according to Bankoff's assumptions, indicates that the agreement of the Bankoff's assumptions with the results obtained at boiling on solid surfaces has the elements of fortuitousness.

P. Griffith and J. Wallis [14] carried out the experiments of boiling water on a copper surface with artificially made nucleation centres. They obtained superheating of 2°C only, 5°C lower than the superheating (7°C) at which on the mercury surface the first nucleation centres were made active without any roughness elements [1]. The authors explain the increase of the number of nucleation centres by the activation of centres on roughness element of smaller and smaller dimensions and they state that the whole copper surface is covered with nucleation centres at the superheating of 18°C, while the same can be said for the boiling on the mercury surface at the superheating of 24°C.

This leads to the conclusion that Bankoff, as well as many other researchers, overstresses the influence of the heating surface roughness.

The problem of slight superheating on solid heating surfaces is explained by many authors with absorbed air or some other gas in liquid state. But C. W. Gates and P. H. Sabersky [15] showed that by degasing a platinum wire (from which water was boiled) the superheating of the wire is increased only by 5.5°C. This increase of the superheating disappears after a shorter boiling. Dean in his work [10] expresses the opinion that the absorbed gas makes easier the appearance of the bubble, but that such nucleation centres are not steady and have no practical value. It can be concluded also on the basis of the work [15] if the achieved superheating increase of 5.5°C is compared with the theoretical superheating which (for water on atmospherical pressure) is calculated to amount of about 200°C.

Dean [10] explains the appearance of the bubbles, with relatively slight superheatings in real conditions, by means of the pressure decrease in the vortex centres created by the liquid movement. But when we take into account that the liquid in the vicinity of the vortex axis rotates as a solid body, the contribution of the vortex liquid movement to the decrease of the superheating would be comparatively slight and of the same value as the contribution of the roughness and gas absorbed.

Many authors have noticed or experimentally proved that some physico-chemical qualities of material in contact with boiling surfaces have considerable influence on the superheating at boiling.

Thus, for example, J. C. Pease and L. R. Blinks [16] state the influence of the intermolecular forces on the cavitation, i.e. that weak inter-molecular forces between wall and liquid make the cavitation easier. They realize as well that the surfaces with unprotected non-polar molecules make the cavitation easier as well as the solid crystal surfaces. However, they stated that the surfaces with unstable boundaries (crystals that are melting) make the cavitation more difficult.

E. N. Harvey and his collaborators [17] stated a more difficult phenomenon of cavitation when the liquid wets the surface than when it does not do it.

The authors of the paper [15] suppose that a material of solid surface can overtake such qualities that make possible that the bubble nucleus generates on the surface with considerably lower energy. They state that the knowledge about the surface energy is not sufficient for the nucleus arising to be explained.

The fact that at the research of the drop superheating in sulphur acid by the same experimental procedure, about four times stronger liquid superheating have been achieved than in Wood's metal [5], shows that metals influence the decrease of the overheating.

Wishing to investigate experimentally the statement about the influence of metals on the superheating in Heat Transfer Department of the Boris Kidrič Institute. boiling experiments of N-heptan and N-pentan from the sulphur acid surface have been carried out. The experiments were carried out with the same equipment as that used for work [1], with the exception that, instead of mercury, sulphur acid has been used as a heating surface; other conditions of the experiment remained unchanged.

N-heptan started to boil only at the superheating of $\Delta t = 56°C$ at the heat flux $= 1.83 \cdot 10^4 \frac{k\,cal}{m^2}$, while with the same heat flux from the mercury surface it started to boil at the superheating of $\Delta t = 26°C$.

N-pentan started to boil from the sulphur acid surface only at the superheating $\Delta t = 59°C$ at the heat flux $q = 1.35 \cdot 10^4$ kcal/m², while with the same heat flux it started to boil from the mercury surface at $\Delta t = 26°C$.

4. Conclusions

It can be concluded that the influence of the roughness, absorbed air and vortex on the boiling (superheating at the boiling) was overestimated, that mild convective streams and temperature gradients in the system do not cause any deviations from the results that the kinetic theory has given starting from idealized conditions. However, intensive convective streams and great temperature gradients as appear in the boundary layer beside the heating surface show already a considerable influence. These conditions may be comprehended by an analysis, such as Kagan [20] made. Besides, the influence of metal and solid surfaces on the boiling shows that there are some other factors, too, influencing the process in the boundary layer, which up to the present date have not been examined. It is supposed that, by applying methods of thermodynamical non-reversible processes, the thermo-dynamical system could be better defined, on which one should apply the methods of the kinetic theory with a much rigorous definition of the boundary conditions and that in this way could come to the results that would be in better agreement with experimental results.

Acknowledgement

The author wishes to express his thanks during the work to Dr Naim Afgan for useful advice and suggestions.

NOMENCLATURE

Ai	generalized force
ai	external parameters
bi	internal parameters
F	thermodynamic potential
g	number of molecules in bubble
g_{kr}	number of molecules in critical bubble
h	Planck's constant
k	Boltzmann's constant
m	molecular weigth
N	Avogadro's number
n	number of bubbles in system
n_g	number of bubbles in g molecules
P	pressure
P_s	saturation pressure
P_k	critical pressure
q	heat flux
r	radius of bubble
r_{kr}	critical radius of bubble
S	entropy
S_g	bubble surface area
T	temperature
T_s	boiling temperature

T_k	critical temperature
t	time
U	internal energy
V	volume
v	molar volume
W	reversible work of isothermal vapor
α_g	mean number of liquid molecules, evaporated for 1 sec/cm²
β_g	mean number of vapor molecules condensed for 1 sec/cm².
Θ	contact angle
λ	molar latent heat of vaporisation
μ_1	chemical potential of liquid
μ_2	chemical potential of vapor
Π	$=P/P_k$
ρ_G	vapor density
ρ_L	liquid density
σ	surface tension
τ_n	$=T/T_k$
τ_s	$=T_s/T_k$

REFERENCES

1. M. Novaković, M. Stefanović, Int. J. Heat Mass Transfer, 7, p. 801—807 (1964).
2. M. Novaković, Lj. Jovanović, M. Stefanović, N. Ninić, Proc. Third Int. Heat Transfer Conf., vol. III p. 212—219, Chicago, 1966.
3. V. P. Skripov, Teplo i massoperenos, vol. 2, p. 60 Minsk, 1960.
4. H. Vakeshima, K. Takata, J. Phys. Soc. Japan, 13, 678, (1958).
5. Contacts with Leningrad, politech. Institute.
6. S. Lazić, Mag. debree thesis; Mechanical Engineering Faculty of Beograd, IBK—444, 1966.
7. F. B. Kenrick, et. al., J. Phys. Chem., 28, 1308 (1924).
8. W. Döring, Z. F. physik Chemie, 36B (1937); 38B, 292 (1937).
9. J. C. F'scher, J. Appl. Phys., 19, 1062 (1948).
10. R. B. Dean, J. Appl. Phys., 15, 5, p. 446 (1944).
11. S. G. Bankoff, Trans. ASME, 79, 4, 735 (1957).
12. M. Volmer, Z. f. Elektrochemie, 35, 555 (1929).
13. H. B. Clark et al., Chem Eng. Progr. Symp. Series, 55, 29, p. 103 (1959).
14. P. Griffith, J. Wallis, Chem. Eng. Progr. Symp. Series, 56 (90), 49 (1960).
15. R. H. Sabersky, C. W. Cates, Jet Propulsion, 25, 2, 67 (1952).
16. D. C. Pease, L. R. Blinks, J. Phys. Coll. Chem., 51, 2, p. 556 (1947).
17. E. N. Harvey et al., J. Appl. Phys., 18, 2, 162 (1947).
18. M. Stefanović, N. Ninić, Teplo i massoperenos, Vol. 2, Minsk, 1968.
19. E. I. Nesis, UFN, Vol. 87, vip. 4, p. 615 (1965).
20. Iu. Kagan, Zh. F. H., 34, 1, p. 92 (1960).
21. V. P. Skripov, G. V. Ermakov, Zh. F. H., 38, 2, p. 396 (1964).
22. A. I. Rusanov, »Fazovie raznovesiya i poverhnostn'e ravlenia«, Izd. »Himia«, Leningrad, 1967.

THE MECHANISM OF NUCLEATE BOILING HEAT TRANSFER

Lj. JOVANOVIĆ and M. STEFANOVIĆ

Boris Kidrič Institute of Nuclear Sciences, Beograd — Vinča, Yugoslavia

In order to study the nucleation process in boiling heat transfer, a photographic investigation was made of nucleation from a mercury surface maintained as clean as possible, thus avoiding the influence of surface roughness. In the first stage the heat transfer process was investigated for four different liquids and the results are correlated [1, 2]. The further investigation was an extension in order to study the kinetics of vapor bubbles.

1. Results and Discussion

The previous obtained boiling heat transfer curves for water and ethyl alcohol are given in Fig. 1. with fluxes on which photography was made. In previous reports [1, 2] is shown that the first part of the curve cover the region of convective boiling and after the transition the region of nucleate boiling follows.

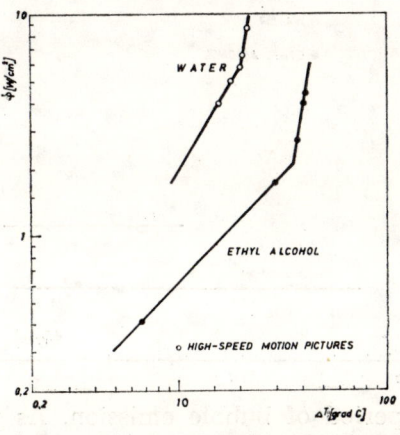

Fig. 1. Boiling heat transfer curve.

The number of nucleation centres versus heat flux is given in Fig. 2. In the region of convective boiling the number of nucleation centres is nearly constant. After the transition the number of nucleation centres rapidly increase with the increasing heat flux. On the diagram are given as well the experimental results of Cumo et all [6] and Gaertner [7]. Cumo and the others boiled from the wire. Gaertner boiled from a smooth metal surface and his results are in good agreement with ours.

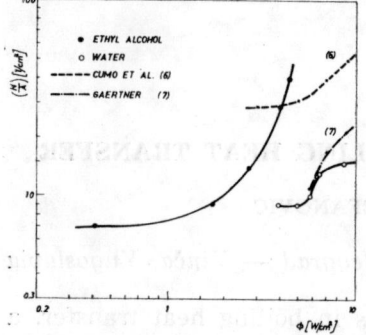

Fig. 2. Variation of bubble source concentration with heat flux.

Bubbles of various diameter size are formed at varying frequency from the same nucleation centre and at the same heat flux. Therefore, it is necessary to make many measurements of the measured magnitude and because of its statistical nature to form a distribution curve. For analysis the most probable and mean values of measured magnitude, taken from distribution curves, were used [3, 4]. The most probable bubble diameter at departure versus heat flux is given for water and alcohol in Fig. 3. For alcohol, bubble diameter at departure calculated by Fritz relation is in good agreement with our results. For water, because of the strong dependence of diameter at departure of heat flux our experimental results diverge strongly from Fritz relation.

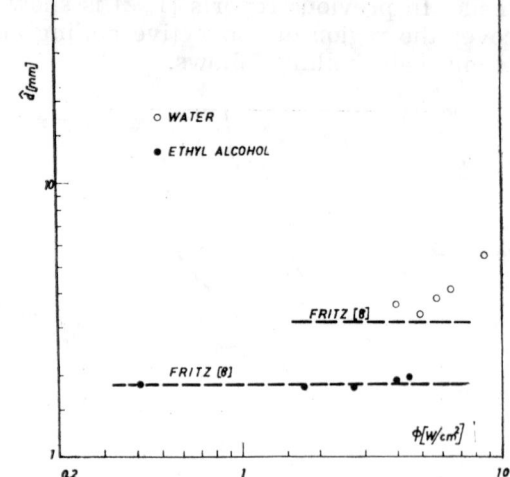

Fig. 3. Most probable bubble diameter at departure as a function of heat flux

A great attention is paid to the period of bubble emission. As the other investigators noticed too, the period of emission consists of delay time and growth period of bubble. Our results show that the mean delay time depends very much of heat flux, i.e. with the increasing heat flux the delay time decreases. The growth period is approximately constant and independent of the heat flux (Fig. 4). Bobrovich and Mamontova [13] (for heat fluxes over 20 W/cm² and boiling from a horizontal surface) have found that the frequency of emission is constant and independent of heat flux (there nearly does not exist delay time) and the bubble diameter at departure increases with increasing heat flux. It can be seen that the period of bubble emission depends on heat flux as the delay time depends on it. Jakob's assumption that the delay time is equal to the growth period, agrees only for low heat fluxes (less than 1 W/cm^2 for alcohol).

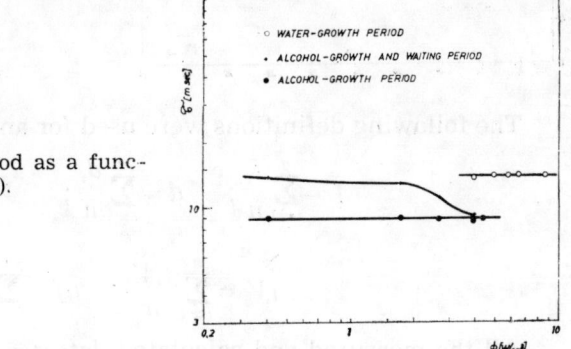

Fig. 4. Growth and waiting period as a function of heat flux (4).

It is obvious that the delay time is connected with the thermal boundary layer formation after the bubble departure from the heating surface [5]. It is noticed in our experiments that bubbles on the periphery of heating surface have a small and irregular frequency [3, 9]. This is because of the great delay time which comes out as the result of surrounding colder liquid.

2. Relation Between Bubble Frequency and Bubble Diameter at Departure

The product of bubble frequency and bubble diameter at departure enters into many analyses of nucleate boiling. In early investigations it came out that the diameter at departure and frequency product is constant and independent of heat flux [8]. In further investigations the same results are obtained. But the recent investigations show that it perhaps can be only for a small region of fluxes. Conclusion that »at fixed flux (and pressure) the product $f_i \cdot d_i$ is the same for each bubble source within reasonable statistical scatter« [10] seems invalid as well (Fig. 5). A number of investigators tried to correlate d_i and f_i. There is a great number of factors which influence the boiling, and there are some problems in the definition of the mean values of the variables used in product. In some cases a limited number of bubbles, choosen freely, were used in analysis. In our investigations are used about 200 data per one measured flux. One film took place for less than one second, and due to this on every flux five films were taken in order to avoid the influence of the time fluctuations. Rallis and Jawurek [10] have pointed out to the difference which appear when for the product $f \cdot d$ is taken the arithmetic mean value for

d_i and f_i and when the product is the arithmetic mean of $f_i \cdot d_i$ for individual bubble. Shortly $\overline{fd} \neq \overline{df}$.

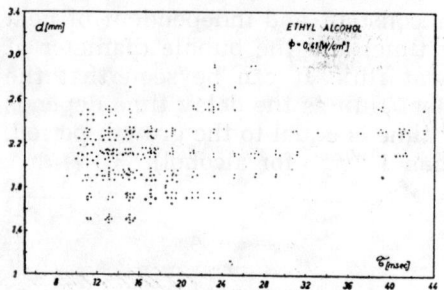

Fig. 5. Bubble diametar versus bubble period for individual bubbles for 0,41 (W/cm²) [4]

The following definitions were used for analyses:

$$\overline{f} = \sum_{i=1}^{n} \frac{f_i}{n}; \quad \overline{d} = \sum_{i=1}^{n} \frac{d_i}{n}; \quad \overline{V} = \sum_{i=1}^{n} \frac{V_i}{n}; \quad \overline{fV} = \sum_{i=1}^{n} \frac{f_i V_i}{n}; \quad \overline{df} = \sum_{i=1}^{n} \frac{d_i f_i}{n} \quad (1)$$

All the measured and calculated data are in Tab. 1 and Tab. 2. There is not the frequency of emission for water because we have not been in position to measure the delay time [4]. The difference for alcohol between $\overline{d} \cdot \overline{f}$ and $\overline{df}$ is bellow 5% and can be taken as scattering of experimental results. The difference is bigger between $\overline{Vf}$ and $\overline{Vf}$ but only for the first and third heat flux where the difference is about 15%. This difference is understandable if the curve distribution of »d« is observed [4]. The curves for the first and third heat flux are very asymetric with a great number of big bubbles influencing the product Vf. The big bubble phenomenon on the first and third flux can be the characteristic of these experiments. It is interesting to point out that disagreement is minimum for higher heat fluxes. From all these the conclusion is that the difference between $\overline{fd}$ and $\overline{fd}$ ($\overline{Vf}$ and $\overline{Vf}$) is small and it should not be the reason of different correlations in the literature. We suppose (between the others), the difficulties arise if the results of boiling from the wire are used for study.

3. Proposed Heat Transfer Mechanism

Different boiling regions in nucleate boiling from a horizontal surface, defined by a number of investigators for boiling from a mercury surface were also distinguished [1, 2, 4]. In [4] we reported about two regions: convective boiling and nucleate boiling region.

In convection boiling, where the convective process is dominant in heat transfer, nucleation centers appear in the »dead« zones of convection [4]. In the »dead« zones the water streams leave the heating surface and the highest superheating may be expected. The vapor generated at the nucleation centres was in the form of isolated bubbles. The energy carried away from a surface by bubbles may be formulated as:

$$q_L = r\rho_v \cdot \overline{fV} \cdot \frac{N}{S} \tag{2}$$

The latent heat transport defined by such way accounts for the heat that leaves the surface by conduction and convection to form bubbles up to their departure. Other part of heat is transferred by natural convection from the heated surface. Increasing the heat flux the flow pattern is not changed, number of nucleation centres are constant and only increased the quantity of heat, transferred by natural convection. In transition when flow pattern changes the number of nucleation centres grows with increasing heat flux. New centres appear near the existing centres and only a few nucleation centres appear outside of these zones in the higher convective current. Occasionally coalescence of bubbles appears above some nucleation centres. In nucleate boiling region, where process of nucleation and its secondary effects are predominant, latent heat transport is dominant on natural convection heat transport. The whole surface is covered uniformly by bubbles. With such an increase in the number of nucleation centers the boiling became more chaotic. The coalescence of bubbles occurred more frequently. The convective currents rising from surface are destroyed.

Many variables in nucleate boiling, specially different nucleating characteristics of heating surface, are the reason for a lot of correlations.

Zuber [14] reported about the similarity of nucleate boiling in the region of isolated bubbles with natural convection. Some investigators [7, 14,] suggested that the latent heat transport played great part in nucleate boiling heat transfer. Rallis and Jawurek [10] found that »latent heat transport and convective together account for the total heat flux in saturated boiling«. In previous report using idea of convective heat transport and latent heat transport sum, good agreement with experimental results is shown [4]. In present study for natural convection correlation $Nu = c\,(GrPr)^n$ was used. Average bubble spacing for characteristic geometrical magnitude is taken and defined as:

$$L = \left(\frac{S}{N}\right)^{1/2} \tag{3}$$

Latent heat transport is calculated using Eq. (2).

The results are reported in Fig. 6 for alcohol. Good agreement is clear in the region of convective boiling which is expected. In the nucleate region the mechanism of heat transfer is more complicated and such simple assumption is not valid. Latent heat transport is predominant in that region. In Fig. 7 change of latent and convective heat transport versus total heat flux is reported. In Fig. 8 per cent of latent heat transport in total heat flux is shown.

For water, this analysis is not done as we could not measure the delay time.

In previous article [2] in which heat transfer is treated and the correlation established the transition from convective to nucleate region is reported that it takes place at $K_p{}^{0.7}\ Pe \approx 2.10^5$. The same conclusion is reached by analysing the relation between latent heat transport and total heat flux which has the minimum at the $K_p{}^{0.7}\ Pe \approx 2.10^5$ too. (Fig. 9). Ex-

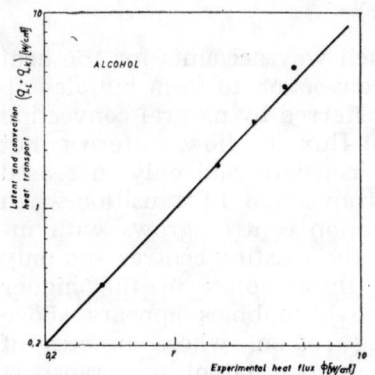

Fig. 6. Latent and convection heat transport versus experimental heat flux.

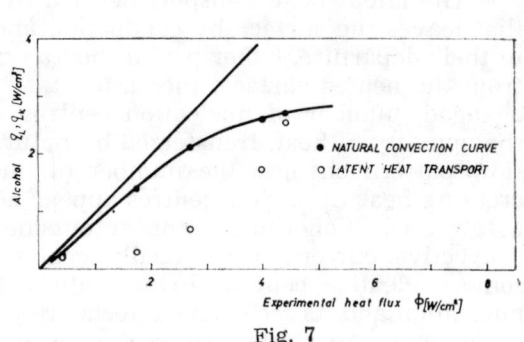

Fig. 7

planation for such behaviour is the following. At the beginning the heat is transported only by convection. When the boiling started, bubbles carried away a lot of heat. Increasing heat flux increases the heat transferred

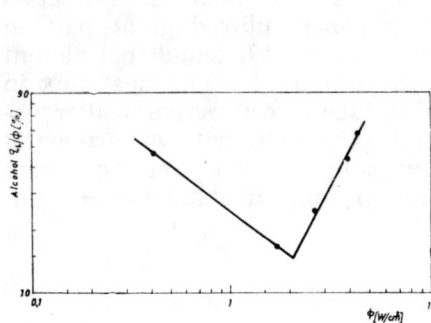

Fig. 8. Latent heat transport as percentage of measured heat flux

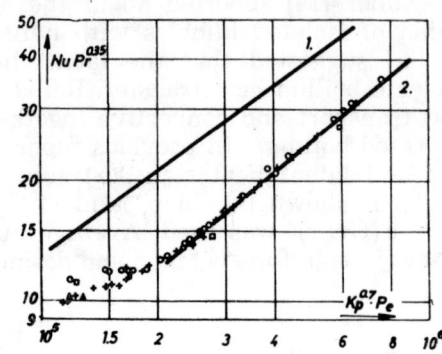

Fig. 9. D'mensionsless correlations for boiling (1). Experimental results for boiling from mercury surface (Novaković and Stefanović). + water; ○ ethyl alcochol; △ benzene; ☐ pentane.
1. Kutateladze's dependence; 2. Correlation Novaković and Stefanović (2)

by convection. The part of heat carried by the bubbles is nearly constant up to the change of flow pattern, because of the constant number of nucleation centres and constant bubble diameter at departure and slow increase of bubble diameter. After the transition the latent heat transport is dominant in total heat transport.

4. Conclusion

The experimental results support the assumption that the heat transport rate in the region of convective boiling can be attributed to convection and latent heat transport.

Latent heat transport is significant for measured heat fluxes in nucleate boiling.

Transition from the region of convective boiling to nucleate boiling region is defined for $K_p^{0.7} Pe \approx 2.10^5$.

The average bubble growth period is approximately constant troughout both boiling regions and independent of the heat flux.

The average product of bubble frequency and departing volume $\overline{fV}$ increases throughout measured heat fluxes.

TABLE 1

Water

Φ w/cm²	3,94	4,99	5,8	6,52	8,61
ΔT grad C	15,7	18,1	20	20,5	21,5
$\overline{d}$ mm	3,53	3,77	4,02	4,14	5,22
$\hat{d}$ mm	3,6	3,33	3,8	4,1	5,45
$\overline{\tau}_{02}$ msec	14,5	14,16	14,7	14,4	14,1
$\hat{\tau}_{02}$ msec	13,5	13,9	13,7	13,75	13,8
$\overline{d} \cdot \overline{f}$ cm/sec	24,3	26,8	27,2	28,8	37,0
$\hat{d} \cdot \hat{f}$ cm/sec	26,7	24,0	27,7	29,8	39,5
$\overline{d \cdot f}$ cm/sec	27,7	—	—	—	37,2
$\overline{V} \cdot \overline{f}$ cm³/sec	1,58	1,985	2,32	2,59	5,3
$\overline{Vf}$ cm³/sec	1,96	—	—	—	5,31

TABLE

Alcohol

Φ w/cm²	0,41	1,74	2,7	3,98	4,41
ΔT grad C	6,7	29,5	37,5	40,3	41
$\overline{d}$ mm	1,925	1,845	1,91	2,03	2,02
$\hat{d}$ mm	1,85	1,78	1,79	1,9	1,93
$\overline{\tau}_{02}$ msec	9,2	9,1	9,1	9,05	9,1
$\hat{\tau}_{02}$ msec	8,85	8,8	8,75	8,7	8,7
$\overline{\tau}$ msec	16,82	14,45	12,0	10,45	9,3
$\hat{\tau}$ msec	13,7	12,8	11,0	9,0	8,8
$\hat{f}$ l/sec	73,0	78,1	90,9	111,1	113,6
$\overline{d} \cdot \overline{f}$ cm/sec	11,45	12,7	15,9	19,5	21,7
$\hat{d} \cdot \hat{f}$ cm/sec	13,5	13,9	16,3	21,1	21,9
$\overline{d \cdot f}$ cm/sec	10,86	12,9	16,0	19,1	21,7
$\overline{V} \cdot \overline{f}$ cm³/sec	0,223	0,227	0,303	0,42	0,464
$\overline{Vf}$ cm³/sec	0,199	0,23	0,348	0,434	0,464

d	bubble diameter at the departure from the heating surface
$\bar{d}$	mean bubble diameter at departure
$\hat{d}$	most probable diameter at departure
f_i	bubble frequency of individual bubble
$\bar{f}$	mean bubble frequency
$\hat{f}$	most probable bubble frequency
τ	bubble period of emission, $1/f$
τ_{01}	delay time
τ_{02}	bubble growth period
r	latent heat of vaporisation
Φ	heat flux
q_L	latent heat transport
q_k	natural convection heat flux
Δt	temperature difference
ρ_v	vapor density
S	heating surface
L	average bubble spacing
V	bubble volume at departure
n	number of bubbles on one heat flux
N	number of nucleation centres
K_p	Kutateladze number
Pe	Peclet number

REFERENCES

1. M. Novaković and M. Stefanović, A/Conf. 28 /P/ 600, Geneva, 1964.
2. M. Novaković and M. Stefanović, Int. J. Heat Mass Transfer, 7, pp. 801 (1964).
3. Lj. Jovanović, M. Novaković and M. Stefanović, Bull. Boris Kidrič Inst. Nucl. Sci., 16, 2, (1965).
4. M. Novaković, Lj. Jovanović, M. Stefanović and N. Ninić, Third Int. Heat Transfer Conf. paper No. 102, Vol. III, Chicago (1966).
5. N. Zuber, A.E.C.U. Report № 4439, 1959.
6. M. Cumo, G. Farello and Pinchera, Third Int. Heat Transfer Conf., paper № 104, Vol. III, Chicago, 1966.
7. R. Gaertner, J. Heat Transfer, ASME 87, 1 (1965).
8. *Voprosi fiziki kipenija*, Moskva, 1964.
9. R. Séméria: Proc. Sym. Two-phase Fluid Flow, London, Feb., 1962.
10. C. Rallis and H. Jawurek, Int. J. Heat Mass Transfer, 7, p. 1051 (1964).
11. R. Cole, AIChE Journal, 13, 4, pp. 779 (1967).
12. H. Ivey, Int. J. Heat Mass Transfer, 10, 1023 (1967).
13. G. Bobrovich and N. Mamontova, Int. J. Heat Transfer, 8, 1421 (1965).
14. N. Zuber, Int. J. Heat Mass Transfer, 6, 53 (1963).
15. Y. Hsu and R. Graham, NASA Report TN-D-594, 1961.
16. F. Moore and R. Mesler: AIChE Journal, 7, 620 (1961).

BOILING HEAT TRANSFER OF THE BINARY MIXTURES

N. H. AFGAN

Boris Kidrič Institute of Nuclear Sciences, Beograd, Yugoslavia

1. Introduction

A series of contemporary technological apparatuses and plants in modern technics operate in conditions of high heat fluxes, so that the need for more intense heat removal from the heating surfaces of these apparatuses has been more and more frequent. Satisfying the present-day technological limitations of modern materials, one of the most efficient way to remove heat from the surface is heat transfer by phase change. This is why an exceptional attention has been for a long time devoted to the study of the mechanisms of boiling heat transfer.

In nuclear reactors this method of heat transport from fuel element surfaces has found wide application. However, it should be pointed out that the upper limit of the permissible heat flux which maw be reached without allowing the superheating of the fuel element has remained an open question. There is no less interest in the use of this method of heat transport in modern space aircraft (flying machines) in which this method of heat removal is also used. Modern electronic equipment in which a large amount of heat is generated in a small space also uses this method of heat removal. The desire for the most compact constructions of the technological apparatuses in modern chemical processes has imposed the need for the study of the mechanism of phase transformation as well as the structure of two-phase flows in different technological operations upon which the efficiency of these apparatuses depends. This has focussed the attention especially of these investigations to the problem of boiling multicomponent mixtures.

The many investigations of the heat transfer mechanism by boiling have to a great extent elucidated this very complex process. Without considering the background of this problem, it is necessary to point out that a complete phenomenological picture of the boiling process is not sufficiently clear in a great number of cases, which has been pointed out by several lectures at this School. Therefore, my wish in this lecture is to try to give little light to the mechanism of boiling binary mixtures. Since this field of research still has not been a subject of extensive development, my attention will be focussed to the basic mechanisms of heat and mass transport in two-phase boundary layer by boiling binary liquids in natural convection.

The dependence of the heat flux on the difference between the temperature of the heating surface and the liquid saturation temperature is called the »boiling curve«. (Fig. 1). The part (AB) of this curve represents

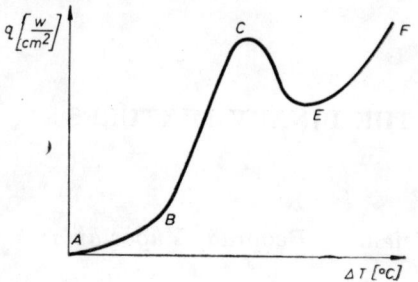

Fig. 1. Typical boiling curve.

the convective transport of heat from the heating wall, while the part (BC) represents buble boiling. This form of heat transport is characterized by the existence of a sufficient number of active nucleation centers, which depends on the superheating of the liquid. Increase in the heat flux brings about increase in the temperature of the wall, which accelerates the formation of bubbles and their mutual interaction. Under these conditions the heat flux in bubble boiling reaches its maximum (C). The heating surface is then covered by a uniform vapour layer. This leads to the thermal isolation of the heating surface and reduction of the heat flux and increase in its temperature. Further increase of the heat flux is followed by a considerable increase in the temperature of the heating surface, which appears as a result of the increased thickness of the weakly conductable vapour layer. The field (CE) is called transition boiling, and the part of the curve (EF) is called film boiling. This descriptive figure of the boiling process is well known and it is presented in order to define the basic ideas which we will discuss later. According to the so far verified experimental facts the boiling curve has the same character in pure liquids and their mixtures.

2. Mechanism of Boiling Binary Mixtures

The mathematical formulation of the growth of spherical bubbles in binary mixtures is possible on the basis of Rayleigh's theory of the development of spherical surfaces in liquid, based on the mechanical equilibrium of forces on the interface. In a general case, this equation has the following form:

$$\frac{P_1+P_2-P_x}{\varepsilon\,\rho'} = \frac{2\,\sigma}{\varepsilon\,\rho'\,R} + R\frac{\partial^2 R}{\partial \tau^2} + \frac{3}{2}\left(\frac{\partial R}{\partial \tau}\right)^2 + \frac{4\,\mu}{R}\frac{\partial R}{\partial \tau} \qquad (1)$$

The heat required for the formation of a new phase is supplied by conducting through a liquid phase. From conditions of heat equilibrium one obtains that

$$\frac{\partial T}{\partial \tau} = \frac{K}{\rho' Cp'}\left(\frac{\partial^2 T}{\partial R^2} + \frac{2}{r}\frac{\partial T}{\partial r}\right) + \frac{Q_o}{\rho' Cp'} \tag{2}$$

On the assumption that mass transfer proceeds only by diffusion we have the following equation:

$$\frac{\partial C}{\partial \tau} = D\left(\frac{\partial^2 C}{\partial r^2} + \frac{2}{r}\frac{\partial C}{\partial r}\right) \tag{3}$$

Besides the fact that spherical symmetry, infinity of the environment, uncompressibility of the liquid, absence of external forces, absence of energy dissipation by viscosity, and the existence of only two components in the mixture are assumed for these three equations, this problem is impossible to solve analytically. Therefore, further simplifications are necessary which will allow analytical solution without loosing the physical characteristic of the process observed.

Based on the assumption that mechanical forces do not exert influence on bubble growth, Scriven [1] obtained an analytical solution of this problem. This means that the influence of the surface tension, the inertial forces and the frictional force is neglected. Regarding that surface tension is greatly important only in the case of very small bubble diameters, the first assumption can be considered permissible. If the bubble growth is not so fast, the assumption about the neglectance of the inertial forces may be considered justified. The friction forces are very small so that they too can be neglected without remarkable effect on the process itself. Mathematical solution of this problem in the case of isothermal bubble growth means that the bubble growth rate depends only on the diffusion rate in the mixture.

In boundary conditions

$$\tau = 0 \quad \text{for} \quad \frac{\partial R}{\partial \tau} = 0$$

$C = C_\alpha$ for all values of r with $\tau = 0$

$C = C_\alpha$ for all values of τ with $r = r_\alpha$

and assuming that the solution of equation (3) has the form

$$R = 2\beta\sqrt{D\tau}$$

one obtains that the coefficient β has the following dependence on density and composition of both phases

$$\emptyset = 2\beta \exp(\beta^2 + 2\varepsilon\beta^2) \int_\beta^\alpha x^{-2} \text{ekp}(-x^2 - 2\varepsilon\beta^3 x^{-1}) dx \tag{4}$$

where $\quad \emptyset = \dfrac{C_\alpha - C_{sat}}{C_L - C_{sat}}$

$\varepsilon = \dfrac{\rho' - \rho''}{\rho''}$

An experimental check on this analytical solution was made for a mixture of water and ethylglycol [2]. Boiling was carried out on the surface with previously defined roughness. Assuming that the bubble diameter is expressed by

$$R = a\tau^n \tag{5}$$

it is possible to determine the constant of bubble growth as a function of the mixture concentration. Figure 2 shows the results of these measurements. The same figure also shows the results obtained with an

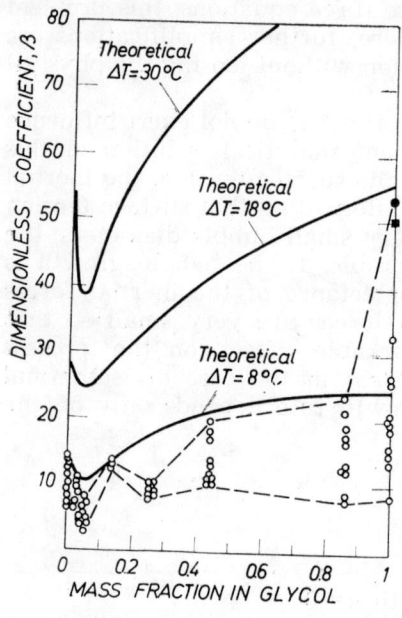

Fig. 2. Experimental values for dimensionless coefficient, for glycol-water mixtures.

analytical expression (4). Although the absolute values differ, the same trend of the curves is obvious. This can be ascribed to the fact that relation (4) does not take into account the mutual influence of heat conductance and mass diffusion and that it is derived for an infinite medium and uniform superheating.

3. Formation of Nucleation Centers by Boiling Binary Liquid Mixtures

It is known that roughness of the surface on which boiling proceeds affects the number of nucleation centers and the intensity of heat removal from the heating surface. According to the theory of boiling one-component liquids the process of phase formation begins in the microscopic pores of roughness. The radius of curvature of the smallest roughness pore which can affect the center for bubble formation depends on the superheated liquid.

The smallest radius of curvature which can act as the nucleation center for pure liquids is given by the expression

$$R_o = \frac{2\tau}{p'\Delta T} \qquad (6)$$

where

$$p' = \left(\frac{dp}{dT}\right) = \frac{r\gamma'\gamma''}{\Delta T(\gamma'-\gamma'')}$$

This expression can also be used to analyze the roughness effect even in binary mixtures where the pressure is a function not only of temperature but of the concentration of the components in the mixture as well.

Taking into account Van der Walls' equations for binary mixtures, Grigoriev [3] derived an equation for the radius of the minimum roughness pore for binary mixtures

$$R_o = \frac{2\sigma}{\left[\dfrac{s''-s'}{v''-v'} + \left|\dfrac{x''-x'}{v''-v'}\left(\dfrac{\partial^2\mu}{\partial x^2}\right)_{PT}\dfrac{dx}{dT}\right|\right]} \qquad (7)$$

The qualitative analysis of this expression shows that R_0 depends first of all on the complex

$$\frac{x''-x'}{v''-v'}\left(\frac{\partial^2\mu}{\partial x^2}\right)_{PT}\frac{dx}{dT}$$

If this one is positive, then with increasing $(x''-x')$, R_0 will decrease. The number of centers for bubble formation will increase, because more and more pores become potential nuclei for the nucleation. If this complex is negative, then with increasing $(x''-x')$, R_0 increases and the number of nucleation centers decreases. The sign in front of this complex depends on the sign in front of $\dfrac{x''-x'}{v''-v'}$ and $\dfrac{\partial^2\mu}{\partial x^2}$. For states far from the critical, $(v''-v')$ is always positive, while $(x''-x')$ can be both positive and negative, depending on the mixture. For example, for mixtures having an azeotrope, for concentrations on one side of the azeotrope it is positive and on the other it is negative. The term in front of the expression $\dfrac{\partial^2\mu}{\partial x^2}\dfrac{\partial x}{\partial T}$ depends on the heat of evaporation.

$$\left(\frac{\partial^2\mu}{\partial x^2}\right)_{PT}\frac{dx}{dT} \cong \frac{h_4 - h_2}{T}$$

The component 1 represents the component whose boiling temperature has a lower value. For most of the mixturaes the sign in front of this term is negative.

4. Heat Transfer by Boiling

As distinct from pure liquids for which we have a rich experimental material on boiling heat transfer, the extent of the so far investigations of binary mixtures has been much less, and not until recently has more work been done along these lines. This is why the present approach to this problem is mainly based on the qualitative description of the process, i.e. attempts to find a similarity between the assumed physical model and the obtained experimental results. Taking into account the basic mechanisms of mass and heat transfer by boiling binary mixtures presented earlier it is possible to distinguish two phsical models of this process. The first one is based on the assumption that the diffusion process on the boundary between the phases defines the dynamic parameters of vapor phase formation. The second model is based on the assumption that the change of the minimum roughness pore radius depends on the concentration and the change of $\dfrac{\partial P}{\partial T}$ defines the structure of the two-phase boundary layer.

In order to get an insight, on the basis of the experimental results obtained so far, into which of the above mentioned models most closely corresponds to the real physical conditions of boiling binary mixtures, we shall deal with the groups of mixtures which help us observe some of the characteristics of mass and heat transport by boiling binary mixtures.

4.1. MIXTURES OF WATER AND ETHYLALCOHOL

Mixtures of water and ethylalcohol make a group which is characteristic for the existence of a positive difference in concentration between the liquid and vapor phase in equilibrium conditions. The boiling tempe-

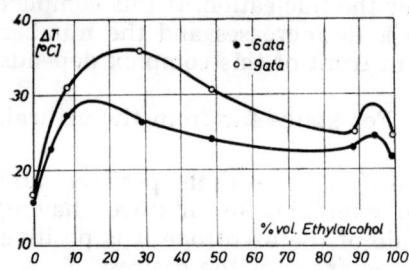

Fig. 3. Water-ethylalcohol mixture $\Delta T = T_w - T_{sat}$ as a function of composition for $q = 20$ w/cm² rough surface

rature of ethylalcohol is lower than that of water so it appears as a more volatile liquid. Figure 3 shows our results of measuring $\Delta T = f(C)$ on boiling these mixtures at a thermal flux of $20\, \dfrac{w}{cm^2}$. The main characteristic of these measurements is the existence of two maxima. If one knows that this mixture has an azeotrope at 89.9% vol. of ethylalcohol, it is obvious that there must be two peaks on the curve $\Delta C = \varphi(c')$. Figure 4 shows this dependence only for the part concerning the positive difference in concentration between the liquid and the vapor phase. The bubble departure diameter is one of the basic characteristics of the two-phase boundary layer on boiling which defines the mechanism of mass and

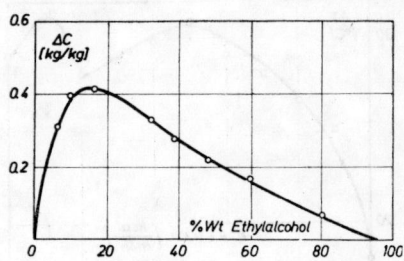

Fig. 4. Water-ethylalcohol mixture ΔC as a function of concentration [4].

heat transport. For this group of mixture Fig. 5 shows the measuring results [4] of the diameter as a function of the concentration of more volatile liquids. Thus it is seen that two pronounced minima exist which correspond to the same concentrations at which there exist maxima of ΔT, i.e. maxima of the absolute differences in concentration between the liquid and vapor phase. All the measurements shown here have been made on a rough surface, so that the existence of nucleation centers is determined by the surface roughness. The radius of the minimum pore of roughness for the azeotropic mixture is equal to the same value for pure liquids of the corresponding physical characteristics. For mixtures whose concentration is greater than the azeotropic concentration, this value is smaller than the corresponding value for ideal mixtures, that is, it is higher with a smaller concentration than the azeotropic. This means that for the mixture water-ethylalcohol the number of nucleation centers for concentrations smaller than the azeotropic is smaller than the number which would correspond to pure liquids having the same physical properties. For concentrations larger than the azeotropic the number of nucleation centers is greater. Our results show that this is true only for concentrations smaller than the azeotropic, while for concentrations larger than the azeotropic it is quite opposite.

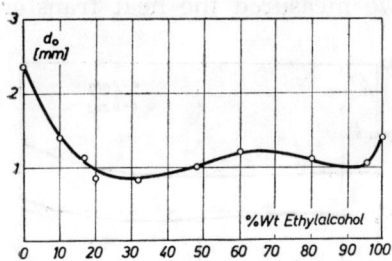

Fig. 5. Water-ethylalcohol mixture d_0 as a function of concentration [4].

In order to make a check as to what extent roughness of the heating surface influences heat transfer by boiling, we measured the heat transfer coefficients on an ideally smooth surface, i.e. on a liquid surface. The results of these investigations for the mixture water-ethylalcohol are given in Fig. 6. It is characteristic in these measurements that there appears both minimum and maximum which are not correlated at all with the difference in concentration between the liquid and vapor phase. These investigations are still under way, so it is difficult to draw any explicit

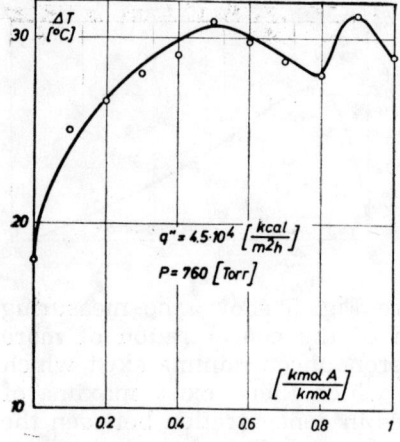

Fig. 6. Water-ethylalcohol mixture $\Delta T = T_w - T_{sat}$ as function of concentration for $q = 4,5 \cdot 10^4$ kcal/m²h, liquid surface.

conclusions at present. However, it may be concluded that the mechanism of heat and mass transport by boiling shows definite characteristics, especially at small concentrations of the components.

The appearance of a pronounced maximum at a concentration which corresponds to the concentration of the minimum of the surface tension implies that surface forces also play an important part in boiling binary mixtures. The existence of maximum at small concentrations of alcohol can be ascribed to the less probability of the formation of nucleation centers.

4.2. MIXTURE OF WATER-GLYCERINE

In order to investigate the boiling of such mixtures that correspond as closely as possible to ideal mixtures, we measured the heat transfer

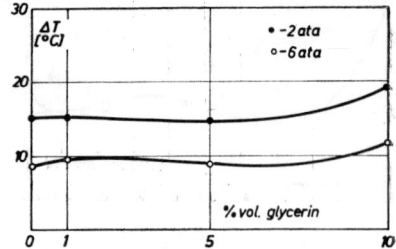

Fig. 7. Water-glycerin mixture $\Delta T = T_w - T_{sat}$ as a functional of composition for $q = 20$ w/cm² rough surface.

coefficient by boiling the mixture water-glycerine. The boiling temperature of glycerine is considerably higher than the corresponding temperature of water, so that water appears as a component with a lower boiling temperature, i.e. as a more volatile component. Figure 7 shows the results of measuring ΔT as a function of the glycerine concentration at $q''=20$ W/cm². These results imply that ideal mixtures do not have any characteristich compared with pure liquids. Particular attention has

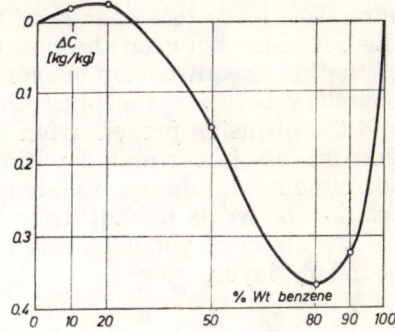

Fig. 8. Ethylalcohol-benzene mixture ΔC as a function of concentration.

been paid to mixtures with low concentrations of glycerine which possess all characteristics of ideal mixtures.

4.3. MIXTURE OF ETHYLALCOHOL-BENZENE

The ethylalcohol-benzene mixture has an azeotrope at a concentration of about 30% vol. of benzene. This characteristic makes this group of mixtures very interesting for the study of the effect of the difference in concentration between the liquid and the vapor phase. Figure 8 shows the curve $\Delta C = \varphi(C)$ for the ethylalcohol-benzene mixture at $p=6$ at [5]. On one side of the azeotrope the difference in concentration between the liquid and the vapor phase is positive, while on the other side it is negative. The results of measuring $\Delta T = f(C)$ for this mixture are shown

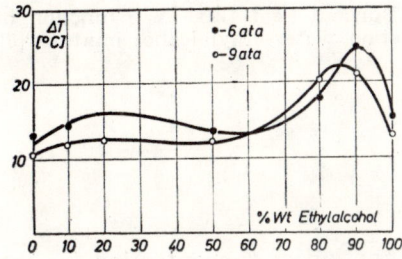

Fig. 9. Ethylalcohol benzene mixture $\Delta T = T_w - T_{sat}$ as a function of concentration for $q = 20$ w/cm² rough surface.

in Fig. 9. The existence of two maxima on this curve whose concentration corresponds to minimum or maximum of the difference in concentration between the liquid and the vapor phase, confirms a similar conclusion drawn for the water-ethylalcohol mixture. The fact is more obvious here that only the absolute value of the difference in concentration between the liquid and the vapor phase exerts influence on heat transfer in boiling binary mixtures. Hence it follows that the diffusion process which depends on the absolute difference in concentration between the liquid and the vapor phase determines the dynamics and structure of the two-phase boundary layer by boiling binary mixtures. This confirms the hypothesis that the evaporation of mere volatile component on the interface of the liquid and vapor increases the concentration of the component in relation to the equilibrium concentration for a given temperature and pressure. This leads to a change in the boiling temperature of the mixture on the boundary between the phases and reduction of the rate of conduction of the heat required for evaporation. The change in concentration on the boundary between the phases is determined by the evaporation process and the diffusion process from the surrounding liquid, while the evaporation process is correlated with the rate of heat conduction from the surrounding liquid on the boundary between the phases. This mutual relation between the diffusion process and heat conduction determines the dynamics of vapor bubble growth and the structure of the two-phase boundary layer.

5. Burn-out Heat Flux

It can be accepted with certainty the thesis that transfer from bubble to film boiling is a hydrodynamic phenomenon. Although this conclusion allows the development of the mathematical model whose analytical solutions satisfy the experimental results, it remains an open question as to how some geometrical parameters exert influence on this phenomenon. There are no special reasons for not adopting this basic assumption even when binary mixtures are concerned. Figure 10 shows results

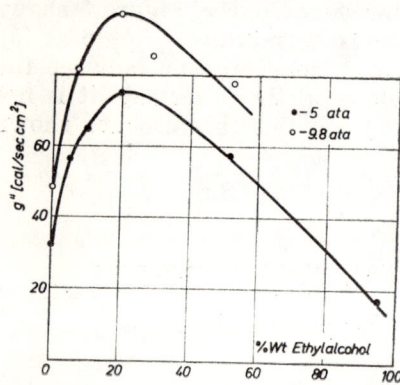

Fig. 10. Burnout heat flux as a function of concentration water-ethylalcohol mixture [6].

of burn-out heat flux measurement as a function of concentration for the water-ethylalcohol mixture [6]. The maximum burn-out heat flux corresponds to the concentration of the mixture at which there is maximum difference in concentration between the liquid and vapor phase. This is confirmed by the fact that even at a heat flux which corresponds to the boiling crisis the structure of the two-phase boundary layer depends on the difference in concentration between the liquid and the vapor phase.

Figure 11 shows the dependence of the burn-out heat flux for ethylalcohol-benzene mixtures [5]. The two maxima on this curve correspond to the concentrations of the liquids having maximal absolute differences in concentrations between the liquid and the vapor phase. This is also in agreement with the earlier statement regarding the dependence of the structure of the two-phase boundary layer on the difference in concentration of the liquid and the vapor phase. Kutateladze—Zuber's criterion

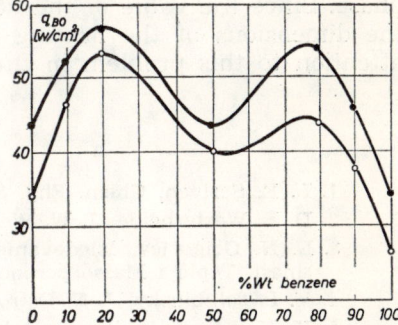

Fig. 11. Burnout heat flux as a function of concentration ethylalcohol-benzene mixture.

on the stability of the two-phase boundary layer, which shows very little sensitivity for pure liquids and has an approximately constant value in a wide range of physical parameters, also shows dependence on the difference in the concentration between the liquid and the vapor phase. However, it should be pointed out that this dependence is not unique and more experimental results will be necessary before a more general form of this dependence is obtained.

6. Conclusion

The investigations of heat and mass transfer by boiling binary mixtures carried out so far have a limited extent and there are insufficient experimental results which confirm some basic characteristics of the model which describe this phenomenon. From the so far available measurements one can draw conclusions which provide grounds for a better definition of the mechanism of heat and mass transport by boiling mixtures.

1. The bubble growth rate in boiling mixtures is determined both by heat transport and by mass diffusion in a boundary layer very close to the interface. This mutual relation between mass and heat transfer is determined by the difference in concentration between he liquid and the vapor phase. The quantitative determination of these dependences requires further experimental evidence.

2. The effect of roughness of the surface on boiling binary mixtures has been experimentally verified, but it is difficult to point out its importance. However, from the so far results it can be concluded that this influence is smaller than one could expect according to the first predictions.

3. The heat transfer coefficient by boiling mixtures depends on the concentration of the components in the mixture. For most of the liquids investigated so far it has been shown that this dependence has a minimum

value with a maximum value of the absolute difference in concentration between the liquid and the vapour phase.

4. The burn-out heat flux for most of the mixtures is higher than the same value for pure components. This value reaches a maximum at a maximum difference in concentration between the liquid and the vapor phase. Since the value of the heat flux for the boiling curve depends on the dimensions of the heating surface, it is necessary to pay special attention to this problem in further investigations.

REFERENCES

1. L. E. Scriven, Chem. Eng. Science, **10**, 1/2 (1959).
2. D. E. Westerheide, J. W. Westwater, A. I. Ch. E. Journal **7**, 3 (1961).
3. L. N. Grigorjev, Isledovanie teploobmena pri kipenii dvuh komponentnih smesi. Teplo i Masso perenos, 1962, Minsk.
4. V. I. Tolubinsky, J. N. Ostrovsky, J. Heat Mass Transfer, Vol. 9 (1966).
5. H. N. Afgan, Boiling heat transfer and burnout heat flux of ethylalcohol-benzene mixtures. Proc. III Int. Heat Transfer Conf., Chicago, 1966.
6. S. J. D. Van Stralen, British Chem. Eng., Vol 8 (1959).

LIST OF PARTICIPANTS

AFGAN Naim
Boris Kidrič Institute of Nuclear Sciences,
Beograd, Yugoslavia

ANASTASIJEVIĆ Predrag
Boris Kidrič Institute of Nuclear Sciences,
Beograd, Yugoslavia

BAEV V. K.
Pure and Applied Mechanics Institute,
U.S.S.R.

BATISTA Jože
J. Stefan Nuclear Institute,
Ljubljana, Yugoslavia

BOBROVICH G. I.
Thermal Physics Institute,
Novosibirsk, U.S.S.R.

BRUN Edmond-Antoine
Université de Paris,
Paris, France

BURDUKOV A. P.
Thermal Physics Institute,
Novosibirsk, U.S.S.R.

CVIJOVIĆ Mihailo
Boris Kidrič Institute of Nuclear Sciences,
Beograd, Yugoslavia

DAVIS R. Michael
Institute of Sound and Vibration Research,
The University,
Southampton, U.K.

DEVOLD Ivar
AB Atomenergi
Studsvik, Sweden

ĐORĐEVIĆ Bojan
Faculty of Technology and Metallurgy,
Beograd, Yugoslavia

ĐORĐEVIĆ Radivoje
Faculty of Mechanical Engineering,
Beograd, Yugoslavia

ĐORĐEVIĆ Vladan
Faculty of Mechanical Engineering,
Beograd, Yugoslavia

DRUKER I. G.
Pure and Applied Mechanics Institute,
Novosib rsk, U.S.S.R.

ERIKSSON Stig Olof
AB Atomenergi,
Studsvik, Sweden

FEDOROV B. I.
Heat and Mass Transfer Institute,
Minsk, U.S.S.R.

FORTIER André
Faculté des Sciences de Paris,
Paris, France

FRAGNAUD Fernand
Faculté des Sciences de Nantes,
Nantes, France

GAŠPERŠIĆ Branko
J. Stefan Nuclear Institute,
Ljubljana, Yugoslavia

GELIN Paul
Centre d'Etudes Nucléaires de Saclay,
Saclay, France

GOUGAT Pierre
Centre National de la Recherche Scientifique,
Meudon, France

GRŽELJ Andrej
Faculty of Mechanical Eng'neering,
Ljubljana, Yugoslavia

HAHNE Erich
Technische Hochschule,
München, Germany

HANJALIĆ Kemal
Faculty of Mechanical Engineering,
Sarajevo, Yugoslavia

HARTNETT J. P.
Dept. of Energy Eng., University of Illinois,
Chicago, U.S.A.

HEAD M. R.
University of Cambridge,
Cambridge, U.K.

IRVINE T. F.
Dept. of Engineering, State Un'versity of New York,
New York, U.S.A.

JOKSIMOVIĆ-TJAPKIN Slobodanka
Faculty of Technology and Metallurgy,
Beograd, Yugoslavia

JONSSON K. Valdimar
Dept. of Mech. Eng., Imperial College,
London, U.K.

JOVANOVIĆ Ljubomir
Boris Kidrič Institute of Nuclear Sciences,
Beorad, Yugoslavia

JOVAŠEVIĆ Vladimir
Boris Kidrič Institute of Nuclear Sciences,
Beograd, Yugoslavia

KHABAKHPASHEVA E. M.
Thermal Physics Institute,
Novosibirsk, U.S.S.R.

KJELLSTRÖM Björn
AB Atomenergi,
Studsvik, Sweden

KOLOVANDIN B. A.
Heat and Mass Transfer Institute,
Minsk, U.S.S.R.

KOVALEV S. A.
Institute of High Temperatures,
Moskva, U.S.S.R.

KONDIĆ Nenad
Boris Kidrič Institute of Nuclear Sciences,
Beograd, Yugoslavia

KOSTELIĆ Aurel
Steam Boiler Factory,
Zagreb, Yugoslavia

KOSTIĆ Života
Boris Kidrič Institute of Nuclear Sciences,
Beograd, Yugoslavia

KUMAR R. N.
Institut für Holzforschung und Holztechnik der Universität München,
München, Germany

KUTATELADZE S. S.
Thermal Physics Institute,
Novosibirsk, U.S.S.R.

LASSAU Gérard
Centre National de la Recherche Scientifique,
Meudon, France

LAUNDER E. Brian
Dept. of Mech. Eng., Imperial College
London, U.K.

LEONT'EV A. I.
Thermal Physics Institute,
Novosibirsk, U.S.S.R.

LOCKWOOD Ch. Frederick
Dept. of Mech. Eng., Imperial College
London, U.K.

MAKSIMOVIĆ Ljiljana
Boris Kidrič Institute of Nuclear Sciences,
Beograd, Yugoslavia

MALÁK Jaroslav
Institute of Nuclear Research,
Praha, Czechoslovakia

MALIĆ Dragomir,
Faculty of Technology and Metallurgy,
Beograd, Yugoslavia

MARKOVIĆ Slavimir
Boris Kidrič Institute of Nuclear Sciences,
Beograd, Yugoslavia

MATHIEU Jean
École Centrale Lyonnaise, Université de Lyon,
Lyon, France

MAYE Jean-Pierre
C.E.A., Faculté des Sciences,
Poitiers, France

MESAROVIĆ Miodrag
Energoprojekt,
Beograd, Yugoslavia

MEYER Guy
Laboratoire de Mécanique Expérimentale des Fluides, Orsay,
Faculté des Sciences,
Paris, France

MIRONOV B. P.
Thermal Physics Institute,
Novosibirsk, U.S.S.R.

MOSTINSKY I. L.
Institute of High Temperatures,
Moskva, U.S.S.R.

NAKORIAKOV V. E.
Thermal Physics Institute,
Novosibirsk, U.S.S.R.

NEVSTRUEVA E. J.
Institute of High Temperatures,

NICOLL W. B.
University of Waterloo,
Waterloo, Canada

NINIĆ Neven
Boris Kidrič Institute of Nuclear Sciences,
Beograd, Yugoslavia

OBROVIĆ Branko
Faculty of Mechanical Engineering,
Kragujevac, Yugoslavia

OBSIGER Vilka
Faculty of Mechanical Engineering
Rijeka, Yugoslavia

OKA S'meon
Boris Kidrič Institute of Nuclear Sciences,
Beograd, Yugoslavia

OPARA Mirko
J. Stefan Nuclear Institute,
Ljubljana, Yugoslavia

OPREŠNIK Miran,
Faculty of Mechanical Engineering,
Ljubljana, Yugoslavia

PAVLETIĆ Radislav
Faculty of Mechanical Engineering,
Ljubljana, Yugoslavia

PAVLOVIĆ Pavle
Bor's Kidrič Institute of Nuclear Sciences,
Beograd, Yugoslavia

PETUKHOV B. S.
Institute of High Temperatures,
Moskva, U.S.S.R.

PEUBE Jean-Laurent
Laboratoire de Dynamique des Fluides,
Faculté des Sciences,
Poitiers, France

PIŠLAR Vladislav
Boris Kidrič Institute of Nuclear Sciences,
Beograd, Yugoslavia

POLIAKOV A. F.
Institute of High Temperatures,
Moskva, U.S.S.R.

POPOV V. P.
Heat and Mass Transfer Institute,
M'nsk, U.S.S.R.

POPOVIĆ Gordana
Faculty of Technology and Metallurgy,
Beograd, Yugoslavia

PUSTYNTSEV G. N.
Heat and Mass Transfer Institute,
Minsk, U.S.S.R.

RADOVANOVIĆ Milan
Faculty of Mechanical Engineering,
Beograd, Yugoslavia

RISTIĆ Momčilo
Boris Kidrič Institute of Nuclear Sciences
Beograd, Yugoslavia

ROHSENOW W. M.
Massachusetts Institute of Technology
Cambridge, U.S.A.

RONAT Jean
Electricité de France,
Chatou, France

RUBTSOV N. A.
Thermal Physics Institute,
Novosibirsk, U.S.S.R.

SALA Ricardo
A.R.S.,
Milano, Italy

SALJNIKOV Viktor,
Faculty of Mechanical Engineering,
Beograd, Yugoslavia

SÉMÉRIA Roger
Centre d'Etudes Nucleaires de Grenoble,
Grenoble, France

SPALDING D. B.
Dept. of Mech. Eng., Imperial College
London, U.K.

SPASOJEVIĆ Dušan
Boris Kidrič Institute of Nuclear Sciences
Beograd, Yugoslavia

STAMENKOVIĆ Ivan
Boris Kidrč Institute of Nuclear Sciences,
Beograd, Yugoslavia

STEFANOVIĆ Miodrag
Boris Kidrič Institute of Nuclear Sciences,
Beograd, Yugoslavia

STIEFEL Ulrich
Swiss Federal Institute for Reactor Research,
Würenlingen, Switzerland

STOS P.
S.N.E.C.M.A.,
Villaroche, France

STUDOVIĆ Milovan
Boris Kidrič Institute of Nuclear Sciences,
Beograd, Yugoslavia

STYRIKOVICH M. A.
Institute of High Temperatures,
Moskva, U.S.S.R.

ŠANDOR Mario
Energoinvest,
Sarajevo, Yugoslavia

ŠAŠIĆ Mane
Faculty of Mechanical Engineering
Beograd, Yugoslavia

ŠIKMANOVIĆ Slobodan
Boris Kidrič Institute of Nuclear Sciences,
Beograd, Yugoslavia

TACCOEN Lionel
Electricité de France,
Chatou, France

TALMOR E.
Rocketdyne Div. of North American Rockwell Corporation,
Canoga Park,
California, U.S.A.

TASIĆ Aleksandar
Faculty of Technology and Metallurgy,
Beograd, Yugoslavia

TOŠIĆ Dragica
Boris Kidrič Institute of Nuclear Sciences,
Beograd, Yugoslavia

TURK Ivan
Faculty of Mechanical Engineering,
Zagreb, Yugoslavia

VAN THINH N.
Laboratoire de Méchanique Expérimentale Fluides,
Faculté des Sciences,
Paris, France

VASIL'EV L. L.
Heat and Mass Transfer Institute,
Minsk, U.S.S.R.

VEHAUC Aleksandar
Boris Kidrič Institute of Nuclear Sciences,
Beograd, Yugoslavia

VEROLLET Edmond
Institut de Mécanique Statistique de la Turbulence,
Marseille, France

VILENSKII V. D.
Institute of High Temperatures,
Moskva, U.S.S.R.

VILIČIĆ Milan
Faculty of Mechanical Engineering,
Zagreb, Yugoslavia

VOJ Peter
Interatom Internationale Atomreaktorbau,
Bensbreg/Köln, Germany

VORONJEC Konstantin
Faculty of Mechanical Engineering,
Beograd, Yugoslavia

WHITELAW J. H.
Dept. of Mech. Eng., Imperial College,
London, U.K.

ZARIĆ Zoran
Faculty of Mechanical Engineering and
Boris Kidrič Institute of Nuclear Sciences,
Beograd, Yugoslavia

ZDRAVKOVIĆ Momčilo
Faculty of Mechanical Engineering
Beograd, Yugoslavia

ZHILIN V. G.
Institute of High Temperatures,
Moskva, U.S.S.R.

ZUBER N.
New York University,
New York, U.S.A.

ZUPANČIČ Jože
Faculty of Mechanical Engineering,
Ljubljana, Yugoslavia

ŽIVANOVIĆ Branislav
Boris Kidrič Institute of Nuclear Sciences,
Beograd, Yugoslavia

REPRESENTATIVES

SERBIAN ACADEMY OF SCIENCES, Beograd
VELIČKOVIĆ Dušan

FEDERAL COMMISION FOR NUCLEAR ENERGY, Beograd
GUZINA Vojin, President

INTERNATIONAL ATOMIC ENERGY AGENCY, Vienna
RISTIĆ Milorad

ORGANIZING COMMITTEE

Chairman
Naim AFGAN, Boris Kidrič Institute of Nuclear Sciences

Members
Predrag ANASTASIJEVIĆ, Boris Kidrič Institute of Nuclear Sciences
Dobrosav MILINČIĆ, University of Beograd
Miran OPREŠNIK, University of Ljubljana
Muhamed RIĐANOVIĆ, University of Sarajevo
Viktor SALJNIKOV, University of Beograd
Ivan TURK, University of Zagreb
Zoran ZARIĆ, Boris Kidrič Institute of Nuclear Sciences

Executive Secretary
Branko JOVIĆ, Boris Kidrič Institute of Nuclear Sciences

HEAT AND MASS TRANSFER
IN FLOWS WITH
SEPARATED REGIONS

DYNAMICS AND THERMODYNAMICS OF SEPARATED FLOWS

H. H. KORST

University of Illinois, Urbana, U.S.A.

1. Introduction

The effects of viscosity in flows characterized by relatively high Reynolds numbers may be classified in terms of the modifications they introduce to the ideal inviscid flow field about prescribed boundaries. While viscosity is always essential near solid walls when the no-slip condition must be satisfied, its influence on the overall flow pattern can range from small corrections (displacement effects of the attached boundary layer) to weak and strong interactions. At present we shall be especially concerned with such conditions which lead to flow patterns in which streamlines break away from guiding walls, and either by re-attachment at the wall (separation bubble) or by realignment with another separation streamline at the end of the wake, enclose, together with that portion of the wall from which the streamlines have departed, a region of separated flow.

If such is the case, the basic assumptions underlying the classical boundary layer theory would be impeded and the parabolic nature of the interaction mechanism is obviously invalidated by the feedback of downstream conditions through the wake.

In contrast to methods of streamwise development of attached boundary layers under prescribed (or iterated) free stream conditions, at most towards a separation point, fully separated flows must be treated by considering simultaneously conditions of separation, wake closure and the viscid-inviscid flow interaction along the boundary of the recirculation region.

The practical importance of flow problems involving separation had been well established long before a reasonable understanding of essential mechanisms had been gained. This led to extensive, but generally not useful, analytical studies based on inviscid concepts (free streamlines, discontinuous potentials) on one side and essentially empirical approaches to produce needed design information.

Consideration of the viscid-inviscid interaction problem and recognizing the importance of introducing a condition for wake closure led to the beginning of a better understanding in the early 1950's.

Since then two major roads have been developed and followed in dealing with both basic and intricate problems involving separated flows: Firstly, the boundary layer-like multi-moment methods proposed originally by Crocco and Lees [1] and expanded further by Lees and Reeves [2] which include closure conditions as a critical (saddle) point for their system of integral conservation equations and secondly, a highly idealized, but physically perceptible flow model distinguishing among and subsequently synthesizing individual flow components (such as the attached boundary layer before separation, its modification during separation, the developing and fully developed free shear layer, reattachment, flow recirculation within, flow development along and outside the separated flow region). This concept is now often referred to in the literature as the Chapman-Korst model [3, 4, 5].

It appears now that the multi-moment integral methods are especially well suited in dealing with such problems where the mixing zone along and the recirculating flow within the separated flow region are closely merged. Such is the case for either shallow separation bubbles or narrow wakes. On the other hand, the flow component approach will be reasonable only if individual mechanisms remain well distinguished and can be quantitatively treated.

Of further interest is the fact that the integral methods always require explicit information on the viscosity coefficient (which is a critical problem in dealing with turbulent separation [6]), while some, even complicated turbulent flow problems involving very large wakes, can be treated with the Chapman-Korst model without specific empirical input.

Due to the better physical insight that can be gained (into both its capabilities and shortcomings) we shall now proceed with an analysis of dynamic and thermodynamic mechanisms involved with the flow components (Chapman-Korst) model.

2. Conservation Equations for Flows with Fully Separated Regions

2.1. DYNAMIC RELATIONS — MECHANICAL ENERGY EQUATION

The mechanical energy equation, obtained directly by integration from the full equation of motion, has gained much prestige in dealing with viscous flows, and, in particular, in its application to boundary layer analysis.

We shall here discuss some important features of two integrations, namely, along streamlines (with additional attention given to thermodynamic implications) and to steady flows within closed stream surfaces, starting out with the equation of motion

$$\frac{D\vec{v}}{Dt} = \vec{F} + \frac{1}{\rho} \nabla \cdot \overline{\overline{p}} \tag{1}$$

where

$\vec{v}$ is the velocity vector,

$\vec{F}$ is the acceleration due to field forces,

ρ is the mass density of the fluid,

$\bar{\bar{p}}$ is the stress tensor composed of hydrostatic and viscous terms as

$$\bar{\bar{p}} = -\bar{\bar{I}}p + \bar{\bar{p}}'$$

∇ is the Hamilton operator for field differentiation, and

$\dfrac{D}{Dt}$ indicates substantial differentation such that in operative form

$$\frac{D}{Dt} = \frac{\partial}{\partial t} + \vec{v} \cdot \nabla$$

Forms of the mechanical energy equation are obtained by scalar multiplication of each term in Eq. (1) by the velocity vector $\vec{v}$, and integration in space.

2.1.1. Integration along a Streamline

Scalar multiplication of Eq. (1) by $\vec{v}$ gives, as we represent the field acceleration $\vec{F} = \vec{g} = -\nabla(gz)$ as the steady gravitational effect

$$\frac{D}{Dt}\frac{v^2}{2} + gz = \frac{\vec{v} \cdot \nabla \bar{\bar{p}}}{\rho} = -\frac{\vec{v} \cdot \nabla p}{\rho} + \frac{1}{\rho}\nabla \cdot (\vec{v} \cdot \bar{\bar{p}}') - \frac{\bar{\bar{p}}' : (\nabla, \vec{v})}{\rho} \tag{2}$$

Now, since $\nabla \cdot (\vec{v} \cdot \bar{\bar{p}}') = -DW'/Dt$ the rate of shear work per unit volume received by the streamline, and the double scalar product of the viscous portion of the stress tensor and the local velocity dyad

$$\bar{\bar{p}}' : (\nabla, \vec{v}) = \frac{D\Phi}{Dt}$$

represents the (positive definite) rate of the mechanical energy dissipated per unit volume (Φ is Rayleigh's dissipation function), we obtain

$$\frac{D}{Dt}\frac{v^2}{2} + gz + \frac{\vec{v} \cdot \nabla p}{\rho} + \frac{1}{\rho}\frac{DW'}{Dt} = -\frac{1}{\rho}\frac{D\Phi}{Dt} \tag{3}$$

As we specialize Eq. (3) to steady flow along a streamline determining the direction »l« while »n« is now to be considered as the perpendicular direction in a two-dimensional flow field, we write, in first approximation (also assuming that $\partial \mu / \partial n \simeq 0$)

$$v\frac{\partial v}{\partial l} + \frac{1}{\rho} + g\frac{\partial z}{\partial l} = \frac{\mu}{\rho}\frac{\partial^2 v}{\partial n^2} \tag{4}$$

The term accounting for available mechanical energy (local value of »Bernoulli constant«)

$$B = \frac{v^2}{2} + \int_{p_{ref}}^{p} \frac{dp}{\rho} + gz$$

is seen to be directly related to the second derivative of the velocity profile

$$\frac{\partial B}{\partial l} = \frac{\mu}{\rho} \frac{\partial^2 v}{\partial n^2} \tag{5}$$

so that the curvature of the velocity profile determines the streamwise change of mechanical energy. In particular, there is a net increase in mechanical energy if $\partial^2 v / \partial n^2 > 0$; that is, where the profile is locally concave. This behavior constitutes an important feature of viscous layers which must negotiate adverse pressure gradients.

For a wall streamline satisfying the no-slip condition $(v \equiv 0)$ will, therefore, be capable of overcoming an adverse pressure gradient by virtue of a profile concavity and the dynamic viscosity according to Eq. (4)

$$\frac{\partial p}{\partial l} = \mu \frac{\partial^2 v}{\partial n^2} \tag{6}$$

which is, of course, a well-known condition of compatibility at the wall.

After separation, the free shear layer develops first in a region of rather insignificant presssure gradients, so that neglecting small streamwise pressure changes, as well as gravitational effects, Eq. (4) now assumes the form

$$v \frac{\partial v}{\partial l} = \frac{\mu}{\rho} \frac{\partial^2 v}{\partial n^2} \tag{7}$$

which relates streamwise velocity changes to the profile curvature. Hence, concave parts of free shear layer (mixing) profiles will produce accelerations, while convex parts of the mixing profile will be further decelerated.

It is of importance to point out that the »weight factor« on the profile curvature, namely, the kinematic viscosity $\varepsilon_t = \mu t / \rho$, assumes very high values for free turbulent shear layers, as can be illustrated by drawing from information developed in more detail in section 3. As an example, let us consider incompressible, isoenergetic, single-stream mixing, and restrict ourselves to fully developed, similarity profiles. We find then a ratio of

$$\frac{\varepsilon_t}{v_{l_{am}}} = 1.92 \times 10^{-3} \, Re_x$$

where Re_x is the Reynolds number based on the length of the mixing region and the free stream velocity, u_{2a}

$$Re_x = \frac{u_{2a} \, x}{v_{l_{am}}}$$

The high effectiveness of such turbulent free shear layers in transferring mechanical energy to streamlines in the concave part of the profile is well borne out. The streamline through the inflexion point of a viscous flow profile separates that part of the profile which receives, from that which supplies, mechanical energy. Since such a streamline will assume an important role in the recompression process, we shall direct attention to the (path) integral $\int dp/\rho$ which appears in the integration of Eq. (4).

$$B_2 - B_1 = \int_{l_1}^{l_2} \frac{\mu}{\rho} \frac{\partial^2 v}{\partial n^2} dl = \frac{V_2^2 - V_1^2}{2} + \int_{p_1}^{p_2} \frac{dp}{\rho} + g(z_2 - z_1) \tag{8}$$

Utilizing the Second Law of Thermodynamics and specifically the definition of entropy, we may write along a streamline.

$$T \frac{Ds}{Dt} = \frac{Dh}{Dt} - \frac{1}{\rho} \frac{Dp}{Dt} = \frac{Dh_0}{Dt} - v \frac{Dv}{Dt} - \frac{1}{\rho} \frac{Dp}{Dt} \tag{9}$$

as h is the enthalpy and T the absolute temperature with subscript o referring to the local stagnation condition.

Again, in first approximation, and for steady flow, the energy equation yields

$$\frac{Dh_0}{Dt} = \vec{v}_0 \nabla h_0 = \frac{v}{\rho} \left[\frac{\partial}{\partial n} \lambda \frac{\partial T}{\partial n} + \frac{\partial}{\partial n} \left(\mu v \frac{\partial v}{\partial n} \right) \right] \tag{10}$$

and for $Pr_t = \mu c_p / \lambda = 1$, while $\partial \mu / \partial n = 0$, c_p = constant

$$T \frac{Ds}{Dt} = \frac{v \mu}{\rho} \left[\frac{\partial^2}{\partial n^2} \left(c_p T + \frac{v^2}{2} \right) - \frac{\partial^2 v}{\partial n^2} \right] \tag{11}$$

One recognizes that if the relation $T_0 = av + b = T + v^2/2c_p$ holds, where a and b are constants, the condition $\partial^2 v / \partial n^2 = 0$ would be sufficient to yield $Ds/Dt = 0$ and a streamline through an inflection point or in an essentially linear portion of a velocity profile in viscous shear flow would exhibit isentropic (although non-adiabatic, irreversible) relations.

2.1.2. Volume Integration for Mechanical Energy

Scalar multiplication of Eq. (1) by $\rho \vec{v}$ yields

$$\frac{\partial}{\partial t}\left(\rho \frac{v^2}{2}\right) + \nabla \cdot \vec{v} \frac{\rho v^2}{2} = \rho \vec{v} \cdot \vec{F} - \nabla \cdot (v p) + p(\nabla \cdot \vec{v})$$

$$+ \nabla \cdot (\vec{v} \; \overline{\overline{p'}}) - \overline{\overline{p}} : (\nabla, \vec{v}) \tag{10}$$

Integration over the volume V with the surface S completely composed of stream surfaces, so that for any (outside) surface normal element

$\vec{ds}$, $\vec{v} \cdot \vec{ds} = 0$ yields, after utilizing the Gauss theorem and specializing here to steady state conditions,

$$0 = \int_v \rho \vec{v} \cdot \vec{F} \, dV + \int_v p(\nabla \cdot \vec{v}) \, dV + \int_s \vec{v} \cdot \bar{\bar{p}}' \, \vec{ds} - \int \bar{\bar{p}}' : (\nabla, \vec{v}) \, dV \quad (11)$$

$$\text{(I)} \qquad \text{(II)} \qquad \text{(III)} \qquad \text{(IV)}$$

In order, these integrals are recognized as representing (i) the rate of work done on the system by field forces (zero if field forces are conservative), (ii) the rate of increase of elastic energy in the system, (iii) the rate of thermodynamic work received by the system through viscous stresses over (moving parts of) the surface, and (iv) the rate of dissipation of mechanical energy over the volume. Since the transport theorem states that for any function P

$$\frac{D}{Dt} \int_{v(t)} P \, dV = \int_{v(t)} \left[\frac{DP}{Dt} + P(\nabla \cdot \vec{v}) \right] dV$$

then, for a fixed volume V and steady state

$$\int_v P(\nabla \cdot \vec{v}) \, dV = 0$$

Hence, we can now conclude that, for a closed steady state system

$$\int_s (\vec{v} \cdot \bar{\bar{p}}') \, \vec{ds} = \int_v \bar{\bar{P}}' : (\nabla, \vec{v}) \, dV = \int_v \Phi \, dV \quad (12)$$

The rate of thermodynamic work done on the system over moving parts of the surface by viscous (shear) forces equals the rate of dissipation of mechanical energy over the entire volume. In the following we shall extend this relation to open system.

2.2. OPEN SYSTEM ANALYSIS

To deal with conservation principles as they apply to open systems (with mass added and extracted) we select, for reasons of convenience, a two-dimensional cavity configuration (see Fig. 1). Despite the geometrical simplicity of the external flow field, it allows us to discuss all pertinent conservation relations [7].*

---

* In addition to the global continuity equation, conservation of species has to be satisfied if mass transfer and/or chemical reactions are present. Such cases are cited in section 3.2.

2.2.1. Conservation of Mass

If mass is added into the cavity at the rate G_B, steady state continuity requires that

$$G_B + \int_j^d \rho u \, dy = 0 \qquad (13)$$

In this form, Eq. (13) applies to a two-dimensional cavity and each term refers to unit depth. Integration of the mass flow leaving near the point of reattachment is carried out between the separating streamline (jet

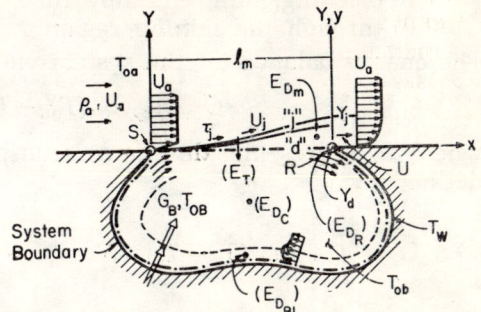

Fig. 1. Cavity flow model. System analysis for mechanical energy balance.

boundary streamline) j, and the reattaching streamline (discriminating streamline) d. In the absence of mass addition $G_B = 0$ and streamlines of j and d coincide. When reattachment to a wall is replaced by confluence of different mixing regions, investigations must be carried out for each individually and added algebraically.

2.2.2. Mechanical Energy Balance

In addition to the shear work done on the moving boundaries of the system, mechanical energy also crosses the system boundaries if $j \neq d$. We account for this in the present case through the kinetic energy, so that, for unit depth

$$\int_0^{l_M} \tau_j u_j \, dx - \frac{1}{2} \int_{y_d}^{x_j} \rho u^3 \, dy = \int \Phi \, dV \qquad (14)$$

2.2.3. First Law of Thermodynamics for Steady State Condition of the Open System Described in Fig. 2

The selection of the system boundaries is the same as in Fig. 1, but, as shown in Fig. 2, three components of energy transfer must now be considered:

(1) Heat transfer by convection from the cavity walls into the thermal boundary-layer developing along these walls—designated here as Q_w.*

(2) Diffusion of the thermal energy carried in this boundary layer into the jet mixing region near the point of confluence S of the wake flow and the approaching free stream. Part of this energy, denoted as Q_d, is diffused into the main stream, outside the streamline »d«, which, in case of zero mass bleed, would coincide with streamline j. The remaining part of the thermal energy is diffused into the bulk of the wake.

(3) Transfer of energy (heat Q_m, shear work W'_m, and in case of mass bleeding, total enthalpy flux between the streamlines d and j) through the mixing region.

The energy balance for the system yields

$$Q_w + c_p\, G_B\, (T_{0B} - T_{0a})\, \Omega_d + Q_d = 0 \tag{15}$$

where Ω_d represents the energy transport rate in the jet mixing region; defined by

$$\Omega_d \equiv \int_{Y_d}^{Y_{Ra}} \rho\, u c_p\, (T_{0a} - T_0)\, dY \equiv Q_m - W'_m + \int_{Y_d}^{Y_j} \rho\, u c_p\, (T_{0a} - T_0)\, dY$$

so after introducing the partition function for the heat source diffusion problem

$$\pi_d = -Q_d/Q_w \tag{16}$$

one may rewrite Eq. (15) in the form

$$\Omega_d = -Q_w(1 - \pi_d) + c_p\, C_B(T_{oa} - T_{oB}) \tag{17}$$

It is obvious that a convective mechanism is needed to sustain the heat transfer across the solid walls of the cavity (or solid wall portion

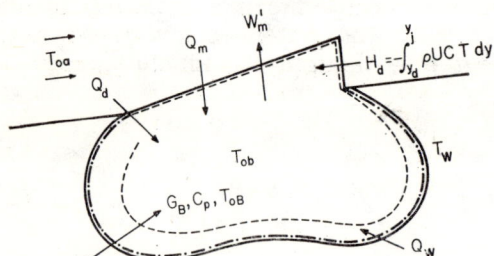

Fig. 2. Cavity heat transfer model. Thermodynamic systems analysis.

of a separated flow region). Some practical problems could be successfully treated under the assumption of a quiescent wake by introducing the

---

* Note that all arrows shown in Fig. 2 would indicate heat added to the system so that the sign of individual contributions will account for the actual flux direction in a specific situation. This is in agreement with the convention to consider heat added to a system, and work delivered by a system, as positive.

quantity Q_w as a lumped parameter, especially when multi-stream interaction leads to a predominance of energy exchange through mixing. To gain insight into the full mechanism controlling heat transfer to and across separated flow regions, finite wake velocities must, however, be considered. These are, in turn, sustained by the dissipative mechanism outlined in section 2.2.2.

For a comprehensive exploration of the system equations, we must now turn to an analysis of its flow components.

3. The Free Turbulent Shear Layer

As pointed out before in section 2.1.1, the dynamics of the viscous free shear layer is central for the understanding of separated flows.

Its development (first by expansion or compression of the approaching—attached—boundary layer near the point of separation, then by a process of turbulent mixing with, and mass entrainment from the wake region) and its contribution to the closure condition at the recompression zone at the end of the wake are, therefore, of greatest importance.

3.1. TRANSITION FROM AN ATTACHED BOUNDARY LAYER TO A FREE SHEAR LAYER

The history of the approaching attached flows (generally producing viscous, thermal and concentration boundary layers) determines the flow profiles upstream of any expansion or compression associated with separation. Initial conditions for the free jet mixing problem are, however, to be formulated by specifying flow profiles downstream of the separation: That is, at the pressure level representative for the following (constant pressure) region along the wake boundary.

This problem is by no means simple in view of the fact that rather large pressure gradients, both in the streamwise and in the transverse direction, are present. In addition, the highly rotational nature of the viscous layer suggests that even turning of a supersonic portion will not result in the appearance of a single family of characteristics, thus eliminating, at least in principle, the simple wave concept expressed by the Prandtl-Meyer relations. Yet, conceding lack of rigor, the problem of profile changes due to either expansion or compression can be treated in different, but nevertheless, rational and obviously adequate ways.

One may argue that transport phenomena can be neglected over the relative short lengths of individual streamlines in an abrupt expansion (as it occurs in base flows) and that, therefore, reversible and adiabatic conditions prevail along such portions of the streamlines. Thus, the change in velocity as each streamline expands through a given pressure ratio can be determined (as well as the flow profile by use of continuity equation).

It is immediately clear that this method, and the very concept of isentropic relations along individual streamlines, will fail to give solutions for the case of compression. We shall, therefore, first treat the expansion case which yields profiles in detail before describing a unified approach to the expansion or compression case, which gives information on (nevertheless important) integral quantities only.

3.1.1. Expansion of a Shear Layer

Restricting ourselves to the isoenergetic case, so that $u_{max}=\text{const}=(2c_pT_o)^{1/2}$ may be used to non-dimensionalize all velocities by introducing the Crocco number, $C=u/u_{max}$, we relate any velocity u, in cross section

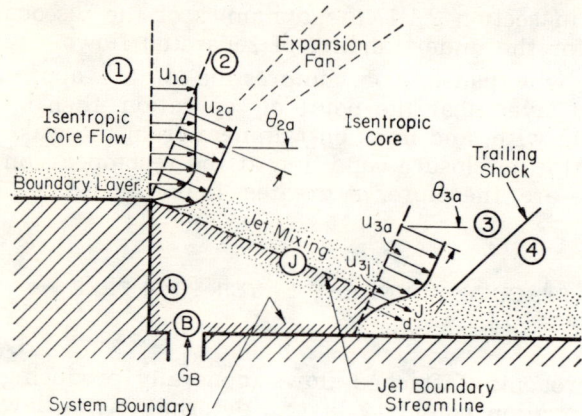

Fig. 3. Supersonic backstep.

1 (see Fig. 3), pressure p_1, to a velocity u_2 along the same streamline at cross section 2, pressure p_2, by the reversible adiabatic relation

$$C_2^2 = 1 - [1 - (C_1^2/C_{1a}^2)\,C_{1a}^2](p_2/p_1)^{(K-1)/K} \tag{18}$$

The continuity equation for a streamtube

$$\rho_1 u_1 \, dy_1 = \rho_2 u_2 \, dy_2 \tag{19}$$

can be utilized for finding

$$y_2 = \int_0^{y_1} \frac{\rho_1 u_1}{\rho_2 u_2} \, dy_1 \tag{20}$$

since

$$\frac{u_1}{u_2} = \frac{C_1}{C_2} = C_1 \cdot \{1 - [1 - C_{1a}^2 (C_1^2/C_{1a}^2)]\,(p_2/p_1)^{(K-1)/K}\}^{-1/2} \tag{21}$$

and

$$\rho_1/\rho_2 = (p_1/p_2)^{-1/K} \tag{22}$$

so that, with $\varsigma_1 = y_1/\delta_1$

$$\frac{y_2}{\delta_1} = \left(\frac{p_2}{p_1}\right)^{1/K} \int_0^{\varsigma_1} \frac{C_{1a}(C_1/C_{1a})}{\{1-[1-C_{1a}^2(C_1^2/C_{1a}^2)](p_2/p_1)^{(K-1)/K}\}^{1/2}} d\varsigma_1 \quad (23)$$

and the profile in cross section 2,

$$C_2/C_{2a} = (C_2/C_{2a})(y_2/\delta_1)$$

can now be found after integration over the known functions in cross section 1, for specific heat ratio K and the pressure ratio p_2/p_1.

The actual integration is conveniently carried out with the help of digital computers. It is, nevertheless, of interest to note that a closed solution can be given for any power law profile for the boundary layer in cross section 1,

$$C_1/C_{1a} = \varsigma_1^{1/n} \text{ for } 0 < \varsigma_1 < 1 \quad (24)$$

(n discrete for finite number of terms) gives

$$\frac{y_2}{\delta_1} = \left(\frac{P_2}{P_1}\right)^{-(K+1)/2K} [\varsigma_1^{2/n} + b^2]^{1/2} \sum_{\nu=0}^{(n-1)/2} (-1)^\nu \frac{(n-1;-2;\nu)}{(n;-2;\nu+1)} b^{2\nu} \varsigma_1^{(n-2\nu-1)/n} \quad (25)$$

with

$$b = \{[1-(p_2/p_1)^{(K-1)/K}]/[C_{1a}^2(p_2/p_1)^{(K-1)/K}]\}^{1/2}$$

where $(m;-d;\nu) = d^\nu [\Gamma(m/d+1)]/[\Gamma(m/d-\nu+1)]$ can be expressed in terms of (tabulated) gamma functions.

By integration of the profile in cross section 2, displacement and momentum thickness can also be obtained, see Fig. 4.

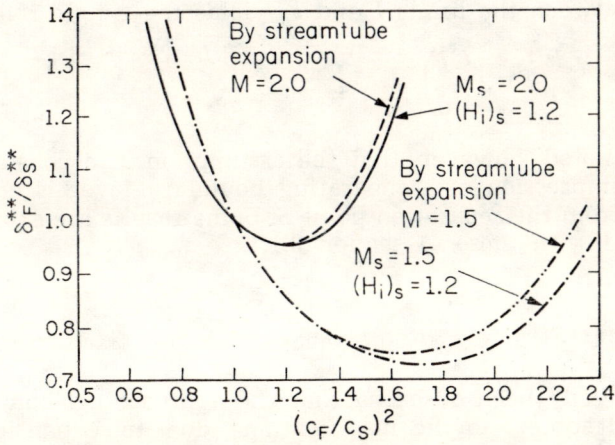

Fig. 4. Momentum thickness ratio across separations with compressions or expansions.

Other approximate methods suggested by Nash [8], Kirk [9], and by Carrière and Sirieix [10] yield similar results.

3.1.2. Unified Treatment of Expansions and Compressions by a Momentum Integral Method

Based on the observation [11] that the shape factor of fully separated flow regions is almost identical in magnitude, and dependence upon Mach number, with that of an attached boundary layer near separation (this happens to be true for separation due to an adverse pressure gradient, but is also satisfied in a more trivial way for expanding shear layers), White [12] developed a unified method for determining the change in momentum thickness of an attached boundary layer as it undergoes an expansion or compression.

He utilized the momentum integral equation for the change in momentum thickness δ^{**} of a compressible shear layer, which, in absence of wall shear forces, assumes the form

$$\frac{d\delta^{**}}{dx} + \frac{\delta^{**}}{u_a}\frac{du_a}{dx}[2+H-M_a^2] = 0 \tag{26}$$

and introduces the Crocco number as $C = u_a/(u_a)_{max}$ and $C^2 = (K-1)M^2/[2+(K-1)M^2]$ so that

$$\frac{d\delta^{**}}{\delta^{**}} + \frac{dC}{C}\{2+H-[2/(K-1)][C^2/(1-C^2)]\} = 0 \tag{27}$$

Noting that the relation between the compressible and equivalent incompressible shape factor for both attached [13] and free isoenergetic shear layers [11] can be represented by

$$H+1 = (H_i+1)[1/(1-C^2)] \tag{28}$$

where H_i can be considered to be practically constant during the expansion $[(H_i)_{exp} \simeq 1.2]$ or separation-compression process $[(H_i)_{sep} \simeq 1.9]$ integration between the limits I and F, yields the simple relation

$$\frac{\delta_F^{**}}{\delta_I^{**}} = \left(\frac{C_I}{C_F}\right)^{(H_i+2)}\left(\frac{C_F^2-1}{C_I^2-1}\right)^{[H_i+1-2/(K-1)]/2} \tag{29}$$

It should be noted, however, that the assumption of a constant value of H_i for the compression of a separating boundary layer is only justified downstream from the separation point S. Some results are shown in Fig. 4 and compared with those of section 3.1.1.

3.1.3. Turbulent Structure of the Separating Shear Layer

It has already been observed that the approaching boundary layer provides (still subject to the modifications due to expansion or compression) the initial conditions for the jet-mixing process. So far, only some details of the kinematic (profile flow) and integral formulations for the dynamic (momentum thickness) aspects of the initial transition to the free shear layer have been considered.

Of equal importance, however, is the question about the initial viscous mechanism in the separated free shear layer. The turbulent structure in the attached boundary layer approaching separation can be illustrated by the apparent kinematic eddy viscosity distribution obtained by Maise and McDonald [14] (see Fig. 5). Peak values are shown to correlate well with boundary layer thickness Reynolds number over a wide range of Mach numbers, Fig. 6 (where representative levels of ε_t/V_{lam} for free shear layers are also indicated).

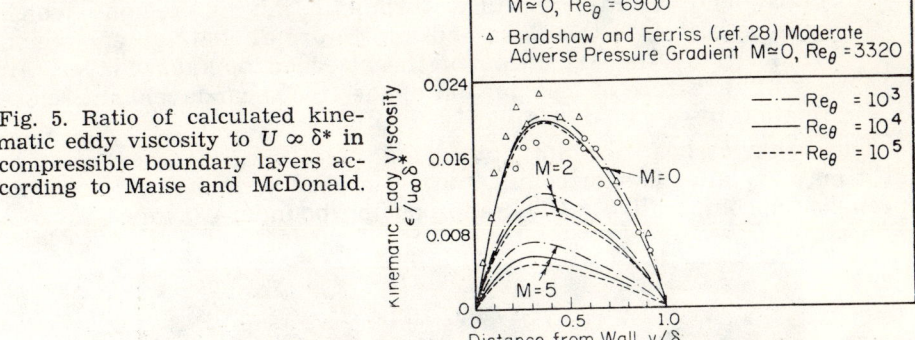

Fig. 5. Ratio of calculated kinematic eddy viscosity to $U_\infty \delta^*$ in compressible boundary layers according to Maise and McDonald.

A developing free shear layer has, then, to provide the link between these entirely different structures. Figure 7 shows the expansion of a turbulent boundary layer around a corner as taken through a Schileren system. An extremely short exposure time of 1.5×10^{-8} seconds, produced by the flashing of a ruby laser light source* freezes the random turbulent disturbances and supports the following interpretation:

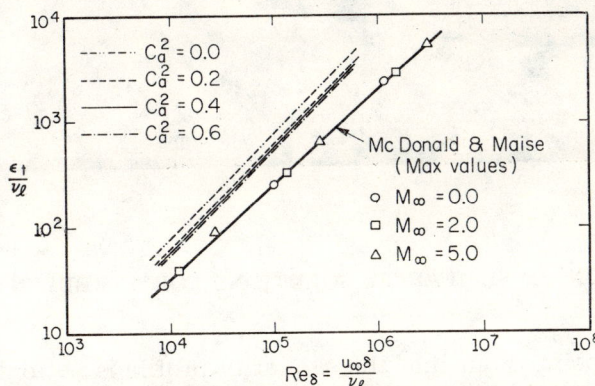

Fig. 6. Comparison of representative peak values for kinematic eddy diffusivity in attached and free shear layer.

---

* The author is indebted to Dr. R. H. Page from Rutgers University for his permission to use this picture.

(i) Strong expansion quenches the turbulence originally contained in the approaching boundary layer.
(ii) The effect of the approaching boundary layer is, therefore, practically restricted to the rotational (kinematic—see section 3.1.1.) aspects of the velocity profile after expansion.
(iii) Expansion of the wall streamline through the pressure ratio p_2/p_1 produces a discontinuity (step) in the velocity profile for $y^2=0$ of the magnitude $C_2\,(y_2=0)=[1-p_2/p_1)^{k-1/k}]^{1/2}$
iv) This discontinuity becomes a most important facet of the initial condition for the subsequent development of the free shear layer. It causes the development of an initially laminar, then transitional, and ultimately fully turbulent mixing region which spreads both into the wake and into the rotational »outer« region produced by the expansion of the attached boundary layer. An essential feature of this portion of the free shear layer is its close relation to so-called »similarity« profiles.

This behavior has been anticipated for some time in the expectation that it may explain the shortcomings of often used »origin shift« concept, especially for comparatively thick approaching boundary layers.

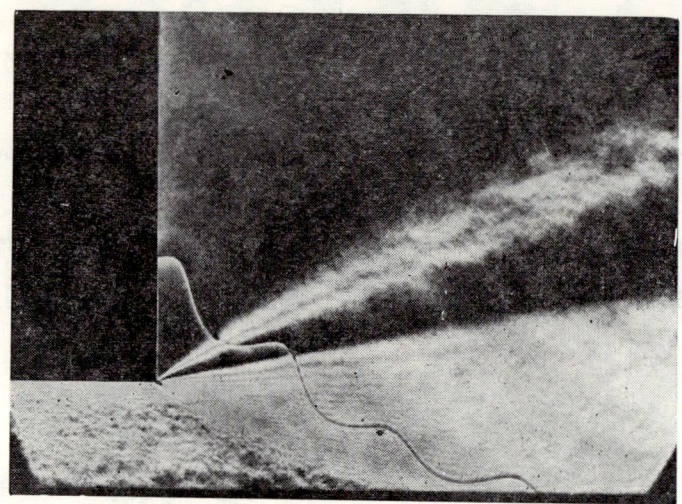

Fig. 7.

3.2. UTILIZATION OF SIMILARITY SOLUTIONS FOR TURBULENT FREE SHEAR LAYERS

We have pointed out that the similarity profile is not only of interest for the asymptotic solution where any initial disturbance in the mixing region has subsided, but that it also has significance in the developing stages when a mixing process follows a strong expansion. The selection of a momentum integral method for dealing with the mixing problem, together with the observation that a nonsimilar mixing profile resembles closely in shape a similar profile well before it attains its dynamic characteristics (e.g., measured by the velocity ratio of the separating streamline) gives added weight to the study of similarity solutions.

Comprehensive surveys on free jet mixing analyses, together with experimental evidence, are given in the books of Pai [15] and Abramovich [16] and numerous monographs have been published since the basic studies conducted by Tollmien [17], Görtler [18], and Reichardt [19]. A momentum integral approach based on a greatly simplified solution of the equation of motion for given initial flow profiles [20] has found much utilization in dealing with free jet mixing because of the availability of tabulated auxiliary functions for both the single stream (half jet) and two-stream problem [21], see Fig. 8.

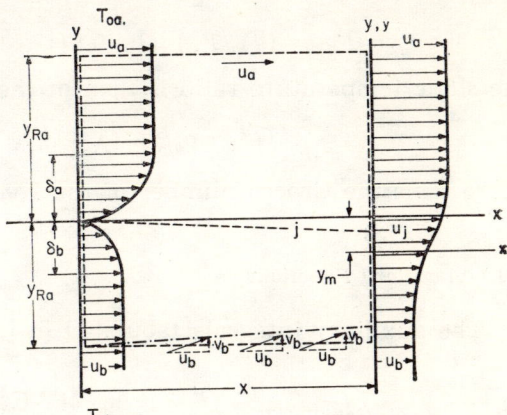

Fig. 8. The jet mixing region.

Asymptotic similarity solutions for the velocity profile conform with Grötler's first approximation which is represented by an error function distribution. Although these and other (e. g., [22]) similarity solutions would be expected to hold for very thin initial boundary layers, or very large lengths of a mixing region, it has already been pointed out that such profiles can be interpreted by judicious use of lateral and longitudinal (origin-shift) [23] displacements-as close approximations for not fully developed jet boundaries.

Details of the analysis and tabulated auxiliary integrals, which, due to the use of Crocco number instead of Mach number, hold for homogeneous streams of any specific heat ratio are given in [21].

Extensions to deal with the mixing of gases with different compositions [24], including reaction [25] and ignition delay [26] have since been made and computer programs developed.

3.2.1. Similar Flow Profiles

It is of interest to note that similarity solutions contain but one empirical constant which can be interpreted as a proportionality factor for the homogeneous independent similarity coordinate $y = \sigma(y/x)$. This parameter

σ is directly related to the spread rate of the velocity profile (such as, e.g., the error function distribution used in [21])

$$\phi = \frac{u}{u_a} = [(1+\phi_b)/2] + [(1-\phi_b)/2]\, erf\, \eta \tag{30}$$

where u_a is the velocity of the faster, u_b the velocity of the slower stream outside of their mixing region, and $\Phi_b = u_b/u_a$ so that single stream mixing will be the special case for which $\Phi_b = 0$.

For fluids of unity (turbulent) Prandtl number, Crocco's energy integral relationship is applicable and the stagnation temperature distribution can be uniquely related to the velocity field throughout such a jet mixing region. Accordingly, one may present the stagnation temperature profile by

$$\Lambda = T_o/_{oa} = (T_{ob}/T_{oa})\,[(1-\phi)/(1-\phi_b)] + (\phi - \phi_b)/(1-\phi_b)$$

The static temperature ratio is then of the form

$$T/T_a = \rho_a/\rho = (\Lambda - C_a^2 \phi^2)/(1 - C_a^2)$$

where C_a is the Crocco number of the free stream.

3.2.1.1. Auxiliary Functions

The auxiliary integrals tabulated in [21] which are defined as

$$I_1(\eta, C_a^2, T_{ob}/T_{oa}, \phi_b) = \frac{\eta_{Rb}\,(1-C_a^2)\,\phi_b}{(T_{ob}/T_{oa}) - C_a^2\,\phi_b^2} + \int_{\eta_{Rb}}^{\eta} \frac{(1-C_a^2)\,\phi}{\Lambda - C_a^2\,\phi^2}\, d\eta$$

$$I_2(\eta, C_a^2, T_{ob}/T_{oa}, \phi_b) = \frac{\eta_{Rb}\,(1-C_a^2)\,\phi_b^2}{(T_{ob}/T_{oa}) - C_a^2\,\phi_b^2} + \int_{\eta_{Rb}}^{\eta} \frac{(1-C_a^2)\,\phi^2}{\Lambda - C_a^2\,\phi^2}\, d\eta$$

$$I_3(\eta, C_a^2, T_{ob}/T_{oa}, \phi_b) = \frac{\eta_{Rb}\,(1-C_a^2)\,(T_{ob}/T_{oa})\,\phi_b}{(T_{ob}/T_{oa}) - C_a^2\,\phi_b^2} +$$

$$+ \int_{\eta_{Rb}}^{\eta} \frac{(1-C_a^2)\,\Lambda\,\phi}{\Lambda - C_a^2\,\phi^2}\, d\eta$$

$$I_4(\eta, C_a^2, T_{ob}/T_{oa}, \phi_b) = \int_{\eta_{Rb}}^{\eta} \frac{(1-C_a^2)\,\phi\,(\phi^2 - \phi_b^2)}{(\Lambda - C_a^2\,\phi^2)}\, d\eta$$

(where η_{RA} is large, positive and η_{RB} large, negative) are helpful for producing much detailed information on the geometry, kinematics, dynamics within and energy transfer across mixing regions. For fixed values of C_a^2 T_{ob}/T_{oa}, and Φ_b these integrals may be represented in short by $I_1(\eta)$, $I_2(\eta)$, $I_3(\eta)$, and $I_4(\eta)$, respectively.

(i) Since the solution for the velocity profile is to fe interpreted in the intrinsic system of coordinates x, y, the reference system of coordinates X, Y, following the »corresponding inviscid jet boundary« [4] will be determined by $X \simeq x$ and $Y = y - y_m(x)$ with $y_m(0) = 0$ when the shift $y_m = \sigma [y_m(x)/x]$ is found from the momentum equation

$$\eta_{jm} = \eta_{RA} - [1/(1-\Phi_b)] [I_2(\eta_{RA}) - \Phi_b I_1(\eta_{RA})] \tag{31}$$

(ii) The jet boundary streamline will satisfy the relation

$$I_1(\eta_j) = [I_1(\eta_{RA}) - I_2(\eta_{RA})]/(1-\Phi_b) \tag{32}$$

which does not require information on σ and the mass flow between the two streamlines η_j and η_d is given by

$$G = \int_{y_j}^{y_d} u \rho \, dy = \frac{x u_a \rho_a}{\sigma} [I_1(\eta_d) - I_1(\eta_j)] \tag{33}$$

(iii) Energy transfer from the main stream into the secondary stream, across streamline η_j can be expressed by

$$\Omega_j = \frac{x \rho_a u_a c_p T_{oa}}{\sigma} [I_1(\eta_{RB}) - I_1(\eta_j) - I_3(\eta_{RA}) + I_3(\eta_j)]$$

$$= \frac{x \rho_a u_a c_p T_{oa}}{\sigma} \left(\frac{1 - T_{ob}/T_{oa}}{1-\Phi_b} \right) [I_2(\eta_j) - \Phi I_1(\eta_j)] \tag{34}$$

so that the Stanton number for the jet boundary streamline will be given by

$$\sigma St = \frac{I_2(\eta_j) - \Phi_b(\eta_j)}{1-\Phi_b} \tag{35}$$

(iv) The local friction coefficient is found to be

$$\sigma C_f/2 = I_2(\eta_j) - \Phi_b I_1(\eta_j) \tag{36}$$

so that $St = (C_f/2) [1/(1-\Phi_b)]$ as a modification of Reynolds analogy for the two-stream mixing region.

(v) Net gain in mechanical energy of the secondary stream due to mixing is then

$$\int_{y_{RB}}^{y_j} \rho u (u^2/2 - u_b^2/2) \, dy = \frac{\rho_a u_a^3}{2} \frac{x}{\sigma} I_4(\eta_j) \tag{37}$$

(vi) The distribution of shear stress, eddy diffusivity, dissipation rates, vertical flow components, displacement and momentum thickness can also be determined with the help of the tabulated auxiliary functions [21].

(vii) Mass transfer coefficient for the mixing region between a half-infinite jet and a stagnant gas of different composition have been determined under the assumptions of perfect, non-reacting gases with effective turbulent Prandtl number and turbulent Schmidt number of unity, for a large range of molecular weight ratios and temperature ratios, by Page and Dixon [24, 27].

All information that can be derived from similarity profiles depends for its interpretation in terms of physical quantities on the knowledge of a single empirical parameter, namely, σ.

3.2.1.2. The Similarity Parameter σ

The similarity parameter σ originally introduced by Görtler [18] has been well established to be twelve for the incompressible mixing between a uniform stream and a quiescent fluid (one-stream mixing), but its dependence upon free stream Mach number and stagnation pressure ratio T_{ob}/T_{oa}, are still subject to much speculation despite the extensive work reported by Abramovich [16] and others [28].

More complicated, and even less settled, are questions concerning the similarity parameter σ_{II} for the two-stream mixing case. Several attempts have been made to achieve a correlation between σ_{II} and σ_I where the latter would be referring to an »equivalent one-stream mixing case« [21, 29], but much is left to be done.

The tie-in between σ and the way it is used to relate a specific analytical solution to experimentally determined flow profiles is another source for disagreement. There appears now to be a preference for using the maximum velocity profile slopes (inflection point tangents), so that for an error function profile one would obtain

$$\sigma = \frac{\partial (u/u_a)}{\partial y}\bigg|_{max} x \, [\pi^{1/2}/(1-\phi_b)] \tag{38}$$

It is here appropriate to recall that σ is, strictly speaking—as a similarity parameter—defined for a specifically selected similarity profile only. Nevertheless, Eq. (38) has been used to follow the growth of σ towards its asymptotic value through the development regions of free jet boundaries. With reference to the remarks made concerning the initial development of free layers following a strong expansion of an attached turbulent boundary layer (section 3.1.3), one can re-evaluate such an apparent growth of σ by considering instead the growth of a locally similar »inner« mixing region within the frozen rotational »outer« region.

Shown in Fig. 9 are the results of such an interpretation as applied to (previously unpublished) experimental data obtained by A. F. Charwat. Details of the actual (iterative) numerical procedure for obtaining the apparent σ from their asymptotic value (or for obtaining their adjusted, asymptotic values from the apparent ones) are here omitted. The agreement indicated that the laterally inhomogeneous nature of developing

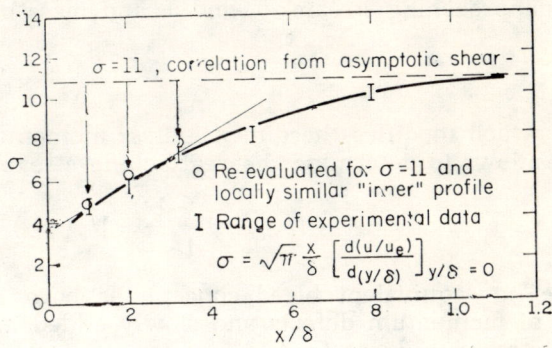

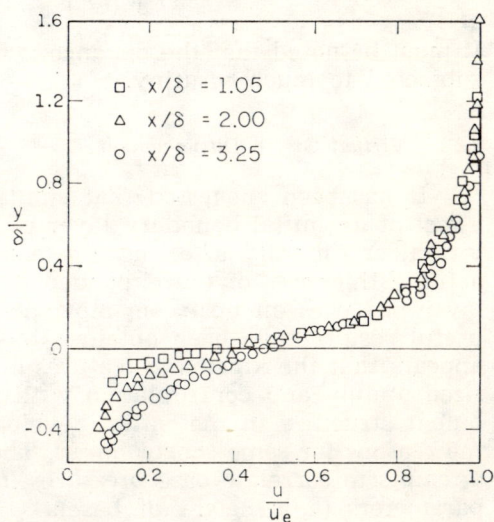

Fig. 9. Apparent and adjusted values for the similarity parameter σ for developing mixing profiles (experimental data by A. F. Charwat).

free turbulence—especially after a strong initial expansion is an important fact which must be carefully weighed when dealing with developing (rather than nearly developed) free shear layers.

3.2.2. Nearly Developed Free Shear Layers

Characterized by flow profiles which resemble closely asymptotic ones, this regime is of much practical importance, and, in addition, lends itself to a rather simple analysis with the help of results obtained and tabulated [21] for similar solutions. The momentum principle together with the idea of displacing the intrinsic coordinate system with respect to the (physical) reference coordinate system is here utilized.

3.2.2.1. Equivalent Mass Bleed Concept

Referring to Fig. 8, one can establish a dimensionless displacement parameter [7, 8] including the effects of momentum thicknesses in both approaching streams θ_a and θ_b and mass bleed G_B,

$$\Delta = \frac{\theta_a}{x} + \frac{\rho_b \theta_b \phi_a^2}{\rho_a x} + \frac{(1-\phi_b) G_B}{\rho_a u_a x} \tag{39}$$

which modifies the dimensionless momentum shift given by Eq. (31) and allows to determine the reattaching streamline η_d from

$$I_1(\eta_d) = \frac{1}{1-\phi_o} [I_1(\eta_{RA}) - I_2(\eta_{RA}) - \sigma \Delta] \tag{40}$$

The »equivalent bleed concept« is borne out by the interchangeability of momentum defects and slowly added mass bleed which, for a single stream, reduces to

$$(G_B)_{qu} = \theta_a \rho_a u_a \tag{41}$$

It must be noted that the distance x which appears in Eq. (31) must be subjected to much scrutiny.

3.2.2.2. Virtual Origin Displacement

It has been suggested that similarity solutions can account for the effect of an initial boundary layer of finite thickness by introducing a virtual origin shift, affecting the meaning of x [8, 9, 23, 30]. Invariably, a finite thickness of the expanded or compressed separating boundary layer produces an upstream movement of the initial origin and many useful results have been obtained by application of this scheme. Yet, it appears that the kinematic features of the initial disturbance are emphasized unduly and confrontation with experimental evidence on the turbulent structure in the initial development of free shear layers reveals the reasons for some shortcomings. These are quite pronounced when one attempts to correlate base pressures for back steps having large δ^{**}/h parameters (h is height of base).

The interesting agreement between theoretical and experimental data on virtual origin shift obtained by Sirieix and Solignac [31] does not distract from this dilemma, since the crucial issue of profile expansion was not touched in their investigation.

At this time we anticipate that a better understanding of the limitations imposed on the use of similarity solutions for developing profiles must come from more detailed studies of the nonsimilar regime.

3.2.3. Developing Free Shear Layers

The availability of high speed digital computers has stimulated the study of systems of nonlinear partial differential equations by numerical integration of difference equations. Powerful methods are now available to improve the economy of computations by iterative-implicit methods applicable to parabolic problems [32].

This makes it attractive to pursue the development of free shear layers from prescribed initial conditions by direct numerical integrations of the fundamental equations for mass momentum and energy conservation. An obvious difficulty arises from the presence of turbulent exchange coefficients. If one allows, however, the crude approximation (which certainly is not justified for the initial development length where the »inner« and »outer« regions in the sense of section 3.1.3 have not yet merged) that either one of $\rho\mu_t$ or μ_t or ε_t are functions of the streamwise coordinate x only, it is possible to obtain velocity profiles from which the viscosity law has been eliminated by a streamwise coordinate distortion [7].

Comparison with experimental profiles allows, then, to establish a relationship between the physical and the distorted coordinate, from which the viscosity law can be derived. Shown in Fig. 10 is the rapid rise of the turbulent eddy diffusivity represented by ε_t/v_{lam} from levels associated with an attached boundary layer (the experiment did not include an initial expansion of the attached boundary layers in the two streams) to those representative for fully developed profiles.

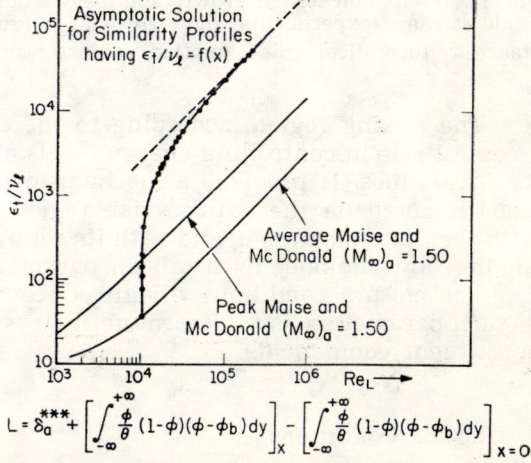

Fig. 10. Turbulent eddy viscosity development in the mixing region between two compressible streams with initial boundary layers.

$$L = \delta_a^{***} + \left[\int_{-\infty}^{+\infty} \frac{\phi}{\theta}(1-\phi)(\phi-\phi_b)dy\right]_x - \left[\int_{-\infty}^{+\infty} \frac{\phi}{\theta}(1-\phi)(\phi-\phi_b)dy\right]_{x=0}$$

Correlation between physical and transformed coordinates was, then, chosen on the basis of the monotonically increasing dissipation length Reynolds number for the viscous region. Study of the initial acceleration of the minimum velocity reveals the compensating effects of an initial carry-over and subsequent breakup of the laminar sublayers, which again underscores the inhomogeneity of the turbulent mechanism in the initial phases of free shear layer development.

Examination of Fig. 11 reveals that the similarity portion of the experimental profile, when matched on the basis of identical initial dissipation levels, point toward a positive origin shift. In addition, one could expect that dynamic similarity would not be present and the meaningful application of momentum shift processed similarity profiles would not be possible before the 45° straight-line portion for the experimental points has been reached. Again, the serious implications of dissimilarity in viscous behavior in »inner« and »outer« regions of expanding (and compressing) separating boundary layers shed much doubt on these and other methods [33] trying to cope analytically with the initial development phases of initially disturbed free shear layers.

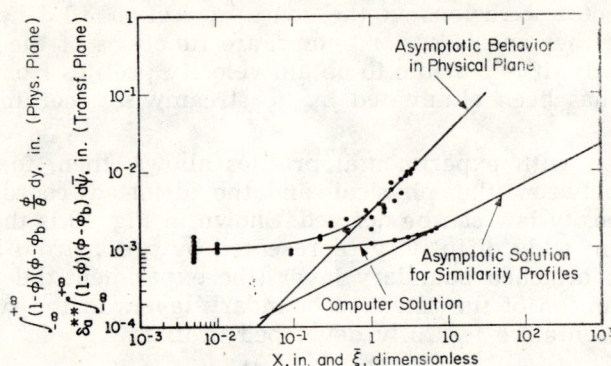

Fig. 11. Dissipation length growth for the viscous shear layer between two compressible streams (experimental data given as a function of physical downstream distance x, theoretical values plotted against transformed downstream coordinate ξ).

The mixing region, according to the analysis given in section 2, has a central role in controlling energy levels along and energy transfer across its streamlines. It provides a mechanism to meet certain physical conditions imposed on the entire wake region to sustain itself in interaction with the external stream, and with its solid boundaries. So far, the problem for the flow components has been parametric.

The closure condition will introduce the needed overall dependency of such parameters as to determine the system solution by synthesis of its constituent components.

4. Closure Conditions

4.1. CRITERIA APPLIED TO THE STAGNATING STREAMLINE

Treating the wake as quiescent and imposing an »escape criterion« on reattaching or realigning streamlines has formed the basis for attacking a large variety of problems involving flows with separation* by utilization of the so-called Chapman-Korst model. Effects of base bleed (actual or equivalent), base heating, base burning or reactive mixing were exclusively observed as selecting the reattaching streamline to assure conservation of mass, to determine the temperature level in the wake to satisfy energy conservation and to establish the energy level for this streamline to allow closure.

---

* An account of the many problems in internal and external flow, including applications to airfoil theory, propulsion-airframe integration and thrust augmentation schemes in jet and rocket propulsion is given in [7].

4.1.1. Point Closure

The closure condition was simply formulated by equating the stagnation pressure of the stagnating streamline as being equal to the static pressure in the realigned or reattached flow [34, 3]). Although the incorrectness of the original assumption, namely, to equate the stagnation pressure with the static pressure of the fully aligned flow after recompression at the end of the wake was recognized at the outset [35] the compensating character of incomplete terminal pressure rise and stagnation pressure reduction along the reattaching streamline due to the always present initial boundary layer effects led to an initial preference to use this point closure. Being simple, free of any empirical corrections, and producing results it provided an unadulterated basis for comparison with experimental evidence. While the influences of a large variety of parameters affecting geometry, external interference, bleed, convective and reactive heating were generally well predicted in a qualitative sense [7], quantitative inaccuracies were easily traced to the recompression scheme employed.

Nash [36] proposed an empirical correction applied to the pressure rise ratio at the end of the wake, which, however, proved to be difficult to relate to as pertinent recognized variables of separated flows.

4.1.2. Distributed Recompression

The observation that the recompression process is a gradual one, the stagnation point of the reattaching streamline being imbedded in the region of still rising pressures, attracted interest to the streamwise variation of the external inviscid flow near the end of the wake. Supersonic flow, with its simple pressure-streamline-direction relationship immediately led Carrière and Sirieix to the suggestion to re-examine the closure condition in the sense of an angular correction for the incomplete turning of the external flow at the streamline closure location [37]. Similar methods for establishing angular reattachment criteria have been proposed by others [38, 39].

4.2. INTEGRAL CLOSURE CONDITIONS

4.2.1. Boundary Layer Rehabilitation Concepts

Boundary layer rehabilitation concepts have been proposed by McDonald [40] and Roberts [41] retaining within their analysis the consideration of a stagnating streamline), and applied to the solution of base pressure problems.

4.2.2. Control Volume Analyses

Instead of associating the closure condition with the mechanical energy level of a single streamline, the conservation of momentum is satisfied for suitably chosen control surfaces enclosing all, or parts, of the reattachment region. Certain assumptions have always to be made which require and introduce some educated guesses. Lamb and Hood [42] use a force balance in both the x- and y-direction, but have to associate their control volume with a highly simplified reattachment streamline configuration, in addition to other simplifications. A conjecture has to be made that the expected solution be unique by exhibiting a double root feature for a vanishing residue of the momentum.

Results obtained by this method can be judged by the reasonable values which can be obtained for the empirical reattachment pressure rise correction suggested by Nash [39]. It is further noteworthy that a pressure distribution along the reattachment region is produced by the assumption of a linear pressure rise to the reattachment point followed by a boundary layer rehabilitation process which accounts for the terminal portion of the pressure distribution. A somewhat more comprehensive control volume approach presently pursued by computer oriented studies at the University of Illinois at Urbana-Champaign, see Fig. 12, is essentially free from assumptions and shows good promise.

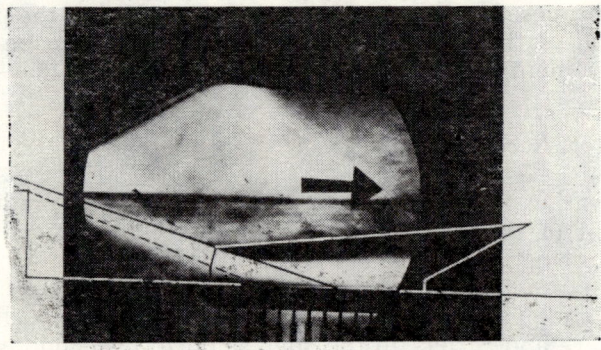

Fig. 12. Control volume configuration and calculated flow field compared with schlieren photo for supersonic flow past a backstep (University of Illinois at Urbana--Champaign, 1969).

Work reported by L. W. Gogish [43] falls into the same category since it introduced empirical information on the pressure-cross section relationship in the viscous flow region.

5. Separated Flows with Finite Wake Velocities

While problems of wake dynamics, such as determination of base pressures, could be attacked on the basis of energy analysis of a single streamline, thermodynamic aspects of wake flows required a complete systems approach. Analyses of wake temperatures and their control by mass bleed [44] and prescribed rates of heat addition through convection [45] or reactions [25] are earlier examples for the systems approach. Yet, with one notable exception [46], no explicit interest was shown in the wake flow field or in introducing mechanisms for convective processes along the wake boundaries outside of the mixing region. The latter remained specified as resulting from the viscid interaction between a uniform free stream (having possibly an initial profile disturbance due to a viscous and thermal boundary layer) and a quiescent wake. This has been one of the reasons for the shortcomings of earlier attempts to analyze theoretically heat transfer in regions of separated flows [47] and the lack of a meaningful formulation for convective heat transfer mechanisms in addition to the jet mixing component further obscured the situation. Only a complete thermodynamic analysis of the entire separated flow region, considered to be a system, and a meaningful treatment of all transfer mechanisms across its boundaries can truly be expected to lead to a better understanding of the problem, which must correctly be defined as »heat transfer to and across regions of separated flows«. Finite wake velocities are needed to account properly for convective mechanisms along the wake boundaries, both along the mixing regions and along solid walls.

Therefore, it will be necessary to discard the assumption of a quiescent wake and to deal with a wake flow field in order to provide a physically meaningful basis for a thermodynamic systems analysis which includes all the significant energy transfer components of a complete flow-and-heat transfer model for separated flows.

5.1. CAVITY FLOW MODEL

While the proposed concepts should prove to be useful in the study of separated flows in a most general sense, application is here restricted to steady flow past cavities, so that the problem of flow separation and flow reattachment remains out of consideration. The analytical treatment specializes throughout to turbulent flow components with effective Prandtl number of $Pr_t = 1$ and also to such cavity configurations (nearly circular or square) for which the dissipative mechanisms can be rationally treated and a wake flow field easily described. This appears most attractive since the merits of the new proposed flow model had to be evaluated by comparison with experimenttal evidence.

The wake flow model is here to be approached in a new way, namely, by formulating consecutive systems approaches for the »dissipative flow model« and the »heat transfer model«, the former being based on a concept involving transfer of mechanical energy to and its dissipation within the system, and yielding information on finite wake velocities; the latter being established with the help of a complete thermodynamic analysis and utilizing the finite wake velocities to deal with the essential convective mechanisms.

5.1.1. The Dissipative Flow Model—Systems Analysis Based on Balance of Mechanical Energy and Conservation of Mass

The separated flow region is here represented as a system (see Fig. 1), the boundaries of which consist of solid walls forming the cavity, the jet boundary streamline j originating at the point of separation S, and a cross section near the point of reattachment R, which closes the gap between the streamlines j and d, the discriminating streamline reattaching at R

($j=d$ for the case of no mass bleed into the wake). Two-dimensional geometry is assumed, although subsequent approximate extension to other suitable geometries is possible.

Restricting our analysis to steady-flow conditions, we recall Eq. (14), section 2.2.2.

$$\int_0^{l_m} \tau_j u_j \, dx - \frac{1}{2} \int_{Y_d}^{Y_j} \rho u^3 \, dY = \int_v e_D \, dV \qquad (14)$$

where the left-hand terms represent the net influx of mechanical energy into the system, while the right-hand term describes the total rate of dissipation of mechanical energy within the system. In particular, the first left-hand term evaluates the rate of shear work transferred from the external stream to the system. The second left-hand term accounts for the outflow of kinetic energy from the system in the case of mass bleeding (an effective means of influencing drag and heat transfer for separated flow regions), which, for mass bleed into the wake at a rate G_B, requires on the basis of conservation of mass (section 2.2.1)

$$G_B + \int_{Y_j}^{Y_d} \rho u \, dY = 0 \qquad (13)$$

In the right-hand integral, e_D represents the local rate of mechanical energy dissipation per unit volume of the system.

Since Eq. (14) reflects a general law, it will be applicable to any separated flow region regardless of wall shape; yet, the quantitative evaluation of the integrals, in particular the dissipation integral, will be possible only if the flow geometry is such that the various mechanisms of dissipation can be delineated and become accessible to quantitative treatment.

If the analysis is restricted to only such wake geometries for which the interaction between the different dissipative regions does not affect their basic mechanisms, the dissipation integral in Eq. (14) can be represented as the sum of individual parts, in the form

$$\int_v e_D \, dV = E_{D_m} + E_{D_R} + E_{D_{BL}} + E_{D_c} \qquad (42)$$

with the right-hand terms now expressing, in order, the total rates of dissipation in the jet-mixing region, the recompression zone, the cavity-wall boundary-layer, and the wake core region. Such local dissipation rates will depend on the kinematics and dynamics of the entire wake flow field.

Our theoretical studies will be concerned with quantitative exploitations of a flow model which allows a discussion of the major and essential mechanisms of separated flows, yet is simple enough in wall geometry and flow field kinematics to give meaning to both theory and comparison with experiment.

The logical choice is a nearly circular cavity. For such a geometry, the core region will exhibit nearly solid body rotation, hence eliminating the dissipative contribution E_{D_c}. The sharp leading edge at point R will reduce the effect of reattachment on the approaching flow, thus giving weight for equating the geometrical length l_m to the effective length of the jet mixing region, while eliminating most of our concern about E_{D_c}.*
(A sharp trailing edge at S will provide a well-defined flow field of confluence, again supporting the physical significance of l_m).

The core flow region will, by itself, produce a nearly constant flow velocity along the edges of the dissipative flow regions (jet mixing region and cavity wall boundary layer) which may be assumed to be thin when compared to the radius of the cavity. We would, then, expect that the jet mixing region can be treated in reasonable approximation by using the results obtained for compressible, non-isoenergetic turbulent ($Pr_t = 1$) jet mixing between two compressible streams at constant pressures [21] since we want to specialize our present investigation to fully turbulent separated flows.

5.1.1.1. Two-stream Mixing

It should be recalled (see section 3.2.1) that solutions to the two-stream mixing problem have been based on a momentum integral method applied to an assumed error-function-profile velocity distribution

$$\phi = \frac{1}{2}\left[(1+\phi_t) + (1-\phi_o)\ erf\ \eta\right] \tag{30}$$

where $\Phi = u/u_a$, $\Phi_b = u_b/u_a$, $\eta = \sigma_{II}(y/x)$, and $\sigma_{II} = f(C_a^2, T_{ob}/T_{oa}, \Phi_t)$ is the empirical spreading parameter for the two-stream mixing profile

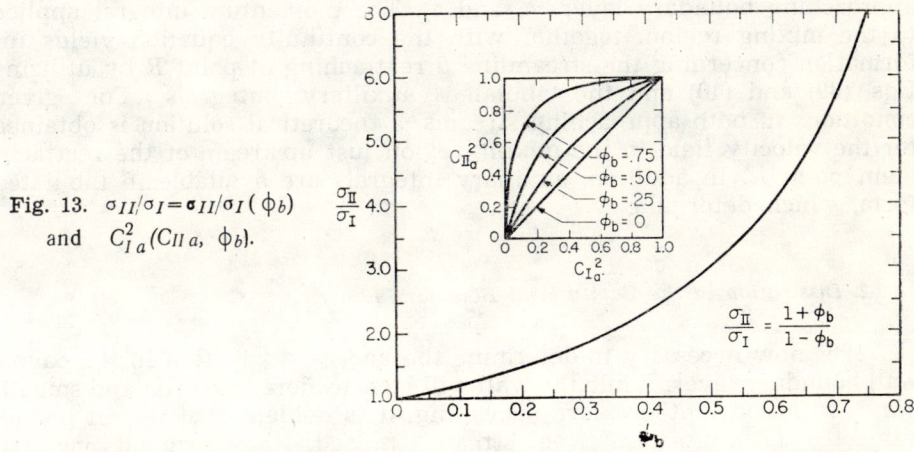

Fig. 13. $\sigma_{II}/\sigma_I = \sigma_{II}/\sigma_I(\phi_b)$ and $C_{Ia}^2(C_{IIa}, \phi_b)$.

---

* The high level of eddy viscosity in the jet mixing region [48] will, however, manifest itself in increased dissipation and heat transfer rates in the wall boundary layers on both sides of R.

For supersonic flow past cavities, the dissipation of mechanical energy in the bow shock wave near the reattachment point R may be of importance [49].

which can be related to the somewhat better known values for the mixing region between a uniform jet and a quiescent wake $\sigma_I = f(C'_o, T_{ob}/T_{oa})$ (see section 3.2.1.2 and Fig. 13), with $C'_a = u^2_a/(2c_p T_{oa})$, denoting the free stream Crocco number. The error function velocity distribution is interpreted for an intrinsic orthogonal system of coordinates x, y, which originates at point S and is slightly rotated with respect to the reference system of coordinates X, Y which also originates at S, but has the X-axis aligned with the direction $S—R$ (see Fig. 14).

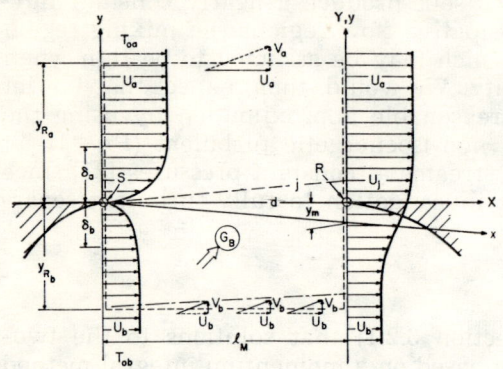

Fig. 14. Two-stream mixing component of the cavity flow model.

While the velocity distribution $\Phi(\eta)$, a similarity solution in terms of η, would normally apply to zero thickness approaching boundary layers, or asymptotically to far-downstream mixing cross sections only, the remaining degree of freedom, i.e., the still unspecified rotation between the two coordinate systems, expressed by y_m/l_m, see section 3.2.2, allows us to account (approximately) for the effects of finite thicknesses of the approaching boundary layers δ_a and δ_b. The momentum integral applied to the mixing region, together with the continuity equation yields information concerning the streamline d reattaching at point R by utilizing Eqs. (39) and (40) and the tabulated auxiliary integrals. For given conditions in both approaching streams, a theoretical solution is obtained for the velocity field in the mixing region just upstream of the reattachment point R. In addition, auxiliary integrals are available in tabulated form, which determine E_{D_m}.

5.1.1.2. Dissipation in the Cavity Wall Boundary Layer

It is now necessary to determine the energy dissipation in the cavity wall boundary layer. While the wall will be considered as a flat and smooth plate, with constant pressure prevailing, it is evident that it will not be subjected to a uniform free stream. Indeed, the terminal velocity distribution of the two-jet mixing problem will become the initial shear flow profile for the flow past the cavity walls. We have, therefore, the problem of energy dissipation in a boundary layer developing in shear flow.

We solve this problem approximately by assuming that the external shear flow retains its velocity distribution [50] and that the wall shear stress is locally controlled by the thickness of the wall shear layer and by the velocity at its edge where it joins the external profile.

A momentum integral method (here restricted to incompressible wake flow), was utilized and applied to an external shear profile resulting from the error profile velocity distribution in the mixing region, and to a wall shear layer $1/n$-th power profile. The resulting set of two simultaneous ordinary differential equations is

$$d\left(\frac{\delta}{l_m}\right)\left[\phi_\delta^2\left(\frac{2}{n+2}\right) - \phi_\delta \phi_b \left(\frac{1}{n+1}\right)\right]$$

$$+ d\phi_\delta \frac{\delta}{l_m}\left(\frac{n}{n+1}\right)\left[\phi_b - \phi_\delta \frac{2(n+1)}{(n+2)}\right]$$

$$= \alpha \phi_\delta^{2n/(n+1)} (\delta/l_m)^{-2/(n+1)} Re_m^{-2/(n+1)} d(x/l_m) \quad (43)$$

$$d\phi_\delta = -(1-\phi_b)\pi^{-1/2} exp[-(\eta_d - \sigma\delta/l_m)^2] \sigma_{II} d(x/l_m) \quad (44)$$

where $Re_m = (u_a l_a \rho_a)/\mu_a$. Selecting $\alpha = 0.0225$ (see [51]) these equations were solved numerically for a wide variety of wake flow conditions with the initial values

$$\frac{x}{l_m} = 0, \quad \frac{\delta}{l_m}(0) = 0, \quad \phi_\delta(0) = \phi(\eta_d)$$

and yielded

$$\delta = \delta\left(\frac{x}{l_m}, Re_m, n, \phi_b, \phi_d\right)$$

The dimensionless dissipation integral $\vec{E}_{D_{BL}}$ was also evaluated as

$$\bar{E}_{D_{BL}} \equiv \frac{E_{D_{BL}}}{1/2 \rho_a u_a^3 l_m} = \frac{I_4(\eta_{RA}) - I_4(\eta_{i\delta})}{\sigma_{II}}$$

$$+ \frac{\delta}{l_m} \phi_\delta^3 \left(\frac{n}{n+1}\right)\left[\frac{\phi_b^2}{\phi_\delta} - \frac{n+1}{n+3}\right] \quad (45)$$

with $I_4(\eta, C_a^2, T_{ob}/T_o, \Phi_b)$, see section 3.2.1.1, available in tabulated form [21].

5.1.1.3. Net Mechanical Energy Transfer through the Mixing Region

We still must find the transfer rate of mechanical energy to the wake, which is again achieved with the help of auxiliary integrals tabulated for the two-stream mixing problem [21].

For better convenience, this can be done by combining

$$\int_0^{l_m} \tau_j u_j dx - \frac{1}{2}\int_{Y_d}^{Y_j} \rho u^3 dY - E_{D_m} = E_T \quad (46)$$

where E_T now represents the net transfer of mechanical energy to the wake, already excluding the portion which has become dissipated in the jet mixing region itself.

The dimensionless expression

$$\vec{E}_T \equiv \frac{E_T}{1/2\, \rho_a\, u_b^3\, l_m} = \frac{I_4(\eta_d)}{\sigma_{II}} \tag{47}$$

can be determined by using the tabulated integrals [21].

5.1.1.4. Determination of the Wake Reference Velocity

The balance of mechanical energy, as expressed by Eq. (14) now can be written in the dimensionless form

$$\frac{E_T}{1/2\, \rho_a\, u_a^3\, l_n} = \frac{E_{D_{BL}}}{1/2\, \rho_a\, u_a^3\, l_m} \tag{48}$$

Noticing that for low speed flow ($C_a^2 = 0$) and nearly isothermal wakes

$$T_{ob}/T_{oa} = 1, \quad \frac{E_T}{1/2\, \rho_a\, u_a^3\, l_m} = f(\phi_b, \Delta)$$

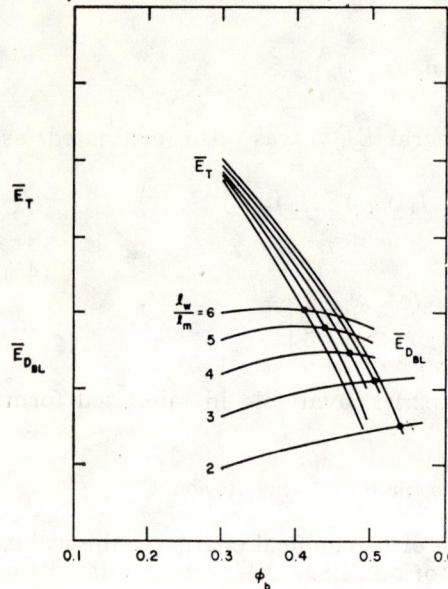

Fig. 15. Determination of wake reference velocity.

and

$$\frac{E_{D_{BL}}}{1/2\, \rho_a\, u_a^3\, l_m} = f(Re_m, \phi_b, l_w/l_m, \Delta)$$

where

$$\Delta = f(Re_m, \delta_a/l_m, \delta_b/l_m, G_B, \phi_b)$$

and
$$\delta_b/l_m = f(Re_m, l_u/l_m, \phi_b)$$

one may use the unknown reference wake velocity Φ_b as the independent variable to find, for given values of Re, δ_a/l_m and l_w/l_m, a value of Φ_b to satisfy the condition for balancing the mechanical energy, Eq. (48). Figure 15 illustrates the matching procedure for $Re_m = 10^6$, $\Delta_a = 0$ and the various wall length ratios $2 \leq (l_w/l_m) \leq 6$. It is then possible to represent a solution for the reference wake velocity applicable to low speed flow past nearly isothermal cavities in the form

$$\phi_b = f(C_a^2 = 0, T_{ot}/T_{oa} = 1, Re_m, \Delta, l_w/l_m) \tag{49}$$

as shown in Fig. 16.

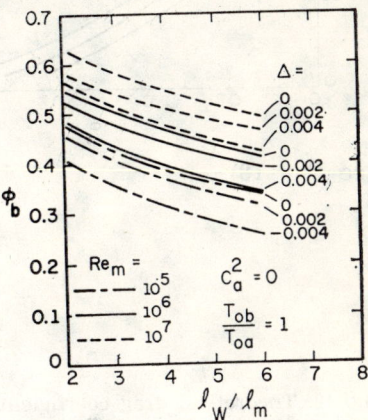

Fig. 16. Wake reference velocity vs. wall-to--mixing length ratio.

5.1.1.5. Drag Coefficient of a Cavity

The drag of a cavity can be related to the jet-mixing flow component (see Figs. 1 and 14) in the following manner,

$$D = \int_0^{l_m} \tau_j \, dx - \int_{Y_d}^{Y_j} \rho u^2 \, dY + G_B u_b \tag{50}$$

when the bleed mass is added to the uniform secondary stream having the velocity u_b. The drag coefficient

$$C_f = \frac{2D}{l_m u_a^2 \rho_a} \tag{51}$$

can be expressed by

$$C_f = \frac{2}{\sigma_{II}} [I_1(\eta_{Ra}) - I_1(\eta_d) + I_2(\eta_d) - I_2(\eta_{Ra}) - \sigma_{II} \Delta] \tag{52}$$

In Fig. 17, C_f has been plotted as a function of Φ_b with Δ as a parameter.

By utilizing the theoretical values for Φ_b [Eq. (49)] (Fig. 16), one can calculate the theoretical drag coefficient for nearly isothermal cavities in low speed flow.

$$C_f = f\,(C_a^2 = 0, T_{ob}/T_{oa},\ Re_m,\ \Delta,\ l_w/l_m) \tag{53}$$

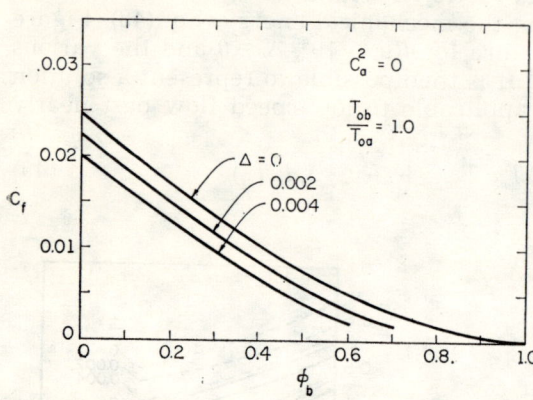

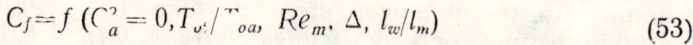

Fig. 17. Theoretical cavity drag coefficient as a function of wake reference velocity and mass bleed.

(see Fig. 18).

Fig. 18. Theoretical drag coefficients for low-speed turbulent flow past adiabatic cavities.

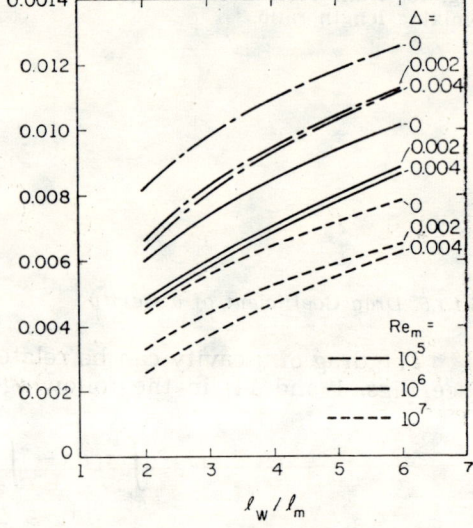

5.1.2. The Heat Transfer Model

The dissipative cavity-flow model based on a system balance for mechanical energy (section 2.2.2.) has furnished information on finite wake velocities (specifically the wake reference velocity Φ_b and the Φ distribution along the cavity wall) needed to account for convective heat transfer mechanisms associated with cavity flows. By the thermodynamic systems analysis (section 2.2.3.), individual components of energy transfer can now be delineated and integrated into a physical perceptive model for heat transfer to and across separated flow regions. We recall Eq. (15).

$$Q_w + c_p G_B (T_{oB} - T_{oa}) + \Omega_d + Q_d = 0$$

and define the *Cavity Stanton number* as

$$St_c = Q_w / [l_m u_a \rho_a c_p (T_w - T_{oa})] \qquad (54)$$

and the *Average Cavity Wall Stanton number** as

$$\overline{St}_w = Q_w / [l_w u_b \rho_b c_p (T_w - T_{ob})] \qquad (55)$$

while the dimensionless group

$$\frac{\Omega_d}{l_m u_a \rho_a c_p (T_{oa} - T_{ob})}$$

for the case of zero bleed, $G_B = 0$, $d = j$, can be identified as the *Mixing Region Stanton number*.

In addition, we define the ratio

$$\pi_d = -Q_d / Q_w \qquad (56)$$

as the partition function.

5.1.2.1. Stanton Number for the Mixing Region

Tabulated auxiliary functions [21] serve conveniently to determine the energy transport through the mixing region as one may express

$$\frac{\Omega_d}{l_m u_a \rho_a c_p (T_{oa} - T_{ob})} = \frac{I_1(\eta_{Ra}) - I_1(\eta_d) - I_3(\eta_{Ra}) + I_3(\eta_d)}{\sigma_{II}(1 - T_{ob}/T_{oa})} \qquad (57)$$

while for zero bleed,

$$St_m = \frac{1 - C_a^2}{\sigma_{II}(T_{ob}/T_{oa} - 1)} \int_{\eta_j}^{\infty} \frac{\phi(\Lambda - 1)}{\Lambda - C_a^2 \phi^2} d\eta$$

$$= \frac{I_1(\eta_{Ra}) - I_1(\eta_j) - I_3(\eta_{Ra}) + I_3(\eta_j)}{\sigma_{II}(1 - T_{ob}/T_{oa})} \qquad (58)$$

which approaches, as a limit for nearly isothermal wakes, $(T_{ob}/T_{oa}) \to 1$.

$$\sigma_{II} St_m = \frac{1}{(1 - \phi_b)} \{[I_1(\eta_{Ra}) - I_1(\eta_j)] - [I_2(\eta_{Ra}) - I_2(\eta_j)]\} \qquad (59)$$

5.1.2.2. Heat Diffusion in the Mixing Region-Partition Function π

This function can be evaluated independently from other heat transfer mechanisms as long as we restrict ourselves to cases which involve small temperature differences. Then, changes in density due to the heat source

---

* The case of non-uniform wall temperature requires introduction and utilization of a calculated average wall temperature [52].

temperature field may be neglected and the latter linearly superimposed by virtue of the now linearized problem which assumes the velocity and density field as given by the jet mixing problem.

Thus, the problem is attacked [53, 54] as describing the diffusion of the thermal energy carried in the thermal boundary along the cavity wall to the point of confluence S, by seeking the temperature field of a line source, having the strength Q_w and diffusing into the mixing region represented by the error function velocity distribution of Eq. (30). This temperature field T_{ex} is being superimposed on the temperature field of the jet mixing region T_m by

$$T = (T_m + T_{ex})*$$

The boundary conditions for T_{ex} are, then, homogeneous. Starting with the diffusion equation

$$\rho u \frac{\partial T_{ex}}{\partial x} + \rho v \frac{\partial T_{ex}}{\partial y} = \varepsilon \rho \frac{\partial^2 T_{ex}}{\partial y^2} + \frac{\partial \rho}{\partial y} \frac{\partial T_{ex}}{\partial y} \tag{60}$$

we express the eddy diffusivity by

$$\varepsilon = (x \mu_a)/(4 \sigma_{II}^2) \tag{61}$$

and utilize the continuity equation

$$\frac{\partial}{\partial x}(u\rho) + \frac{\partial}{\partial y}(v\rho) = 0 \tag{62}$$

together with the similarity condition for the velocity field in the mixing region $\Phi = u/u_a = \Phi(\eta)$, to find

$$\frac{\sigma_{II} v}{u_a} = \eta \phi - \left[\int_{\eta_j}^{\eta} (\rho/\rho_a) \phi \, d\eta\right] / (\rho/\rho_a) \tag{63}$$

Since the concentrated heat source at $S=0$ suggests a similarity solution for the temperature field T_{ex} of the form

$$x \, T_{ex} = \theta(\eta) \tag{64}$$

we may rewrite Eq. (60) in the form

$$\frac{d^2 \theta}{d\eta^2} + \frac{d\theta}{d\eta} \left[\frac{2 C_a^2 (d\phi/d\eta)}{1 - C_a^2 \phi^2} + 4 \frac{\rho_a}{\rho} \int_{\eta_j}^{\eta} \frac{\rho}{\rho_a} \phi \, d\eta\right] + 4 \phi \, \theta = 0 \tag{65}$$

where $\Phi = \Phi(\eta)$ and $\rho/\rho_a(\eta) = (1-C_a^2)/(1-C_a^2 \Phi^2)$ are known solutions of the jet mixing problem, hence given functions of the independent variable η so that Eq. (65) is a linear second-order ordinary differential equation.

---

* As an alternative, the thermal boundary layer approaching the point S can be considered as an initial condition for the temperature fields in two-stream mixing, see section 3.2.3. An iterative procedure for combining all flow and heat transfer components in a comprehensive computer program has been carried out in [52].

Solutions were obtained by numerical integration (Runge-Kutta method) on a digital computer [53]. Absolute levels of θ are not of interest, as we determine the partition function from

$$\pi_d = \left[\int_{\eta}^{\infty}(\rho/\rho_a)\,\phi\,\theta\,d\eta\right] \bigg/ \left[\int_{-\infty}^{+\infty}(\rho/\rho_a)\,\phi\,\theta\,d\eta\right] \qquad (66)$$

and represent results for π_j in the form

$$\pi_j = f\,(T_{ob}/T_{oa} = 1,\ C_a^2,\ \phi_b) \qquad (67)$$

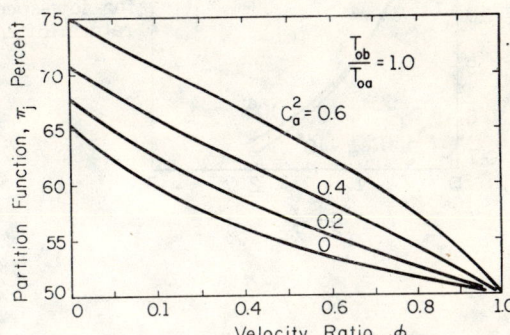

Fig. 19. Partition function π_j.

5.1.2.3. Wall Stanton Number

The average cavity wall Stanton number can be derived from the analysis of the wall boundary layer and, after utilizing Reynolds analogy to define a local Stanton number,

$$St\,(y) = \tau_w/(\rho_\delta\,u_\delta^2) = 0.0225\,[(u_\delta\,\rho_\delta^\delta)/\mu_\delta]^{-1/4} \qquad (68)$$

one expresses for a wall heated over the entire length l_w

$$\phi_b\,\overline{St}_w = 0.0225 Re_m^{-1/4}(l_m/l_w)\int_0^{l_m/l_w}\phi_\delta^{3/4}\,(\delta/l_m)^{-1/4}\,d\,(x/l_m) \qquad (69)$$

It is, therefore, possible to establish, as the result of numerical integration,

$$\phi_b\,\overline{St}_w = f\,(Re_m,\ \phi_b,\ l_w/l_m) \qquad (70)$$

The local convective heat transfer coefficient, which is variable along the cavity wall is determined by

$$h\,(\varsigma)/h = (l_w/l_m)\left[\frac{d}{d\varsigma}\int_0^{\varsigma} St\,(\varsigma)\,u_\delta\,d\varsigma\right] \bigg/ \left[\int_0^{l_w/l_m} St\,(\varsigma)\,u_\delta\,d\varsigma\right] \qquad (71)$$

(where $\zeta = x/l_m$, $x = 0$ at R) showing a monotone decrease as one progresses from point R toward point S, see Fig. 20.

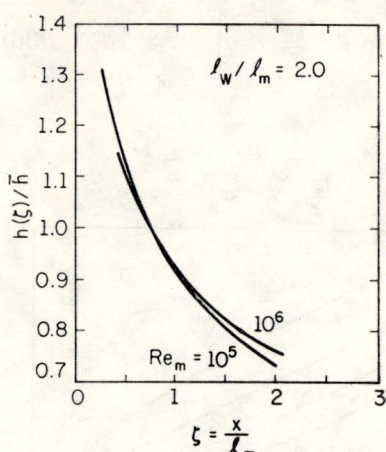

Fig. 20. Theoretical cavity Stanton number St_c for low-speed turbulent flow at small temperature differences.

5.1.2.4. Cavity Stanton Number

Based on on the energy balance, Eq. (15) and the defining equations for dimensionless groups identifying individual energy transfer processes, Eqs. (54), (55), (56), (57), (66), and (70), we may express the cavity Stanton number in the form

$$St_c = \frac{I_3(\eta_{Ra}) - I_s(\eta_d) - [I_1(\eta_{Ra}) - I_1(\eta_d)]}{\sigma_{II}[(T_{\bar{u}}/T_{oa}) - 1](1 - \pi_d)}$$

$$- \frac{G_B}{l_m u_a \rho_a} \frac{[(T_{oB} - T_{oa}) - 1]}{[T_v/T_{oa}) - 1](1 - \pi_d)} \tag{72}$$

For the case with zero bleed, we can also express

$$St_c = \frac{St_m}{1 - \pi_j + (St_n/St_w)(1/\phi_b)(\rho_a/\rho_t)(l/l_w)} \tag{73}$$

and

$$St_c = St_m \frac{(T_{ob}/T_{oa}) - 1}{(1 - \pi_j)[(T_w/T_{oa}) - 1]} \tag{74}$$

Theoretical cavity Stanton number (for zero bleed, low speed turbulent flow, and small temperature differences) are shown in Fig. 21 as a function of the wall length at ratio l_w/l_m for parametric values of Re_m of 10^5, 10^6, and 10^7. Also indicated in Fig. 21 are the regions for which the heat transfer rates across cavities exceed $[(St_c/St_{att}) > 1]$, or are less than $[(St_c/St_{att}) < 1]$, that for attached boundary-layer flow past a flat plate.

Inspection of Fig. 21 indicates that both Reynolds number and cavity wall ratio l_w/l_m (besides the modifying influences of Δ, which are not shown) have a strong effect on the cavity Stanton number. This is to be expected because of the importance of the dissipative and convective mechanisms in the boundary-layer along the cavity walls. It is also noteworthy that the present theory suggests that turbulent cavity Stanton numbers can be lower or higher than those of corresponding attached flat plate flows. The primary criterion for this is the wall length ratio l_w/l_m.

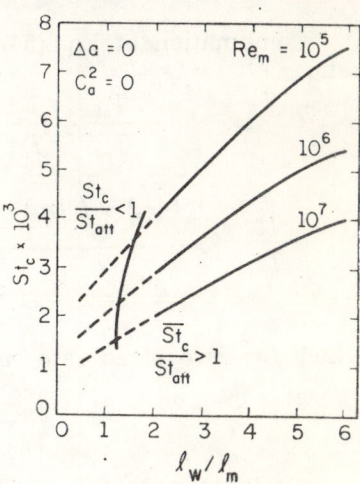

Fig. 21. Local heat transfer coefficient along cavity walls.

In contrast to these findings are the results (for turbulent flow of the theoretical analysis of heat transfer in regions of separated flow by D. R. Chapman [47], according to which, for turbulent flow, there is no influence of Reynolds number, or wall length ratio, and theoretical cavity Stanton numbers for low speed flow are of one order magnitude higher than those for the attached flow past a flat plate. As we inspect the basic assumptions made in Chapman's analysis, we notice that he stipulates (in our notation) $T_w = T_{ob}$, and since he does not account for any other heat transfer mechanisms but the one due to jet mixing, in his case, $\pi_j = 0$ and $\overline{St_w} \to \infty$. Consequently, he equates the Stanton number for the cavity to the Stanton number for the mixing region, $St_c = St_m$, for both the laminar and turbulent regimes.

While the model is incorrect for both regimes, and does not really represent a physically possible mechanism for heat transfer, as has already been observed by Charwat, et al., it is interesting to note that the quantitative consequences are quite different whether one considers laminar or turbulent regimes, as can be shown easily at the hand of Eq. (73) applicable to the case of zero bleed.

In the case of fully laminar flows past and within the cavity, and restricting oneself to configurations where $(l_w/l_m) \simeq 1$, it is by coincidence that the two terms in the denominator of Eq. (73) which are neglected in Chapman's analysis, namely, $-\pi_j + (St_m l_m \rho_a)/(\Phi_b \overline{St_w} l_w \rho_b)$, tend to cancel each other out. For turbulent cavity flows, however, where $St_w \ll St_m$, the denominator in Eq. (73) is much larger than unity and, therefore, the cavity Stanton number St_c is much smaller, than the mixing region Stanton number St_m.

Much additional insight can be gained from the theoretical analysis of wake temperatures.

5.1.2.5. Wake Temperature

Combination of Eq. (54) and Eq. (55) yields the wake temperature ratio

$$\frac{T_{ob}}{T_{oa}} = \frac{T_w}{T_{oa}} - \frac{St_c}{St_w}\left(\frac{T_w}{T_{oa}}-1\right)\frac{l_m}{l_w}\frac{\rho_a}{\rho_b}\frac{1}{\phi_b}$$

$$= \frac{T_w}{T_{oa}} - \frac{I_3(\eta_{Ra}) - I_3(\eta_d) - I_1(\eta_{Ra}) + I_1(\eta_d) - \dfrac{G_b\,\sigma_{II}}{l_m\,\rho_a\,u_a}\left(\dfrac{T_{oB}}{T_{oa}}-1\right)}{\sigma_{II}\,\overline{St}_w\,\dfrac{l_w}{l_m}\,\dfrac{\rho_b}{\rho_a}\,\phi_b(1-\pi_d)} \quad (75)^*$$

which for zero bleed rate can be brought into the form

$$\frac{T_{ob}-T_{oa}}{T_w-T_{oa}} = (1-\pi_j)\frac{St_c}{St_m} \quad (76)$$

A critical evaluation of Chapman's theory can be based on a discussion of Eq. (76).

Control of the adiabatic wall temperature by mass bleed into the cavity is here chosen to illustrate the quantitative aspects of the presented theory. Since the assumption of $Pr_t = 1$ leads to the special case $T_w = T_{ob}$ (as $Q_w = 0$), Eq. (75) can be utilized, together with the results obtained

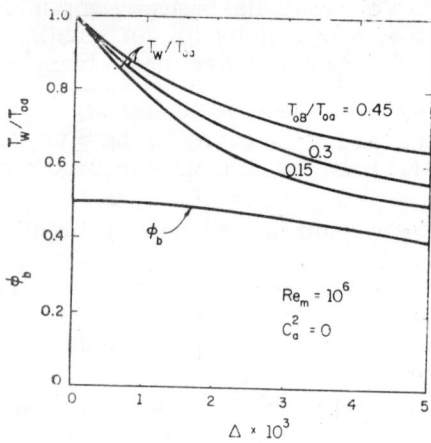

Fig. 22. Control of adiabatic cavity wall temperature by mass injection into the core.

---

* A modification in the expression for η_d in Eq. (75) in the sense of Eq. (40) is necessary to account for the effect of approaching boundary layers.

for the dissipative flow model, to establish a relationship of the form

$$\frac{T_w}{T_{oa}} = f(C_a^2, l_w/l_m, Re_m, T_{oB}/T_{oa}, \Delta)$$

Shown in Fig. 22 are the results of theoretical calculations for $C_a^2 = 0$, $l_w/l_m = 3$, $Re_m = 10^6$ [52] which show the influence of mass bleed into the wake, expressed here by

$$\Delta = (1 - \phi_b) \frac{G_B}{l_m \rho_a u_a}$$

(in absence of »equivalent bleed«) for various parametric values of the mass bleed temperature ratio T_{oB}/T_{oa}.

5.2. EXPERIMENTAL RESULTS

A critical evaluation of the merits of the present theory has to be based on experimental evidence, part of which was available in the literature (such as [55]). Additional tests have been carried out at the University of Illinois with the special aim of clarifying the validity of the new concepts in their overall as well as component manifestations.

Since most of Larson's [55] tests were conducted with axially symmetric cavity configurations, one has first to interpret his models in terms of our length ratio l_w/l_m.

It appears logical to equate the ratio of wall cavity surface to jet mixing surface as equivalent to l_w/l_m. On this basis, the model used by Larson for low speed experiments had an equivalent length ratio of approximately 0.66.

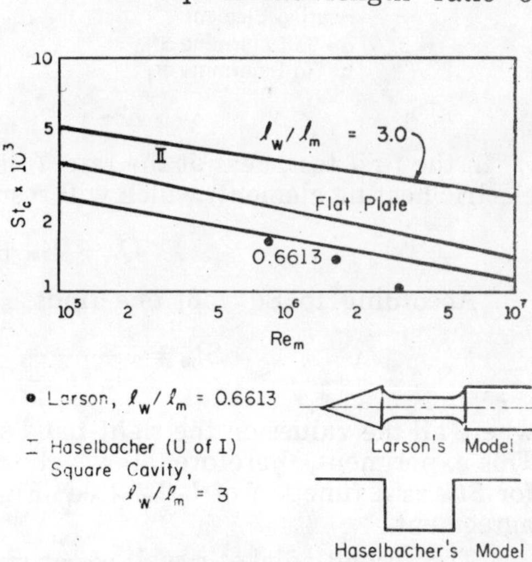

Fig. 23. Theoretical and experimental cavity Stanton numbers.

By cross plotting the results of Fig. 20 for this length ratio, one obtains the theoretical Reynolds number dependency for Larson's configuration as shown in Fig. 23 and can compare it with his experimental data, which results in a fair agreement. Also plotted in Fig. 23 is the theoretical curve for $l_w/l_m = 3$, which could be compared with the experimental data obtained at the University of Illinois for a square heated cavity, again indicating reasonable agreement.

Also of interest is a wake bulk temperature measurement reported by Larson. In searching for a reason for the large discrepancy in the turbulent regime between the results of Chapman's analysis and the experimentally obtained heat transfer data, he noticed that T_{ob} differed much from T_w, and much less from T_{oa}.

He did not, however, report any measurement of the bulk temperature in the laminar flow regime, which would have been revealing in view of Eq. (76).

Additional tests were conducted by Miles [53] in order to explore the validity of assumptions made in establishing the new cavity heat transfer model. Two series of tests were carried out with a nearly circular cavity having insulated walls (see Fig. 24). These experiments were restricted to the case of zero mass bleed.

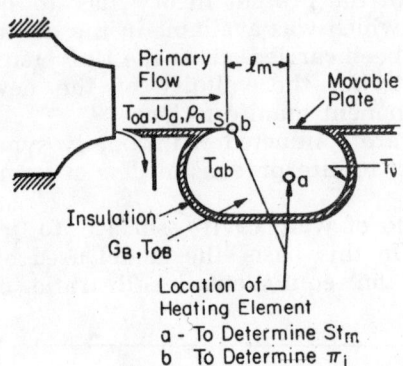

Fig. 24. Heat transfer experiments with insulated cavity.

In the first test, heat at the rate E, is supplied to the core region (by an electric heating element) which will result in an energy balance of the form

$$Q_m - W'_m + E = 0 \tag{77}$$

According to Eq. (56) one finds

$$St_m = \frac{E}{l_m (T_{ob} - T_{oa}) u_a \rho_a c_p} \tag{78}$$

where all the values on the right-hand side are obtained in the experiment. This experiment, therefore, allows checking directly the theoretical values for St_m as a function of l_w/l_m (see Fig. 25), apparently showing good agreement.

The second adiabatic wall experiment had the heating element placed near the point of confluence S. If heat is supplied at the rate E, the energy balance for the system now becomes

$$Q_m - W_m = -E(1 - \pi_j) \tag{79}$$

and solving for π_j with the help of Eq. (56).

$$\pi_j = \frac{St_m \, l_m \, (T_{ob} - T_{oa}) \, u_a \, \rho_a \, c_p}{E} \tag{80}$$

Again, all the terms on the right-hand side can be determined experimentally. Comparison between theoretical and measured values of π_j is

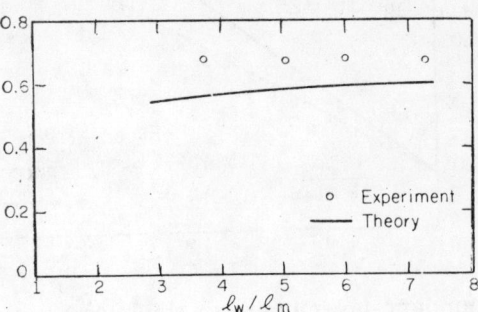

Fig. 25. Insulated cavity: Comparison of theoretical and experimental values at the Stanton number of the mixing region St_m (heating element at location a in Fig. 24, $G_B = 0$).

only fair (Fig. 26), but high sensitivity of this experiment to the exact location of the heating element can be expected.

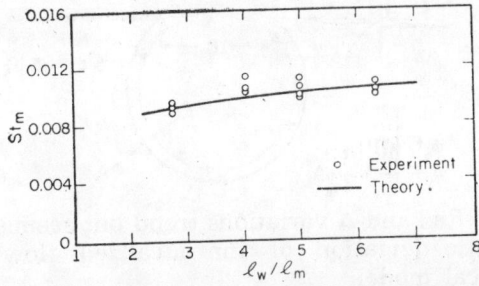

Fig. 26. Insulated cavity: Comparison of theoretical and experimental values for the partition function π_j (heating element at location b in Fig. 24, $G_B = 0$).

Tests were also conducted [52] to check the accuracy of predicting wake temperatures as controlled by mass bleed. Figure 27 gives a comparison between theoretically predicted and measured wake temperatures as determined in low-speed ($C_a = 0$) experiments with an insulated cavity having a heat source placed at the point of confluence S to simulate wall heat transfer.*

With that much evidence for the validity of the cavity model coming from heat transfer phenomena, it is of interest to seek direct support for the physical reality behind the idealized flow model by comparing theoretical and experimental wake velocities.

Experiments conducted by Golik [49] with a nearly circular cavity (see Fig. 28) for which the length ratio was varied by extending a flat plate with a sharp leading edge into the mixing region upstream of the original

---

* The influence of the approaching boundary layer ($\Delta_a \neq 0$) required special attention for the correct thermodynamic interpretation of the equivalent bleed concept, see [55] for further details.

recompression point R, were conducted at a constant Reynolds number of $10^6/ft$ which resulted in a variation of Re_m and Δ (due to an approaching

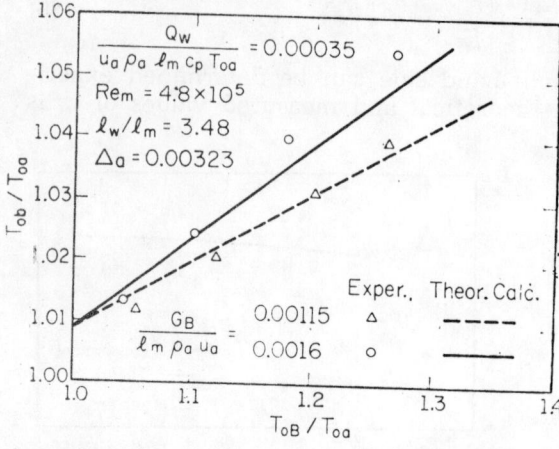

Fig. 27. Insulated cavity: Comparison of theoretical and experimental values for wake bulk temperatures (heating element at location b in Fig. 24, $G_B \neq 0$).

boundary-layer of given thickness) with changing l_w/l_m. Shown in Fig. 29 is a comparison between measured and theoretical values of Φ_b which were

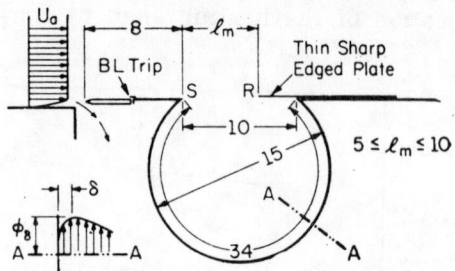

Fig. 28. Experimental determination of wake reference velocities model.

calculated on the basis of the proper Re_m and Δ variations trend but seems to bear out a typical and systematic deviation of the idealized flow geometry, assumed for the theoretical model.

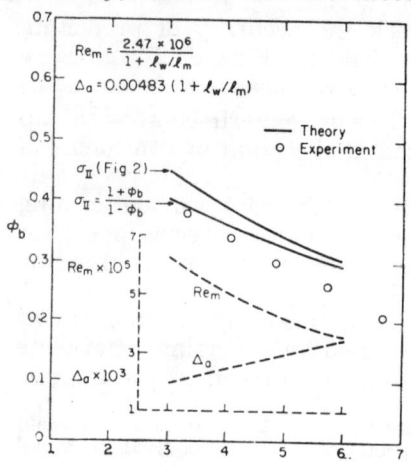

Fig. 29. Comparison of experimental and theoretical wake reference velocities.

5.3. APPLICABILITY TO NONCIRCULAR CAVITIES AND WAKES

The flow regions near the points of confluence S and reattachment R are decreasing the effectiveness of the mixing region, while increasing the weight of dissipative mechanisms. Such a tendency would be reflected by characterizing actual cavities through effective wall ratios l_w/l_m larger than those given by the cavity geometry. Assigning effective wall ratios to cavities or wakes, having geometries differing from the nearly circular shapes (which were considered essential for satisfying the assumptions made in our theoretical analysis) apparently offers a sensible and reasonable approach to extending the usefulness of the present theory.

Such an expedient concept should, however, not be accepted without strong reservations, since the specific cavity geometries can decisively affect the flow pattern in the separated flow region. Depending upon geometry and / or flow conditions past the cavity, flow instabilities [56] or cavity resonance can greatly alter the character of the core flow; which only for nearly circular wakes closely approaches that of solid body rotation.

REFERENCES

1. Crocco, L., and Lees, L., »A Mixing Theory for the Interaction between Dissipative Flows and Nearly Isentropic Streams«, J. Aero. Sci., **19**, 10, pp. 649—676 (October 1952).
2. Lees, L., and Reeves, B. L., »Supersonic Separated and Reattaching Laminar Flows: I. General Theory and Application to Adiabatic Boundary Layer-Shock Wave Interactions«, AIAA J., 2 : 11, pp. 1907—1920 (1964).
3. Korst, H. H., »Comments on the Effect of Boundary Layer on Sonic Flow through an Abrupt Cross-section Area Change«, J. Aero. Sci., **21**, pp. 568—569 (1954).
4. Korst, H. H., »A Theory for Base Pressures in Transonic and Supersonic Flow«, J. Appl. Mech., 23, pp. 593—600 (1956).
5. Chapman, D. R., Kuehn, D. M., and Larson, H. K., »Investigation of Separated Flows in Supersonic and Subsonic Streams with Emphasis on the Effect of Transition«, NACA Report 1356, 1958.
6. Alber, I. E., and Lees, L., »Integral Theory for Supersonic Turbulent Base Flows«, AIAA J., **6**, pp. 1343—1351 (July 1968).
7. Korst, H. H., »Turbulent Separated Flows«, VKI. CN 666, von Karman Institute for Fluid Dynamics, Rhode-Saint-Genese, Belgium, 1967.
8. Nash, J. F., »An Analysis of Two Dimensional Turbulent Base Flow Including the Effect of the Approaching Boundary Layer«, NPL Aero. Rep. 1036, ARC, 1962.
9. Kirk, F. N., »An Approximate Theory of Base Pressure in Two-dimensional Flow at Supersonic Speeds«, R.A.E. TN Aero 2377, March 1954.
10. Carrière, P., and Sirieix, M., »Facteurs d'influence du recollement d'un écoulement supersonique«, ONERA Memo Techn. 20, 1961.
11. White, R. A., »Turbulent Boundary Layer Separation from Smooth-Convex Surfaces in Supersonic Two Dimensional Flow«, Ph. D. Thesis, Dept. of Mech. Eng., University of Illinois, 1963.
12. White, R. A., »Effect of Sudden Expansions or Compressions on Turbulent Boundary Layer«, AIAA J., **4**, pp. 2232—2234 (1966).
13. Culick, R. E. C., and Hill, J. A. F., »A Turbulent Analog of the Stewartson-Illingworth Transformation«, J. Aero. Sci., 25, pp. 259—262 (1958).
14. Maise, G., and McDonald, H., »Mixing Length and Kinematic Eddy Viscosity in a Compressible Boundary Layer«, AIAA Paper 67—199, presented at the AIAA 5th Aerospace Sciences Meeting, New York, January 23—26, 1967.
15. Pai, S. I., *Fluid Dynamics of Jets*, D. Van Nostrand Co., New York, 1954.
16. Abramovich, G. N., *The Theory of Turbulent Jets* (translation), MIT Press, Cambridge, Mass., 1963.

17. Tollmien, W., »Berechnung turbulenter Ausbreitungsvorgänge«, ZAMM, 6, pp. 468—478 (1926).
18. Görtler, H., »Berechnung von Aufgaben der freien Turbulenz auf Grund eines neuen Näherungsansatzes«, ZAMM, 22, pp. 244—254 (1942).
19. Reichardt, H., Ü»ber eine neue Theorie der freien Turbulenz«, ZAMM, 21, pp. 257 ff (1941).
20. Korst, H. H., »Auflösung eines ebenen Freistrahlrandes bei Berücksichtigung der Ursprünglichen Grenzschichtströmung«, Oesterr. Ing. Arch., 7, pp. 152—157 (1954).
21. Korst, H. H., and Chow, W. L., »Non-isoenergetic Turbulent ($Pr_t=1$) Jet Mixing between Two Compressible Streams at Constant Pressure«, NASA CR-49, 1966.
22. Jacques, R., and Gailly, A., »Mélange supersonique turbulent et application aux problèmes de recollement«, AGARD CR-4, Part I, pp. 271—301.
23. Page, R. H., and Dixon, R. J., »Base Heat Transfer in a Turbulent Separated Flow«, Proc. 4th Int. Symp. on Space Technology and Science, Tokyo, AGNE Corp., Tokyo, Japan, pp. 295—308, 1963.
24. Page, R. H., and Dixon, R. J., »Computer Evaluation of an Integral Treatment of Gas Mixing«, Third Conference on Performance of High Temperature Systems, Pasadena, California (proceedings in press), 1964.
25. Davis, L. R., »Experimental and Theoretical Determination of Flow Properties in a Reacting Near Wake«, AIAA J., 6, pp. 843—847 (1968).
26. Da-Riva, L., and Urrutia, J. L., »Ignition Delay in Diffusive Supersonic Combustion«, AIAA J., 6, pp. 2095—2101, (1968).
27. Page, R. H., »Heat and Mass Transfer in Separated Flows«, Advanced Heat Transfer, B. T. Chao, ed., University of Illinois Press, Urbana, Chicago, London, pp. 277—300, 1969.
28. Bauer, R. C., »Another Estimate of the Similarity Parameter for Turbulent Mixing«, AIAA J., 6, pp. 925—927 (1968).
29. Miles, J. B., and Shih, J. S., »Similarity Parameter for Two-Stream Turbulent Jet-Mixing Region«, AIAA J., 6, pp. 1429—1430 (1968).
30. Hill, W. G., Jr., »Initial Development of Compressible Turbulent Free Shear Layers«, Ph. D. Thesis, Rutgers—The State University, New Brunswick, New Jersey, May 1966.
31. Sirieix, M., and Solignac, J. L., »Contribution à l'étude expérimentale de la couche limite de mélange turbulent isobare d'un écolement supersonique«, AGARD CP 4, Part I, p. 241.
32. Clausing, A. M., »Numerical Methods in Heat Transfer«, Advanced Heat Transfer, B. T. Chao, ed., University of Illinois Press, Urbana, Chicago, London, pp. 157—216, 1969.
33. Lamb, J. P., »An Approximate Theory for Developing Turbulent Free Shear Layers«, Paper 66-WA/FE-17, presented at the ASME Winter Annual Meeting, New York, 1966.
34. Korst, H. H., »Research on Transonic and Supersonic Flow of a Real Fluid at Abrupt Increases in Cross Section«, Technical Status Report A-PR-4, AE 18 (600)—392 Dept. of Mech. Eng., University of Illinois, Urbana, Illinois, July 1953.
35. Chapman, D. R., and Korst, H. H., Discussion of »Theory for Base Pressures in Transonic and Supersonic Flow«, J. Appl. Mech., 24, p. 484 (1957).
36. Nash, J. F., »A Discussion of Two Dimensional Turbulent Base Flow«, N.P.L. Aero R 1162 (ARC 27 175), July 1965.
37. Carrière, P., and Sirieix, M., »Facteurs d'influence du recollement d'un écoulement supersonique«, 10th Congrès International de Mécanique Appliquée, Stresa, 1960.
38. Page, R. H., Hill, W. G., and Kessler, T. J., »Reattachment of Two-dimensional Supersonic Turbulent Flows«, ASME Paper 67-FE-20, 1967.
39. Sirieix, M. J., Delery, J., and Mirande, J., »Recherches expérimentales fondamentales sur les écoulements réparés et applications«, ONERA TP 520, Paris, 1967.
40. McDonald, H., »Turbulent Shear Layer Reattachment with Special Emphasis on the Base Pressure Problem«, Aeronautical Quarterly, 15, p. 247 (1964).
41. Roberts, J. B., »On the Prediction of Base Pressure in Two-dimensional Supersonic Turbulent Flow«, ARC R and M 3434, 1966.

42. Lamb, J. P., and Hood, C. G., »An Integral Analysis of Turbulent Reattachment Applied to Plane Supersonic Base Flows«, Trans. ASME, Ser. C., **90**, pp. 553—560, 1968.
43. Gogish, L. W., »The Approximate Calculation of the Critical Pressure Ratio for Turbulent Boundary Layer Attachment and Separation in Supersonic Flow«, Mechanics of Liquids and Gases, Academy of Sciences of U.S.S.R., **4**, pp. 103—108 (1968).
44. Page, R. H., and Korst, H. H., »Non-isoenergetic Compressible Jet Mixing with Consideration of Its Influence on the Base Pressure Problems«, Proc. 4th Midwestern Conf. on Fluid Mechanics, Purdue University, pp. 45—68, 1955.
45. Chow, W. L., and Korst, H. H., »Influence on Base Pressures by Heat and Mass Addition«, ARS J., **32**, p. 1094 (1962).
46. Baum, E., et al., »Recent Studies of the Laminar Base-Flow Region«, AIAA J., **2**, pp. 1527—1534 (1964).
47. Chapman, D. R., »A Theoretical Analysis of Heat Transfer in Regions of Separated Flows«, NACA TN 3792, 1956.
48. Mueller, T. J., »An Experimental Investigation of the Reattachment of Compressible Two Dimensional Jets«, paper presented at the Fluids Engineering Conference of the ASME, Philadelphia, Pa., May 18—21, 1964.
49. Golik, R. J., »On the Dissipative Mechanisms within Separated Flow Regions«, Ph. D. Thesis, Dept. of Mech. Eng., University of Illinois, Urbana, Illinois, 1962.
50. Ting, L., »Boundary Layer over a Flat Plate in the Presence of Shear Flow«, Phys. Fluids, **3**, pp. 78—81 (1960).
51. Schlichting, H., *Boundary Layer Theory*, Chap. XXIII, McGraw-Hill Book Co., Inc., 1960.
52. Bales, E. L., »On Heat Transfer within and across Nearly Circular Cavities Including the Effects of Variable Wall Temperature and Mass Bleed«, Ph. D. Thesis, Dept. of Mech. Eng., University of Illinois, Urbana, Illinois, 1967.
53. Miles, J. B., »Stanton Number for Separated Turbulent Flow Past Relatively Deep Cavities«, Ph. D. Thesis, Dept. of Mech. Eng., University of Illinois, 1963.
54. Miles, J. B., »Heat Diffusion from Line Source into Mixing Region of Two Parallel Streams«, AIAA J., **2**, pp. 2038—2040 (1964).
55. Larson, H. K., »Heat Transfer in Separated Flow«, J. Aero Sci., **26**, 11, pp. 731—737 (1959).
56. Charwat, A., et al., »An Investigation of Separated Flows«, J. Aero. Sci., **28**, 7 (1961).

GENERALIZED TURBULENCE TRANSPORT EQUATIONS

C. W. HIRT

University of California, Los Alamos Scientific Laboratory,
Los Alamos, New Mexico, U.S.A.

1. Introduction

Most investigations of fluid turbulence have concentrated on the development of statistical theories that describe the underlying structure of turbulence [1]. These theories are necessarily complex and do not lend themselves to the solution of practical problems. Fortunately, in most problems it is only the mean turbulent behavior that is of interest, in which case the details of a complete statistical theory are superfluous. This situation may be likened to the motion of a fluid that represents the average motion of a large number of interacting molecules. The molecular dynamics could be studied directly, but that is difficult when there are on the order of 10^{23} molecules, and such detail is usually unnecessary. Instead, the average molecular motion is described by a set of fluid equations with the underlying molecular structure contained in a few transport coefficients.

In the same way, it would useful if the average behavior of a turbulent fluid could be described by equations containing turbulent transport coefficients. For general application, equations of this kind must be applicable to free turbulence and wall generated turbulence, and must cover all levels of turbulence intensity, since these cases often occur simultaneously within the same problem. Whether or not equations of this type can actually be found, for all problems of interest, is open to discussion. Certainly there are special cases, such as steady channel flow, where the average turbulence behavior can be approximated [1, 2, 3]. This paper describes the derivation of such turbulence transport equations. The work condensed here summarizes devolopments contained in [4—7], and suggests that the turbulence energy decay, E, be used to characterize the turbulence in place of the previously used scale function S.

2. Basic Equations

This study is restricted to incompressible fluids, and it is assumed that an adequate description of the fluid dynamics is given by the Navier-Stokes equations,

$$\frac{\partial V_i}{\partial t} + \frac{\partial}{\partial x_k} V_k V_i = -\frac{\partial \Pi}{\partial x_i} + \nu \frac{\partial^2 V_i}{\partial x_k^2}, \qquad (1)$$

and the incompressibility condition

$$\frac{\partial V_k}{\partial x_k} = 0. \qquad (2)$$

A summation convention for repeated indices is implied in (1)—(2), ν is the kinematic viscosity, and Π is the ratio of pressure to constant density, ρ.

Consider an ensemble of identical fluid systems defined in identical spatial regions and satisfying identical boundary conditions, but each having slightly different initial conditions. By the average behavior of a fluid flow we shall mean an average with respect to the ensemble. In this way, difficulties associated with space or time averages in transient flow situations are avoided.

The pressure and velocity can be separated into mean and fluctuating parts,

$$V_i = U_i + u_i$$
$$\Pi = \Phi + \phi,$$

where the capital letters, U_i and Φ, denote the ensemble averages of V_i and Π, and the lower case letters, u_i and ϕ, denote the fluctuations of individual members of the ensemble from the average. Accordingly, the ensemble average of u_i or ϕ, denoted by a bar over the quantity as $\bar{u}_i$ or $\bar{\phi}$, must vanish.

Inserting the mean and fluctuating quantities into (1)—(2) and taking an ensemble average leads to the familiar equation of Reynolds [8] for the mean flow,

$$\frac{\partial U_i}{\partial t} + \frac{\partial}{\partial x_k} U_k U_i = -\frac{\partial \Phi}{\partial x_i} + \nu \frac{\partial^2 U_i}{\partial x_k^2} - \frac{\partial}{\partial x_k} \overline{u_k u_i}, \qquad (3)$$

$$\frac{\partial U_k}{\partial x_k} = 0 \qquad (4)$$

The quantity $\overline{u_k u_i}$ is the ensemble average of the product of two fluctuating velocity components. The product $\rho \overline{u_k u_i}$ represents a flux of kth component momentum in the ith direction as produced by correlations between the fluctuating velocity components. This quantity, called the Reynolds stress tensor, is given a special symbol

$$R_{ki} = \overline{u_k u_i}, \qquad (5)$$

In (3)—(4) it is apparent that the only influence of the turbulent fluctuations on the mean flow is through the Reynolds stress, which contributes to an additional momentum transfer.

3. Turbulence Viscosity Concept

The simplest direction in which to proceed is to assume that the transfer of momentum by fluctuating velocities is similar to the transfer of momentum by molecular processes,

$$R_{ij} = \frac{2}{3} q \, \delta_{ij} - \sigma \left(\frac{\partial U_j}{\partial x_i} + \frac{\partial U_i}{\partial x_j} \right) \quad (6)$$

where $q = 1/2 \overline{u_k u_k}$ is the turbulence kinetic energy per unit mass, and σ is a turbulence viscosity coefficient. Relation (6) is the most general tensor expression that is linear in the gradients of the mean flow, Galilean invariant and symmetric in the indices. Through (6) the six independent components of R_{ij} are replaced by two variables, q and σ. Various assumptions can be introduced to express q and σ in terms of the local mean flow and its gradients [1]. Although an approach of this sort is highly attractive, because of its simplicity, it is severely limited. If the flow is inhomogeneous q and σ cannot be determined solely on the basis of local properties, but must be influenced by transport from other points in the flow. Relation (6) also assumes the turbulence is isotropic, which is far from the truth in many problems of interest.

In general, it is necessary to work with all the Reynolds stress components, so that different turbulence energies can be associated with different directions, and to use transport equations for the transient behavior of the stress components.

4. Reynolds Stress Equation

An equation for R_{ij} can be derived directly from the Navier-Stokes equations. Subtraction of (3) from (1) gives an equation for the fluctuating velocity u_i,

$$\frac{\partial u_i}{\partial t} + U_k \frac{\partial u_i}{\partial x_k} + u_k \frac{\partial U_i}{\partial x_k} + u_k \frac{\partial u_i}{\partial x_k} = -\frac{\partial \phi}{\partial x_i} + \frac{\partial}{\partial x_k} \left(\nu \frac{\partial u_i}{\partial x_k} + R_{ik} \right). \quad (7)$$

Subtraction of (4) from (2) also gives

$$\frac{\partial u_k}{\partial x_k} = 0. \quad (8)$$

Multiplying (7) by u_j, adding the corresponding equation with i, j interchanged, and taking the ensemble average,

$$\frac{\partial R_{ij}}{\partial t} + U_k \frac{\partial R_{ij}}{\partial x_k} + R_{ik} \frac{\partial U_j}{\partial x_k} + R_{jk} \frac{\partial U_i}{\partial x_k} + \frac{\partial}{\partial x_k} \overline{u_i u_j u_k} = \quad (9)$$

$$- \left(\overline{u_i \frac{\partial \phi}{\partial x_j}} + \overline{u_j \frac{\partial \phi}{\partial x_i}} \right) + \nu \left(\overline{u_i \frac{\partial^2 u_j}{\partial x_k^2}} + \overline{u_j \frac{\partial^2 u_i}{\partial x_k^2}} \right).$$

Three additional unknown correlations are present in (9): the triple velocity correlation, $\overline{u_i u_j u_k}$, the velocity-pressure correlation, $\overline{u_i \, \partial \phi / \partial x_j}$, and the velocity-velocity derivative correlation, $\overline{u_i \frac{\partial^2 u_j}{\partial x_k^2}}$. Additional equations could be derived for these correlations by manipulating Eqs. (3) and (7), but more unknown correlations would then appear and nothing is gained by this process alone. The chain of equatons for the higher correlations must be broken at some point. One method for accomplishing this is described here, in which approximations are introduced for the unknown correlations appearing in (9). Before investigating these correlations, however, it is convenient to rewrite the right side of equation (9),

$$\frac{\partial R_j}{\partial t} + U_k \frac{\partial R_{ij}}{\partial x_k} + R_{ik}\frac{\partial U_j}{\partial x_k} + R_{jk}\frac{\partial U_i}{\partial x_k} + \frac{\partial}{\partial x_k} \overline{u_i u_j u_k} =$$

$$- \left(\frac{\partial}{\partial x_i}\overline{u_j \phi} + \frac{\partial}{\partial x_j}\overline{u_i \phi} \right) + \overline{\phi \left(\frac{\partial u_j}{\partial x_i} + \frac{\partial u_i}{\partial x_j} \right)} \qquad (10)$$

$$+ \nu \frac{\partial^2 R_{ij}}{\partial x_k^2} - 2\nu \overline{\left(\frac{\partial u_i}{\partial x_k} \frac{\partial u_j}{\partial x_k} \right)}.$$

5. The Viscous Terms

The viscous term has been separated into two parts. The first part (next to last term in (10)) represents the molecular diffusion of R_{ij}, and need no further discussion. The second part (last term in (10)) is the dissipation of R_{ij} by molecular processes. Strictly speaking, this term is purely dissipative only for the diagonal elements of R_{ij}, but we shall assume it is dissipative for the off diagonal elements as well. Let E_{ij} denote this dissipation

$$E_{ij} = 2\nu \overline{\frac{\partial u_i}{\partial x_k} \frac{\partial u_j}{\partial x_k}}. \qquad (11)$$

As a first guess let the viscous decay of R_{ij} be assumed proportional to R_{ij},

$$E_{ij} = \frac{E}{q} R_{ij}. \qquad (12)$$

The scalar E is the energy decay rate for the total turbulence energy, and is defined as

$$E = \nu \overline{\frac{\partial u_m}{\partial x_k} \frac{\partial u_m}{\partial x_k}}. \qquad (13)$$

Relation (12) replaces the six independent components of E_{ij} with the one unknown variable E. The price paid for this simplification is that all components of R_{ij} decay with the same rate, E/q. The scalar E plays a central role in the remainder of this paper as it is used, together with the Reynolds stress, R_{ij}, to characterize the turbulence.

In references [4—7] a scale function S was used to characterize the turbulence in place of E. The scale is introduced according to the relation

$$E = \frac{2 \nu \Delta q}{S^2} \tag{14}$$

where Δ is a function of $S\sqrt{2q}/\nu$ and is chosen to give the correct decay rates in the limits of high and low levels of turbulence intensity. When E is used as a dependent variable, however, the Δ function is eliminated as an unknown, and the correct decay rates are assured by suitably choosing the equation for E.

6. The Triple Velocity Correlation

The triple velocity correlation $\overline{u_i u_j u_k}$ is considered next. The divergence of this correlation appears in (10) and represents the convection of $u_i u_j$ with fluctuating velocity u_k. To gain some perspective on how to treat this term it is helpful to first consider the correlation of u_k with a transportable scalar quantity such as heat or solute concentration. As a first approximation the fluctuation in the scalar, $w(\vec{r})$, at position $\vec{r}$ is due to the turbulent convection of the mean value $W(\vec{r} - u_m \delta t)$ from position $\vec{r} - u_m \delta t$ to position $\vec{r}$ in the time interval δt. That is,

$$w(\vec{r}) = W(\vec{r} - u_m \delta t) - W(\vec{r}). \tag{15}$$

Expanding the right side to first order in δt,

$$w(\vec{r}) = -\delta t\, u_m \frac{\partial W}{\partial x_m}$$

Multiplying by u_k and averaging

$$\overline{w u_k} = -\delta t\, \overline{u_k u_m} \frac{\partial W}{\partial x_m}$$

The time interval, δt, is proportional to the average transit time of the convecting eddies, and for dimensional reasons must be proportional to q/E. Hence,

$$\overline{w u_k} = -\tau_w \frac{q}{E} R_{km} \frac{\partial W}{\partial x_m} \tag{16}$$

where τ_w is a constant of order unity. Equation (16) relates the correlation $\overline{w u_k}$ to the gradient of the mean value W.

Approximations like (16) were first introduced by Boussinesq [1] and are often used to approximate the correlations between various kinds of fluctuating quantities. Although (16) should be restricted to transportable scalar quantities, suppose we apply it to the triple correlation $\overline{u_i u_j u_k}$. Allowing for the necessary symmetry in indices,

$$\overline{u_i u_j u_k} = -\frac{1}{3}\frac{q}{E}\left(R_{im}\frac{\partial R_{jk}}{\partial x_m} + R_{jm}\frac{\partial R_{ki}}{\partial x_m} + R_{km}\frac{\partial R_{ij}}{\partial x_m}\right). \tag{17}$$

Unfortunately the contraction of (17) is not consistent with the gradient approximation (16). Contracting the i, j indices, and defining $q' = \frac{1}{2} u_i u_i$, (17) reduces to

$$\overline{q' u_k} = -\frac{1}{3}\frac{q}{E}\left(R_{km}\frac{\partial q}{\partial x_m} + R_{im}\frac{\partial R_{ik}}{\partial x_m}\right), \tag{18}$$

which does not have the form of (16). In reference [6] additional terms were added to (17) to make its contraction reduce to

$$\overline{q' u_k} = -\frac{q}{E} R_{km}\frac{\partial q}{\partial x_m}. \tag{19}$$

However, in applying the resulting equations to turbulent channel flow it was discovered that undesirable effects were produced by the added terms. Since q' is not a transportable quantity in the same sense as a solute concentration, it really is not necessary to make the contraction of (17) have the form of (16).

An alternative approach to the triple correlation is to derive an exact equation using (7) in the same way that (9) was derived. Of course this equation contains the quadruple correlation $\overline{u_i u_j u_k u_m}$. A reasonable approximation for this is given by [2]

$$\overline{u_i u_j u_k u_m} = R_{ij} R_{km} + R_{ik} R_{jm} + R_{im} R_{jk}, \tag{20}$$

and then the triple correlation equation is

$$\frac{\partial \overline{u_i u_j u_k}}{\partial t} + U_m \frac{\partial \overline{u_i u_j u_k}}{\partial x_m} + \overline{u_i u_j u_m}\frac{\partial U_k}{\partial x_m} + \overline{u_i u_k u_m}\frac{\partial U_j}{\partial x_i}$$

$$+ \overline{u_j u_k u_m}\frac{\partial U_i}{\partial x_m} = -\left(\overline{u_i u_j \frac{\partial \phi}{\partial x_k}} + \overline{u_i u_k \frac{\partial \phi}{\partial x_j}} + \overline{u_j u_k \frac{\partial \phi}{\partial x_m}}\right)$$

$$+ \nu \frac{\partial^2}{\partial x^2}\overline{u_i u_j u_k} - \left(R_{mj}\frac{\partial R_{ik}}{\partial x_k} + R_{km}\frac{\partial R_{ij}}{\partial x_m} + R_{im}\frac{\partial R_{jk}}{\partial x_m}\right)$$

$$-2\nu\left(\overline{u_i \frac{\partial u_j}{\partial x_m}\frac{\partial u_k}{\partial x_m}} + \overline{u_j \frac{\partial u_i}{\partial x_m}\frac{\partial u_k}{\partial x_m}} + \overline{u_k \frac{\partial u_i}{\partial x_m}\frac{\partial u_j}{\partial x_m}}\right) \tag{21}$$

The last term in (21), which represents the viscous decay of the triple correlation, could be approximated by

$$-\frac{E}{q}\overline{u_i u_j u_k},$$

in the same way that the decay of R_{ij} was treated. Then, once the pressure-velocity correlations are specified, (20) could be solved simultaneously with the equations for U_i and R_{ij}. There are ten independent components in the triple correlation, so the use of (20) more than doubles the number of dependent variables. For this reason relation (17) will be retained as a first approximation for $\overline{u_i u_j u_k}$. If it is later found necessary to have a more exact representation, Eq. (20) can then be coupled to the other equations.

7. The Pressure-Velocity Correlations

The most difficult correlation terms to approximate are those involving the pressure. The pressure does not depend solely on local fluid properties, but is influenced by the entire region occupied by fluid. To see this it is only necessary to recall that ϕ satisfies a Poisson equation whose formal solution is

$$\phi = \frac{1}{2\pi}\iiint\left(\frac{\partial U_m}{\partial x_k}\frac{\partial u_k}{\partial x_m}\right)\frac{1}{r}dV' - \frac{1}{4\pi}\iiint\frac{\partial^2}{\partial x_m \partial x_k}(\overline{u_m u_k} - u_m u_k)\frac{1}{r}dV' \qquad (22)$$
$$+ \frac{1}{4\pi}\iint\left[\frac{1}{r}\left(\frac{\partial \phi}{\partial n}\right) - \phi \frac{\partial}{\partial n}\left(\frac{1}{r}\right)\right]dS',$$

where the volume integrations are over the entire fluid with respect to a point P', and the surface integral is over the surface bounding the fluid. All quantities in the integrands are evaluated at P', except r, which is the distance from P to P'.

Some authors [2, 3] have tried to estimate pressure-velocity correlations using (22). In these attempts the surface integral is neglected as unimportant, except near a boundary, and the mean velocity appearing in the first volume integral is developed in a Taylor series about P. The coefficients in this expansion, and contributions from the second volume integral, are left as constants to be determined by comparison with experimental data.

A different approach is followed here. Approximations are introduced to represent the physical consequences expected from the pressure-velocity correlations. There are two kinds of pressure correlations appearing in (10). The gradient of $\overline{\phi u_k}$ representing the rate at which work is done by the fluctuating pressure, and $\overline{\phi\left(\frac{\partial u_j}{\partial x_i} + \frac{\partial u_i}{\partial x_j}\right)}$, which controls the transfer of turbulence energy between the components u_1^2, u_2^2, and u_3^2.

In earlier works [4—7] it was argued that $\overline{\phi u_k}$ should satisfy a gradient approximation like (16),

$$\overline{\phi u_k} = -\frac{q}{E} R_{km} \frac{\partial \Phi}{\partial x_m}. \tag{23}$$

However, in an application to channel flow this choice led to a negative diffusion of energy, and hence, to unstable equations. If, on the other hand, we accept the interpretation of $\rho \overline{\phi u_k}$ as the flux of work done by the fluctuating pressure in the kth direction then it is reasonable to suppose that this correlation should be proportional to the to the gradient of the kth component of turbulence energy. In tensor invarient form this might be written as

$$\overline{\phi u_k} = -\frac{q^2}{E} \frac{\partial R_{kn}}{\partial x_m}, \tag{24}$$

where the coefficient has been inserted for dimensional reasons. The negative sign directs the flux from high to low regions of turbulence energy. This choice agrees with that used by Donaldson [9], except that he uses as a coefficient the macroscale times the square root of $\overline{u_k^2}$. This choice cannot be written in tensor invarient form, and is restricted to high levels of turbulence intensity.

The second pressure correlation, $\overline{\phi \left(\frac{\partial u_j}{\partial x_i} + \frac{\partial u_i}{\partial x_j} \right)}$, primarily accounts for the exchange of energy between the diagonal components of R_{ij}. Since this correlation vanishes upon contraction of its indices it does not influence the total turbulence energy. It is this term that drives the turbulence toward isotropy. A simple relaxation approximation is used here,

$$\overline{\phi \left(\frac{\partial u_j}{\partial x_i} + \frac{\partial u_i}{\partial x_j} \right)} = -\omega \frac{E}{2q} \left(R_{ij} - \frac{2}{3} q \delta_{ij} \right), \tag{25}$$

where ω is a constant of order unity. This approximation is the same as Rotta's [3], but it differs from that used by Donaldson [9] in that he replaces E/q with $(A_1 U^2 + A_2 q)^{1/2}/L$, where L is the macroscale. Donaldson's choice is not Galilean invariant. In [6] an additional effect was added to the right side of (25) to account for departures from isotropy near rigid walls. This aspect of the pressure-velocity correlation is considered later.

8. Transport Equation for the Reynolds Stress

With expressions (12), (17), (24), and (25) substituted into (10) the Reynolds stress equation becomes

$$\frac{\partial R_{ij}}{\partial t} + U_k \frac{\partial R_{ij}}{\partial x_k} = -\left(R_{ik} \frac{\partial U_j}{\partial x_k} + R_{jk} \frac{\partial U_i}{\partial x_k} \right)$$

$$+ \frac{\partial}{\partial x_k}\left[\frac{q}{3E}\left(R_{im}\frac{\partial R_{jk}}{\partial x_m}+R_{jm}\frac{\partial R_{ki}}{\partial x_m}+R_{km}\frac{\partial R_{j}}{\partial x_m}\right)\right]$$

$$+\frac{\partial}{\partial x_i}\left(\frac{q^2}{E}\frac{\partial R_{jk}}{\partial x_k}\right)+\frac{\partial}{\partial x_j}\left(\frac{q^2}{E}\frac{\partial R_{ik}}{\partial x_k}\right) \quad (26)$$

$$-\omega\frac{E}{q}\left(R_{ij}-\frac{2}{3}q\,\delta_{ij}\right)+\nu\frac{\partial^2}{\partial x_k^2}R_{ij}-\frac{E}{q}R_{ij}.$$

The corresponding equation for the turbulence energy, q, is obtained by contracting (26),

$$\frac{\partial q}{\partial t}+U_k\frac{\partial q}{\partial x_k}=-R_{mh}\frac{\partial U_m}{\partial x_k}+\frac{\partial}{\partial x_k}\left[\frac{q}{3E}\left(R_{km}\frac{\partial q}{\partial x_m}+R_{jm}\frac{\partial R_{jk}}{\partial x_m}+3q\frac{\partial R_{km}}{\partial x_m}\right)\right] \quad (27)$$

$$+\nu\frac{\partial^2 q}{\partial x_k^2}-E.$$

9. Transport Equation for the Energy Decay Rate

To complete this theory an equation for E must be constructed. In most previous theories a scale function S has been used in place of E to characterize the turbulence. S was then specified by a mixing length postulate or similarity argument. The disadvantage of these methods is that S is, in general, a non-local function, and should be governed by a transport equation. Previous attempts in this direction were made by Chou [2], who employed a vorticity decay equation, and by Rotta [3], who introduced an energy spectrum equation. In this paper an equation is derived for the turbulence energy decay rate, E. Using the definition of E given by (13), the fluctuating velocity Eq. (7) can be manipulated to give

$$\frac{\partial E}{\partial t}+U_k\frac{\partial E}{\partial x_k}=-2\nu\overline{\left(\frac{\partial u_k}{\partial x_j}\frac{\partial u_n}{\partial x_j}\right)}\frac{\partial U_m}{\partial x_k}-2\nu\overline{\left(\frac{\partial u_j}{\partial x_k}\frac{\partial u_j}{\partial x_m}\right)}\frac{\partial U_m}{\partial x_k}$$

$$-2\nu\overline{\left(u_k\frac{\partial u_j}{\partial x_m}\right)}\frac{\partial^2 U_j}{\partial x_n\partial x_k}-2\nu\overline{\frac{\partial u_j}{\partial x_m}\frac{\partial u_k}{\partial x_m}\frac{\partial u_j}{\partial x_k}}$$

$$-\nu\frac{\partial}{\partial x_k}\overline{\left[u_k\left(\frac{\partial u_i}{\partial x_m}\right)^2\right]}-2\nu\overline{\frac{\partial u_j}{\partial x_m}\frac{\partial^2\phi}{\partial x_m\partial x_j}} \quad (28)$$

$$+\nu\frac{\partial^2 E}{\partial x_k^2}-2\nu^2\overline{\left(\frac{\partial}{\partial x_k}\frac{\partial u_j}{\partial x_m}\right)^2}$$

This is the D equation in references [5—7] multiplied by 2ν.

The higher correlations appearing in (28) must be approximated in terms of U_i, R_{ij}, and E to close the set of turbulence transport equations. This was done in [5, 6], where physical intuition served as a guide in sorting out the physical processes expected to influence the evolution of E. Except for the absence of a Φ diffusion term, and a q diffusion term,, the equation used here is identical to that in [6],

$$\frac{\partial E}{\partial t}+U_k\frac{\partial E}{\partial x_k}=-\Omega\frac{E}{q}R_{mk}\frac{\partial U_m}{\partial x_k}+\frac{\partial}{\partial x_k}\left(\frac{q}{E}R_{km}\frac{\partial E}{\partial x_m}\right)+\nu\frac{\partial^2 E}{\partial x_k^2}+\Lambda\frac{E^2}{q}. \quad (29)$$

The terms on the right side of (29) represent, respectively, creation, turbulent diffusion, molecular diffusion, and decay terms. The quantities Ω and Λ are functions of the nondimensional variable $\eta = q^2/\nu E$. This variable is proportional to the ratio of a turbulence viscosity, q^2/E, and the molecular viscosity, ν. It is equal to one forth the turbulence Reynolds number defined by Batchelor and Townsend [10]. For most turbulent flows η is much larger than unity; it is less than unity only for very low intensity turbulence.

10. Determination of the Functions Ω and Λ

Functions Ω and Λ are determined by requiring that our equations agree with known theoretical and experimental results. For example, in the limit of homogeneous isotropic turbulence Eqs. (27) and (29) are

$$\frac{\partial q}{\partial t}=-E$$

$$\frac{\partial E}{\partial t}=-\Lambda\frac{E^2}{q}.$$

(30)

In this limit the turbulence is subject only to decay processes. According to Hinze [1] (p. 208) E is related to the Taylor micro-scale, λ, by

$$E=\frac{10\,\nu q}{\lambda^2}. \quad (31)$$

Combining (30) and (31) it is easy to show that

$$\frac{\partial \lambda^2}{\partial t}=10\,\nu\,(\Lambda-1). \quad (32)$$

Again referring to Hinze [1] (p. 209) Eq. (32) implies that $\Lambda=2.0$, except in the limit of vanishing turbulence intensity when $\Lambda=1.4$. The experimental data of Batchelor and Townsend [10] suggests a rapid transition between these two Λ values at a value of η of approximately 1.25. Thus, as a first approximation

$$\Lambda=\begin{cases} 2, & \text{for } \eta\geq 1.25 \\ 1.5, & \text{for } \eta<1.25 \end{cases} \quad (33)$$

Now consider a homogeneous shear flow for which Eqs. (27) and (29) reduce to

$$\frac{\partial q}{\partial t} = -R_{12}\frac{\partial U}{\partial y} - E$$

$$\frac{\partial E}{\partial t} = -\Omega \frac{E}{q} R_{12}\frac{\partial U}{\partial y} - \Lambda \frac{E^2}{q}.$$
(34)

An equilibrium state, other than the trivial one of no turbulence, exists only if $\Omega = \Lambda$. However, this choice has not been successful in applications to channel flow. A value of Ω equal to Λ is too large, and forces q to decay away. A better value for Ω is unity, which is approximately one half Λ.

A partial understanding of these results can be gained from the experimental investigations of Laufer [11] into turbulent channel flow. In the center of the channel there is a balance between the viscous decay of turbulence energy and the diffusion of energy into the center from the wall regions. Thus, near the walls there must be an excess of creation versus decay to supply the energy needed in the center. Otherwise, the turbulence must decay away, which is precisely what is observed in numerical studies when Ω is equal to Λ.

11. Determination of ω

One variable, ω, remains unaccounted for in Eqs. (3), (26), and (29), but it must be kept in mind that additional numerical factors of order unity may be necessary in some of the diffusion terms (the τ-factors in (16)). It may also be necessary to add or adjust terms in (26) and (29) as more applications of these equations are attempted.

The turbulence transport equations have been successfully applied to the passage of homogeneous turbulence through a contracting duct. Agreement was obtained with the experimental data of Uberoi [12] with a value of ω equal to 0.5. In this application, it was also necessary to add to the R_{ij} equation a term equal to

$$\frac{2}{5}\mu q \left(\frac{\partial U_j}{\partial x_i} + \frac{\partial U_i}{\partial x_j}\right).$$
(35)

where μ was found to be 0.45. This term was proposed in [6] as arising from the pressure-velocity correlations. Similar terms were obtained by Chou [2] and Rotta [3], although the correspondence is not exact.

12. Application to Two-Dimensional Channel Flow

As another application of the transport equations consider the stationary flow of turbulence in a two-dimensional channel. Let the flow direction be aligned with x_1 (or x) and the normal direction between the channel walls with x_2 (or y). All quantities are functions of y only. Letting

Γ denote the constant pressure gradient, $-\partial \Phi/\partial x$, and assuming R_{23}, R_{13} are zero, Eqs. (3), (26) and (29) reduce to

$$\nu \frac{\partial^2 U}{\partial y^2} - \frac{\partial R_{12}}{\partial y} = -\Gamma \tag{36}$$

$$\frac{\partial \Phi}{\partial y} + \frac{\partial R_{22}}{\partial y} = 0 \tag{37}$$

$$\frac{E}{q} R_{11} + 2 R_{12} \frac{\partial U}{\partial y} = \frac{\partial}{\partial y} \left[\frac{q}{3E} \left(2 R_{12} \frac{\partial R_{12}}{\partial y} + R_{22} \frac{\partial R_{11}}{\partial y} \right) + \nu \frac{\partial R_{11}}{\partial y} \right]$$
$$- \frac{\omega E}{\partial q} (2 R_{11} - R_{22} - R_{33}) \tag{38}$$

$$\frac{E}{q} R_{22} = \frac{\partial}{\partial y} \left[\frac{q}{3E} (2 R_{22} + R_{11} + R_{33}) \frac{\partial R_{22}}{\partial y} + \nu \frac{\partial R_{22}}{\partial y} \right]$$
$$- \frac{\omega E}{6q} (2 R_{22} - R_{11} - R_{33}) \tag{39}$$

$$\frac{E}{q} R_{33} = \frac{\partial}{\partial y} \left[\frac{q}{3E} R_{22} \frac{\partial R_{33}}{\partial y} + \nu \frac{\partial R_{33}}{\partial y} \right] - \frac{\omega E}{6q} (2 R_{33} - R_{11} - R_{22}) \tag{40}$$

$$\left(1 + \frac{\omega}{2}\right) \frac{E}{q} R_{12} + R_{22} \frac{\partial U}{\partial y} = \frac{\partial}{\partial y} \left[\frac{q}{3E} R_{11} \frac{\partial R_{22}}{\partial y} \right.$$
$$\left. + \frac{q}{6E} (7 R_{22} + 3 R_{11} + 3 R_{33}) \frac{\partial R_{12}}{\partial y} + \nu \frac{\partial R_{12}}{\partial y} \right] \tag{41}$$

$$\frac{\Lambda E^2}{q} + \Omega \frac{E}{q} R_{12} \frac{\partial U}{\partial y} = \frac{\partial}{\partial y} \left[\left(\nu + \frac{q}{E} R_{22} \right) \frac{\partial E}{\partial y} \right] \tag{42}$$

The μ term mentioned previously has been omitted from these equations, since it contributes an unwanted decay in the R_{12} equation. This discrepancy can be resolved only through further applications of the basic equations.

Equations (36)—(42) are seven coupled, nonlinear equations for stationary turbulence in a two-dimensional channel. Solutions of these equations can be obtained numerically. Choosing $\omega=3.0$ and $\Omega=1.4$, for example, these equations exhibit many of the flow characteristics observed in the experimental data of Laufer [11] (Figs. 1—3). Data in these figures are nondimensionalized with one-half the separation distance between the channel walls, d, and the maximum mean flow velocity, U_o.

Although the calculated results are remarkably good, there are two places where some refinement is needed. The most obvious discrepancy is the low mean velocity profile near the wall (Fig. 1). This is caused by a too rapid rise in R_{12} at the wall (Fig. 2), which smooths the velocity profile in this region. Secondly, the turbulence energy components at the wall

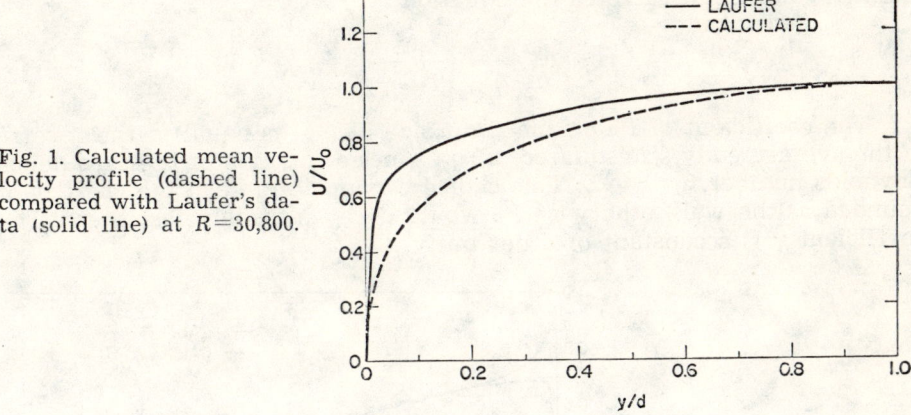

Fig. 1. Calculated mean velocity profile (dashed line) compared with Laufer's data (solid line) at $R=30,800$.

do not exhibit the observed behavior (Fig. 3). There are discrepancies in the sharp peak in R_{11} and the fact that R_{22} is too large while R_{11} and R_{33} are too small. In the latter case it might be supposed that energy associated

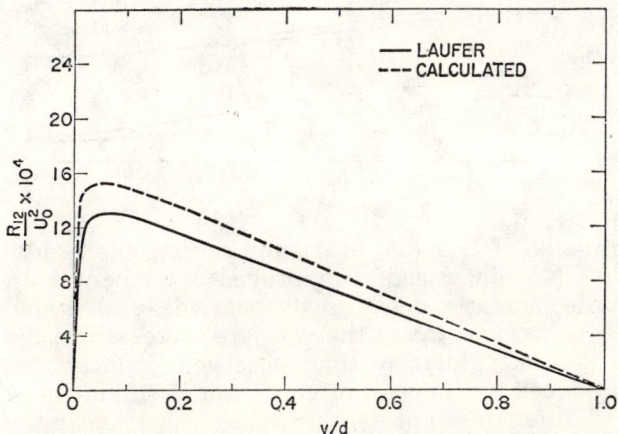

Fig. 2. Calculated turbulence stress R_{12} (dashed line) compared with Laufer's data (solid line) at $R=30,800$.

with motions normal to the wall should be transferred to the tangential components because of the impenetrability of the wall. The presence of a wall effect term that transfers normal energy to tangential energy could conceivably arise from the pressure-velocity correlations. Assuming that the transfer rate should be proportional to the energy itself, we have experimented with the introduction into the R_{ij} equation of a term equal to

$$-\zeta \frac{E}{q} \left(P_{im} R_{mj} + P_{jn} R_{ni} - \frac{2}{3} \delta_{j} P_{nm} R_{nm} \right) \qquad (43)$$

where ζ is a constant of order unity. P_{ij} is a wall effect tensor defined at a wall as

$$P_{ij} = n_i n_j, \qquad (44)$$

where n_i is the unit normal to the wall. The value of P_{ij} decays rapidly away from the wall. This decay should be nearly complete in a distance of the order of the integral scale of turbulence, $q^{3/2}/E$. As a first guess one might try a P_{ij} that satisfies the equation

$$\nabla^2 P_j = -\frac{\Upsilon E^2}{q^3}\left(\frac{\eta^2}{1+\eta^2}\right)/P_{ij}. \tag{45}$$

The coefficient of P_{ij} on the right side of (45) is equal to the inverse of the average eddy size squared, E^2/q^3, times a function of the turbulence Reynolds number, $\eta = q^2/\nu E$. This choice was made to keep the coefficient bounded at the wall, otherwise P_{ij} would decay infinitely fast there. The coefficient Υ is a constant of order unity.

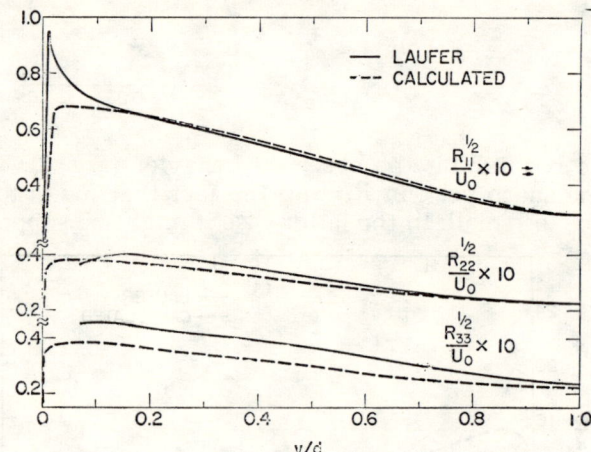

Fig. 3. Calculated turbulence energy components (dashed lines) compared with Laufer's data (solid lines) at $R = 30,800$.

Adding (43) to the right side of Eqs. (38)—(41), and solving the resulting equations together with (45) does lead to improved agreement with Laufer's data. The R_{12} profile increases more slowly near the wall, which raises the mean velocity. Also R_{11} and R_{33} at the wall are increased at the expense of R_{22}. Although a peak similar to that observed in $R_{11}^{1/2}$ at $y/d \approx 0.012$, in Fig. 3, is produced by the wall effect tensor, a similar peak is also produced in $R_{33}^{1/2}$. Laufer does not exhibit data for $R_{33}^{1/2}$ near the wall ($y/d < 0.075$), but the trend of his data suggests that this peak should not occur. Further experimenting with the wall effect tensor is necessary before its usefulness can be fully determined.

13. Acknowledgment

I should like to acknowledge my indebtedness to Francis H. Harlow for initiating the turbulence transport equations, and for his continuing interest in their development, and to Bart J. Daly for the channel flow results described in Section 12.

This work was performed under the auspices of the United States Atomic Energy Commission.

REFERENCES

1. Hinze J. O., *Turbulence*, McGraw-Hill Book Company, N.Y., 1959.
2. Chou P. Y., Quart. Appl. Math. **3**, 38 and 3, 198 (1945).
3. Rotta J., Z. Physik, **129**, 547 and **131**, 51 (1951).
4. Harlwo F. M. and Nakayama P. I., Phys. Fluids **10**, 2323 (1967); Los Alamos Scientific Laboratory Report, LA-3854, 1968.
5. Harlow F. H., Los Alamos Scientific Laboratory Report, LA-3947, 1968.
6. Harlow F. H. and Hirt C. W., Los Alamos Scientific Laboratory Report, LA-4086, 1969.
7. Hirt C. W, Suppl. Phys. Fluids, »Proc. Int. Symp. on High-Speed Computing in Fluid Dynamics«, Montery, Califonia, August, 1968 (to be published).
8. Reynolds O., Phil. Trans., A186, 123 (1895).
9. Donaldson C. duP., AIAA J., **7**, 271 (1969).
10. Batchelor G. K. and Townsend A. A., Proc. Roy, Soc. (London) A194, 527 (1948).
11. Laufer J., »Investigation of Turbulent Flow in a Two-Dimensional Channel«, National Advisory Committee for Aeronautics Report 1053, 1951.
12. Uberoi M. S., J. Aero. Sci., **23**, 754 (1956).

HEAT TRANSFER FROM SINGLE TUBES AND BANKS OF TUBES IN CROSSFLOW

A. ZHUKAUSKAS

Academy of Sciences of the Lithuanian SSR, Institute of Physical and Technical Problems of Energetics, Vilnius, USSR

1. Introduction

Heat transfer surfaces are mostly made of tubes due to the simplicity of their manufacture and installation. Therefore, studies of heat transfer mostly concentrate on heat transfer from tube surfaces.

Developments in the field of steam boilers have led to investigations of heat transfer from single tubes and banks of tubes in gas flows. A number of investigations has been devoted to this problem, mostly in air flows, its physical properties scarcely differing from those of combustion gases.

Influences of bank arrangement and relative longitudinal ($a = s_1/d$) and transverse pitch ($b = s_2/d$) on heat transfer have been analyzed thoroughly (Fig. 1). Local heat transfer along the tube perimeter also has been studied.

As a result of numerous investigations and generalizations V. Avtufyev, E. Grimison, M. Mikheyev, N. Kuznetsov et al. [1—4] proposed relations for heat transfer predictions from single tubes and banks of tubes of various arrangement in gas crossflow.

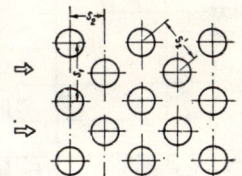

Fig. 1. Tube arrangement in banks.

Nevertheless, most of the previous investigations are limited to lower Re numbers. Consequently, the proposed relations are valid only for lower Re and Pr numbers. Due to the recent developments of numerous new fields of engineering, heat transfer in fluid flows at higher Re and Pr numbers has become of interest.

The present report is devoted mainly to heat transfer from single tubes and banks of tubes to fluids flowing across the tubes. The influence of the fluid properties, temperature differences, heat flux direction, flow pattern and other factors on heat transfer have been studied comprehensively. The report includes the results of recent investigations of heat transfer from banks of tubes in gas crossflow at high Re numbers. Extensive experimental data analyzed include various arrangements of tube banks, as well as of single tubes in crossflow in the range of Pr numbers from 0.7 to 500 and Re numbers from 1 to $2 \cdot 10^6$.

2. Flow Across a Tube

The flow regime — laminar or turbulent — in the boundary layer has a considerable effect on heat transfer. Therefore, before passing to the problems of heat transfer let us consider some details of fluid flow past single tubes and banks of tubes.

When a viscous fluid flows across a tube, a laminar boundary layer is formed in the front part of the tube. Its thickness gradually increases. Due to shear stresses in the boundary layer, some energy dissipates. At the same time pressure in the main flow decreases ($dp/dx < 0$), velocity in the direction of the flow increases, and particles of fluid from the boundary layer are drawn into the main flow, continuing to move along the surface of the cylinder in spite of friction. In the rear portion of the cylinder main flow pressure increases ($dp/dx > 0$). As velocity in the direction of flow decreases, the velocity of the fluid particles in the boundary layer decreases to zero and eventually starts moving in the opposite direction.

The existence of layers in which fluid flows in opposite directions gives rise to a vortex. As soon as a vortex separates from the cylinder, another vortex is formed. In this way the rear portion of the cylinder is in the region of a mixed vortex flow.

The laminar boundary layer separates from the sides of the cylinder at an angle $\varphi = 82°$. This separation point has negligible dependence on the Re number. On only 45 percent of the cylinder surface the boundary layer is attached to the surface. This flow pattern exists from $Re = 9$ to the critical Re numbers in the order of $2 \cdot 10^5 - 5 \cdot 10^5$.

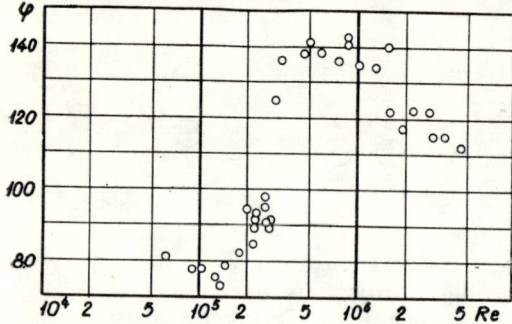

Fig. 2. Position of separation point on a cylinder depending on Re.

As the Re number increases, the boundary layer gradually becomes turbulent. It receives additional kinetic energy from the main flow through turbulent fluctuations, and the separation point is displaced downstream. In this case, according to measurements performed by Achenbach [5], in air flow the separation point may be displaced to the angles ranging from $\varphi=95°$ to $\varphi=140°$. However, as is seen in Fig. 2, at $Re=2 \cdot 10^6$ the separation angle starts to decrease, and the separation point is displaced towards the front part of the cylinder.

The change of the friction coefficient C_w in function of the Re number corresponds to the described flow pattern across a circular cylinder. Figure 3 suggests that the dependence of C_w on the Re number is approximately constant up to $Re=2 \cdot 10^5$. At this Re number the friction

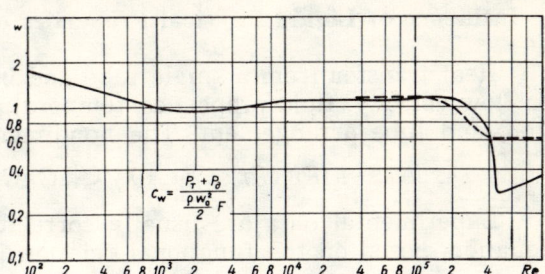

Fig. 3. Cylinder drag coefficient as a function of Re. Broken line — Achenbach's curve [5].

coefficient decreases sharply. This phenomenon is explained by the fact that the displacement of the separation point towards the rear part of the tube decreases the area of the vortex region.

Flow across a tube in a bank of tubes is influenced by the neighboring tubes. Tubes in a row form narrow flow passages, causing flow acceleration and corresponding pressure gradients. Both the velocity distribution in the boundary layer and the flow pattern in the vortex region at the rear part of the tube change accordingly. The separation point of the boundary layer displaces more or less downstream, depending on the transverse pitch. In staggered tube banks the boundary layer separates at approximately 135—145°, depending on the transverse and longitudinal pitch and the Re number.

At low Re numbers the flow across tube banks is laminar. The effect of relatively large vortices in the recirculation zone on the boundary layer in the front portion of the tube is made negligible by viscous forces and the negative pressure gradient. The boundary layer at the tube is laminar, and at the rear portion a vortex flow is formed. Such a pattern of flow across tubes at $Re<10^3$ may be described as predominantly laminar.

With Re numbers greater than 10^3 the flow pattern across a tube bank considerably changes. Between tubes the vortex flow appears with a correspondingly higher degree of turbulence. Although the flow at the front part of the tube is influenced by the turbulence, a laminar boundary layer still exists. Such a flow pattern may be defined as mixed. Turbulence and its intensity between the tubes depend on the tube arrangement and the Re number.

The mixed flow pattern covers a wide range of Re numbers and changes only at $Re>2 \cdot 10^5$. At higher Re numbers the flow between the tubes becomes intensively turbulent, causing turbulization of the boundary layer, and the laminar flow exists only on very restricted areas of the front parts of the tubes. The flow pattern is predominantly turbulent.

This is indicated by the fact that the variation of the total friction coefficient in the tube bank acquires a corresponding character.

Consequently, we may distinguish three different flow patterns across tubes with respect to the Re number: predominantly laminar flow at $Re<10^3$; mixed flow at $5 \cdot 10^2 < Re < 2 \cdot 10^5$; and predominantly turbulent flow at $Re > 2 \cdot 10^5$.

3. Influence of Liquid Physical Properties on Heat Transfer

Heat transfer from a single tube and banks of tubes depends mainly on flow velocity, fluid properties, temperature difference, heat flux direction, and tube arrangement. The nondimensional relation is as follows:

$$Nu = f(Re, Pr, \mu_f/\mu_w, \lambda_f/\lambda_w, C_f/C_w, \rho_f/\rho_w, S_1/d_1, S_2/d_2) \qquad (1)$$

Experimental data are usually correlated by the following power equation, based on the functional relation Eq. 1.

$$Nu = c Re^m Pr^n, \qquad (2)$$

where the exponent n is determined by the physical properties of the fluid.

Experiments performed by Makarevichius, Shlanchiauskas and the author [6, 7] on heat transfer from single tubes and banks of tubes in different fluids enable a fuller analysis of the physical properties and their influence on heat transfer. The influence of fluid properties is mainly due to the variations of the properties with temperature in the boundary layer. The choice of the characteristic temperature for the evaluation of the physical properties is therefore important.

The influence of the fluid properties may be established by two different methods. By the first method the physical properties are evaluated on the basis of the mean flow temperature, and an additional parameter is introduced in Eq. 2. By the second method a certain characteristic temperature is chosen laying between the mean flow and the wall temperature, by which the influence of the fluid properties on heat transfer is taken into account. The relation is the same as in the case of the constant physical properties. From the analysis of experimental data we concluded that the best choice of the characteristic temperature is the main flow temperature t_f and that an additional parameter (Pr_f/Pr_w) has to be introduced.

In literature the exponent n in the range from 0.31 to 0.33 is often quoted for tubes in crossflow. This is suggested by theoretical solutions of heat transfer from a plate in a laminar boundary layer flow. Our extensive theoretical and experimental investigations [8—10] suggest that the exponent n over the Pr number depends on the flow pattern across tubes. For the heat transfer from a plate in a laminar boundary layer flow $n=0.33$, but in a turbulent boundary layer it becomes $n=0.43$. A study of the relation between the velocity and the temperature field in semi-empirical theories of turbulent heat transfer showed that the exponent n also varies to a cretain extent with the Pr number.

On the other hand, E. Eckert [11] has deduced the following analytical relation for the calculations of local heat transfer from various wedge-shaped bodies in laminar boundary layer flows:

$$Nu_c = 0.56\,(\beta+0.2)^{0.1}/(2-\beta)^{0.5}\,Re_x^{0.5}\,P_r^{0.333} + 0.067\,\beta - 0.026\,\beta^2 \qquad (3)$$

It is seen that the exponent n over the Pr number in a laminar boundary layer flow depends on the wedge angle.

In the case of mean heat transfer calculations from a cylinder the exponent n over the Pr number is assumed to have a value between 0.33 and 0.43. Detailed studies of heat transfer from a single tube in crossflow of transformer oil, water and air [6], suggest that the exponent n over the Pr number has a value between 0.37 and 0.38.

Investigations of heat transfer [7] in 27 banks of in-line and staggered tubes in crossflow of different fluids in the range of Pr numbers from 0.7 to 500 suggest that for the mean heat transfer calculations in all sorts of tube banks the exponent n over the Pr number has the value of 0.36. This

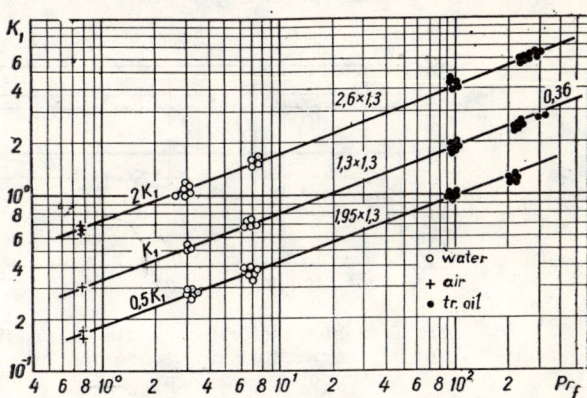

Fig. 4. The evaluation of power index for Pr for staggered banks.

is supported by Fig. 4, in which some experimental data for banks of staggered tubes are correlated in the form:

$$Nu_f\,Re_f^{-0.60} = f(Pr_f).$$

Another important problem is presented by the influence of heat flux direction on heat transfer. Experiments suggest that the heat transfer coefficient has a higher value for heating than for cooling and that its value increases with the temperature difference.

Mikheyev [3] has proposed to account for the influence of sudden changes in the fluid properties in the boundary layer by introducing a ratio Pr_f/Pr_w in the power of 0.25. Another parameter often used is (μ_f/μ_w). In viscous fluids like water and oil, it is mainly viscosity that changes with temperature across the boundary layer, therefore $(Pr_f/Pr_w) \approx (\mu_f/\mu_w)$. The analytical calculations of the laminar boundary layer heat from a flat plate in the flows of various fluids have shown that the influence of the changes of the other physical properties on heat transfer still constitute up to 7 percent of the whole influence.

Figure 5 presents data [8 to 12] from the variation of the ratio Pr_f/Pr_w for a flat plate streamlined by a laminar boundary layer flow. For heating

$k = 0.25$ and for cooling $k = 0.19$. The curves are drawn in accordance with the theoretical calculations [12], and the points coincide with our experimental data [8]. Fig. 6 presents experimental data [10] from the variations

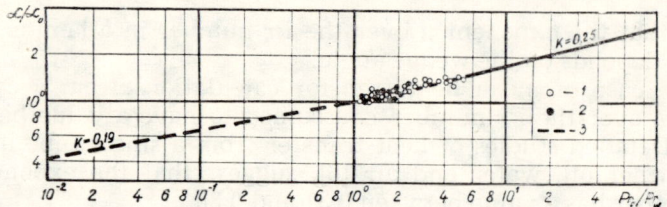

Fig. 5. Local heat transfer from a plate in laminar flow as a function of temperature difference and heat flux direction: 1, 2 — heating of water and transformer oil respectively, 3 — theoretical curve [12] α_0 — heat transfer coefficient for constant physical properties.

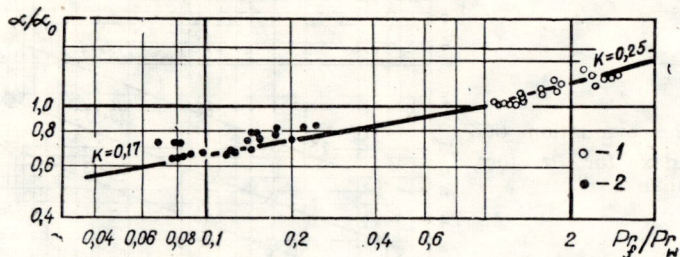

Fig. 6. Heat transfer from plate in turbulent flow as a function of temperature difference and heat flux direction. 1 — transformer oil, heating. 2 — glycerine, cooling.

of Pr_f/Pr_w for a flat plate streamlined in a turbulent boundary layer. For heating $k = 0.25$ and for cooling $k = 0.17$.

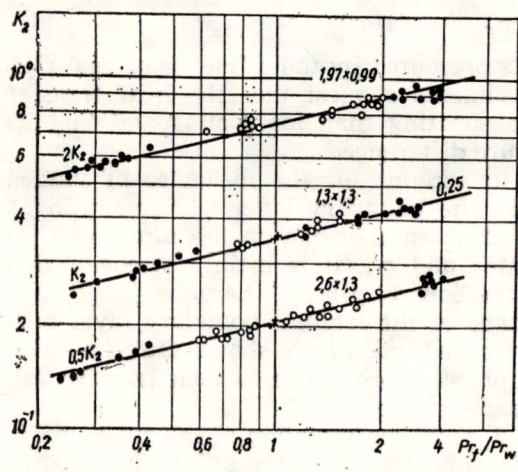

Fig. 7. Evaluation of power index for the ratio Pr_f/Pr_w for staggered banks.

Thus, in the case of cooling the power index of the ratio Pr_f/Pr_w is lower than in the case of heating. For practical purposes, 0.25 may be chosen for both cases. In the case of heat transfer from single tubes [6] and banks of tubes [7] in crossflow of various fluids we have chosen $K = 0.25$ as the value of the exponent over the ratio Pr_f/Pr_w.

Figure 7 presents experimental heat transfer results as a function of Pr_f/Pr_w in banks of staggered tubes. In Fig. 7 the characteristic temperature is that of the main flow. The variation of physical properties with temperature may well be accounted for by the introduction of $(Pr_f/Pr_w)^{0.25}$.

4. Mean Heat Transfer From a Single Tube

Resluts of detailed studies of heat transfer from single wires and tubes in crossflow of air, water and transformer oil are presented in the above-mentioned form in Fig. 8. On the abscissa the values of the Re number and on the ordinate are given as:

$$Nu_f Pr_f^{-0.38} (Pr_f/Pr_w)^{-0.25}$$

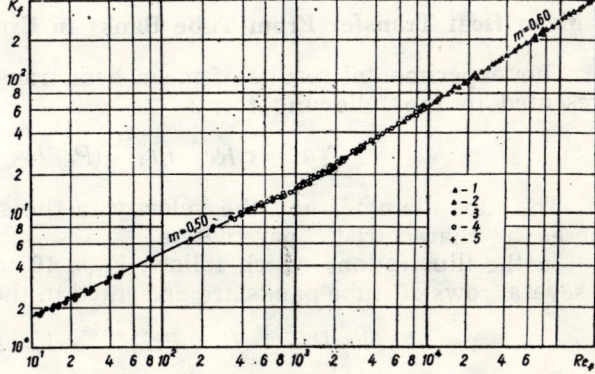

Fig. 8. Heat transfer from single wires and tubes: 1 — transformer oil, heating, 2 — transformer oil, cooling, 3 — water, heating, 4 — water, cooling, 5 — air.

The experimental data are well-correlated both for cooling and for heating. The proposed correlation is then:

for: $Re_f = 1 \cdot 10^1 - 1 \cdot 10^3$.

$$Nu_f = 0.50 \, Re_f^{0.5} \, Pr_f^{0.38} \, (Pr_f/Pr_w)^{0.25} \qquad (4)$$

and for: $Re_f = 1000$ to $200,000$

$$Nu_f = 0.25 \, Re_f^{0.6} \, Pr_f^{0.38} \, (Pr_f/Pr_w)^{0.25} \qquad (5)$$

In the case of air flow relations Eq. 4 and Eq. 5 are simplified:
for: $Re_f = 10$ to 1000

$$Nu_f = 0.44 \, Re_f^{0.5} \qquad (6)$$

and for: $Re_f = 1 \cdot 10^3 \div 2 \cdot 10^5$

$$Nu_f = 0.22\, Re_f^{0.6} \tag{7}$$

In Fig. 9 our relations (Eq. 4 and Eq. 5) are compared with the curves of the other authors.

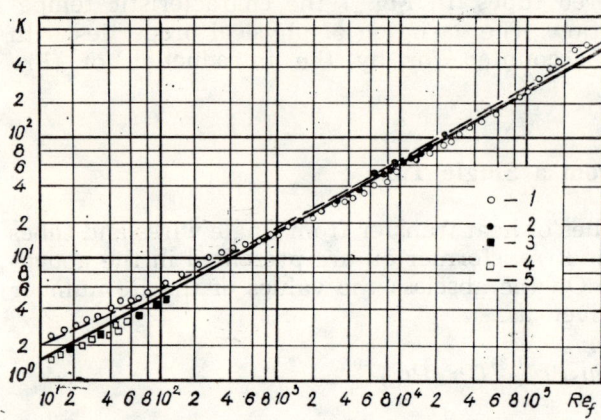

Fig. 9. Comparison of the results of different authors: 1 — experimental data by Hilpert [13], 2 — Mikheyev, air, 3, 4 — experimental data by Davis water and paraffin [14], 5 — McAdams curve [15]. $K = Nu_f Pr_f^{-0,38} (Pr_f/Pr_w)^{-0,25}$

5. Mean Heat Transfer From Tube Banks in Crossflow

Our experimental results of mean heat transfer from tube banks are presented in the following form:

$$Nu_f = c\, Re_f^m\, Pr_f^{0.36}\, (Pr_f/Pr_w)^{0.25} \tag{8}$$

The tube diameter and the velocity in the smallest cross section are chosen as characteristic parameters.

In the illustrations which follow, Figs. 10 to 24, mean heat transfer of several rows of tube banks are presented in the following general form:

$$Nu_f\, Pr_f^{-0.36}\, (Pr_f/Pr_w)^{-0.25} = K = f(Re_f). \tag{9}$$

5.1. PREDOMINANTLY LAMINAR ZONE

For $Re = 10^2$ to 10^3 the heat transfer coefficient is proportional to the fluid velocity on the power 0.5 ($m = 0.5$). In this range of Re numbers heat transfer from tube banks with large and medium longitudinal pitches is the same as in the case of single tubes.

With decreasing longitudinal pitch, heat transfer in in-line tube banks decreases because of the »shading« effect. As the longitudinal pitch increases, heat transfer decreases in comparison with the heat transfer from a single tube, because the velocity profiles in the fluid passages of the tube bank are analogous to that in a channel flow.

Figure 10 presents three in-line tube banks in the range of low Re numbers. In this case the heat transfer coefficient is proportional to the velocity on the power 0.3 ($m = 0.3$).

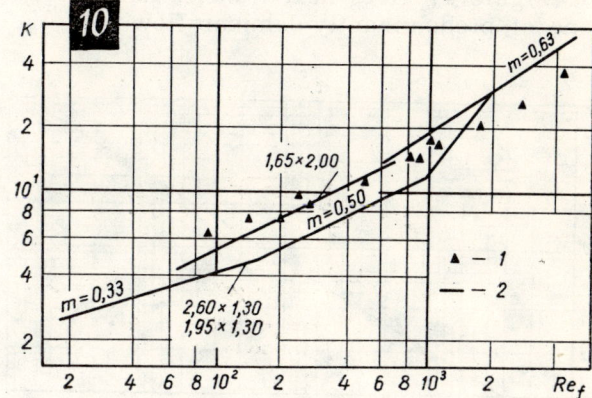

Fig. 10. Heat transfer from in-line banks in predominantly laminar flow 1 — first row, 2 — inner row.

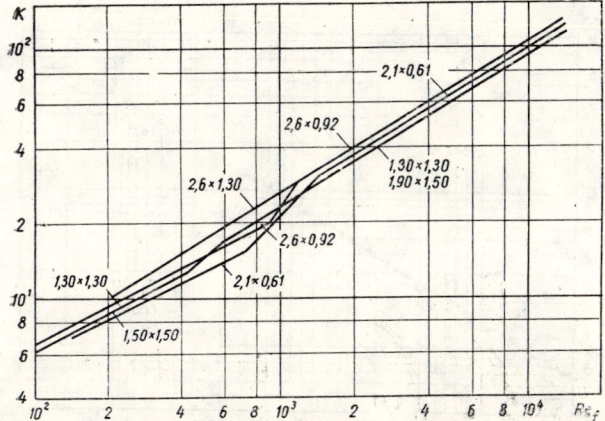

Fig. 11. Heat transfer from staggered banks in predominantly laminar flow.

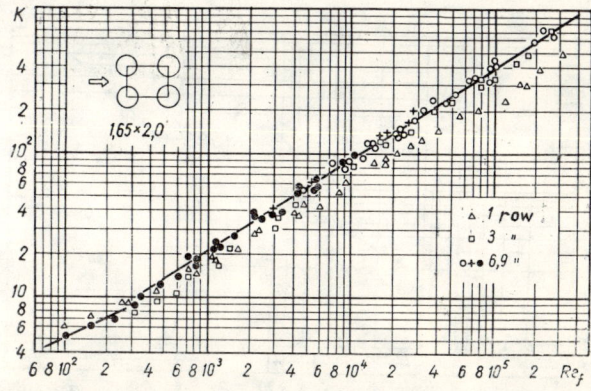

Fig. 12. Heat transfer from in-line bank 1.65×2.00.

Figure 11 gives heat transfer data for staggered tube banks in the zone of predominantly laminar flow.

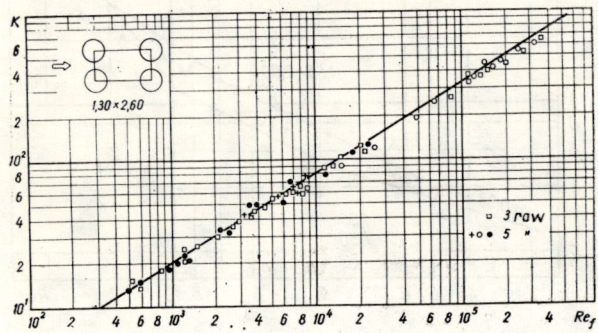

Fig. 13. Heat transfer from in-line bank 1.30×2.00.

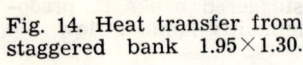

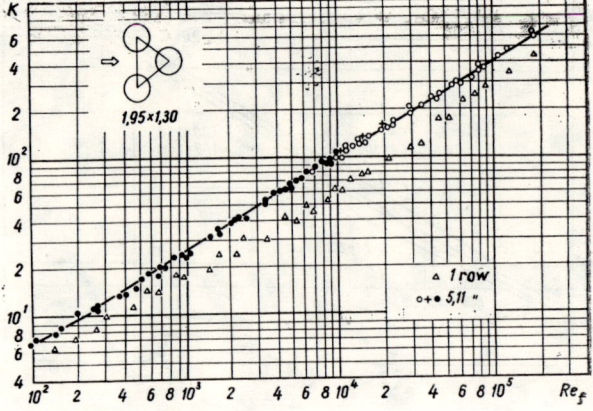

Fig. 14. Heat transfer from staggered bank 1.95×1.30.

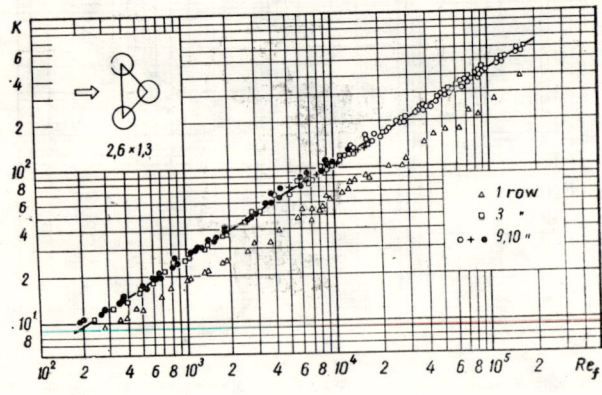

Fig. 15. Heat transfer from staggered bank 2.60×1.30.

Figures 12 to 17 give further results of heat transfer from banks of tubes in crossflows of water, air and transformer oil.

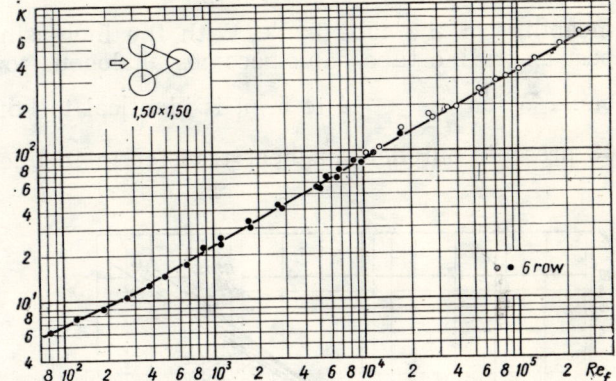

Fig. 16. Heat transfer from staggered bank 1.50×1.50.

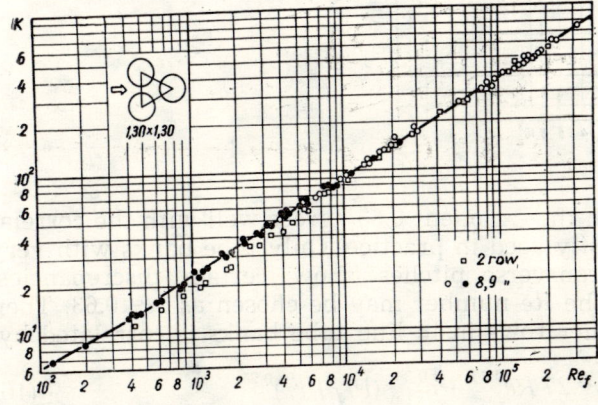

Fig. 17. Heat transfer from staggered bank 1.30×1.30.

5.2. MIXED FLOW ZONE

For $Re > 10^3$ heat transfer from the first row is lower by 30—40 percent than heat transfer from subsequent rows and is approximately equal to heat transfer from a single tube. In the first two or three rows the flow becomes fully turbulent. From the fourth row on heat transfer is stable.

For $Re > 10^3$ a certain portion of the surface in the inner rows has a laminar boundary layer, but most of the surface is streamlined by vortex flow, and the boundary layer is predominantly turbulent. The local distribution of heat transfer along the perimeter of a tube in a bank at $Re > 10^3$ is determined by the presence of the zones of different flow pattern. Because of this, the exponent m over the Re number varies from 0.55 to 0.73, depending on the bank arrangement. With the decreasing transverse pitch a, m decreases, and with the decreasing longitudinal pitch

b, m increases. The exponent depends mainly on the ratio of the longitudinal and transverse spacings between the tubes, $\frac{b-1}{a-1}$, i.e. on the permeability of the tube bank. With the increasing permeability, the flow past an in-line tube bank becomes analogous to that of a staggered tube bank. When $\frac{b-1}{a-1}=4$ to 5, m is identical to both types of banks. Figure 18 presents our data on heat transfer in in-line tube banks at $Re>10^3$.

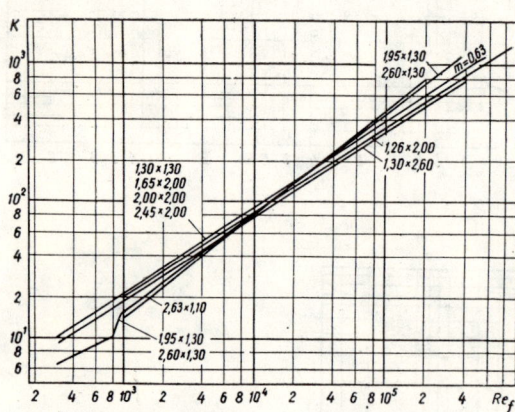

Fig. 18. Comparison of heat transfer of in-line banks of different geometry.

The correlation curve with the exponent 0.63 agrees well with the correlation curves which are mostly used in practice. Only tube banks with very small longitudinal and transverse pitches show certain discrepancies. Thus, the exponent over the Re number may be chosen as $m=0.63$. Then heat transfer from the inner rows in in-line tube banks is correlated by:

$$Nu_f = 0.27 \, Re_f^{0.63} \, Pr_f^{0.36} \, (Pr_f/Pr_w)^{0.25} \qquad (10)$$

The results of our experiments in staggered tube banks of different arrangement are presented in Fig. 20. The influence of the flow pattern on the heat transfer is expressed by a different exponent over the Re number, in this case being $m=0.6$.

The effect of the pitch in staggered tube banks is seen from Fig. 20. Heat transfer increases with the decrease of the longitudinal pitch and, to a lesser extent, with the increase of the transverse pitch a. Variation of the constant c may be correlated by the geometrical parameter a/b on the power of 0.2, when $a/b<2$. For $a/b>2$, the factor c becomes constant, $c=40$, in banks of this type the minimum flow cross section area is in the diagonal direction, and the variations of c are due to the variation in flow pattern. Tube banks of this type are not common in practice, and heat transfer relations for the pitches other than those studied here are of negligible interest. The general relation of heat transfer from tubes in inner rows of staggered tube banks is:

for: $a/b < 2$:
$$Nu_j = 0.35\,(a/b)^{0.2}\,Re_f^{0.60}\,Pr_f^{0.36}\,(Pr/Fr_w)^{0.25} \tag{11}$$

and for: $a/b > 2$:
$$Nu_j = 0.40\,Re_f^{0.60}\,Pr_f^{0.36}\,(Pr_f/Pr_w)^{0.25} \tag{12}$$

Banks of compact in-line tubes must be considered separately. Heat transfer results from tubes of large diameter in in-line banks in water

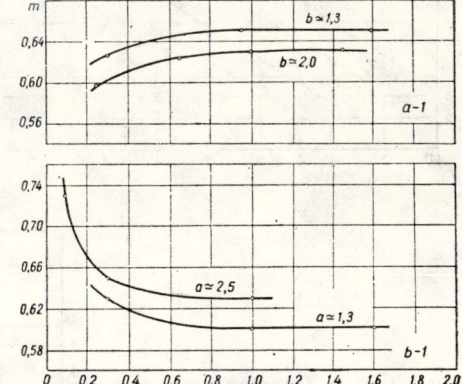

Fig. 19. Power index for Re in banks of in-line arrangement as a function of longitudinal and transverse pitch.

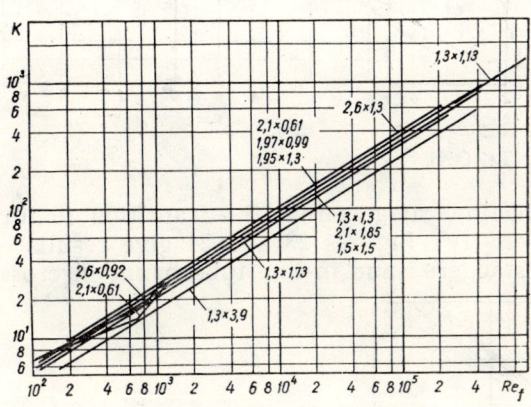

Fig. 20. Comparison of heat transfer from staggered banks of different geometry.

flow [16] are correlated in Fig. 21, with the maximum velocity as the parameter and in Fig. 22 with the average flow velocity as the parameter.

When the spacings in a tube bank are small, maximum velocity affects only a small portion of the tube surface, so that the mean velocity is a better choice for the characteristic parameter. That is why the heat transfer results correlated on the basis of the maximum velocity are widely scattered. Generalized results of different tube banks on the basis of the mean flow velocity correlate well and may be compared with other curves.

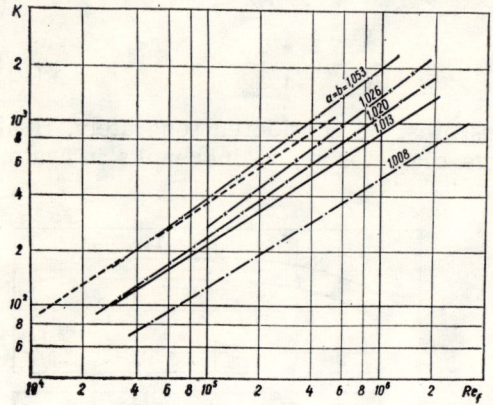

Fig. 21. Heat transfer from compact in-line banks with respect to maximum velocity. Broken line-heat transfer from bank 1.30×1.30.

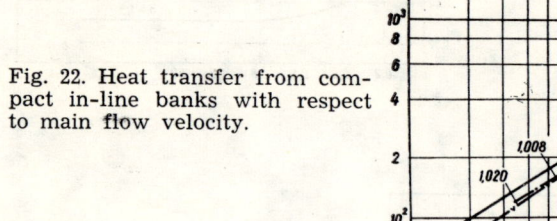

Fig. 22. Heat transfer from compact in-line banks with respect to main flow velocity.

5.3. PREDOMINANTLY TURBULENT ZONE

Experimental results in air flow suggest that the transition to turbulent flow takes place at $Re > 2 \cdot 10^5$. Figures 23 and 24 give results of heat transfer in inner rows of staggered and in-line tube banks in cross-

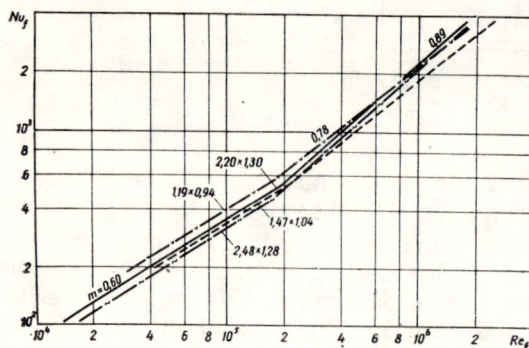

Fig. 23. Heat transfer from staggered banks in the range of large Re [17].

flow of air. These experimental results have been obtained in the Academy of Sciences of the Lithuanian SSR [17].

Heat transfer results from staggered tube banks in the range of $Re < 2 \cdot 10^5$ agree well with our data [7]. The exponent over the Re number in this range of Re numbers is $m = 0.6$. In Fig. 23 the zone of predominantly turbulent flow begins at about $Re = 2 \cdot 10^5$, which is indicated by an increase of the exponent n from 0.60 to between 0.8 and 0.9.

Differences in heat transfer from staggered tube banks of different arrangements in the studied range of pitch ratios n over exceed 30 percent. Heat transfer is more intensive in banks with larger tube spacings (large values of a/b).

The results for in-line tube banks are analogous. In the range of $Re > 1 \cdot 10^5$ heat transfer results of in-line tube banks in crossflow of air correlate well with our data for liquids, and at $Re > 1 \cdot 10^5$ heat transfer increases sharply. The exponent over the Re number increases from 0.63 to between 0.8 and 0.9, depending on the bank arrangement (Fig. 24).

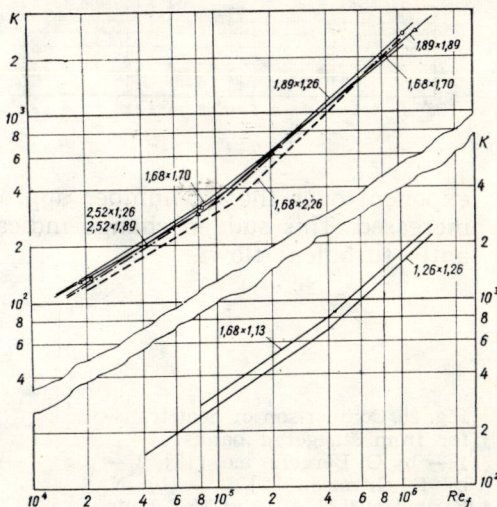

Fig. 24. Heat transfer from in-line banks in the range of large Rr [17].

6. Comparison With the Data of Other Authors

Heat transfer results of various authors in in-line tube banks with pitch ratios $1.3 \cdot 1.3$ and $2.0 \cdot 2.0$ are given in Fig. 25. The broken line corresponds to our data, 1 — experimental data by O. Bergelin et al. [18], 2 — V. Kays, A. London [19], 3 — N. Kuznetsov [4] — Lyapin [20], 5 — J. Stasiulevichius [17], 6 — V. Isachenko [21], 7 — E. Grimison [2], 8 — [4], 9 — [17], 10 — F. Hammeke et al. [22].

In Fig. 26 heat transfer results of various authors are compared for staggered tube banks with equilateral triangles having transverse pitch ratios from 1.3 to 2.0 in the stable heat transfer region.

Our results agree well with those of other authors. Our heat transfer results in liquids correlate satisfactorily with the data of O. Bergelin [18] for tube banks in oil and with the data of Isachenko [21] for tube banks in water.

In a number of papers [17, 20, 22, 24] a sudden change in the heat transfer correlation curve is noted in the range of $Re > 2 \cdot 10^5$, where the

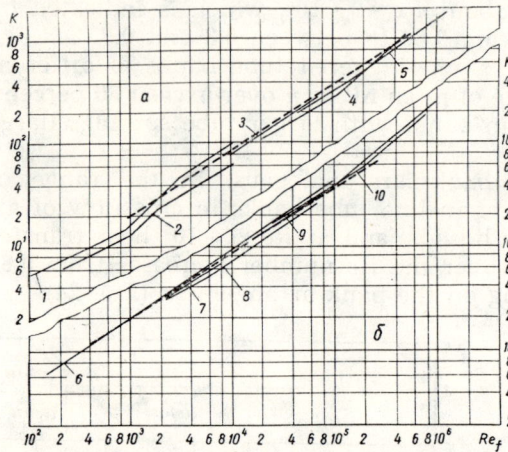

Fig. 25. Comparison of heat transfer from in-line banks:
1 — the curve of data by O. Bergelin a.o., 2 — by W. Kays and A. London [19], 3 — by N. Kuznetsov [4], 4 — by M. Lyapin [20], 5 — by J. Stasiulevichius [17], 6 — by V. Isachenko [21], 7 — by E. Grimison [2], 8 — [4], 9 — [17], 10 — by K. Hammeke a.o. [22].

exponent over the Re number approaches 0.8 and the heat transfer is increased. This sudden change indicates the transition to the predominantly turbulent flow.

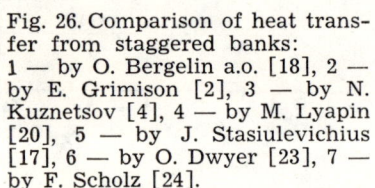

Fig. 26. Comparison of heat transfer from staggered banks:
1 — by O. Bergelin a.o. [18], 2 — by E. Grimison [2], 3 — by N. Kuznetsov [4], 4 — by M. Lyapin [20], 5 — by J. Stasiulevichius [17], 6 — by O. Dwyer [23], 7 — by F. Scholz [24].

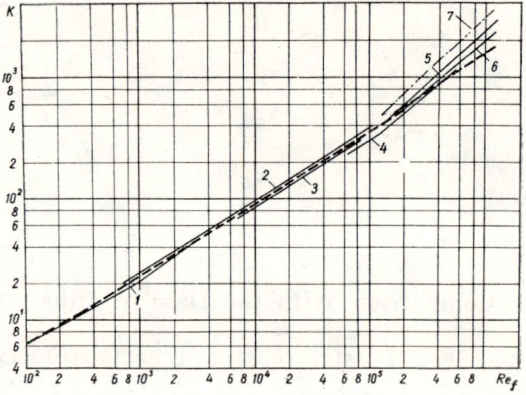

7. Proposals on Mean Heat Transfer Calculations in Banks

In Fig. 27 recommended correlations for various Re number ranges differing in the flow pattern are given for the calculation of heat transfer in inner rows of in-line tube banks. In Fig. 28 correlations are given for heat transfer calculations in inner rows of staggered tube banks.

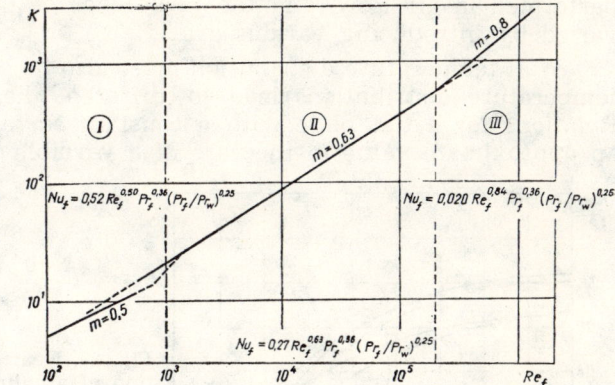

Fig. 27. Heat transfer calculation for in-line banks.

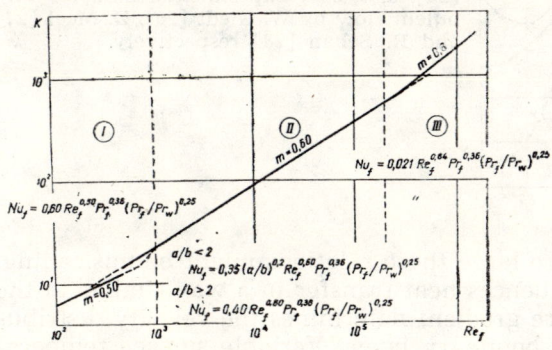

Fig. 28. Heat transfer calculation for staggered banks.

8. Local Heat Transfer

Distributions of the local heat transfer coefficients along the perimeter of a single cylinder closely follow the flow patterns described in the beginning of the report.

Because the boundary layer in the front part of the cylinder is laminar, heat transfer may be evaluated by analytical methods. It should be noted that heat transfer from the front part of a cylinder depends on whether the main flow is laminar or turbulent. Theoretical results may be obtained for the laminar main flow, and if the main flow is undisturbed or only slightly disturbed, the experimental results for local heat transfer are in agreement with the calculations. Experimental results are compared in Fig. 29 with the theoretical predictions. The experimental results corresponding to a turbulent main flow lie higher than the predictions.

J. Kestin [29] studied the influence of turbulence of the main flow on heat transfer from the front part of a cylinder and has found that the heat transfer increases with the intensity of turbulence of the main flow. The increase of turbulence of the main flow from 0.8 to 2.7 percent causes a 25 percent increase in heat transfer rates. The experiments have been performed in air flow, and the turbulence of the main flow has been varied by introducing various grids.

Wall temperature distribution must also be considered. Different wall temperature distributions lead to different heat transfer coefficients. Relations for a flat plate with a constant surface temperature are only an approximate value in the case of a variable wall temperature, and if

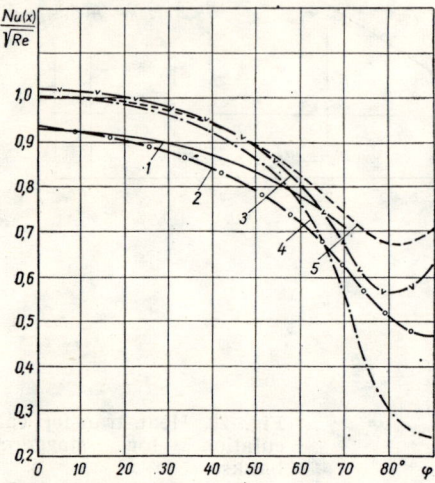

Fig. 29. Heat transfer from the front portion of a cylinder in crossflow: 1 — theoretical calculation by N. Kruzhylin [25], 2 — experimental data for laminear flow [25], 3, 4, 5 — experimental data for turbulent flow by W. Giedt [26], D. Meel [27] and R. Seban [28] respectively.

the temperature differences are large these relations might be misleading. The temperature gradient influences heat transfer in a way similar to the pressure gradient. The pressure gradient determines the velocity distribution and the thickness of the boundary layer. Variable surface temperatures determine the temperature distribution and the thickness of the thermal boundary layer. The thickness of the thermal boundary layer varies with the temperature gradient with a corresponding effect on the heat transfer coefficient, which is inversely proportional to the thickness of the thermal boundary layer. This is confirmed by the analysis [8] of exact and approximate solutions of heat transfer from bodies with randomly varying temperature.

Heat transfer from a surface streamlined by laminar boundary layer is higher in the case of q_w=const. than in the case of t_w=const. The difference in theoretical calculations for the front part of the cylinder is about 30 percent. The previous theoretical calculations refer to t_w=const. Therefore, in calculations of heat transfer and analysis of experimental data the character of surface temperature variations must be considered.

As noted above, theoretical heat transfer calculations in this case are only possible up to the separation point where minimal heat transfer occurs.

The boundary layer separates near the equator, and the rear portion of the cylinder is in the zone of a complex vortex flow. All this limits the possibilities of analytical solution, and heat transfer in this zone can be determined only from experimental data. The experiments show that after the separation point the heat transfer coefficient increases in the rear part, and the character of heat transfer variation depends mainly on the Re number. In Fig. 30 for a low Re number heat transfer is lower in the rear part than in the front part. Heat transfer increases with the Re number, and for $Re > 5 \cdot 10^4$ heat transfer is higher in the rear part than in the front part.

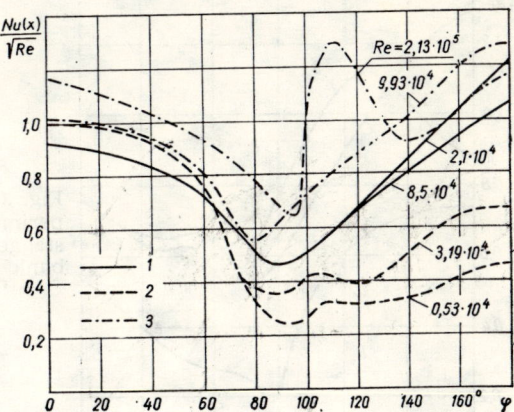

Fig. 30. Heat transfer variation along tube perimeter at different Re: 1 — by N. Kruzhylin [25], 2 — by D. Meel [27], 3 — W. Giedt [26].

Figure 30 also shows that in the range of $Re > 2 \cdot 10^5$ heat transfer variations are of different character. This is obviously connected with the turbulence of the boundary layer. Here, a certain portion of the boundary layer becomes turbulent, and the separation point together with minimum heat transfer is removed downstream. The vortex formed on the separation point enables the access of a cold mass of fluid to the surface, and heat transfer increases sharply. Heat transfer is most intensive at $\varphi = 110°$, with a subsequent decrease.

For a tube in a bank the main relations of local heat transfer are the same as for a single cylinder. Distribution of heat transfer on the surface is determined by the flow pattern around a tube in a bank, which depends to a large extent on bank geometry. In in-line tube banks two points of attack by the main flow on the tube are observed, e.g. two points of maximum heat transfer. Flow around a tube in a staggered tube bank in general is similar to the flow past a single cylinder.

As noted above, flow past a tube in a bank is intensively turbulent, and the boundary layer is truly laminar only in the front part and only for low Re numbers.

In Fig. 31 characteristic differences of heat transfer variations in inner rows of tube banks and on a single cylinder are presented. As a consequence of the flow pattern and the considerable turbulence of the main flow, heat transfer is higher from the front part of a tube in a staggered tube bank than from a single tube. Heat transfer from the rest of the tube surface is also higher in a tube bank because of the higher flow vorticity. In in-line tube banks heat transfer from the front part of a tube is also higher than from a single cylinder. Maximum heat transfer is observed at $\varphi = 50°$, which is caused by the attack of the stream coming from the preceding spacing between tubes.

As the Re number increases, flow past a tube bank becomes more and more turbulent, and local heat transfer becomes equally distributed along the tube perimeter. This is illustrated in Fig. 32 for a water flow.

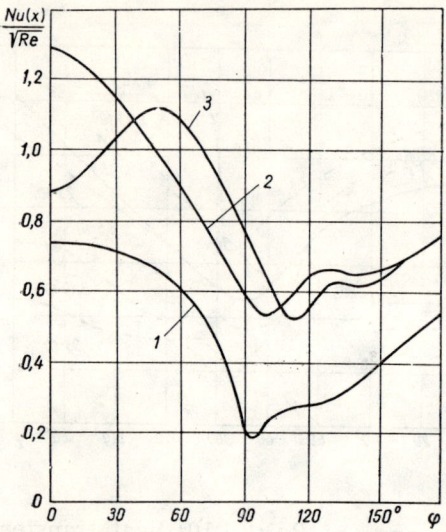

Fig. 31. Heat transfer variation along tube perimeter: 1 — single tube, 2 — tube in a staggered bank 2.00×2.00, 3 — tube in a bank of in-line arrangement 2.00×2.00.

Fig. 32. Local heat transfer from a bank of in-line arrangement. 2.60×1.30 in water flow depending on Re.

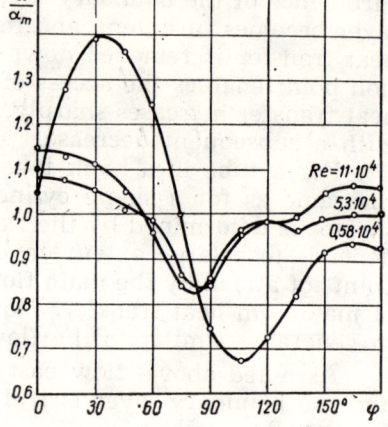

9. Fluid Friction Across Banks of Tubes

Based on our experiments with various liquids and on the results of Bergelin et al. [18], nomograms of fluid friction across tube banks have been compiled (Figs. 33, 34).

It follows from the nomograms that fluid friction depends mainly on relative transverse pitch and is increased with the decreasing pitch.

The variation of fluid friction as a function of the flow pattern can be observed from the curves on Figs. 33 and 34.

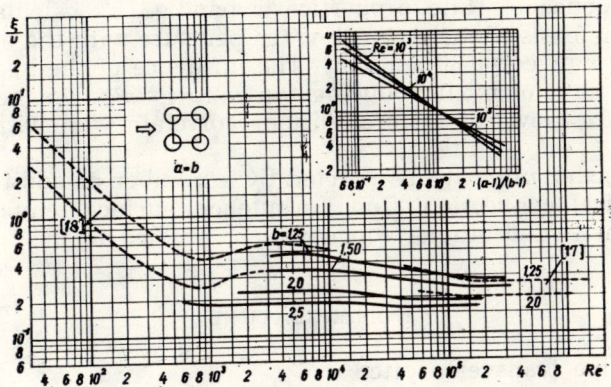

Fig. 33. Evaluation of drag coefficient in banks of in-line arrangement. Relative longitudinal pitch is characteristic in the main plot.

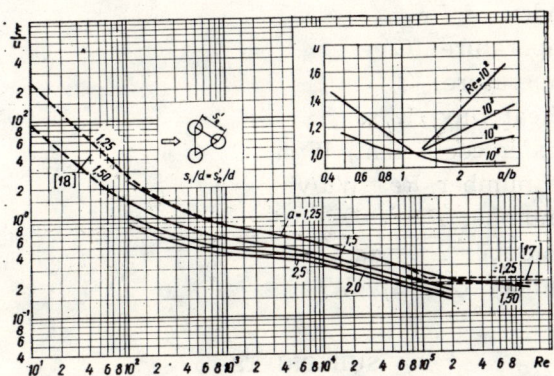

Fig. 34. Evaluation of drag coefficient in banks of staggered arrangement. Relative transverse pitch is characteristic in the main plot.

10. Conclusions

Based on the experimental investigations, performed at the Institute of Physical and Technical Problems of Energetics of the Academy of Sciences of Lithuanian SSR, heat transfer from single tubes and banks of tubes in crossflow of various fluid is described. Experiments with single tubes and tube banks of 43 different arrangements in crossflow of various liquids and gases cover the range of Pr numbers from 0.7 to 500 and Re numbers from 1 to $2 \cdot 10^6$. The results have been compared with the data obtained by other authors.

Heat transfer from single tubes and banks of tubes is analyzed in function of tube arrangement, fluid properties, heat flux direction, and temperature difference.

The influence of fluid properties is handled by calculating the Pr_f number on the basis of the mean flow temperature with the exponent over the Pr_f being between 0.36 and 0.38. The influence of the heat flux direction is taken care of by introducing the ratio Pr_f/Pr_w on the power 0.25. The results for flows of different fluids correlate well with the results for gases.

Heat transfer intensity is affected by a number of additional factors, such as natural convection, flow turbulence, surface temperature variation, roughness, and others, which will be studied in the future.

NOMENCLATURE

a relative transverse pitch
b relative longitudinal pitch
c constant
d diameter, mm
m power index of Re
n power index of Pr
t_f flow temperature, °C
t_w wall temperature, °C
q specific heat flux, W/m^2
α heat transfer coefficient, $W/m^2\,°C$
Re Reynolds number $Re = Wd/\nu$
Pr Prandtl number $Pr = c\mu/\lambda$
Nu Nusselt number $Nu = \alpha d/\lambda$
Eu Euler number $Eu = \Delta p/\rho W^2$

SUBSCRIPTS

f characteristic main flow temperature
w characteristic wall temperature.

REFERENCES

1. Antus'ev V. M., Kozachenko L. S., Sovetskoe kotloturbostroenie, 5 (1937); Sovetskoe kotloturbostroenie, 6 (1937); Teploperedacha i aerodinamicheskie soprotivlenia konvektivn'ikh poverkhnostei nagreva, ONTI, M., 1938.
2. Grimison E. D., Trans. ASME, 59, 7 (1937).
3. Mikheev M. A., Osnov'i teploperedachi, Gosenergoizdat, M., 1956.
4. Kuznetsov N. V., Rabochie protesess'i i vopros'i usovershenstvovaniia konvenktivn'ikh poverkhnostei kotel'n'ikh agregatov, Gosenergoizdat, 1958.
5. Achenbach E., J. Fluid Mech., 34, 4 (1968).
6. Zhukauskas A. A., Teploenergetika, 4 (1954); Teploperedacha i teplovoe modelirovanie, Izd. AN SSSR, 1959.

7. Zhukauskas A., Makariaichius V., Shanchiauskas A., *Teplootdacha puchkov trub v poperechnom potoke zhidkosti,* »Mintis«, Vil'nius, 1968.
8. Zhukauskas A., Zhiugzhda I., Inzh. Fiz. Zh., **4**, 11 (1961).
9. Zhukauskas A., Ambrazyavichyus A., Int. J. Heat Mass Transfer, **3**, 4 (1961).
10. Shanchiauskas A., Ulinskas R., Zhukauskas A., Trud'i AN Litovskoi SSR, seriia B, 4 (59) (1969).
11. Eckert E., VDI-Forschungsheft, 416 (1942).
12. Shenchianas P., Makariavichius V., Tamonis M., Zhukauskas A., Trud'i AN Litovskoi SSR, seriia B, 4 (59) (1969).
13. Hilport R., Forsch. Ing-Wes., 4 (1953).
14. Davis A., Philosophical Magazin. **47**, 282 (1924).
15. McAdams W., *Heat Transmission,* New York, 1954.
16. Samoshka S., Makariavichius V., Shlanchiauskas A., Zhiugzhda I., Zhukauskas A., Trud'i AN Litovskoi SSR, seriia B, 3 (5) (1967).
17. Stasiuliavichius I. K., Samoshka P. S., Trud'i AN Litovskoi SSR, seriia B, 4 (33) 1963); Inzh. Fiz. Zh., **7**, 11 (1964).
18. Bergelin O. P., Davis E. S., Hull H. L., Brown G. A., Sullivan F. W., Doberstein S. L, Trans. ASME, **71**, 4 (1969); Trans. ASME, **72**, 6 (1950); Trans. ASME, **74**, 6 (1952).
19. Kays W. W., London A. I., *Compact Heat Exchangers,* McGraw-Hill, New York, 1958.
20. Liapin M. B., Teploenergetika, 9 (1956).
21. Isachenko V. P., Teploenergetika, 8 (1955); Sb. *Teploperedacha i teplovoe modelirovanie,* Izd. AN SSSR, 1959.
22. Hammeke K., Heinecke E., Scholz F., Int. J. Heat Mass Transfer, 10, p. 427 (1967).
23. Dwyer O. E., Sheeman T. V., Weisman J., Horn F. L., Schomer H. T., Industrial and Engineering Chemistry, **48**, 10, (1956).
24. Scholz F., Chemie-Ingenieur-Technik, 20 (1968).
25. Kruzhilin N. S., Zh. Tekh. Fiz., 6, v'ip. 3 (1936), v'ip. 5 (1936).
26. Giedt W. N., J. Aero. Sci., **18**, 11 (1951).
27. Meel D. S., Int. J. Heat Mass Transfer, 5, p. 715 (1962).
28. Seban R. A., Trans. ASME, **82**, 2 (1960).
29. Kestin J., *Advances in Heat Transfer,* 3, Academic Press, New York, 1966.

GAS CONCENTRATION MEASUREMENTS IN BOUNDARY LAYER

EDMOND A. BRUN

Faculté des Sciences, Université de Paris, France

1. Definitions

Let us recall some wellknown definitions.

1.1. Let a homogeneous mixture of fluids occupy a domain D. In this domain in the volume dv around the point M, at time t, the mass of the mixture is dm and the density, at point M and time t, is:

$$\rho = \frac{dm}{dv}$$

Let dm_i be the mass of species i in the volume dv at time t, the *mass concentration of species* i, *at point* M *and time* t, is:

$$\Gamma_i = \frac{dm_i}{dv}, \quad \text{with} \quad \Sigma \, \Gamma_i = \rho. \tag{1}$$

The mass fraction of species i, *at point* M *and time* t, is:

$$\gamma_j = \frac{dm_i}{dm} = \frac{\Gamma_i}{\rho}, \quad \text{with} \quad \Sigma \, \gamma_i = 1. \tag{2}$$

The molar concentration of species i, *at point* M *and time* t, is:

$$C_i = \frac{dm_i}{M_i} \frac{1}{dv} = \frac{\Gamma_i}{M_i} \tag{3}$$

Let

$$C = \Sigma \, C_i$$

be the molar concentration of the mixture, *the molar fraction of species* i, *at point* M *and time* t, is:

$$c_i = \frac{C_i}{C} \quad \text{with} \quad \Sigma \, c_i = 1 \tag{4}$$

1.2. In the problem of interest, where the mixture is flowing, we also have to define quantities relative to mass transfer: velocities and mass flux. The component u_i along the x axis of the *mass velocity of species i*, at point $M = [x, y, z]$ and time t, is given by:

$$dm_i = \Gamma_i u_i \, dA \, dt$$

where dm_i is the mass of species i flowing, during the time dt, through the element of area dA, centered on M and normal to the x axis.

Therefore, *the mass flux density* $\vec{\varphi}_i$ *of species i, at point M and time t*, is related to the velocity $\vec{U}_i = (u_i, v_i, w_i)$ of species i, at the same point and time, by

$$\vec{\varphi}_i = \Gamma_i \vec{U}_i \tag{5}$$

Here, the reference system is fixed in space.

The mass average velocity of the mixture, or barycentric velocity, at point M and time t, is:

$$\vec{U} = \frac{\Sigma(\Gamma_i \vec{U}_i)}{\rho} = \Sigma(\gamma_i \vec{U}_i) \tag{6}$$

This is a fictitious velocity.

In order to characterize the motion of species i in the flowing fluid, one introduces the relative velocity

$$\vec{V}_i = \vec{U}_i - \vec{U}, \tag{7}$$

which is called *the mass diffusion velocity of species i in the mixture at point M and time t*.

Of course, we have the relation

$$\Sigma(\gamma_i \vec{V}_i) = 0$$

Here, the reference system is fixed on the velocity $\vec{U}$. *The mass diffusion flux density* of species i, at point M and time t, is:

$$\vec{J}_i = \Gamma_i(\vec{U}_i - \vec{U}) = \Gamma_i \vec{V}_i \tag{8}$$

1.3. Similarly, one defines:
— the molar flux density of species i,

$$dn_i = C_i u_i \, dA \, dt; \qquad \vec{\omega}_i^* = C_i \vec{U}_i \tag{9}$$

— the molar average velocity of the mixture,

$$\vec{U}^* = \frac{\Sigma(C_i \vec{U}_i)}{C} = \Sigma(c_i \vec{U}_i^*) \tag{10}$$

— the molar diffusion velocity of species i,

$$\vec{V}_i^* = \vec{U}_i - \vec{U}^* \qquad \text{with} \qquad \Sigma(c_i \vec{V}_i^*) = 0 \tag{11}$$

Here, the reference system is fixed on the velocity $\vec{U}^*$.
— the molar diffusion flux density of species i,

$$\vec{J}^* = C_i(\vec{U_i} - \vec{U}^*) = C_i \vec{V_i}^* \tag{12}$$

2. Mass Transfer in the Boundary Layer

2.1. The aim of this paper is to describe the techniques of concentration measurements in the vicinity of boundaries. Such measurements rest on the principles of the thermodynamics of irreversible phenomena, such as that of the »local state«; to examine them would take me too far and they will be assumed known.

On another hand, I find it useful to show, at the beginning, the interest of concentration measurements in the boundary layer by exposing a simple problem which will be seen again later.

During missile reentry in the atmosphere, the components of the air undergo, when flowing through the shock wave and due to the strong rise in temperature, a number of transformations: dissociation, formation of nitric oxide, NO and ionization. In the vicinity of the wall the temperature, on the contrary, drops sharply, which should lead to the transformations opposite of the above mentioned, but the gaseous mixture does not reach, at each point, the state of chemical equilibrium corresponding to the physical conditions, this, as the time required for a complete chemical reaction in an element of fluid is large compared to the time this element remains in the boundary layer.

However, generally, the wall acts catalytically and increases largely the reaction. Therefore, the species concentrations are not the same on the two frontiers of the boundary layer and diffusion of these species will occur, *mass transfer thus being superimposed to heat transfer* through the boundary layer.

To investigate, on a simple case, this transfer mechanism, G. Lassau [1] let a nitrogen plasma ($N_2 + N + N_+$ + electrons) flow, at a velocity of the order of 250 m/s, on a steel flat plate parallel to the stream. The plasma, at a temperature of the order of 2000°C, is not in equilibrium and, practically, there is no chemical reaction in the free stream and in the boundary layer. However, in contact with the wall, maintained at a temperature of the order of 500°C, it is known by experience that a catalysis occurs, the initial reaction being monoatomic. The mechanism of the catalysis is probably that described by Rideal. An atom of gaseous nitrogen is adsorbed and another impinging gaseous atom, striking the adsorbed one, creates the molecular nitrogen, the adsorption sites on the surface being always saturated in adsorbed atoms [2]:

$$N + S \rightarrow N_{ads} \tag{a}$$

$$N + N_{ads} \rightarrow N_{2\,ads} \tag{b}$$

$$N_{2\,ads} \rightarrow N_2 + S \tag{c}$$

The initial monoatomic reaction occurs only forward; therefore the law of chemical kinetics is written, with C_N standing for the concentration in monoatomic nitrogen, at time t:

$$-\frac{dC_n}{dt} = k\ C_n, \qquad (13)$$

where k is the kinetic reaction constant. Integration leads to consider k as the reciprocal of a time constant:

$$\tau = \frac{C_n}{d\ C_n/dt} \qquad (14)$$

On the plate, the time constant is very small due to catalysis; in the boundary layer, on the contrary, it is very high, the kinetic reaction constant being practically zero.

If, instead of a steel plate, a pyrex plate is placed in the plasma jet, the catalysis does not occur.

2.2. More generally, let us deal with a *chemically reacting boundary layer* and where, therefore, *mass diffusion phenomena occur*.

The diffusion of species i can be characterized by a time τ_m, which will be called simply »*mechanical time*«: it *represents* the time it takes species i to go through the boundary layer:

$$\tau_m = \frac{\delta}{V_i}; \qquad (15)$$

δ is the boundary layer thickness and V_i the average diffusion velocity of species i through the boundary layer.

Let C_i be the concentration of species i, at a given point of the boundary layer, whereas the temperature at this point would lead to an equilibrium concentration C_{ie}. A »*chemical time*« can be defined by:

$$\tau_c = \frac{C_{ie} - C_i}{dC_i/dt}. \qquad (16)$$

It is the ratio of the shift from equilibrium over the reaction rate which brings back to equilibrium. In the particular case of a first order forward reaction, τ_c is no more than the time constant defined in Eq. (14).

Consideration of these two characteristic times allows a classification of the different types of flows.

If $\tau_c < \tau_m$, the reaction is very fast; the *boundary layer is in equilibrium at every point*.

If $\tau_c > \tau_m$ the reaction does not have time to occur in the flow, *the boundary layer is frozen*. This is the case of the fast catalysis at wall, dealt with previously.

If τ_c and τ_m are of the same order of magnitude, *relaxation phenomena occur in the boundary layer*.

The importance of boundary layer profile concentrations investigation is then understandable; it not only allows to know the constitution of the boundary layer, but furthermore it determines the type of flow and the reaction mechanisms which occur.

3. Measurements by Sampling

3.1. The *sampling probe* is, most often, a cylindrical tube, the axis of which is tangent to the stream, a drop in pressure in the tube, due generally to pumping, carries the gas sample to the *analyzer* placed outside of the stream.

The sampling methods have serious drawbacks: the insertion of the probe perturbs the stream and therefore the *analyzed sample is not necessarily that which would exist, without the probe, where the center of the probe aperture is*.

In order to minimize the perturbations due to the insertion of the probe, a number of measures must be taken.

a) of course, *the size of the probe must be as small as possible* [3]. Indeed, the amount of perturbation is related to the ratio of the tube diameter to the boundary layer thickness and this ratio can thus be kept down by increasing the boundary layer thickness: in particular, this is the reason why the above mentioned experiments of Lassau were carried at low pressures.

b) the *tube axis must be as parallel as possible to the stream direction* which existed before insertion of the probe, otherwise a distortion of the field of velocity would be superimposed to the perturbation of the streamlines density due to this insertion.

c) the perturbation of the velocity along the outside wall of the sampling tube will be as small as possible if the sucking rate in the probe is equal to that which crossed the section of the tube before its insertion in the boundary layer (*isokinetic probe*). This measure does not allow to disturb too much the diffusion flux and chemical reaction rate in the vicinity of the tube and mainly where the sampling is performed.

d) it must be made sure that *the probe is not a heat sink or source* in the boundary layer and that thermal equilibrium must be nearly reached between the probe wall and surrounding fluid: a heat source would change, not only the temperature distribution, but also the concentration gradient due to a change in chemical reaction rate.

e) *the probe wall must have no chemical effect on the gas* (catalysis for example in the case of a gas out of equilibrium); furthermore it must not undergo any change (fusion, sublimation, etc.).

f) finally *the sample, while being sucked, must not be submitted to any change:* if a chemical reaction occurs within the tube, the transfer time must be as small a fraction as possible of the chemical time; should an adsorption occur, it must be kept as small as possible by avoiding the use of metals (teflon walls, apertures drilled in sapphire plates) and by using a steady flow allowing the analysis to be carried only after adsorption equilibrium is reached [4].

3.2. We will deal more closely with the last condition, very specific to the problem of interest, and which takes on great importance in the case of a sample not in chemical equilibrium.

Let us deal, for example, with the experiments on the boundary layer of a nitrogen plasma out of equilibrium. The probe is made of a quartz tube: this material has a very low catalytic reaction constant for the transformation of atomic nitrogen; it can be assumed that it changes very little of the concentration gradient in the boundary layer. However, the transfer of atomic nitrogen in the tube up to the analyzer lasts long enough for an appreciable catalytic action to occur and, upon arrival in the analyzer, to have a different sample. The solution is to transform the atomic nitrogen into a stable species before any transfer for analysis.

It is known that hydrocarbons react with activated nitrogen to give cyanhydric acid, traces of CN radicals and cyanogen gas C_2N_2 [5, 6]. With methane, which was the hydrocarbon used, the main reaction is:

$$N + CH_4 \to HCN + 3/2\, H_2 \qquad (d)$$

The reaction rate is high enough to be considered as instantaneous [7]. The idea is to react the atomic nitrogen with the hydrocarbon, then to transport the stable products into the analyzer.

The gases mix at the entrance of the probe and are sucked by a vacuum pump, the analyzer is linked to this vacuum pump. The flow rate of methane compared to the rate of nitrogen must be such that the chemical reaction (d) is complete, but there is evidently no point in diluting too much the sample in methane since the analyzer will yield more accurate measurements as the fractions of nitrogen and cyanhydric acid in the mixture are high. The rate of flow of methane is then increased

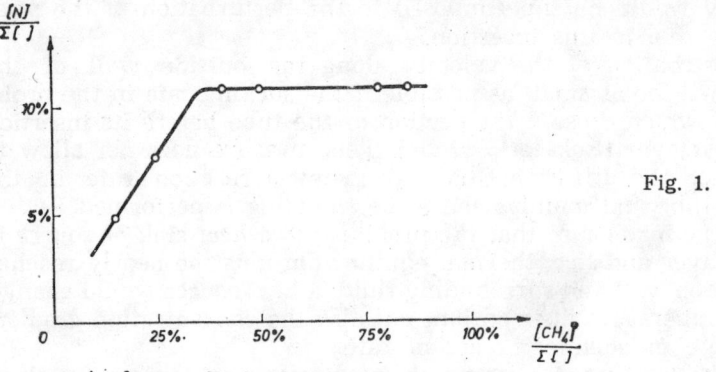

Fig. 1.

progressively: analysis shows that the fraction of atomic nitrogen detected increases first proportionally to the rate of flow of methane until it becomes constant (Fig. 1). At this time the transformation of atomic nitrogen is complete; the results of the analysis corresponding to the minimum flow rate of methane allowing the complete transformation of N into HCN are kept.

This is given as an example but there are, as will be seen later, other cases where *a fast auxiliary chemical reaction can freeze the state of the mixture* when a reaction of the mixture components between the probe and the analyzer is feared.

4. Different Types of Remote Analyzers

It is not intended here to give a lecture on chemical analysis. Different techniques will nevertheless be recalled, because they were used in the problems in interest.

4.1. MASS SPECTROMETER

This is the apparatus most used in our laboratory: indeed, the analysis is carried at a very low pressure, between 10^{-5} and 10^{-11} torr. This is very interesting in the case of our laboratory, where mostly analyses in rarefied gas flows are carried.

The molecules of the mixture, inserted in a very low pressure container (Fig. 2), are ionized, through an electron bombardment for example,

$$M + e \rightarrow 2e + M^+. \qquad (e)$$

The ions thus created flow through a set of apertures in plates raised to electric potentials such that focusing of these ions is realized. The ion beam then enters into the element which allows separation. These separators might involve either a combination of electric and magnetic fields as in the Omegatron, or grids raised to alternating electric potentials as in

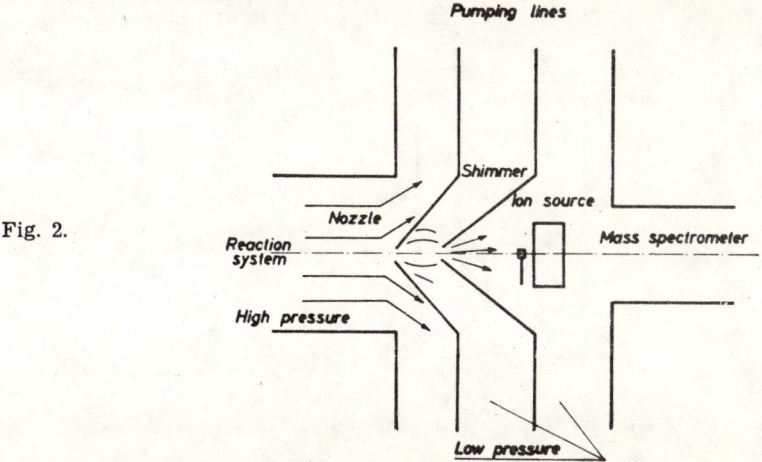

Fig. 2.

the Topatron, or a quadripolar tube, the electrodes of which are submitted to a variable frequency electric potential, as in the Varian. In the latter case, which, in our opinion, is the system best fitted to the measurements of interest, for a given frequency, only those electrons corresponding to a ratio mass/electric charge (M/q) within a narrow band have a regular sinusoidal trajectory, whereas the others oscillate transversally with an increasing amplitude until they reach the walls of the quadripole, allowing their deionization and elimination. The ions, in resonance with the quadripole frequency (Fig. 3) reach an electron multiplicator leading finally to a recording device. By changing the frequency, all values of M/q are scanned; those corresponding to ions contained in the beam create peaks of height proportional to the molar concentrations (Fig. 4).

The resolution of a mass spectrometer is defined by the ratio $M/\Delta M$, where M is the ion mass and ΔM the shift in mass corresponding to the peak width, measured at half height. In the range of molar mass between 1 and 50, the resolution is of the order of 50, whereas for higher molar masses it is of the order of 20.

The sensitivity is very high and a pressure of 10^{-14} torr of a component such as nitrogen can be detected.

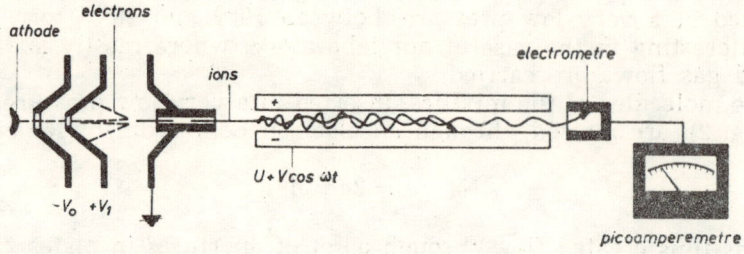

Fig. 3.

The mass spectrometer is now very much used for the investigation of many gas reactions [8, 3]; its use for the determination of concentration profiles in a boundary layer, with or without chemical reactions, seems to

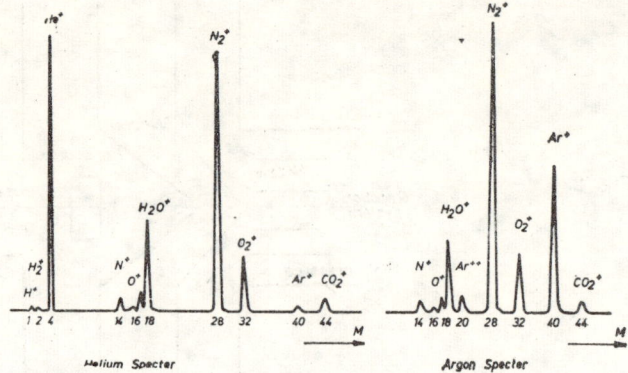

Fig. 4.

be less frequent. Provided the care necessary for all sampling method is taken, it leads however, in most cases, to good results which, in our laboratory, could be checked with other methods.

4.2. CHROMATOGRAPHY

The sample can of course be analyzed by chromatography. It is a method very much used by chemists. However, in the present case, it is less interesting than the method based on the mass spectrometer and this for several reasons.

Firstly, the pressure in the analyzer must be higher than 5 mm Hg and the mass flow rate in the chromatometer must be considerably higher than in the mass spectrometer; this might, in some cases, create large perturbations (case of free diffusion).

Second, the response time is long, considerably longer than in a mass spectrometer.

Finally, the chromatometer must be prepared to work on the mixture components, whereas the mass spectrometer determines all the components while scanning and, in particular, it is very useful in the analysis of gases containing atoms free radicals, etc.

4.3. OTHER TYPES OF ANALYZERS

There are of course many types of analyzers based on different physical properties, but all have the above mentioned drawbacks of chromatography: high flow rates, lengthy experiments (response time of the order of a minute), specificity of components.

a) *The infrared radiation absorption analyzer* is based on the fact that many gases (CO, CO_2, hydrocarbons, SO_2, NH_3) have absorption bands in the infrared: an absorbing component weakens, proportionally to its concentration, the infrared radiation in a given wave-length, which is characteristic of the gas and independent of its temperature and pressure.

The measurement is of the differential type, as the energies involved are weak. For example, two tubes of equal length contain one ❶ dry pure air (reference tube) the other ❷ the gaseous mixture with the component i to be measured. This mixture flows through tube ❷ (Fig. 5). The

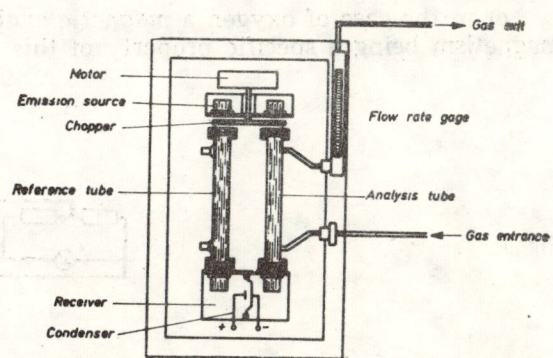

Fig. 5.

same infrared radiation goes through both tubes and is then received in two identical chambers, separated by a thin membrane, containing the component i to be analyzed. The gas of the chambers will absorb the radiations characteristic of component i. As the emission is interrupted five times a second by the rotation of a chopper, the temperatures and, consequently, the pressures in the two chambers are modulated at the frequency of the chopper. With the membrane, one measures the difference in instantaneous pressure which is nearly proportional to the selective weakening that one of the radiations undergoes in tube ❷ where the mixture containing the component to be measured flows.

b) The *catharometer*, an analyzer based on the difference of thermal conductivities of gases, is also a differential apparatus: four wires, of resistance variable with temperature make up a Wheatstone bridge and are set in such a way that two opposite wires are swept by the gas mixture to be studied, the other two by the reference gas (Fig. 6). The difference

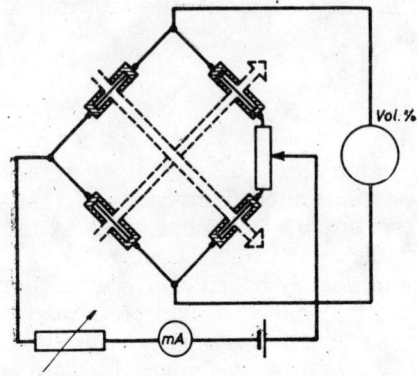

Fig. 6.

of thermal conductivity between the gas mixture and the reference gas are enough to generate, in the bridge diagonal, a current proportional to the concentration to be measured. In order to proscribe corrosion and catalysis, the measuring wires are inserted in perfectly sealed glass capillaries.

c) in the case of oxygen a *magnetic analyzer* can be employed, paramagnetism being a specific property of this component. In the apparatus

Fig. 7.
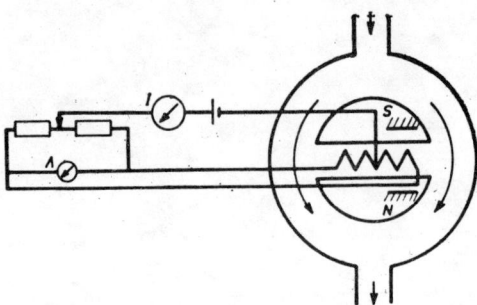

displayed on Fig. 7, due to the heterogeneous magnetic field a thermomagnetic flow, the velocity of which is proportional to the oxygen concentration in the gas under study, occurs in the small electrically heated transversal tube.

5. In Situ Measurements, without Probe, in the Boundary Layer

These measurements are obviously the best, particularly if they do not require the addition of gas in the boundary layer. They usually employ optical methods and thus cannot, unfortunately, be applied to all cases.

5.1. INTENSITY OF EMISSION RAYS

This method is applicable only to two dimensional flow and for a fairly high temperature gas mixture (plasma for example).

The observation plane in the boundary layer is both parallel to the stream and normal to the wall (Fig. 8). The image is produced by the lense L on a spectrometer slit which is parallel to the observation plane and normal to the wall. On the plate, appear spectral rays of the com-

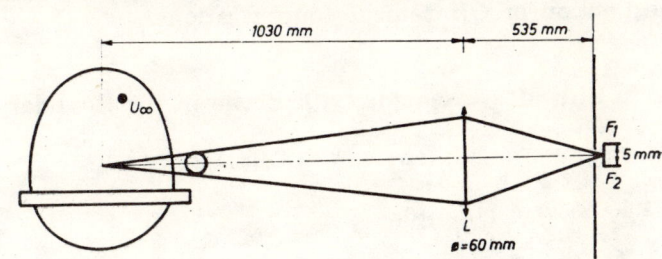

Fig. 8.

ponent in the gas mixture. With a special microdensitometer, the intensity in each point of each ray is determined. As this intensity is proportional to the concentration of the component i of the mixture, the concentration as a function of distance from the plate is thus obtained directly.

The optics of the apparatus must be such that:

a) the entire thickness of the boundary layer is found on the spectrometer slit,

b) not only the observation plane, but practically every point of the boundary layer which illuminates the slit can be assumed practically at infinity and this requires that the width of the boundary layer on one hand and the focal distance of the lense on the other hand be small compared to the distance of the lense to the middle of the boundary layer.

c) however, as the spectrometer integrates all the light coming out of the boundary layer, its width must be large enough compared to the part perturbed by the lateral edges of the plate.

This method is particularly fitting for the study of plasmas [9]. It allows, up to the near vicinity of the wall, to determine the distribution of the electron density in boundary layer along flat walls and also atom and ion concentrations. However the interpretation of the spectra is accurate only for a boundary layer in equilibrium.

Instead of an optical spectrum, a Raman spectrum, or the electronic paramagnetic resonance, can also be used.

5.2. ABSORPTION OF A LIGHT BEAM

This method again fits only a two dimensional flow.

A cylindrical light beam, parallel to the wall and normal to the stream, goes through a boundary layer; the light beam diameter is small compared to the boundary layer thickness. If the component i of the gas mixture absorbs a light ray or an infrared band, the variation of the optical absorption, proportional to the concentration of component i, as a function of the distance to the wall is investigated.

This method is more difficult to carry out than that of emission: it must be made sure that the beam remains rigorously parallel while going through the boundary layer.

As in the case of emission, the length that the light beam goes through must be large enough for the lateral edge effects to be negligible.

This method has been used in particular in the infrared for the investigation of OH radical concentrations.

6. In Situ Measurements with Probe in the Boundary Layer

6.1. DETERMINATION BY PHOTOCHEMISTRY

6.1.1. N Determination

We mentioned previously the complete reaction of *atomic nitrogen on methane*, with formation of the component HCN the molar concentration of which is proportional to that of N. This reaction is luminous and gives out a blue coloration due to the CN bands [10]. If the CN concentration is high enough, the radiated energy, due to the reaction, is proportional to the atomic nitrogen concentration [11].

The probe used is a quartz tube, with axis normal to the wall, where the methane comes out. In order to measure the luminous intensity of the reaction, the image of the area in the vicinity of the probe aperture is formed on a photomultiplier.

However, it must be noticed that the light emitted throughout the plasma located on the system optical axis is also integrated; the measurement will be accurate only if, in the radiation received by the photomultiplier, that part coming out of the chemical reaction dominates; in other words, the ratio of the intensity I obtained with injection over the intensity I_o, without injection, must be large; it is brought close to 6 by a proper choice of the spectral region employed. The relative concentration in atomic nitrogen is then given by:

$$\frac{I-I_o}{I_\infty - I_{o\infty}}, \qquad (17)$$

where I_∞ and $I_{0\infty}$ are the reference luminous intensities, which will be, for example, the intensities outside the boundary layer.

6.1.2. O determination

When a small quantity of nitric oxide NO is added to atomic oxygen, a yellow-green light is emitted; it is due to the set of slow reactions.

$$O + NO + M \rightarrow NO_2^* + M \qquad (f)$$

$$\begin{cases} NO_2^* + M \rightarrow NO_2 + M & \text{(g)} \\ NO_2^* \rightarrow NO_2 + h\nu & \text{(h)} \end{cases}$$
$$\overline{O + NO \rightarrow NO_2 + h\nu}$$

but atomic oxygen reacts *rapidly* on NO_2 to give

$$O + NO_2 \rightarrow NO + O_2 \tag{i}$$

in such a way that nitric oxyde is regenerated. Therefore, with a small quantity of nitric oxyde, the light emitted during the set of reactions is proportional to the atomic oxygen concentration [12]. If the mixture is sucked in a tube after NO addition, it is found a diminution of the luminous intensity I along the tube axis (x axis), such that:

$$-\frac{dI}{dx} = -\frac{dC_o}{dx}. \tag{18}$$

In order to determine the absolute atomic oxygen concentration C_o at a given point of the tube and, consequently, in any other point, it is sufficient to add nitric oxyde in progressively increasing quantity until light disappears at the injection point; the molar concentration in O is then, according to reaction (i) equal to that of NO_2, which is known if the flow rate of NO_2 is measured.

6.1.3. H determination

A very analogous method allows to measure the molar concentration in atomic hydrogen with nitric oxyde, according to the set of slow reactions that produce red light.

$$\begin{cases} H + NO + M \rightarrow HNO^* + M & \text{(j)} \\ M + HNO^* \rightarrow HNO + M & \text{(k)} \\ HNO^* \rightarrow HNO + h\nu. & \text{(l)} \end{cases}$$

Here again, if the quantity of nitric oxyde is low, the reaction rate is proportional to the molar concentration in atomic H and the nitric oxyde is regenerated through the fast reaction

$$H + HNO \rightarrow H_2 + NO. \tag{m}$$

6.2. DETERMINATION BY MICROCALORIMETRY

In principle, with a microcalorimeter inserted in the flow, the amount of heat produced in a complete chemical reaction which leads to the disappearance of one of the components of the mixture can be measured [13]. In this case, only systems out of equilibrium are dealt with. We will consider here only the case of catalytic reactions, for which the probes are of small size and consequently the most fitting for use in boundary layer [14, 15]. Even in this case, the size of the probes is too large for them to be used in the near vicinity of the wall, but we will see that, in the case of thickened boundary layer investigated by Lassau, the microcalorimetric probe has given results in accordance with those of other methods used concurrently. We will see rapidly the principle of this method.

The heat of dissociation of nitrogen N_2 is produced when the nitrogen N recombine. Thus with a sample of the mixture taken at a point of the boundary layer, one can go back owing to the amount of heat due to this recombination by catalysis on the walls, to the concentration of atomic nitrogen in the mixture.

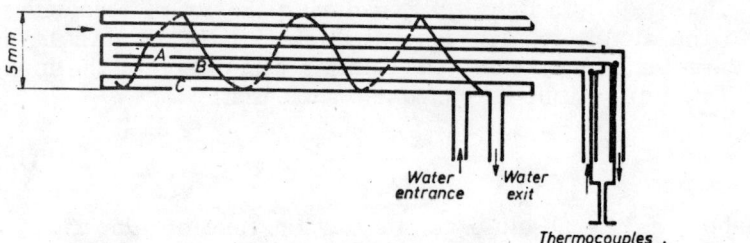

Fig. 9.

The sampling is carried out with an isokinetic probe (Fig. 9): a given mass flow rate of gas is sucked through the annular tube of the probe, and the gas transmits heat to two water flows, one inside, the other outside. By measuring the water and sucked gas, flow rates, as well as the entrance and exit temperatures of the water in the two pipes, with or without pumping in the probe, the heat flux received by the walls of the annular tube can, by difference, be obtained. Of course, by correcting, the heat fluxes transmitted to the water through other mechanisms (convection, radiation) must be taken into account.

6.3. COMPARISON OF THE METHODS

The three methods used by Lassau, mass spectrometer, measurement after sampling, in situ measurement by photochemistry and in situ measurement by microcalorimetry, gave, as shown on the curves of Fig. 10, very consistent results: the elevations represent the relative concentra-

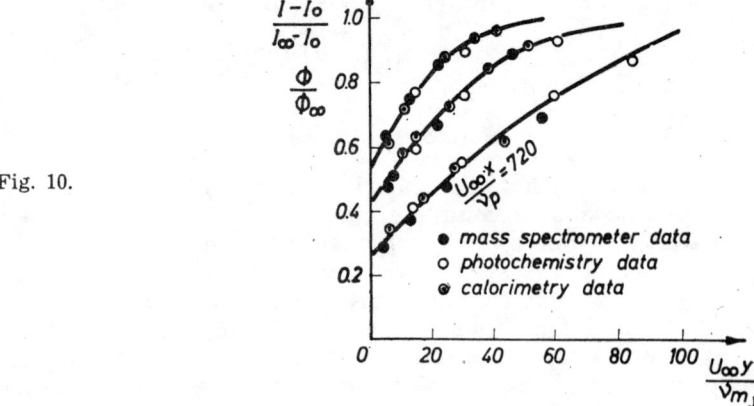

Fig. 10.

tions. The kinematic viscosities, in the Reynolds numbers proportional to the distance to the wall, are computed from the average temperature of the boundary layer (1200°C), whereas those defining the Reynolds numbers of the curves are computed from the wall temperature.

In conclusion, it can be said that the experimental methods are fitting, with good precision, to investigate concentration profiles.

7. Results and Applications

I could not end this lecture without giving some results and applications of the measurements.

7.1. WALL CATALYSIS OF A FROZEN MIXTURE $N+N_2$

In Fig. 10, in the vicinity of the wall, the concentration profiles are practically straight; this quasilinearity is demonstrated by boundary layer theory [16]. It is then easy to obtain, by extrapolating these profiles the relative wall concentration defined, for example, as in Eq. (17). It is found then that the mass flux at the wall $(da/dy)_p$ is proportional to the wall concentration a_p (Fig. 11), which shows that the reaction of atomic

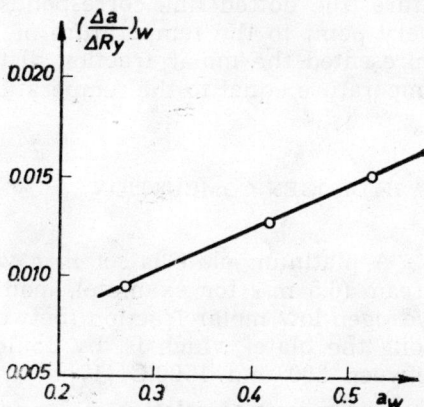

Fig. 11.

nitrogen disappearance is of first order, as was said previously. On Fig. 12 the elevation represents the dimensionless number da/dRy, where $Ry = V_\infty y/\nu$.

The knowledge of the mass flux and the wall concentration allows to determine the values of the catalysis rate constant which is about 45 m/s on the steel plate, under the experimental conditions.

7.2. INVESTIGATION OF AN ARGON PLASMA

The work by Valentin [9] of an argon plasma allowed him to obtain, using the in situ measurement of the luminous intensity of the continuous background, the electron concentration distribution curve C_e/C_{emax} versus the distance normal to the plate, and for a distance from the leading edge of 10 cm (Fig. 12). The value C_{emax} is the electron density in the free stream of $2 \cdot 10^{15}$ e/cm³. The full line assumes a constant electron tempe-

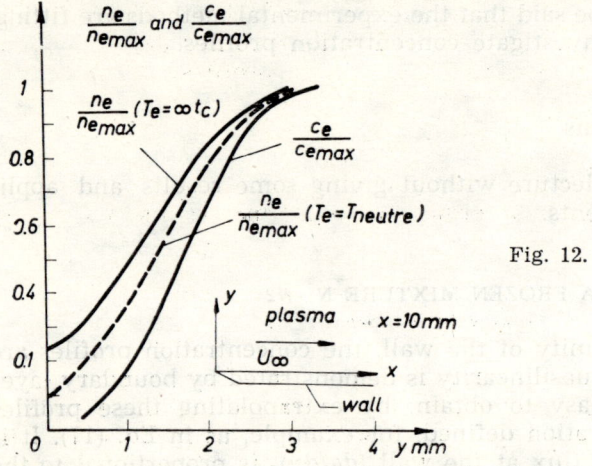

Fig. 12.

rature; the dotted line corresponds to an electron temperature equal in every point to the temperature of the neutrals. Also on the figure, is represented the molar fraction distribution curve, assuming the electron temperature equal to the temperature of the neutrals, in every point.

7.3. HYDROGEN COMBUSTION

A platinum plate is set in a wind tunnel, parallel to a low speed stream (6.5 m/s for example), made of a cold mixture of air and of a hydrogen low molar fraction (between 0.02 and 0.06). The flow is laminar along the plate, which is, by Joule effect, maintained at a temperature between 300 and 1500°C [17].

The curves of relative molar fractions versus distance to the wall are displayed on Fig. 13. If the wall temperature is lower than 600°C (curve 1), a laminar diffusion profile is found: the hydrogen goes to the wall by diffusion and is consumed only on the wall by catalytic wall combustion. Curves 3 and 4, corresponding to temperatures higher than 900°C, have a nul gradient of molar fraction at the wall, which shows that the hydrogen combustion occurs totally in the boundary layer. A mixed process (a fraction consumed in the boundary layer and another on the plate) is displayed in curve 2, corresponding to a temperature of 880°C. In all cases, combustion is always complete.

If the wall is non-catalytic (platinum plate coated with aluminium oxyde or gold), in the case where the reaction is incomplete in the boundary layer, it will not go to completion at the wall. Figure 14 shows, for a temperature of 800°C, a speed of 6.6 m/s and a molar fraction of

0.0185, the molar fraction profiles for different abscissae. The mass flux density is zero at the wall.

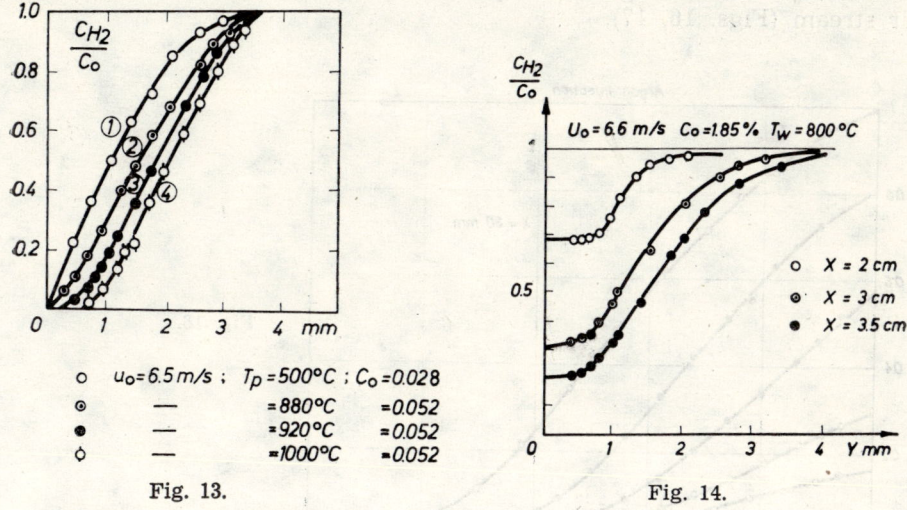

Fig. 13.

Fig. 14.

7.4. METHANE COMBUSTION

The results obtained for methane combustion in the boundary layer are not entirely analogous to those obtained for hydrogen combustion [18]. Added to the combustion in homogeneous phase and the catalytic wall combustion, there seems to be a catalytic combustion in homogeneous phase probably due to the products of platinum evaporation [19]. This is shown by the fact that the magnitude of the homogeneous phase combustion depends upon the nature of the plate, all other factors held constant (Fig. 15). The plate temperature is such (1320°C) that the reaction is complete in the boundary layer, whatever the nature of the plate.

Fig. 15.

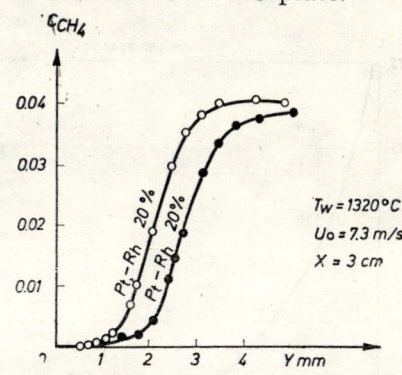

7.5. WALL INJECTION

Investigations of concentration profile in boundary layers are of interest not only when there are chemical reactions close to the wall, but also when there is wall sublimation, evaporation or injection. As an example, I will mention Cornil's experiments [20] concerning argon or helium injection through a flat plate immersed in a hypersonic ($M=5$) air stream (Figs. 16, 17).

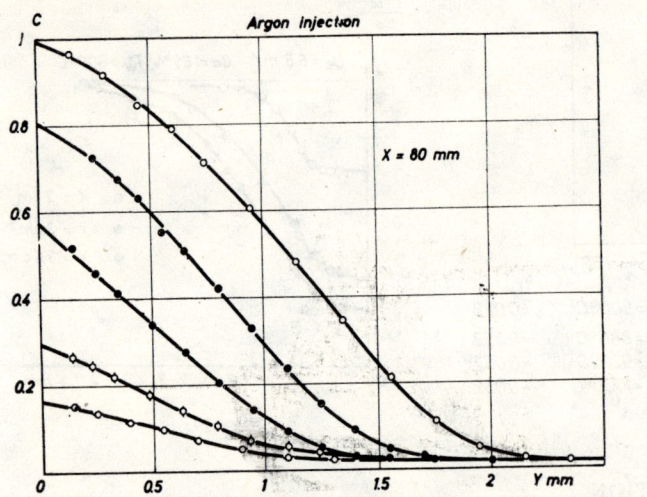

Fig. 16.

The dynamic or diffusion boundary layer thicknesses increase with injection: for argon injection, the dynamic thickness increases from 1.5 to 2.5 when the injection rate $\rho_p v_p/\rho_e u_e$ increases up to $1.58 \cdot 10^{-3}$ and the diffusion thickness increases from 1.2 to 1.6 for argon and from 1.5 to 2.5 for helium as the injection rate changes from 0.11 to $0.495 \cdot 10^{-3}$.

It can be seen that, for the same injection rate, the boundary layer diffusion thickness is larger for helium than for argon; on the contrary, the mass fractions at the wall are smaller in the case of helium.

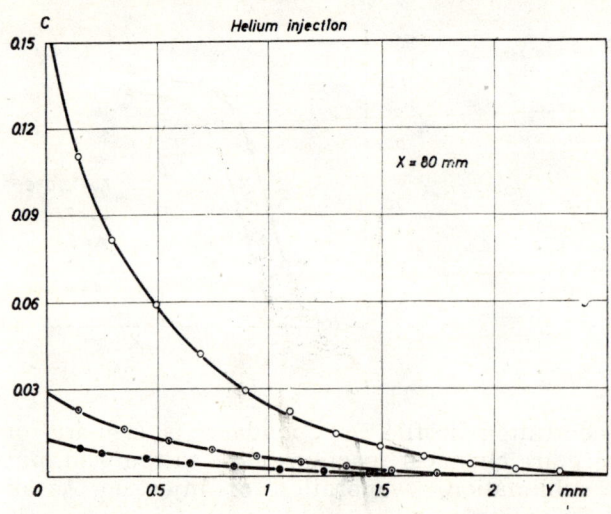

Fig. 17.

REFERENCES

1. Lassau, Chimie et Industrie, Génie chimique, **100**, 7 (1968).
2. Le Goff, Cassuto and Pentenero, Industrie chimique belge, 4 (1964).
3. Tine: Agardograph 47, Pergamon Press, 1961.
4. Fristrom, Grunfelder and Favin, J. Phys. Chem., **65**, 587 (1961).
5. Herron, Franklin and Bradt, Canad. J. Chem., **37**, 579 (1959).
6. Fontijn, Rosner and Kurzius, Aerochem. Res. Lab., 8 (1962).
7. Blades and Winkler, Canad. J. Chem., **29**, 1022 (1951).
8. Stevenson, *Ion-Molecule-Reactions,* Mc Dowell, ed. Mc Graw-Hill, 1963.
9. Valentin, Piar and Lacase, J. Phys., **29**, 4, C.3—44 (1968).
10. Young and Sharpless, J. Chem. Physics, **39**, 4 (1963).
11. Van Tiggelen and Feugier, Revue de l'I.F.P., vol. XX, 7 (1965).
12. Melville and Gowenlock, *Experimental Methods in Gas Reactions,* Mc Millan and Co, 248, 1964.
13. Au and Sprengel, Z. Flugwiss, **14**, 4 (1966).
14. Haenig, A.R.S. J., **29**, 5 (1959).
15. Rosner, AIAA J., **2**, 4, (1964).
16. Lassau, C. R. Acad. Sci., **261**, 4617 (1965).
17. Valentin, Annales Phys. (1961).
18. Cabannes and Valentin, Bull. Société chimique de France, 166 (1962).
19. Devore, Eyraud and Prettre: C.R. Acad. Sci., **248**, 1227 (1958), and **248**, 2345 (1959).
20. Cornil, C.R. Acad. Sci., session. July 16, 1969.

PRESSURE MEASUREMENTS IN HIGHLY TURBULENT FLOWS

MARC BARAT

Centre Technique des Industries Aérauliques et Thermiques, Paris, France

1. Introduction

In every turbulence definition, it is important to underline that any fluctuation of the quantities to be measured is accompanied by direction fluctuations; this character is essential; in every measurement in turbulent flow, the directional properties of the utilized instrument and the pattern of flow directional fluctuations are partly conditioning the validity of experimental results.

Unfortunately, most of the measuring instruments cannot locally distinguish the respective effects of the magnitude fluctuations and of the direction fluctuations, which partly explains the insufficiency of the knowledge regarding this metrology problem.

The working of any measuring device may be disturbed in turbulent flow either simultaneously or separately at different levels, for instance:
— at the probe level,
— at the leads level,
— at the measuring device level.

As regards both the last cases, the conditions leading to a correct working are known, particularly the damping conditions of the manometric device.

We shall examine the turbulence influence at the level of the probe itself and, particularly, the directional fluctuations influence. The errors which originate from this are sometimes, as we shall see it later on, much superior to those which are usually evoked (for instance: damping of the manometrical system, etc.).

Before drawing up correction formulas regarding the measurements effected with the total or static pressure probes, we are going to precisely state certain assumptions regarding the working of such probes in turbulent flow.

2. Influence of Turbulence on the Static Pressure Probes

Since the works of S. Goldstein [1] and A. Fage [2] regarding the turbulence influence on the static pressure probes, it is generally accepted that the indicated pressure exceeds the true static pressure.

S. Goldstein points out the theoretical complexity of this problem and admits that the turbulence influence is evidenced by the appearance of a velocity pressure at the level of each orifice.

S. Goldstein proposes an expression of the following shape:

$$p_m = p_s + c\, \rho\, \overline{U^2}$$

p_{sm} being the measured static pressure
p_s the true static pressure
U the velocity fluctuating component
c being a factor included between 0 and 1/3.

No justification relating to these limits is supplied. Nevertheless, he proposes a probable value $c = \dfrac{1}{6}$ if it is admitted that this pressure is due to the normal turbulent component only.

As a matter of fact, it seems that the following elementary reasoning may be formulated. Let us consider a usual cylindrical probe alined to the mean direction $\overline{U}$ of a turbulent flow (Fig. 1). Let $d\sigma$ denote any surface

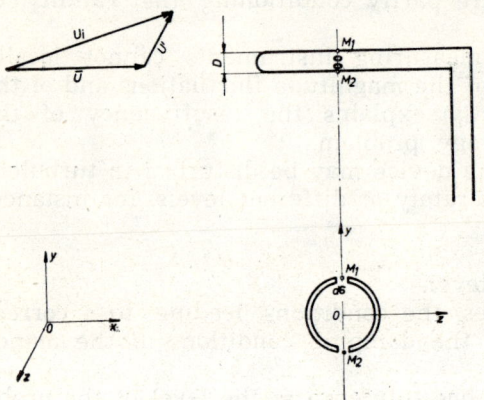

Fig. 1.

element corresponding to a pressure tap orifice. Let us consider particularly a $d\sigma$ element perpendicular to Oy, the dynamic pressure introduced will be ranging about $\dfrac{1}{2}\rho\overline{v^2}$. If there is no correlation between the normal fluctuations at the elements $d\sigma$ an expression of the following form may be admitted:

$$\overline{p_m} = \overline{p_s} + \frac{1}{2}\rho\,\overline{v^2}$$

this when assuming that the turbulence is homogeneous and isotropic, $\overline{p_{sm}}$ being the static pressure measured by means of such a probe, correctly damped.

This expression is obviously similar to:

$$\overline{p_{sm}} = \overline{p_s} + \frac{1}{6} \rho \overline{U'^2}$$

This assessment of the factor c is evidently very arbitrary as if it is admitted that there is no correlation between the fluctuations of the normal component in two diametrically opposed elements, the probe diameter is at least ranging about the scale of sizes of the considered turbulence. In this case the measuring instrument disturbs the flow and particularly modifies in a considerable manner the transverse turbulence intensity.

In a more general way S. Goldstein admits that the measured pressure is of the form:

$$\overline{p_m} = \overline{p_s} + k_s \rho \, (\overline{v^2} + \overline{w^2})$$

When comparing with the preceding expression, one is led to admit a maximum value $k_s = 0{,}25 \left(c = \dfrac{1}{6} \right)$ when $\overline{v^2} = \overline{w^2}$ if one accepts the previous reasoning. This positive coefficient would depend on the geometrical characteristics of the probe and would then be included between 0 and 0.5 $\left(0 < c < \dfrac{1}{3} \right)$.

A. Fage [2] has experimentally confirmed the value $k_s = 0.25$ in flows into long circular and rectangular ducts.

Another more correct reasoning may allow a justification of the value $c = \dfrac{1}{3}$. This value may be estimated in the following manner, starting from the Navier-Stokes equations applied to a statistically steady turbulent flow, the mean velocities and pressures fields of which are axisymmetrical. Let it be in cylindrical co-ordinates:

$$V_r = \overline{V_r} + v$$
$$V_\varphi = \overline{V_\varphi} + w$$
$$U = \overline{U} + u$$
$$p_s = \overline{p_s} + p'_s.$$

the probe axis being confounded with the Ox axis (Fig. 2). Account being taken of the continuity equation, which can be written down:

$$\frac{\partial V_r}{\partial r} + \frac{V_r}{r} + \frac{1}{r} \frac{\partial V_\varphi}{\partial \varphi} + \frac{\partial U}{\partial x} = 0$$

and noticing that one has:
$$\overline{V_\varphi} = 0$$
$$\overline{V_r} = 0$$
$$\overline{\frac{\partial}{\partial \varphi}} = 0$$

one obtains after taking up the time-averaged quantities:
$$\rho \frac{\overline{w^2 - v^2}}{r} - \frac{\partial}{\partial x}(\rho \, \overline{uv}) = \frac{\partial}{\partial r}(\overline{p_s} + \rho \, \overline{v^2})$$

One assumes a weak variation of the turbulent tangential stress in the x direction.

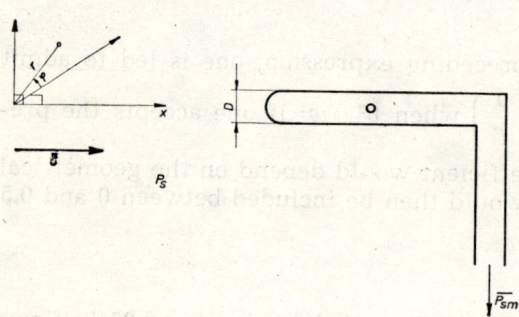

Fig. 2.

Let it be:
$\overline{p_{sm}}$ the mean pressure at $r = \dfrac{D}{2}$
$\overline{p_s}$ the mean pressure at $r = \infty$
obtains:
By integrating the preceding expression between $r = \dfrac{D}{2}$ and $r = \infty$ one obtains

$$\overline{p_{sm}} + \rho \int_{r=\frac{D}{2}}^{r=\infty} \frac{\overline{w^2 + v^2}}{r} dr = \overline{p_s} + \rho \, \overline{v^2}$$

If one admits that the integral is practically null one obtains:
$$\overline{p_m} = \overline{p_s} + \rho \, \overline{v'^2}$$
or still:
$$\overline{p_{sm}} = \overline{p_s} + \frac{1}{3} \rho \, \overline{U'^2} \quad (k_s = 0.5)$$

The static pressure tube behaves itself in some way as a pressure tapping at the wall.

Both these reasonings involve a diameter D of probe large with respect to a turbulence scale.

A more complete interpretation may be supplied by the following theoretical picture [3]:

2.1. THE TURBULENCE IS HOMOGENEOUS AND ISOTROPIC

One adopts for instance the length of transverse correlation L_g to characterize the turbulence scale.

The dimensions of small eddies are defined by λ_g.

D is the probe diameter.

Three cases may occur:

2.1.1. ($D > L_g$) — In two diametrically opposed orifices, situated at a distance D one from the other, and particularly in two orifices M_1 and M_2 (Fig. 1), there is no correlation between the velocity fluctuations.

The mean velocity over-pressure introduced by the turbulent normal component at the level of each orifice is, by instance, of the form:

$$K \rho \overline{v^2}$$

The static pressure measured by the whole of the orifices is then:

$$\overline{p_{sm}} = \overline{p_s} + K \rho \overline{v^2}$$

In this case only, K is always positive; $K = 0.5$ or $K = 1$ according as one adopts either one or the other of the previously set out reasonings.

2.1.2. ($D < \lambda_g$) — At any moment, the fluctuations in two points diametrically opposed M_1 and M_2 are in phase: all happens as if the pressure probe was inclined at each moment in relation to the instantaneous velocity U_i.

In this case, one has $K < 0$.

Introducing $K_s = -K$, one has the relation:

$$\overline{p_{sm}} = \overline{p_s} - K \cdot \rho \overline{v^2}$$

We shall see that this factor receives a simple interpretation in connection with the probe calibration according to the angle of attack, that is to say with the probe directional properties.

2.1.3. ($\lambda_g < D < L_g$) — The measured pressure depends in this case of the energy distribution between the different eddies. One has then an expression of the form

$$\overline{p_{sm}} = \overline{p_s} + K \rho \overline{v^2}$$

with $K < 0$ or $K > 0$ according to the eddies distribution. For the small diameter probes (a few millimeters) we shall see that in fact, one has always in the flows of industrial type: $K < 0$.

2.2. THE TURBULENCE IS ANY WHATEVER

The turbulence scale will be characterized by a correlation length L_g. One will assume that the correlation lengths defined starting from the functions of transverse and longitudinal correlation according to the axis O_y and O_z are ranging about the same value. The same simplification will be admitted as regards the dimension λ_g of the small eddies.

Three cases may occur:

2.2.1. $D > L_g$: The previous interpretation remains valid. The orifices placed on a diameter parallel to the axis O_y supply a pressure:

$$\overline{(p_{sm})}_{Oy} = \overline{p_s} + K \rho \overline{v^2}$$

The orifices placed on a diameter parallel to the axis O_z supply a pressure:

$$\overline{(p_{sm})}_{Oz} = \overline{p_s} + K \rho \overline{w^2}$$

One may admit that the mean measured pressure for the whole of the orifices is then approximately of the form:

$$\overline{p_{sm}} = \overline{p_s} + K \rho \left(\frac{\overline{v^2} + \overline{w^2}}{2} \right)$$

with $K > 0$.

2.2.2. $D < \lambda_g$: By analogy one may admit an expression of the previous form with $K < 0$.

One lays

$$K_s = -K.$$

We shall see that a theoretical interpretation effectively leads to the following expression:

$$\overline{p_{sm}} = \overline{p_s} - K_s \rho \left(\frac{\overline{v^2} + \overline{w^2}}{2} \right)$$

2.2.3. $\lambda_g < D < L_g$: One will likewise admit:

$$\overline{p_{sm}} = \overline{p_s} + K \rho \left(\frac{\overline{v^2} + \overline{w^2}}{2} \right)$$

with $K < 0$ or $K > 0$ according to the eddies distribution.

Note: Strictly speaking, it would be preferable to bring in a correlation length L_l formed from a function of lengthwise correlation.
Account being taken that the inequality $D > L_g$ shall not bring in anything but a comparison between orders of magnitude, it seemed preferable to us to define the turbulence considered with the help of the usual corellation length L_g.
As a matter of fact, this length, defined starting from a transverse correlation function, is more surely accessible from the experimental point of view. Any

way, in the case of a homogeneous and isotropic turbulence, one has the relations

$$L_f = 2 L_g \quad \text{and} \quad \lambda_f = \sqrt{2}\, \lambda_g$$

3. Static Pressures Measured with the Help of Different Diameters Probes in a Flow of the Rectangular Long Duct Type

In order to evidence the influence of the turbulenve scale and to verify the previous interpretation, we have effected some measurements at the outlet of a rectangular duct (Fig. 3).

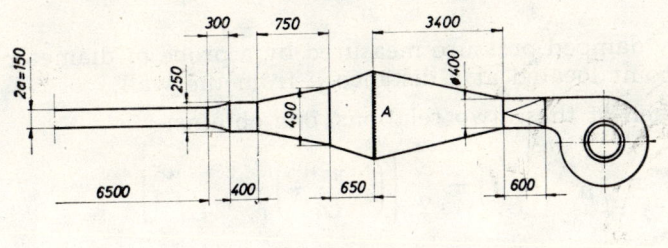

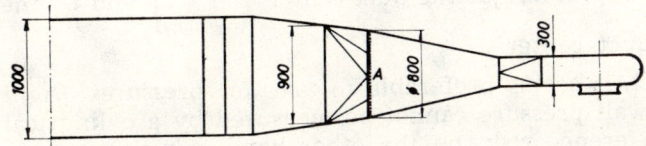

Fig. 3.

The last part of this duct (made in sheet steel) was provided with outside reinforcements so as to obtain a good flatness of the walls in view of avoiding any local divergence or convergence of the flow. A filter involving three expanded metal thicknesses (meshes 8×2 mm) was located in A (Fig. 3).

The ejection velocity at the duct center was ranging about 40 m/sec. The probes of respective diameters $D=2$ mm, $D=8$ mm, $D=16$ mm, were similar between them, the probe diameter $D=2$ mm being in accordance with the diagram of Fig. 4. The measurement of pressures was effected

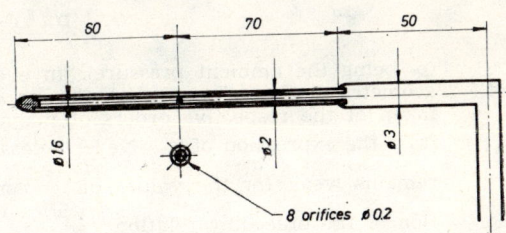

Fig. 4.

at the outlet of the duct, in a median plane, the pressure tapping orifices being placed at the very level of the ejection section.

— Static pressure in a cross section —

In a rectangular duct section, one knows that one has the relation:

$$\overline{p_s} + \rho \overline{v^2} = \text{constant}$$

$\overline{p_s}$ being the actual mean static pressure in one point of the considered section.

By definition, one denotes:

$$\overline{p_{sm}} = \overline{p_s} + K \frac{\rho}{2} \left(\overline{v^2} + \overline{w^2} \right)$$

$\overline{p_{sm}}$ being the correctly damped pressure measured by a probe of diameter D at the considered point located at a distance y from the wall.

Account being taken of these two relations, one obtains:

$$\frac{(\overline{p_{sm}})_1 - \overline{p_{sm}}}{\rho \, \overline{U_1^2}} = \frac{\overline{v^2}}{\overline{U_1^2}} - \left(\frac{\overline{v^2}}{\overline{U_1^2}}\right) + \frac{K}{2}\left[\left(\frac{\overline{v^2}+\overline{w^2}}{\overline{U_1^2}}\right)_1 - \frac{\overline{v^2}+\overline{w^2}}{\overline{U_1^2}}\right]$$

$(\overline{p_{sm}})_1$ being the measured pressure at the duct center $\left(\dfrac{y}{a}=1\right)$ and $\overline{U_1}$ the mean velocity at the duct center.

As a matter of fact, it seems preferable to take for pressures origin the central value; the wall pressure cannot be measured by a cylindrical probe without an interference risk; on the other hand, a wall pressure tapping assumes a perfectly realized duct and introduces in some way a slightly different measuring process.

By excluding the points too near of the wall one may admit: $\overline{v^2} = \overline{w^2}$; in this case, one then has the relation:

$$\frac{(\overline{p_{sm}})_1 - \overline{p_{sm}}}{\rho \, \overline{U_1^2}} = (1-K)\left[\frac{\overline{v^2}}{\overline{U_1^2}} - \left(\frac{\overline{v^2}}{\overline{U_1^2}}\right)_1\right]$$

N o t e: If one admits that at the outlet of the duct, one has the relation:

$$\overline{p_s} + \rho \, \overline{v^2} = p_o$$

(p_o being the ambient pressure), an eventual error, due for instance to the geometrical characteristics of the pressure orifice, forbids, account being taken of the respective orders of magnitude, a determination of K starting from the expression of $\overline{p_{sm}}$. Nevertheless, the variation of this error probably remains weak for the values of $\dfrac{y}{a}$ considered, which justifies the utilization of the preceding relation.

Practically, in an industrial type installation comprising no regulation device with a sonic nozzle, one cannot obtain a sure quantitative value of K, the accuracy degree required for the knowledge of the distribution of the transverse turbulent intensity being in other respects prohibitive.

Figures 5, 6 and 7 supply, in function of different values of K arbitrarily selected, the distribution of the transverse turbulent intensity and this by admitting that $\left(\dfrac{v'}{U_1}\right)_1 = 0.044$ (central value).

The dotted line curve drawn on these three figures represents $\dfrac{v'}{U_1}$ measured with the help of a new process farther on described, and this process precisely supplies the value $\left(\dfrac{v'}{U_1}\right)_1 = 0.044$ for $\dfrac{y}{a} = 1$.

These figures clearly indicate the evolution of the coefficient K which shall be assigned to a given diameter probe in a same turbulent flow.

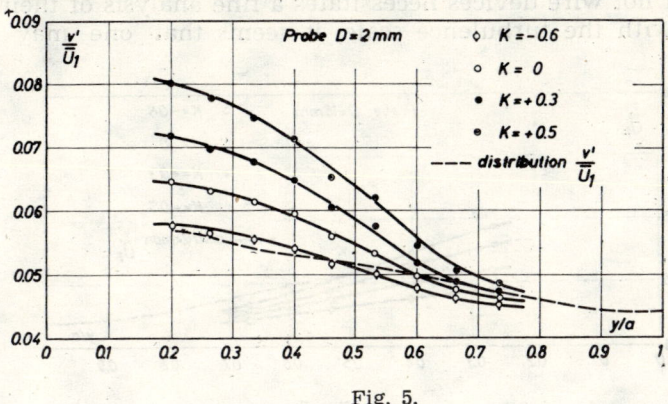

Fig. 5.

One establishes that a probe such as $D=16$ mm shall be utilized so that a value $K>0$ be plausible; on the contrary a probe such as $D=2$ mm leads to $K \approx -0.6$. In particular, one may notice that the curve correspond-

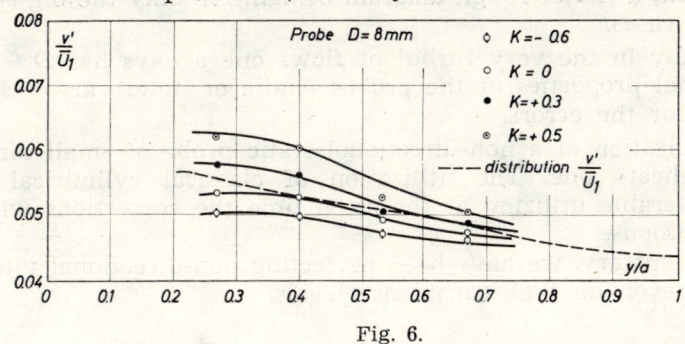

Fig. 6.

ing to $K=-0.6$ (Fig. 5) seems to confirm the distribution of the transverse turbulent intensity (dotted line curve) obtained by the previously mentioned process, this for values $0.2 < \dfrac{y}{a} < 0.6$, these limits approximately fixing the scope wherein this process is applicable.

We shall see that a theoretical interpretation precisely assigns to the coefficient K the value -0.6 ($D \ll L_g$).

These results distinctly confirm the previously said hypothesis.

Remark: The utilized installation showed a small contraction and the turbulent intensity level was distinclty higher than that obtained by J. Laufer [4]. In any way, if one adopts the value obtained by J. Laufer, i.e. $\left(\dfrac{v'}{U_1}\right) = 0.022 \left(\dfrac{y}{a} = 1\right)$, the evolution of the coefficient K remains the same.

Let us recall that in a turbulent flow, the big eddies are the energy bearers. If, for the study of the turbulence degradation (dissipation), the utilisation of hot wire devices necessitates a fine analysis of their working in relation with the turbulence scale, it seems that one may adopt as

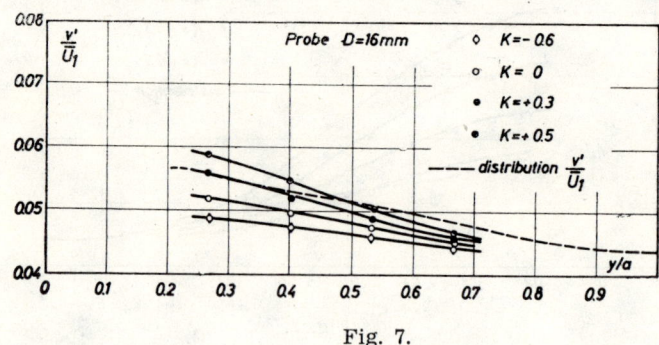

Fig. 7.

regards the problem of the turbulence correction of the pressure probes (static or total) a rather rough diagram bringing in only the big eddies in most of the cases.

Practically in the very turbulent flows one always has $D \ll L_g$ and the directional properties of the probes (static or total) are essentially responsible for the errors.

The realization of a non-directional static probe of small dimension is a very delicate one. The utilization of classical cylindrical probes remains preferable utilizing at the same time the corrections which we farther on propose.

On the contrary, we have been perfecting non-directional total pressure probes, even for high turbulence levels.

4. Influence of the Turbulence on the Total Pressure Probes

Let us consider an »ideal« total pressure probe which would measure the pressure prevailing at the level of a stagnation point whatever be the orientation of the streamlines. Let us place this probe in a turbulent flow (Fig. 8).

At every moment one has:

$$P_i = p_{si} + \frac{1}{2} \rho U_i^2$$

P_i being the stagnation pressure at point A
p_{si} being the static pressure at point A
U_i being the instantaneous velocity at point A.

Generally speaking, it is not the quantity which can be reached, but a mean value. Assuming that the damping of the manometric system is correct, one will measure:

$$\overline{P} = \overline{p_s} + \frac{\rho}{2} \overline{U}^2 + \frac{\rho}{2} (\overline{u^2} + \overline{v^2} + \overline{w^2})$$

$$\left(\overline{P_i} = \overline{P} = \frac{1}{T} \int_0^T P_i \, dt \right)$$

One may see in particular that the determination of the velocity $\overline{U}$ implies:
1) The correct measurement of the mean static pressure in point A or at least in a point where similar flow conditions are prevailing.
2) The knowledge of values $\overline{u^2}$, $\overline{v^2}$, $\overline{w^2}$.
3) The correct measurement of $\overline{P}$.

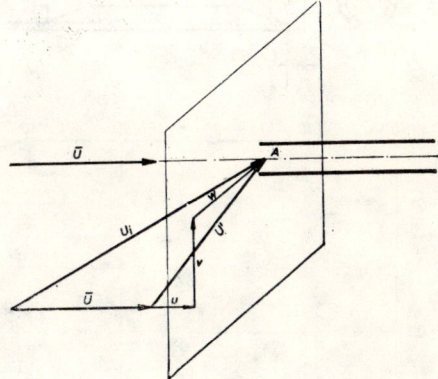

Fig. 8.

The previous expression is valid for an »ideal« probe without any directional properties; in fact most of the probes usually utilized give variable indications according to the angle of attack.

Some authors have tried to check the expression validity of $\overline{P}$ by modifying the turbulence level with the help of screens placed in a duct situated immediately after a convergent, the mean velocity being maintained constant. The static pressure was assumed as constant. Contrarily to the previsions, the turbulence increase produced a decrease of the measured total pressure [5].

This anomaly explains itself, as we shall see later on, starting from the directional properties of the utilized probe.

One can as well as for the static pressure probes, establish a classification of a characteristic dimension of the probe in relation to a turbulence scale L_g. It is probable that the characteristic dimension of the probe is the diameter d of the pressure tapping orifice in which a stagnation point is established.

Practically one has always $d \ll L_g$. Still in this case all happens as if the pressure probe is inclined in relation to the instantaneous velocity U_i and the directional properties of the probe are the ones which introduce an error.

This justifies the perfecting of two practically punctual probes types, showing out a great insensibility to the orientation (45°) (see Figs. 9 and 10). P_θ is the pressure obtained for an inclination θ in relation to the relative wind [6].

One certain correction method drawn up by us imposed in other respects the utilization of such probes [7].

These two probes types are realizable with outside diameters $D = 1$ mm, particularly the model represented on Fig. 9.

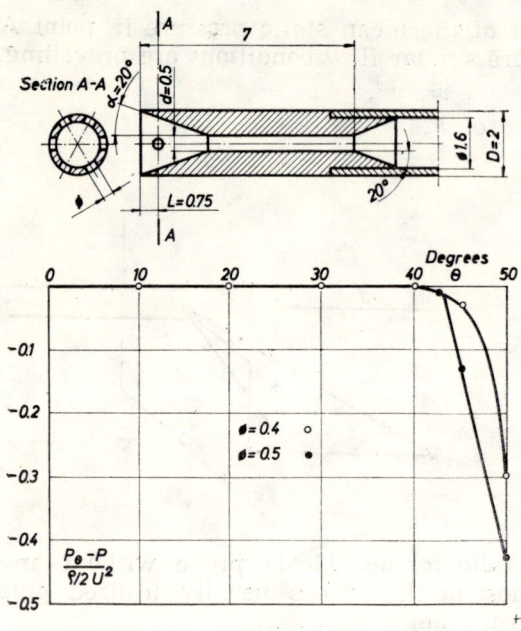

Fig. 9.

Such probes may be considered as »ideal« probes in most of the turbulent flows. In this case the measured value $\overline{P}$ is really equal to P_i.

These probe models permit to evidence, by a direct experimental comparison, the errors which affect certain probes in the turbulent flows and to thus verify certain admitted hypothesis in order to establish the correction methods.

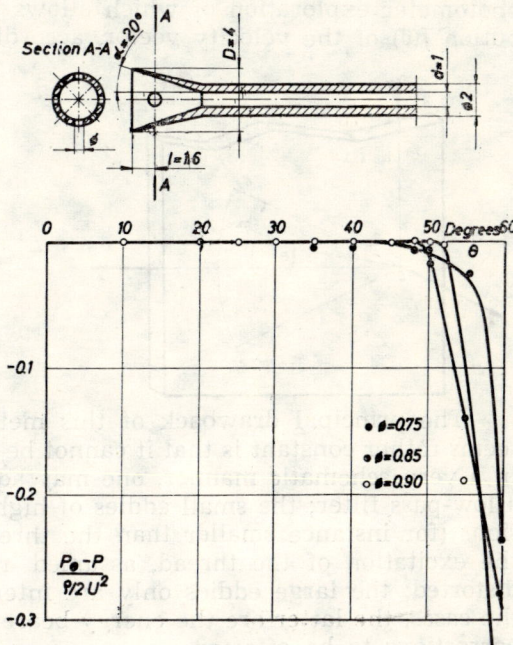

Fig. 10.

5. Correction Method

5.1. ANGULAR DISTRIBUTION $f(\theta)$ OF THE VELOCITY VECTOR IN ONE POINT

We have at first tried to visualize very turbulent flows in the vicinity of a pressure probe.

In order to obtain the direction of the velocity vector one knows that a simple process consists to introduce in the flow a fine and light thread which orients itself according to the local velocity.

If this direction varies in function of time, the thread describes a conical volume the pattern of which gives a qualitative indication on the agitation. This process is more particularly utilized to detect the separation onset.

The utilisation of this process for the succinct examination of the flow in the vicinity of static probes, has prompted us to perfect a photographic method leading to quantitative results.

The experimental fitting up is comfortable to the diagram of the Fig. 11. The axis Ox is oriented according to the mean direction of the considered flow.

The thread is fixed by an extremity to the point O. A luminous plane flux, mingled with the Oxy plane, lights the thread, the length of which is large enough with respect to the luminous plane thickness, in such a way that the volume swept by the thread involves only a small part comprised in the luminous plane.

A camera, the optical axis of which is parallel to the axis, takes plates which present themselves under the form of fuzzy sectors, and the microphotometer exploration of which allows to determine the angular distribution $f(\theta)$ of the velocity vector, according to the Oxy plane.

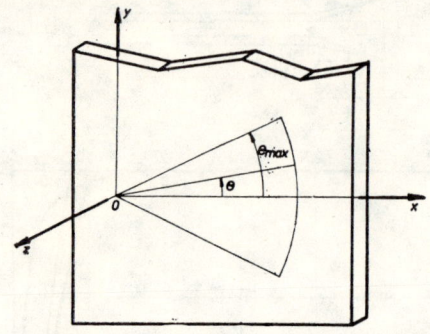

Fig. 11.

The principal drawback of this method which, in other respects, seems rather constant is that it cannot be applied to low scale turbulences. In a very schematic manner, one may admit that this device is acting as a low-pass filter; the small eddies of high frequency and of small dimensions (for instance smaller than the thread length) do not participate to the excitation of the thread, assumed rectilinear and that cannot be distorted; the large eddies only are intervening. But in fact, in most of the cases, the latter are the energy bearers, which are responsible for the corrections to be effected.

The realization of utilizable plates necessitates a certain number of precautions and particularly the elimination of phenomena in connection with the luminations intermittency.

Numerous realized plates, either in free jets or in long conduit, have always supplied angular distributions $f(\theta)$ of the Gauss curve type. For

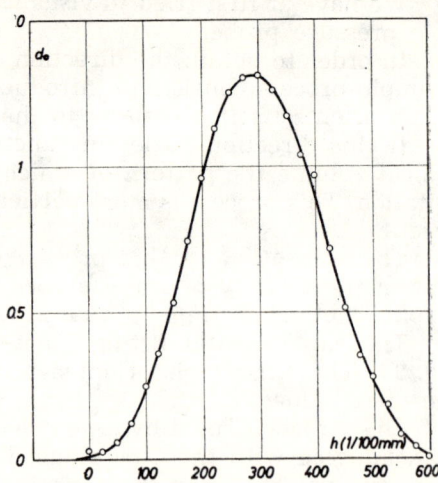

Fig. 12.

instance Fig. 12 is relating to the recording effected on a jet axis of initial diameter $D = 60$ mm at a distance $X = 6D$. The reading of the optical partial density d_o (which is proportional to $f(\theta)\,d\theta$ has been effected along an arc of a circle with a radius $R=10$ mm; h represents a guide mark according to the Oy axis.

This method supplying a diagram of the θ angles distribution of the velocity vector in relation with the average direction has allowed us to establish a theoretical correction regarding the usual cylindrical static pressure probes and some total pressure probes of various forms.

5.2. RELATION BETWEEN THE ANGULAR DISTRIBUTION $f(\theta)$ VARIANCE AND THE CORRESPONDING TRANSVERSE VELOCITY FLUCTUATIONS VARIANCE

At any moment, the instantaneous velocity vector, in a given point presents an angle θ_y with regard to the xOz plane and an angle θ_z with regard to the xOy plane (Fig. 13).

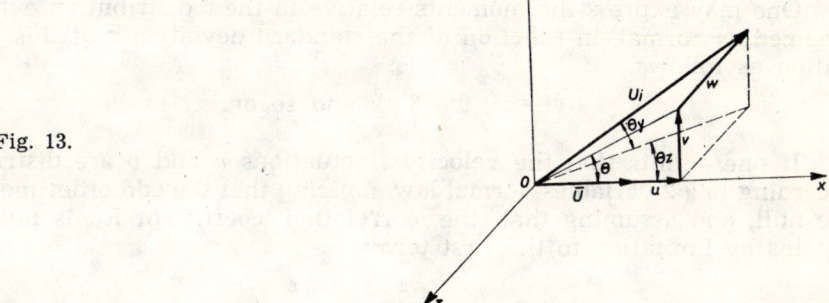

Fig. 13.

We shall admit that these two angles distribution is a 2 variables normal distribution. The application conditions of such a law, which appears as a generalization of the normal law, are often realized.

One may demonstrate in this case that if the linear correlation coefficient

$$R_{\theta_y \theta_z} = \frac{\overline{(\theta_y - \bar{\theta}_y)(\theta_z - \bar{\theta}_z)}}{\sigma_{\theta_y} \cdot \sigma_{\theta_z}}$$

σ_{θ_y} being the marginal standard deviation of the variable θ_y) is null, the marginal distributions laws and the bound distributions laws are parallel and particularly have the same variance. The distribution of θ_y bound by θ_z for a thickness strip $d\theta_z$ and of ordinate $\theta_z = O$ is precisely the distribution law $f(\theta)$ obtained by the photographic process previously described. This latter may be connected with the corresponding transverse velocity fluctuations variance. As a matter of fact, let us consider the transverse velocity fluctuations v and w; we shall admit that these fluctuations distribution is a 2 variables normal distribution; we shall admit that the correlation coefficient R_{vw} is null; in this case, the marginal distribution of v is parallel to the corresponding bound distribution and particularly to that observed in the xOy plane. The variance $\overline{v^2}$ will then be similar to that observed in the xOy plane.

In this xOy plane (confounded with the luminous plane) one has:

$$\text{tg } \theta = \frac{v}{U+u}$$

By writing the following expansion:

$$\text{tg } \theta = \frac{v}{U} - \frac{vu}{U^2} + \frac{vu^2}{U^3} - \frac{vu^3}{U^4} + \ldots \left(\text{with } \frac{u}{U} < 1\right)$$

and by taking the time-averaged value of $tg^2\theta$ one obtains:

$$\overline{\theta^2} + \frac{2}{3}\overline{\theta^4} + \ldots = \frac{\overline{v^2}}{U^2} - 2\frac{\overline{v^2 u}}{U^3} + 3\frac{\overline{v^2 u^2}}{U^4} + \ldots$$

One may express the moments relative to the θ distribution, which is assumed as normal, in function of the standard deviation σ of this distribution as follows:

$$\overline{\theta^2} = \sigma^2; \quad \overline{\theta^4} = 3\sigma^4; \text{ and so on.}$$

If one admits that the velocity fluctuations u and v are distributed according to a 2 variables normal law, noticing that the odd order moments are null, and assuming that the correlation coefficient R_{uv} is null, one obtains by limitation to the first terms:

$$\sigma^2 + 2\sigma^4 + \ldots = \frac{v'^2}{U^2} + 3\frac{u'^2}{U^2} \cdot \frac{v'^2}{U^2} + \ldots$$

One sees that, practically, one can admit:

$$\overline{\theta^2} = \sigma^2 = \frac{v'^2}{U^2}$$

Hence, assuming that the observable maximum deviation θ_{max} is such that $\theta_{max} = 3\sigma$:

$$\frac{\theta_{max}}{3} = \frac{v'}{U}$$

5.3. METHOD I

The photographic method previously described supplied us with an angular distribution diagram of the velocity vector according to a plane. This distribution follows a normal law for the usual turbulent flows. Besides, we only consider the flows in which can be admitted:

$$\overline{v^2} = \overline{w^2}.$$

5.3.1. Let be assumed at first that the indication I_θ of the probe placed in a flow of constant velocity U is such that one has the following relation with the angle θ:

$$\frac{I_\theta - I}{\frac{\rho}{2} U^2} = -K \sin^2 \theta$$

I being the value indicated for $\theta = 0$.

We shall see, as a matter of fact, that the indication of the cylindrical static pressure probes satisfies to that expression, as well as, evidently, the indication of total pressure probes with spheric or hemispheric nose.

Let us consider such a probe placed in a turbulent flow of following pattern: the instantaneous velocity is assumed constant in a plane passing by the Ox axis and distributed according to a normal law.

The contribution, for the probe indication, due to the passages within the elementary solid angle included between the cones of respective angles θ and $\theta + d\theta$ is (Fig. 14):

$$\mu(\theta) \cdot d\theta = -K \cdot 2\pi\theta \cdot f(\theta) \cdot \sin^2 \theta \cdot d\theta$$

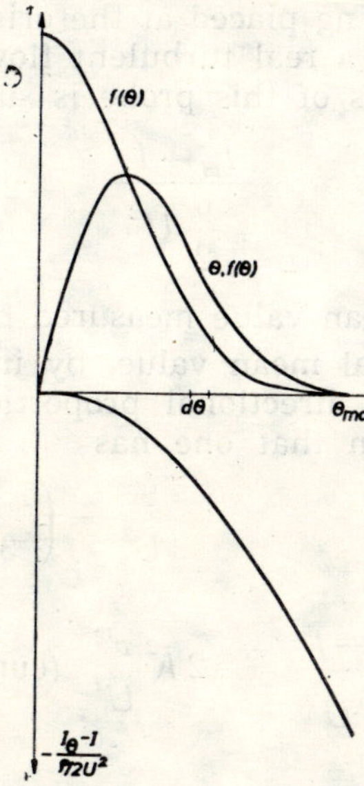

Fig. 14.

The total contribution for the maximum considered angle θ_{max} is:

$$(M(\theta))_0^{\theta_{max}} = -\int_0^{\theta_{max}} K \cdot 2\pi\theta \cdot f(\theta) \cdot \sin^2 \theta \cdot d\theta$$

The considered passages number is:

$$(N(\theta))_0^{\theta_{max}} = \int_0^{\theta_{max}} 2\pi\theta \cdot f(\theta) \cdot d\theta$$

Hence the error $\left(\dfrac{\overline{\Delta I}}{\dfrac{\rho}{2} U^2}\right)_{\theta_{max}}$ corresponding to the considered angle θ_{max} is:

$$\left(\frac{\overline{\Delta I}}{\dfrac{\rho}{2} U^2}\right)_{\theta_{max}} = \frac{(M(\theta))_0^{\theta_{max}}}{(N(\theta))_0^{\theta_{max}}} = -K \frac{\displaystyle\int_0^{\theta_{max}} \theta \cdot f(\theta) \cdot \sin^2 \theta \cdot d\theta}{\displaystyle\int_0^{\theta_{max}} \theta \cdot f(\theta) \cdot d\theta}$$

This expression can be easily calculated if one admits:

$$\sin \theta \approx \theta$$

One finally finds:

$$\left(\frac{\overline{\Delta I}}{\dfrac{\rho}{2} U^2}\right)_{\theta_{max}} = -2 K \sigma^2 = -2 K \left(\frac{\theta_{max}}{3}\right)^2;$$

the mean value $\overline{\theta}$ being placed at the origin (Fig. 14).

Let us consider a real turbulent flow; the error introduced by the directional properties of this probe is such that one approximately has:

$$\frac{\overline{I_m} - \overline{I}}{\dfrac{\rho}{2} \overline{U^2}} = -2 K \left(\frac{\theta_{max}}{2}\right)^2$$

$\overline{I_m}$ being the mean value measured by this probe,
$\overline{I}$ being the real mean value, by instance, measured by a probe without any directional properties.

Account being taken that one has:

$$\frac{\overline{v'^2}}{\overline{U^2}} = \left(\frac{\theta_{max}}{3}\right)^2$$

we may still write:

$$\frac{\overline{I_m} - \overline{I}}{\dfrac{\rho}{2} \overline{U^2}} = -2 K \frac{\overline{v'^2}}{\overline{U^2}} \quad \text{(curve I, Fig. 15)}$$

Remarks:

One may evidently use the series expansion for $sm^2\theta$; by limitation to the two first terms, one is led to the expression:

$$\left(\frac{\overline{\Delta I}}{\dfrac{\rho}{2} U^2}\right) = -K \left[\frac{\displaystyle\int_0^{\theta_{max}} \theta^3 \cdot f(\theta) \cdot d\theta}{\displaystyle\int_0^{\theta_{max}} \theta f(\theta) \cdot d\theta} - \frac{1}{3} \frac{\displaystyle\int_0^{\theta_{max}} \theta^5 \cdot f(\theta) \cdot d\theta}{\displaystyle\int_0^{\theta_{max}} \theta \cdot f(\theta) \cdot d\theta} \right]$$

One finally has:

$$\frac{\overline{I_m} - I}{\frac{\rho}{2}\overline{U^2}} = -K\left(2\frac{\overline{v'^2}}{\overline{U^2}} - \frac{8}{3}\frac{\overline{v'^4}}{\overline{U^2}}\right) \quad \text{(curve (III)} - \text{Fig. 15)}$$

The curve (I) (Fig. 15) represents in most of the cases a sufficient approximation.

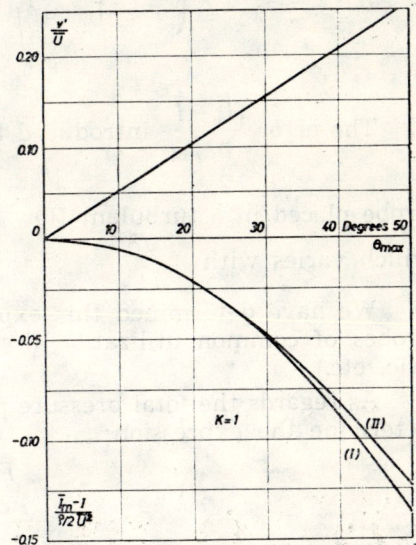

Fig. 15.

Briefly, for a probe, the calibration of which may be put under the form:

$$\frac{I_\theta - I}{\frac{\rho}{2}U^2} = -K \sin^2 \theta.$$

the error introduced by its directional properties, in a turbulent flow $(\overline{v^2} = \overline{w^2})$, is indicated, with a good approximation, by the expression:

$$\frac{\overline{I_m} - \overline{I}}{\frac{\rho}{2}\overline{U^2}} = -K\frac{\overline{v'^2}}{\overline{U^2}}$$

5.3.2. The indication I_θ of the probe placed in a flow of constant velocity U, in function of the angle θ, is of the form:

$$\frac{I'_\theta - I}{\frac{\rho}{2}U^2} = -\varphi(\theta) \quad\quad (\theta(>0)$$

I being the indicated value for $\theta = 0$.

The previously utilized reasoning remains valid, but, generally, the different integrals will not be calculable. One is led to numerically calculate for each value $\theta_{max}=3\sigma$ considered, the expression:

$$\left(\frac{\overline{\Delta I}}{\frac{\rho}{2}U^2}\right) = -\frac{\int_0^{\theta_{max}=3\sigma} \theta \cdot f(\theta) \cdot \varphi(\theta) \cdot d\theta}{\int_0^{\theta_{max}=3\sigma} \theta \cdot f(\theta) \cdot d\theta}$$

The error $\dfrac{\overline{I_m} - \overline{I}}{\frac{\rho}{2}\overline{U^2}}$ introduced by the directional properties of such a probe placed in a turbulent flow such as $v'^2 = w'^2$ is one negative value which varies with $\dfrac{\overline{v'^2}}{\overline{U^2}}$.

We have determined this expression for a whole of total pressure probes of common utilization (hypodermic, with thin edges, hemispheric, etc.).

As regards the total pressure probes, the possibility to experimentally determine the expression:

$$\frac{\overline{P_m} - \overline{P}}{\frac{\rho}{2}\overline{U^2}}$$

in a given turbulent flow, and this due to the probes without orientation sensitivity that we have previously described, allows to verify the validity of calculation hypothesis which we have previously admitted. As a matter of fact, different probes placed in a same turbulent flow shall supply an experimental value:

$$\frac{\overline{P_m} - \overline{P}}{\frac{\rho}{2}\overline{U^2}}$$

corresponding to a same value of θ_{max} $\left(\text{or of } 3f\dfrac{\overline{v'^2}}{\overline{U^2}}\right)$.

We shall see that this is fully verified.

5.4. METHOD II

For a conventional static pressure probe, the expression

$$\frac{\overline{p_{.\theta} - p_s}}{\frac{\rho}{2}U^2} = -K_s \sin^2\theta$$

yields a good approximation of the error introduced by the probe inclination of an angle θ in relation with the velocity U of a uniform flow.

- $p_{s\theta}$ being the measured pressure for an inclination θ
- p_s being the measured pressure for an inclination θ=0 (real static pressure).
- $K_s \approx 0.60$ for the conventional probes (Fig. 16). This coefficient slightly depends on the Reynolds number formed with the probe outside diameter

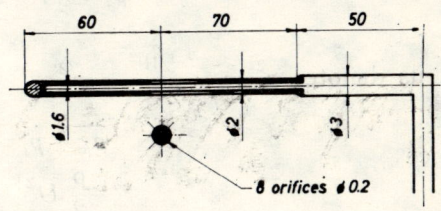

Fig. 16.

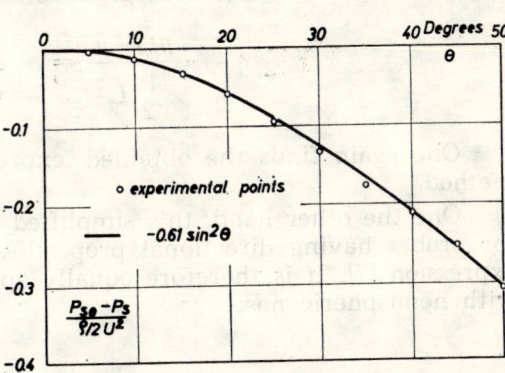

Let us place such a probe in parallel with the mean velocity $\overline{U}$ direction of a statistically-steady turbulent flow.

The instantaneous velocity will be:

$$U_i = \sqrt{(\overline{U}+u)^2 + v^2 + w^2}$$

The angle θ_i formed by the velocity vector U_i and the probe axis is that one has:

$$\sin \theta_i = \frac{\sqrt{v^2+w^2}}{\sqrt{(\overline{U}+u)^2+v^2+w^2}}$$

The instantaneous measured value p_{sm} is such that:

$$\frac{p_{sm}-p_s}{\frac{\rho}{2}U_i^2} = -K_s \sin^2 \theta_i$$

Account being taken of the previous expression, one has:

$$p_{sm} - p = -K_s \frac{\rho}{2}(v^2 + w^2)$$

This relation is true at any moment. If the measuring apparatus is suitably damped, the mean indication $\overline{p_{sm}}$ will be such, that:

$$\overline{p_{sm}} - \overline{p_s} = = K_s \frac{\rho}{2}(\overline{v^2} - \overline{w^2})$$

Let us denote: $\overline{v_m^2} = \dfrac{\overline{v^2} + \overline{w^2}}{2} = \overline{v_m'^2}$

Hence

$$\frac{\overline{p_{sm}} - \overline{p_s}}{\frac{\rho}{2} U^2} = -2 K_s \frac{\overline{v_m'^2}}{U^2}$$

If the turbulence is homogeneous and isotropic, one has:

$$\frac{\overline{p_{sm}} - \overline{p_s}}{\frac{\rho}{2} U^2} = -2 K_c \frac{\overline{v'^2}}{U^2}$$

One again finds the obtained expression by the previous general method.

One the other hand, this simplified reasoning cannot be utilized but for probes having directional properties conformable with the previous expression [7]. It is therefore equally applicable to total pressure probes with hemispheric nose.

5.5. METHOD III

It is known that the measured pressure P_θ through an orifice situated on a sphere placed in a constant velocity flow U where the static presure p_s is prevailing is expressed by:

$$P_\theta = p_s + \rho \frac{U^2}{2} - \frac{9}{4} \rho \frac{U^2}{2} \sin^2 \theta$$

θ being the inclination between the far relative wind and the orifice axis.

If one places in a statistically steady turbulent flow a sphere of diameter D such as $D < \lambda_g$, the orifice of pressure tapping being connected with a suitably damped manometer, one obtains:

$$\overline{P_m} = \overline{p_s} + \overline{U_i^2} - \frac{9}{4} \frac{\rho}{2} \overline{U_i^2 \sin^2 \theta(t)}$$

U_i being the instantaneous velocity at the considered point,
$\theta(t)$ being the angular distribution law of velocity.

The finished dimensions of the pressure tapping orifice situated on the sphere introduce a coefficient $K_t < \dfrac{9}{4}$, coefficient which can be determined by a calibration within a uniform flow. For instance, for a sphere conformable to the Fig. 17, $K_t = 1.90$.

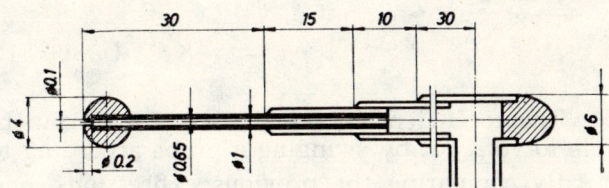

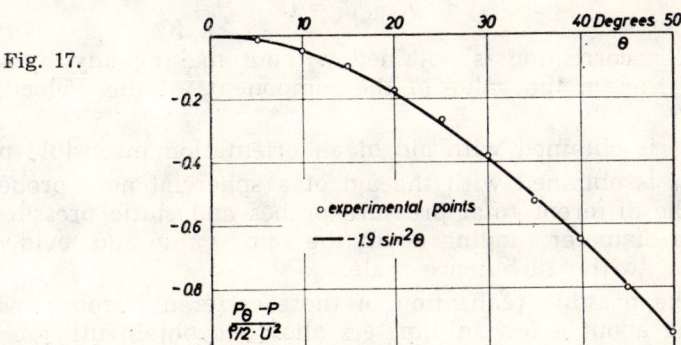

Fig. 17.

We have been able to realize similar probes with a sphere diameter $D = 2{,}5$ mm which have the same coefficient $K_t = 1.90$.

One may admit that this coefficient remains valid in turbulent flow.

This hypothesis, in other respects, has been confirmed by swinging a sphere according to a known law $\theta(t)$.

We have seen that the total pressure probes previously described are practically insensible to the orientation; a probe of this type, placed at the considered point, yields the following indicated pressure:

$$\overline{P} = \overline{p_s} + \frac{\rho}{2}\,\overline{U_i^2}$$

Thus, two successive measurements effected at a same point within the turbulent flow determine the expression:

$$\overline{P} - \overline{P_m} = K_t \cdot \frac{\rho}{2} \cdot \overline{U_i^2 \sin^2 \theta(t)}$$

On the other hand, we have seen that the measured static pressure p_{sm} with a cylindrical static probe, placed in a constant velocity flow U

where the static pressure p_s is prevailing, is such that:

$$p_{sm} = p_s - K_s \frac{\rho}{2} U^2 \sin^2 \theta$$

This probe, being placed at the same point, within the same turbulent flow, and connected with a correctly damped manometer, one obtains:

$$\overline{p_{sm}} = \overline{p_s} - K \cdot \frac{\rho}{2} \overline{U_i^2 \sin^2 \theta (t)}$$

The coefficient K_s remains valid in turbulent flow, as we have been able to verify it by swinging a probe according to a known law.

By comparing the previous expressions, one obtains the following error $(p_s - p_{sm})$ introduced on the static pressures measurements:

$$\overline{p_s} - \overline{p_{sm}} = (\overline{P} - \overline{P_m}) \frac{K_s}{K_t}$$

This correction is obtained without making any hypothesis on the law $\theta(t)$ or on the value of the components of the velocity fluctuation.

$\overline{P}$ is obtained with aid of an orientation insensible probe,
$\overline{P_m}$ is obtained with the aid of a sphercial nose probe.

The different total pressure probes and static pressure probes shall have a diameter ranging about the same value and evidently small in relation to the turbulence scale.

The possible realization of these different probes with diameters ranging about a few millimeters allows to obtain utilizable measures in most of the flows.

5.6. METHOD IV (measuring of the transverse turbulent intensity)

This method directly derives from the Method III.

The total pressure P_θ measured with a spheric or hemispheric nose probe placed in a constant velocity flow U, is such as:

$$\frac{P_\theta - P}{\frac{\rho}{2} U^2} = -K_t \sin^2 \theta,$$

θ being the inclination between the far relative wind and the pressure tapping orifice axis,
P being the actual total presure.

This expression remains valid for some angles θ so much higher as the orifice diameter is small, the coefficient K_t being, in other respects, influenced by this diameter.

The reasoning applied to the case of static pressure probes remains valid.

Let us consider such a probe placed in parallel with the mean velocity direction $\overline{U}$ of a statistically-steady turbulent flow.

The instantaneous value P_m is such that:

$$\frac{P_m - P}{\frac{\rho}{2} U_i^2} = - K_t \sin^2 \theta_i$$

P being the measured instantaneous total pressure, for instance with the aid of an ideal probe without any directional properties.

One will obtain:

$$P_m - P = - K_t \rho \frac{v^2 + w^2}{2}$$

If the measuring device is correctly damped, the mean indication $\overline{P_m}$ will be such that:

$$\frac{\overline{P_m} - \overline{P}}{\frac{\rho}{2} \overline{U}^2} = - 2 K_t \frac{\overline{v_m'^2}}{\overline{U}^2}$$

and, if the turbulence is homogeneous and isotropic,

$$\frac{\overline{P_m} - \overline{P}}{\frac{\rho}{2} \overline{U}^2} = - 2 K_t \frac{\overline{v'^2}}{\overline{U}^2}$$

$\overline{P}$ is obtained with the aid of a probe of Fig. 10 type or Fig. 9 one.
$\overline{P_m}$ is obtained with the aid of a spherical or hemispheric nose probe.

5.7. BRIEFLY, THESE METHODS MAY BE UTILIZED IN DIFFERENT WAYS

5.7.1. *The turbulence components are known*

METHOD I:

$\overline{v^2} = \overline{w^2}$ (central region of a jet for instance).
This method allows to effect a turbulence correction on any static or total pressure probe the directional properties (any whatever) of which are known by calibration in uniform flow.

$\overline{v^2} \neq \overline{w^2}$: It will be possible to effect an approximate correction by adopting a mean value:

$$\overline{v_m^2} = \frac{\overline{v^2} - \overline{w^2}}{2}$$

METHOD II:

$\overline{v^2}=\overline{w^2}$ or $\overline{v^2}\neq\overline{w^2}$, and the calibration curve of the probe is assumed of the form:

$$\frac{I_\theta - I}{\frac{\rho}{2}U'^2} = -K\sin^2\theta$$

This method allows to effect a turbulence correction on any probe the directional properties of which are conformable to the above expression.

5.7.2. *The turbulence components are unknown:*

METHOD III:
This method allows to effect a turbulence correction for the (cylindrical) classical static pressure probes.

METHOD IV:

This method allows to determine an expression:

$$\overline{v_m^2} = \frac{\overline{v^2}+\overline{w^2}}{2}$$

$\overline{v^2}=\overline{w^2}$: This expression allows the rigorous determination of a turbulence correction for whatever directional properties probes, with the aid of the METHOD I.
$\overline{v^2}\neq\overline{w^2}$: It will be possible to effect an approximate correction by adopting a mean value $\overline{v^2}_m$.

6. Application

6.1. CORRECTION OF TOTAL PRESSURE

We have determined with the aid of the general method I the error introduced on most of the models of total pressure probes usually utilized in function of the transverse turbulence level (Figs. 18, 19, 20, 21, 22, 23).

If, for instance, these different probes are placed in a same turbulent flow, utilizing the experimental value

$$\frac{\overline{P_m}-\overline{P}}{\frac{\rho}{2}\overline{U}^2}$$

corresponding to each probe type, one shall obtain, starting from the theoretical curves, $\frac{v'}{U}$ neighbouring values.

Let us recall that:

$\overline{P_m}$ is the measured value by the considered probe,
$\overline{P}$ is the measured value by a non-directional probe Fig. 10 type or Fig. 9 one.

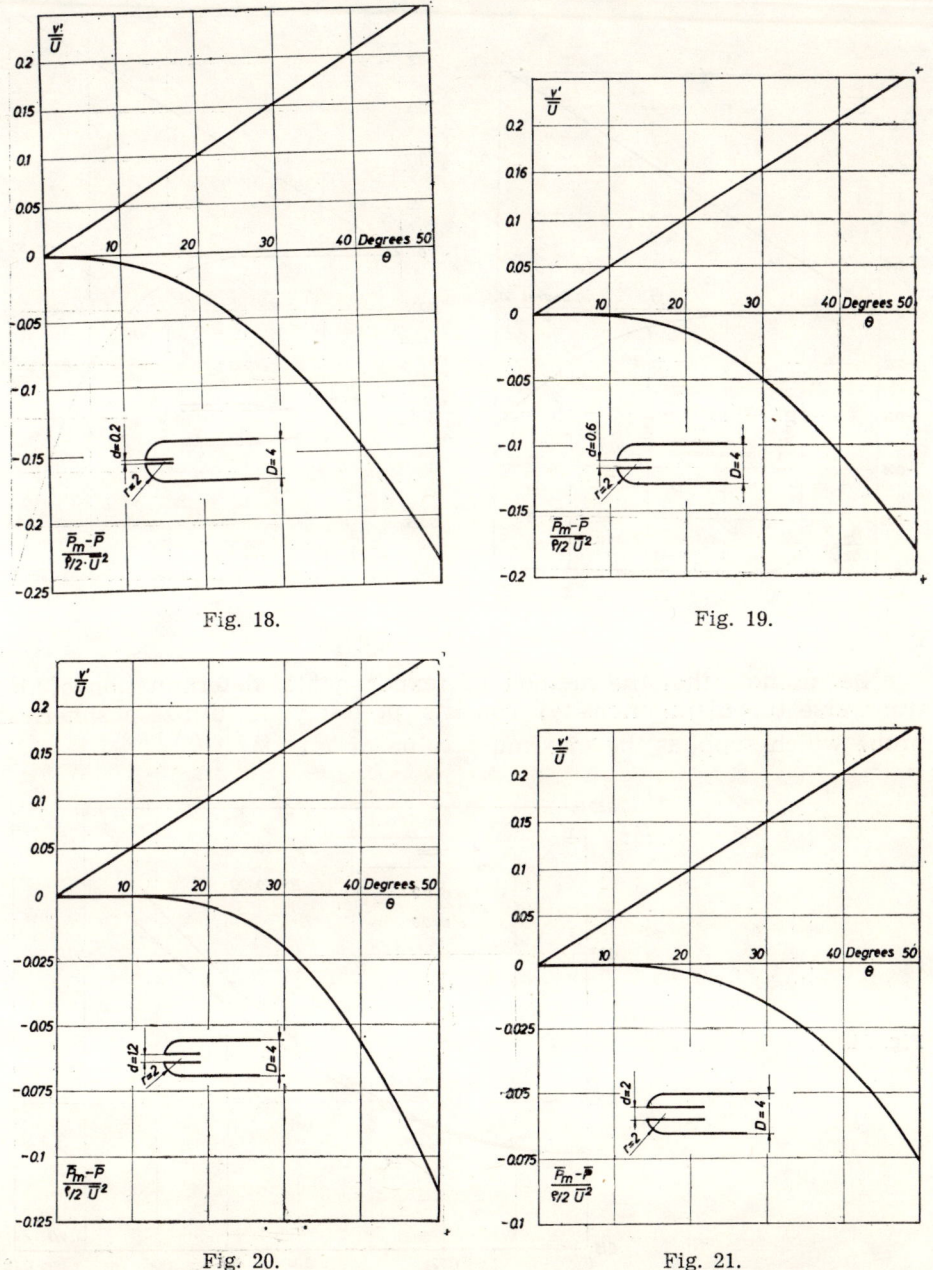

Fig. 18.

Fig. 19.

Fig. 20.

Fig. 21.

It is, for instance, what is evidenced on the Fig. 24, these two probes types having been placed in a plane jet at a distance $\dfrac{x}{2a} = 20$ of the ejection orifice.

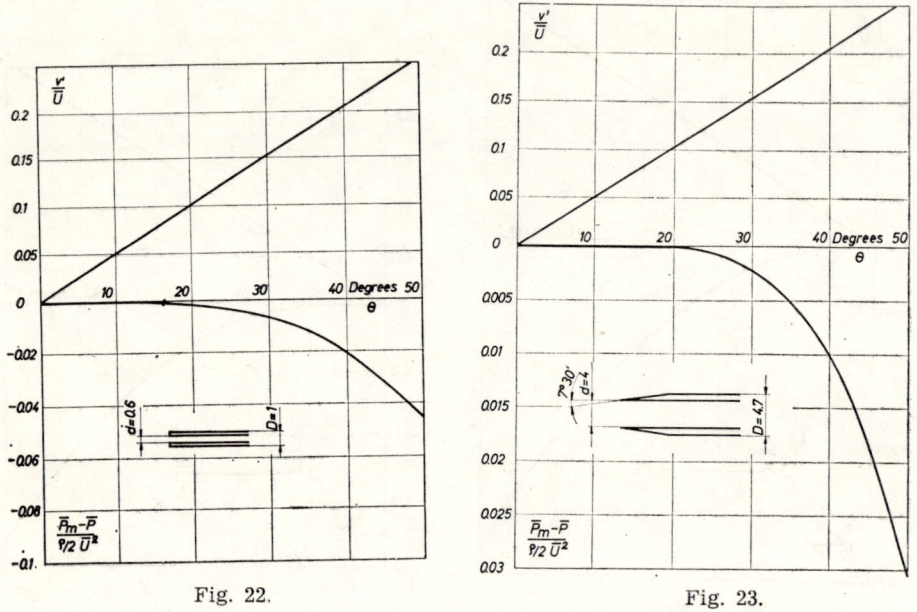

Fig. 22.   Fig. 23.

Let us note that the method IV (experimental determination of the transverse turbulent intensity) consists precisely to utilize a spherical probe which supplies the maximum value of $|\overline{P}_m - \overline{P}|$.

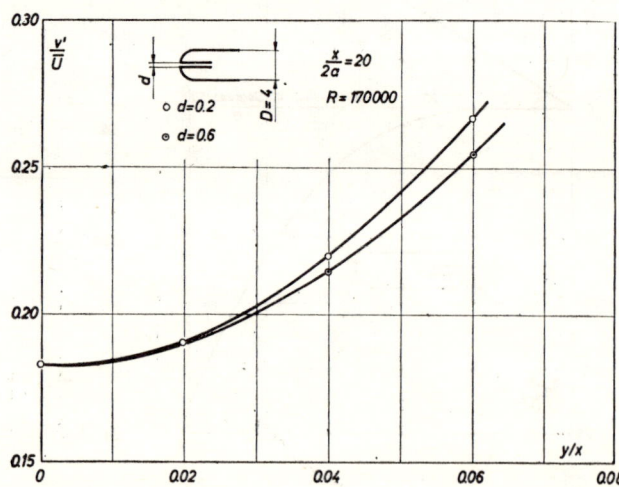

Fig. 24.

6.2. STATIC PRESSURE CORRECTION AND DETERMINATION OF THE TRANSVERSE TURBULENT INTENSITY

We have realized a plane jet with the aid of an installation partly conformable to the Fig. 3. The rectangular duct was suppressed and the final plane convergent had a high contraction ratio. The initial thickness of the jet was $2a = 50$ mm and the initial jet width was 1 000 mm.

The application of the methods III and IV has been effected for different ejection velocities, i.e. for Reynolds numbers $R = 100\,000$ and $R = 170\,000$ $\left(Re = \dfrac{U_0\, 2a}{\nu}, U_o \text{ being the velocity in the jet core}\right)$.

The method III has permitted us to determine the distribution of the actual mean static pressures $\overline{p_s}$ prevailing within a plane jet. The measured mean static pressure $\overline{p_{sm}}$ was supplied by a cylindrical static probe of outside diameter $D = 2$ mm.

The measured values have been corrected with the aid of the formula:

$$\overline{p_s} - \overline{p_{.m}} = (\overline{P} - \overline{P_m})\frac{K_s}{K_t}$$

$\overline{P}$ being the total pressure indicated by a probe of a similar type to the Fig. 10. This probe outside diameter is equal to 2 mm.

$\overline{P_m}$ being the total pressure measured with the aid of a spherical probe similar to that of the Fig. 17. This probe outside diameter was equal to 2.5 mm.

The value $(\overline{P} - \overline{P_m})$ has been obtained by a differential measurement, both the probes being simultaneously placed in the flow.

The coefficient K_s relating to the static pressure probe was equal to 0.61.

The coefficient K_t relating to the total pressure spherical probe was equal to 1.90.

The mechanical realization difficulties have not allowed us to obtain a spherical probe of a diameter equal to 2 mm. Practically such dimensions are already very satisfying since some measurements effected with probes of outside diameter equal to 4 mm have supplied measures similar to those obtained with the aid of the above described probes.

The pressure measurements have been effected for sections such as $5 \leqslant \dfrac{x}{2a} \leqslant 20$, x being the distance comprised between the considered section and the ejection orifice; y represents the distance between the considered point and the jet center. These pressure measurements, for each section, have been made in the middle of the jet width.

So as to allow a confrontation with the method IV, and this by the intermediary of the expression:

$$\overline{p_s} + \rho\, \overline{v^2} = p_o,$$

(p_o being the pressure prevailing outside the jet) we have presented the obtained data under the form:

$\left(\dfrac{p_o - \overline{p_s}}{\rho \, \overline{U}^2} \right)^{\frac{1}{2}}$, $\overline{U}$ being the mean local velocity.

We have also mentioned the corresponding values

$$\left(\dfrac{p_o - \overline{p_s}}{\rho \, \overline{U}^2_{max}} \right)^{\frac{1}{2}} \quad \text{and} \quad \left(\dfrac{p_o - \overline{p_s}}{\rho \, U^2_0} \right)^{\frac{1}{2}},$$

$\overline{U}_{max}$ being the mean maximum velocity in the considered section and $\overline{U}_o$ the velocity in the jet core.

The dotted curves correspond to the value:

$$\dfrac{p_o - \overline{p_s}}{\dfrac{\rho}{2} \, \overline{U}^2}$$

The method IV has permitted us to determine the value of the transverse turbulent intensity $\dfrac{v'}{\overline{U}}$, this in the limit where one admits $\overline{v^2} = \overline{w^2}$, which is the case in the jet central part. These values have also been presented under the corresponding forms:

$$\dfrac{v'}{\overline{U}_{max}} \quad \text{and} \quad \dfrac{v}{U_o}$$

One finds out that a good agreement exists between the obtained values specially on the jet axis (Figs. 25 and 26). The curves dispersion does not exceed that obtained by the processes utilizing a hot wire device.

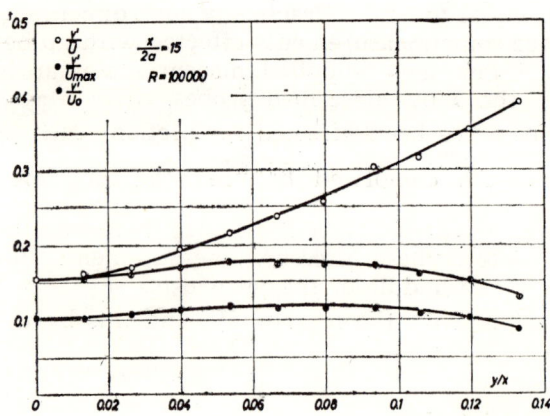

Fig. 25.

Numerous other explorations such as $5 \leqslant \dfrac{x}{2a} \leqslant 20$ have supplied consistent results.

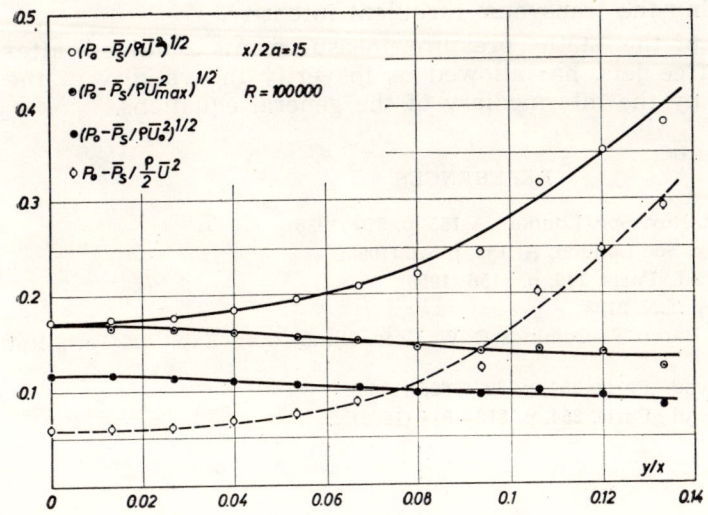

Fig. 26.

Remark: We have represented the values of $\dfrac{v'}{U}$ (or the corresponding values $\left(\dfrac{\overline{p_0 - p_s}}{\rho \overline{U}^2}\right)^{\frac{1}{2}}$ superior to 0.26, although these values correspond to angles θ_{max} such as the measuring of $\overline{P}$ supplied by the non-directional probe is no longer correct. The values of the transverse turbulent intensity, superior to this limit, are obtained by defect. Any way in the central region of the jet, the values obtained by our method are slightly superior to those obtained by the processes utilizing hot wire devices, but in good concordance with those obtained by heat diffusion, for instance. The transverse turbulent intensity thus appears to have a neighbouring value of the longitudinal turbulent intensity in a considerable part of the jet.

One may think that the velocity directional fluctuations are such that the working of the hot wires placed in an X-array is distinctly perturbed. As a matter of fact, for instance, for $\dfrac{v'}{U} \approx 0.25$, θ_{max} is equal to 45°.

6. Conclusion

It is necessary to effect measurements as punctual as possible; this necessitates in most of the flows, small probe dimensions in relation to the turbulence scale; consequently, we have been able to propose correction theoretical methods regarding the classical probes (total pressure probes or static pressure probes). The perfecting of a punctual and non directional total pressure probe has allowed us to put up a correction method when the transverse turbulent intensity is unknown and to experimentally verify the theoretical corrections concerning the total pressure probes.

One of the proposed processes moreover introduces itself as a new method of measuring the transverse turbulent intensity.

The accuracy of the static pressure measurements obtained after correction in the free jets, has allowed us to verify the validity of the proposed methods by the intermediary of the general equations.

REFERENCES

1. Goldstein S., Proc. Roy. Soc. London, A **155**, p. 570 (1936).
2. Fage A., Proc. Roy. Soc London, A **.155**, p 576 (1936).
3. Barat M., C.R. Acad., Paris, **246**, p. 1156 (1958).
4. Laufer J., N.A.C.A. T.N. 2123.
5. Alexander, L. G., Baron T., Comings E. W. Univ. Illinois Eng. Exp. Sta. Tech. Rpt № 8, 1950.
6. Barat M., C. R. Acad., Paris, **254**, p. 623—625 (1962).
7. Barat M., C. R. Acad., Paris, **254**, p. 812—814 (1962).

THE USE OF ELECTROCHEMICAL TECHNIQUES TO STUDY FLOW FIELDS AND MASS TRANSFER RATES

THOMAS J. HANRATTY

Department of Chemistry and Chemical Engineering, University of Illinois, Urbana, Illinois

1. Introduction

A number of years ago Hanratty (1956) presented a paper which proposed a model for the turbulent exchange of mass and momentum with a solid wall. At that time it became apparent that measurements of time averaged temperatures and velocities are not adequate to guide theoretical work and that studies which more directly reflected the character of the turbulence in the immediate vicinity of the wall were needed. Therefore experiments with this goal were initiated in our laboratory and in 1962—63 Reiss published two papers which showed how electrochemical techniques may be used to study flow fluctuations in the viscous sublayer. Since then we have applied electrochemical techniques extensively to study flow fields and concentration fields. Shaw (1964) studied mass transfer fluctuations for a fully developed turbulent concentration field in a one inch diameter pipe. Mitchell (1966) examined in detail the effect of spatial averaging over the test electrode and the effect of the concentration boundary layer on the frequency response. He obtained results on the properties of the component of the fluctuating velocity gradient in the direction of mean flow at the wall of a one inch pipe. More recently Sirkar (1969b) has repeated the experiments of Reiss, Shaw and Mitchell in a 7.625 inch pipe in order to obtain better resolution. He extended previous experimental techniques by measuring the properties of the component of the fluctuating velocity gradient in a direction transverse to that of the mean flow. Dimopoulos (1968) and Son (1969) have applied electrochemical techniques to more general boundary layer flows. They studied the variation of the wall caused by the flow of a fluid past a circular cylinder. By using a specially designed sandwich electrode, Son (1969) was able to detect the separation point quite accurately. Karabelas (1968) has designed an electrode configuration which is capable of measuring the direction of the wall shear stress in a three-dimensional boundary layer flow. Eckelman (1968) has looked at the possibility of using electrochemical techniques to study velocity fluctuations away from the wall. Shaw and Reiss (1963) and Son (1967) have used electrochemical techniques to study the effect of the length of the transfer section on the rate of mass transfer to a pipe wall since it is quite easy to isolate small mass transfer sections.

The extensive experience that we have had with electrochemical methods enables us to use them with enough confidence that they are now a standard technique in our laboratory. They are currently finding application in our studies of flow over solid wavy surfaces, of flow through packed beds, of turbulence close to a wall, of turbulence between rotating cylinder, and of flow around a cylinder.

In this presentation the methods that have been developed to measure local mass transfer rates and local wall velocity gradients will be described.

2. Description of the Technique

In these applications an electrochemical reaction is carried out at the surface of an electrode which is mounted flush with a solid wall. This test electrode may be the entire cathode or part of the cathode depending on whether it is embedded in an inert surface or is part of a large electrode surface. The anode of the electrochemical cell is located downstream of the cathode. It has a much larger area than the cathode so that the electric current flowing in the circuit is controlled by happenings at the cathode surface. If the test electrode is part of a larger cathode surface it can be used to measure local mass transfer rates or fluctuations in the local mass transfer. If it constitutes the entire cathode surface, it can be used to measure the time averaged and fluctuating velocity gradient at the wall.

The electrochemical redox reaction used in most of our experiments is as follows:

$$\text{Cathode} \quad Fe(CN)_6^{-3} + e^- \rightarrow Fe(CN)_6^{-4}$$

$$\text{Anode} \quad Fe(CN)_6^{-4} \rightarrow Fe(CN)_6^{-3} + e^-$$

The electrolyte consists of a dilute solution of potassium ferriand ferrocyanide (0.001 to .01 molar each) in a strong (1 to 2 molar) aqueous solution of sodium hydroxide. The sodium hydroxide acts as a low resistance vehicle for current flow. The bulk concentration of ferricyanide ions C_B remains constant because the electrode reactions destroy and produce equimolar quantities of each species. The flux of ferricyanide ion N to the surface of a test electrode of area A is related to the cell current I by the relation

$$N = (I/AF)(1-T) \tag{1}$$

where F is Faraday's constant. The transferrence numbers T for all of the solutions used are approximately 0.001. Consequently, the current is determined by mass transfer rather than charge transfer.

A mass transfer coefficient K can be defined by the following equation:

$$N = K(C_B - C_W) \tag{2}$$

The concentration of ferricyanide ion at the wall C_W depends on the rate of reaction. At large enough voltages across the two electrodes the reaction rate is fast enough that C_W is esentially zero at the test electrode and the current is controlled by the magnitude of the mass transfer coefficient. The experiments consist of controlling the voltage at a value so that $C_W \simeq 0$ and measuring the current.

For the electrochemical system used in these studies the Schmidt number Sc is about 2400. Therefore, the size of the concentration boundary layer δ_c is quite small and the measured current is related to the flow field in the immediate vicinity of the electrode surface. The velocity field in the concentration boundary layer may be represented by the first term of a Taylor series expansion in y, the distance from the wall. The velocities in the x- and z- directions are therefore related to the velocity gradients at the wall in the following manner:

$$U = S_x y \tag{3}$$

$$W = S_z y \tag{4}$$

From continuity the velocity perpendicular to the wall is given by

$$V = -\left(\frac{\partial S_x}{\partial x} + \frac{\partial S_z}{\partial z}\right) \frac{y^2}{2} \tag{5}$$

3. Measurement of Time Averaged Velocity Gradients

3.1. THE MAGNITUDE OF THE VELOCITY GRADIENT

The time averaged local velocity gradient at a location on a wall can be measured with small circular or rectangular cathodes embedded in an inert surface. The circular electrodes can vary in diameter from 0.005 inches upward. The rectangular electrodes are typically 0.003 in. long and 0.03 in. wide. The test electrodes are formed by inserting a platinum wire or sheet into a small hole drilled into the test section. They are glued in place with epoxy resin and sanded flush with the pipe wall using progressively finer grades emery paper. The anodes can be a sheet of platinum located in the fluid downstream of the test section.

Let us first consider a two-dimensional flow which is not changing in the direction of mean flow and a rectangular test electrode with its long side perpendicular to the direction of mean flow. A mass balance for the ferricyanide ion in the neighborhood of the electrode surface gives the following time averaged equation:

$$\overline{S}_x y \frac{\partial \overline{C}}{\partial x} = D \left(\frac{\partial^2 \overline{C}}{\partial y^2} + \frac{\partial^2 \overline{C}}{\partial x^2} + \frac{\partial^2 \overline{C}}{\partial z^2}\right) \tag{6}$$

Diffusion in the z-direction can be made negligibly small by using an electrode which is wide compared to the thickness of the concentration boundary layer. For reasonably large values of $\bar{S}_x$ the term governing diffusion in the x-direction can be neglected so that (6) becomes

$$\bar{S}_x\, y\, \frac{\partial \bar{C}}{\partial x} = D\, \frac{\partial^2 \bar{C}}{\partial y^2} \qquad (7)$$

An expression relating the mass transfer coefficient to $\bar{S}_x$ can be obtained by solving (7) using the boundary conditions $\bar{C}=C_W$ at $y=0$, $\bar{C}=C_B$ at $y=\infty$, and $\bar{C}=C_B$ at $x=0$. The following equation is obtained:

$$\frac{KL}{D} = 0.807\, Z^{1/3}, \qquad (8)$$

where D is the diffusion coefficient, L is the length of the electrode and Z is a dimensionless wall velocity gradient defined as $(\bar{S}_x L^2/D)$. Reiss (1963) has shown that (8) may be used for circular electrodes provided L is taken as 0.82 times the electrode diameter.

Equation (8) becomes unreliable for small values of Z because of the effects of longitudinal diffusion and because of natural convection caused by density variations in the concentration boundary layer. The additional contribution to the value of $\frac{KL}{D}$ by streamwise diffusion is estimated as $0.19\, Z^{-1/6}$ from a calculation performed by Ling (1962). The effect of natural convection depends on the orientation of the electrode and the magnitude of the group $G_L/R_L^2\, Sc^{1/3}$, where G_L and R_L are the Grashof and Reynolds numbers based on the length of the electrode and Sc is the Schmidt number. For small values of $G_L/R_L^2\, Sc^{1/3}$ the following combined corrections due to natural convection and streamwise diffusion may be used (Dimopoulos, 1968)

$$\frac{KL}{D} = 0.807\, Z^{1/3} + 0.19\, Z^{-1/6} \pm 0.253\, \frac{G_L}{R_L}\left(\frac{Sc}{Z}\right)^{1/3} \sin\theta, \qquad (9)$$

where θ is the angle between a normal to the test electrode and the vertical. Equation (9) has not yet been properly tested so at present it should be used only to estimate when the measurements are being affected by streamwise diffusion and by natural convection.

The application of Eq. (8) to general boundary layer flows where the velocity gradient is varying in the direction of flow has been found to be valid except in the neighborhood of the front stagnation point and the separation point. Near the front stagnation point the variation of $\bar{S}_x$ over the electrode surface could be important so that the effect of the velocity component normal to the surface must be taken into account in the mass balance

$$-\frac{d\bar{S}_x}{dx}\frac{y^2}{2}\frac{\partial \bar{C}}{\partial y} + \bar{S}_x\, y\, \frac{\partial \bar{C}}{\partial x} = D\, \frac{\partial^2 \bar{C}}{\partial y^2} \qquad (10)$$

The solution of (10) is

$$\left(\frac{\partial \overline{C}}{\partial y}\right)_{wall} = \left(\frac{1}{9D}\right)^{1/3} \frac{C_B}{0.893} \frac{\{\overline{S}_x(x)\}^{1/2}}{\left[\int_0^x \{\overline{S}_x(x)\}^{1/2} dx\right]^{1/3}} \quad (11)$$

where x is the distance from the front stagnation point. When the electrode is located right at the front stagnation point the measured mass transfer rate is related only to $\dfrac{d\overline{S}_x}{dx}$, and for some region within about 5° of the front stagnation point it is related to both $\overline{S}_x$ and $\dfrac{d\overline{S}_x}{dx}$. Equation (8) is not valid near a separation point because the velocity field is not adequately represented by the first term in a Taylor series and because natural convection effects could be important. A recent article by Spence and Brown (1968) deals with the first aspect of this problem.

3.2. THE DIRECTION OF WALL VELOCITY GRADIENTS

Measurements from single electrodes are capable of giving the magnitude and not the direction of the velocities close to a solid surface. Son (1969) has shown that a sandwich of two electrodes can be used to determine flow direction in a two-dimensional boundary layer. Two platinum electrodes, 1 in. x 0.005 in. separated by 0.002 — 0.003 in of insulation were embedded in the surface of a cylinder as shown in Fig. 1. The

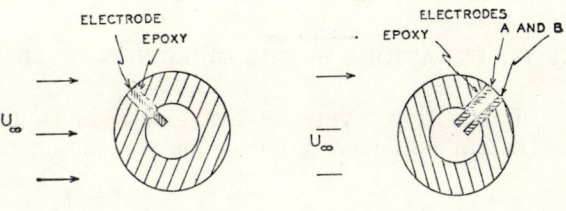

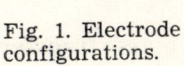
Fig. 1. Electrode configurations.

cylinder could be rotated so that the position of the sandwiched electrodes could vary with respect to the oncoming flow. If the flow is steady and the two electrodes are activated, one of the measurements will be significantly smaller than the other since it is in the wake of the other. If the flow has large fluctuations, it is not easy to determine the flow direction from the time averaged signal. In this case one of the electrodes is operated as the test electrode. If the sudden activation of the other electrode causes a sudden change in the signal from the test electrode, then the test electrode is in the wake of the second electrode. By using this technique it is possible to determine the separation point for flow around a 3/4 in. cylinder to within 1°.

In a three-dimensional boundary layer a circular electrode will give the magnitude of the velocity gradient at a wall. The electrode arrangement shown in Fig. 2 was designed by Karabelas (1968) to determine the

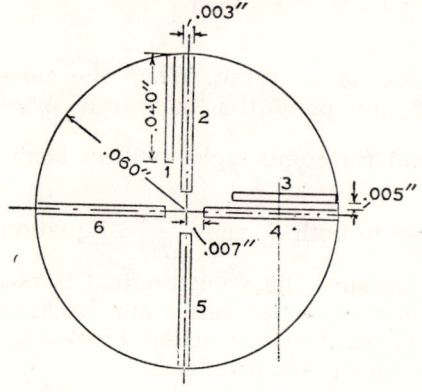

Fig. 2. Probe dimensions.

direction of the wall velocity gradients for a three-dimensional flow. The angle which the flow is making can be calculated from mass transfer rates to either electrode pair 2 and 4 or electrode pair 5 and 6. Two pairs of electrodes are used in order to obtain a better average at the center of the probe. The auxiliary electrodes 1 and 3 determine roughly the direction of flow at the beginning of the measurements.

4. Measurement of Fluctuations in the Velocity Gradient

4.1. FLUCTUATIONS IN THE DIRECTION OF MEAN FLOW

If the flow over the test electrode is unsteady the mass transfer coefficient will have a steady as well as an unsteady component,

$$K = \overline{K} + k \tag{12}$$

If one uses a circular electrode or a rectangular electrode with its long side perpendicular to the direction of mean flow, the fluctuating component k can be related to the component of the fluctuating velocity gradient in the direction of mean flow, s_x. The problems involved in relating s_x to k are similar to those encountered in hot wire anemometry in relating current fluctuations to flow fluctuations (Hinze 1959). The ability of the hot wire to measure directly high frequency velotiy fluctuations is limited by the thermal inertia of the wire. A capacitance effect of the concentration boundary layer over a diffusion controlled electrode limits direct measurement of high frequency velocity gradient fluctuations. Large turbulent intensities can cause non-linear response in both systems. Non-uniform flow over the wire length and over the electrode surface can be a source of error.

Consider a rectangular electrode with its long side perpendicular to the direction of mean flow. The mass balance may be written as

$$\frac{\partial C}{\partial t} + S_x y \frac{\partial C}{\partial x} = D \frac{\partial^2 C}{\partial y^2} \tag{13}$$

If a pseudo-steady state approximation is made, whereby the term $\frac{\partial C}{\partial t}$ is neglected, the relation between the instantaneous mass transfer coefficient and the instantaneous velocity gradient is given as

$$\frac{(\overline{K}+k)L}{D} = 0.807 \left[\frac{(\overline{S}_x + s_x) L^2}{D} \right]^{1/3} \tag{14}$$

If $\frac{s_x}{\overline{S}_x} \ll 1$ the above expression can be simplified to yield the relation

$$\frac{k}{\overline{K}} = \frac{1}{3} \left(\frac{s_x}{\overline{S}_x} \right) \tag{15}$$

The validity of the linearization assumption can be checked by a method similar to that outlined by Hinze (1959) for the hot-wire anemometer. It is found that for $\overline{(s_x^2)}^{1/2}/\overline{S}_x$ less than 0.5 the error in using the linear model is less than 3%.

The pseudo-steady state approximation is valid for small frequencies. For higher frequency velocity fluctuations the mass transfer fluctuations are smaller than predicted by (15). A correction for the pseudo-steady state approximation can be obtained by solving the linearized form of (13),

$$\frac{\partial c}{\partial t} + s_x y \frac{\partial \overline{c}}{\partial x} + \overline{S}_x y \frac{\partial c}{\partial x} + s_x y \frac{\partial \overline{C}}{\partial x} = D \left(\frac{\partial^2 \overline{C}}{\partial y^2} + \frac{\partial^2 c}{\partial y^2} \right) \tag{16}$$

An equation for the fluctuating component of the concentration field is obtained after subtracting the time averaged quantities from (16):

$$\frac{\partial c}{\partial t} + \overline{S}_x y \frac{\partial c}{\partial x} + s_x y \frac{\partial \overline{C}}{\partial x} = D \frac{\partial^2 c}{\partial y^2} \tag{17}$$

This is to be solved using the boundary conditions that $c=0$ at $y=0$, at $y \to \infty$ and at $x=0$. If the response of the concentration boundary layer is sought to a harmonic oscillation in s_x of frequency n, then

$$s_x = \hat{s}_x \, e^{i 2 \pi n t} \tag{18}$$

and

$$c = \hat{c} \, e^{i 2 \pi n t} \tag{19}$$

Mitchell (1966) calculated the amplitude c for different values of n by numerically integrating (17). The results of these calculations can be represented as a correction to the pseudo-steady state approximation.

$$\left| \hat{k} \right|^2 = A^2 \, \hat{k}_s^2, \tag{20}$$

where

$$\frac{\hat{k}_s}{K} = \frac{1}{3}\left(\frac{\hat{s}_x}{\overline{S}_x}\right). \tag{21}$$

The coefficient A is a function of the dimensionless frequency

$$\tilde{n} = (2\pi n\, L^{2/3}/D^{1/3}\, \overline{S}^{2/3}), \tag{22}$$

For small values of $\tilde{n}$ Mitchell found that his results could be represented as

$$A^2 = 1/(1 + 0.060\, \tilde{n}^2) \tag{23}$$

If the concentration boundary layer is causing considerable damping, then the velocity intensity is calculated from the spectral distribution function of the mass transfer fluctuations W_k by the equation

$$\frac{\overline{(s_x^2)}^{1/2}}{\overline{S}_x} = 3\frac{\overline{(k^2)}^{1/2}}{\overline{K}}\left[\int_0^\infty \frac{W_k\, dn}{A^2}\right]. \tag{24}$$

More work is needed on this problem to extend Mitchell's calculations to higher frequencies and to devise instrumental methods for accounting for the frequency response.

The use of (13) implies that the flow field is uniform over the surface of the wall electrode. If the scale of the fluctuating flow is small compared to the size of the electrode then averaging can occur and the value of the mass transfer intensity measured by the electrode $\overline{<k>^2}$ is smaller than the true local intensity $\overline{k^2}$. The magnitude of this averaging depends on the function describing the correlation coefficient R. For a rectangular electrode of with W

$$\overline{<k>^2} = \frac{2\,\overline{k^2}}{W^2}\int_0^W (W-g)\, R(g)\, dg \tag{25}$$

It is usually desirable to design the electrode so that averaging is negligible. This is especially true if frequency spectra of the fluctuating field are desired since it is not known how to correct for the frequency response of the concentration boundary layer under circumstances such that averaging is occurring.

Correlation measurements can be made by measuring the product of the signals from two separate electrodes. The spacing and design of the electrodes depend on the system and on the scale of the fluctuations. For example, Mitchell (1966) measured the transverse correlation function in a 1 in. pipe by embedding a packet of rectangular electrodes in the wall. The electrodes in the packet were oriented with their long side parallel to the flow direction and were separated by insulators which were 0.006 in. wide.

Shaw (1964) used measurements of the effect of the size of the electrode on the intensity and Eq. (25) to calculate a scale for the fluctuations. Sirkar (1969) has recently shown that this precedure can lead to serious error if prior knowledge about the shape of the correlation function is not known.

4.2. FLUCTUATIONS TRANSVERSE TO THE DIRECTION OF MEAN FLOW

Flow fluctuations in a direction transverse to that of the mean flow can be measured with a rectangular electrode embedded in the wall at an angle to the mean flow (Sirkar 1969a). The sensitivity of these electrodes to the transverse fluctuations depend strongly on their orientation. Consider the single slanted electrode shown in Fig. 3. The angle θ between the instantaneous direction of flow and the direction of mean flow is defined by the equation

$$\tan \theta = -\frac{S_z}{\overline{S}_x + S_z}. \qquad (26)$$

The electrode surface can be divided into a number of strips of length q parallel to the instantaneous flow direction. As can be seen from Fig. 3, the value of q will be less near the edges of the electrode than in the

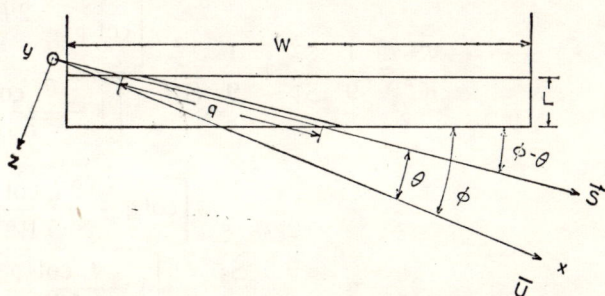

Fig. 3. Electrode configuration.

central regions. If the pseudo-steady state approximation is valid, one can apply Eq. (8) to each of these strips so that

$$K = \sigma \left(\frac{S}{q}\right)^{1/3} \qquad (27)$$

where

$$S = [(\overline{S}_x + s_x)^2 + s_z^2]^{1/2} \qquad (28)$$

As long a $\varphi - \theta \geqslant \psi$ where $\psi = \tan^{-1}(L/W)$ the average mass transfer coefficient over whole electrode surface at a given instant is

$$K = \sigma \left[\frac{S \sin (\varphi - \theta)}{L}\right]^{1/3} \left[1 + \frac{L}{5W} \cot (\varphi - \theta)\right] \qquad (29)$$

The term containing $\dfrac{L}{W}$ appears because q is varying near the edges of the electrode. It can be seen that as $\frac{L}{W} \to 0$ this edge effect can be neglected.

Equation (29) can be expanded under the restriction that $\frac{|S_z|\cot \varphi}{|\overline{S}_x+s_x|}<1$ to give

$$K=\sigma\left(\frac{\overline{S}_x \sin \varphi}{L}\right)^{1/3} \left[1+\frac{L \cot \varphi}{5\,W}+ \right.$$

$$\left. \frac{1}{3}\frac{s_x}{\overline{S}_x}+\frac{s_z \cot \varphi}{3\overline{S}_x}+\frac{s_x\,L \cot \varphi}{15\,\overline{S}_x\,W}-\frac{2\,L\,s_z \cot^2 \varphi}{15\,W\,[\overline{S}_x+s_x]}+\cdots\right] \tag{30}$$

The time averaged mass transfer coefficient is then given as

$$\overline{K}=\sigma\left(\frac{\overline{S}_x \sin \varphi}{L}\right)^{1/3} 1+\left(\frac{L \cot \varphi}{5\,W}\right) \tag{31}$$

The fluctuating component of the mass transfer coefficient, k, can be obtained by subtracting (31) from (30).

$$\frac{\overline{k^2}}{\overline{K}^2}=\frac{1}{9}\frac{\overline{s_x^2}}{\overline{S}_x^2}+\frac{1}{9}\frac{\overline{s_z^2}}{\overline{S}_x^2}\frac{\left[\cot \varphi - \dfrac{2\,L \cot \varphi}{5\,W}\right]^2}{\left[1+\dfrac{L \cot \varphi}{5\,W}\right]^2}$$

$$+\frac{2}{9}\frac{\overline{s_x s_z}}{\overline{S}_x^2}\frac{\left[\cot \varphi - \dfrac{2\,L \cot \varphi}{5\,W}\right]}{\left[1+\dfrac{L \cot \varphi}{5\,W}\right]} \tag{32}$$

Equation (32) relates the measured $\overline{k^2}$ to three statistical properties of the flow field $\overline{s_x^2}$, $\overline{s_z^2}$, and $\overline{s_x s_z}$. Because of symmetry $\overline{s_x s_z}$ may usually be taken as zero. Therefore $\overline{s_x^2}$ can be determined from measurements with the electrode perpendicular to the direction of mean flow and $\overline{s_z^2}$ is calculated from measurements with a slant electrode. Equation (30) is valid only so long as $-\dfrac{s_z}{\overline{S}+s_x}< 0.2$. For larger values of φ or smaller values of $\cot \varphi$ the upper limit on s_z is higher.

The sensitivity of a slant electrode to transverse velocity fluctuations can be examined by considering a field for which $\sqrt{\overline{s_x^2}}/\overline{S}_x = 0.32$ and $\sqrt{\overline{z_z^2}}/\overline{S}_x = 0.10$. The value of $\sqrt{\overline{k^2}}/\overline{K}$ measured with a 45° electrode with $W/L=10$ would only be about 4.5% higher than the value measured with a 90° electrode. For the same velocity field the mass transfer intensity is calculated to be about 35 percent higher for a 12.5° electrode than for a 90° electrode.

The direct measurement of transverse velocity fluctuations can be accomplished with the electrode arrangement shown in Fig. 4. This was

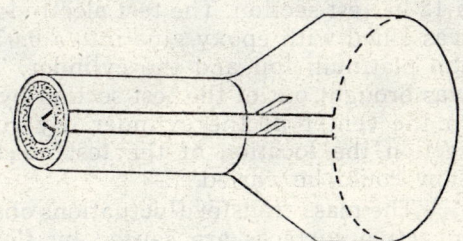

Fig. 4. V electrodes assembly.

recently used by Sirkar (1969b) in his studies of turbulence close to a wall. The instantaneous total mass transfer coefficients of electrode 1 and electrode 2 are

$$K_1 = \overline{K} + \sigma \left(\frac{\overline{S}_x \sin \varphi}{L} \right)^{1/3} \left[\frac{1}{3} \frac{S_x}{\overline{S}_x} \left(1 + \frac{L \cot \varphi}{5W} \right) \right. \tag{33}$$
$$\left. + \frac{1}{3} \frac{S_z}{\overline{S}_x} \left(\cot \varphi - \frac{2L \cot \varphi}{5W} \right) \right]$$

$$K_2 = \overline{K} + \sigma \left(\frac{\overline{S}_x \sin \varphi}{L} \right)^{1/3} \left[\frac{1}{3} \frac{S_x}{\overline{S}_x} \left(1 + \frac{L \cot \varphi}{5W} \right) \right. \tag{34}$$
$$\left. - \frac{1}{3} \frac{S_z}{\overline{S}_x} \left(\cot \varphi - \frac{2L \cot \varphi}{5W} \right) \right]$$

The difference of the signals between the electrodes gives a signal that is proportional to s_z since

$$K_1 - K_2 = \frac{2}{3} \frac{S_z}{\overline{S}_x} \left(\cot \varphi - \frac{2L \cot \varphi}{5W} \right) \tag{35}$$

5. Measurement of Local Mass Transfer

Local mass transfer rates and the properties of the fluctuating mass transfer rate have been measured for a fully developed concentration profile in a pipe (Shaw, 1964; Sirkar, 1969b). The variation of the mass transfer rate has been measured around the surface of a cylinder which is exchanging mass with a flowing stream (Dimopoulos, 1968).

In the pipe studies the anode is either a long section of nickel piping or a large number of nickel sheets. The test electrode consisted of nickel wires having diameters of 0.015 in. and larger. The wires were glued with epoxy resin into holes in the pipe wall that were drilled about 0.006 in. oversize. The epoxy glue served to insulate the test electrode from the rest of the cathode. The inside of the test section was then sanded smooth. In the studies of mass transfer to a cylinder a platinum foil was wrapped around the central 1 in. of a 1 in. diameter plastic cylinder which spanned a 12 in. test section. The test electrode was a 0.020 in. platinum wire which was glued with epoxy glue into a 0.030 in. hole which was drilled through the platinum foil and the cylinder. The wiring to the electrode surfaces was brought out of the test section by leading it out through a hole drilled in the center of the cylinder. By rotating the test cylinder in the test section the location of the test electrode with respect to the oncoming flow could be varied.

The mass transfer fluctuations observed from test electrodes embedded in active surfaces are caused by flow fluctuations perpendicular to the surface and flow fluctuations transverse to the mean flow. However, as yet, we are not able to relate the mass transfer fluctuations to the flow fluctuations in a simple fashion, as is possible for test electrodes embedded in an inert surface.

6. Experimental Details

6.1. DESIGN OF THE EXPERIMENTS

A number of precautions have to be taken in the design of the experiments to avoid contamination of the electrode surface and of the electrolyte and to avoid side reactions. Flow systems are fabricated from stainless steel or from plastic. Before carrying out a run the electrolyte is purged of dissolved oxygen by bubbling nitrogen through the solution and the test section is washed with soap and water. The fluid in the storage tank and flow system is blocked from exposure to light as much as possible because ferrocyanide decomposes slowly according to the reaction

$$Fe(CN)_6^{-4} + 2H_2O \xrightarrow{light} Fe(CN)_5 H_2O^{-3} + OH^- + HCN.$$

A flow loop used in the study of flow around a cylinder by Son (1969) is shown in Fig. 5. It has a maximum capacity of 1900 gallons per minute and holds 900 gallons of fluids. The temperature is controlled to within $\pm 0.05°C$ of $25.00°C$. The 6 in. $\times$ 12 in. test section was constructed from stainless steel supporting members and plastic panels. The remainder of the flow loop including the honeycomb, screens and pump were fabricated from stainless steel. During operation the liquid in the riser is kept in contact with an atmosphere of nitrogen. When not in operation the electrolyte is stored in a black polyethylene tank. The storage system has an auxiliary heat exchanger to remove heat generated during the preparation of the solutions. Nitrogen is bubbled through the liquid in the storage tank before it is pumped into the flow loop for an experiment.

The equipment shown in Fig. 6 was designed to study fully developed turbulent flow in a 7.625 in. pipe up to Reynolds numbers of 100,000. The system occupies four floors. The electrolyte discharges from a stainless

steel storage tank on the fourth floor through a stainless steel pump on
the ground floor to the stainless steel calming section which consists of

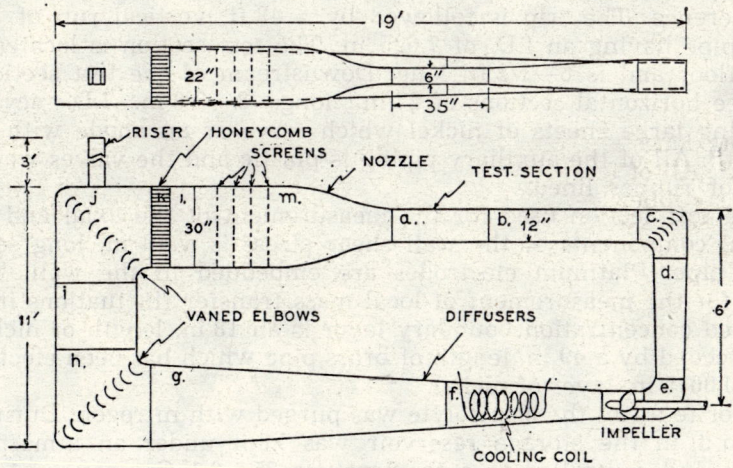

Fig. 5. Water tunnel.

a 6 in. standard 90° elbow with turning vanes, a 3 ft. long diffuser with
a square 22 in. × 22 in. section, a honeycomb, a 9 in. settling chamber, a

Fig. 6.

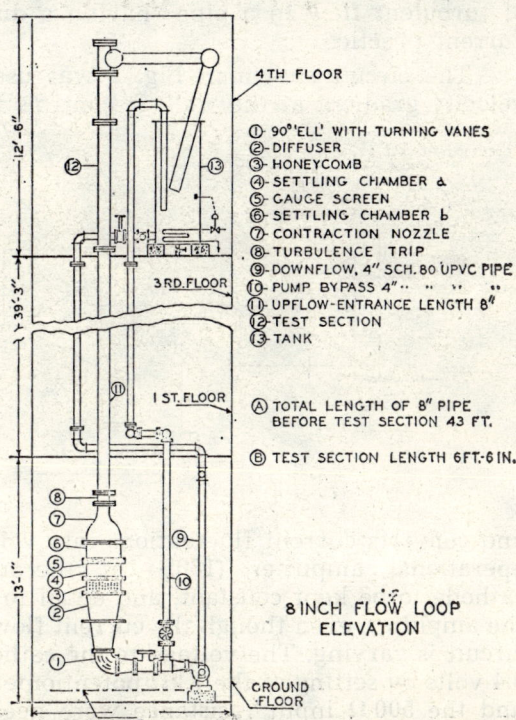

20 mesh wire screen, another 9 in. long settling chamber, and a 20 in. long contraction. The flow at the inlet to the pipe is tripped by a 1/2 in. long ring consisting of a series of 3/8 in. equilateral triangles around the circumference. The trip is followed by a 43 ft. vertical run of Van-Cor UPVC pipe having an I.D. of 7.625 in. The test section is located on the fourth floor and is 6—1/2 ft. long. Downstream of the test section there are three horizontal sections of 27 in. long, 8—3/8 in. I.D. acylic pipe containing large sheets of nickel which serve as an anode with an area of 190 ft.^{2}. All of the auxiliary piping is plastic and the valves were either plastic or rubber lined.

The test section used for the measurement of the axial and circumferential components of the wall shear stress is a 40 in. long section of acrylic pipe. Platinum electrodes are embedded in the wall. The test section for the measurement of local mass transfer fluctuations in a fully developed concentration boundary layer is an 18 in. length of nickel pipe. It is preceded by a 49 in. length of brass pipe which has been electroplated with a 0.0015 in. layer of nickel.

Prior to a run the electrolyte was purged with nitrogen. During a run the liquid in the storage reservoir was kept under an atmosphere of nitrogen and controlled to a temperature $25 \pm 0.1°C$.

6.2. DESIGN OF THE ELECTRONIC CIRCUITS

The electronic circuitry used by Sirkar (1969b) in his recent studies of turbulent flow in a pipe having a diameter of 7.625 in. is typical of current practice.

The circuit shown in Fig. 7 was used for the measurement of the velocity gradient at the wall. It controls the voltage of the test electrode

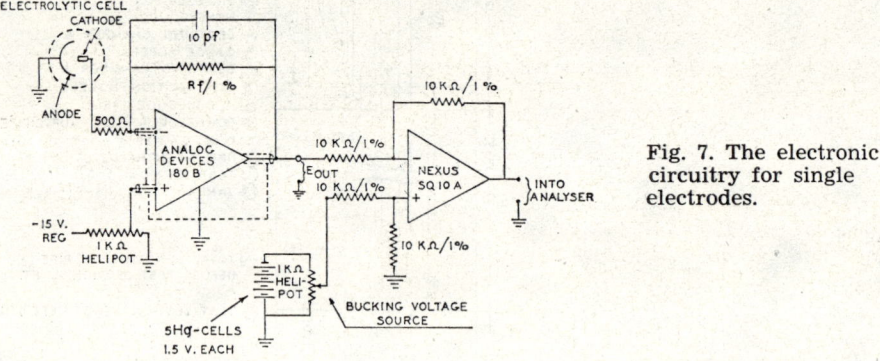

Fig. 7. The electronic circuitry for single electrodes.

and converts current fluctuations into voltage fluctuations. The use of an operational amplifier (180B) at the output enables the voltage of the cathode to be kept constant, and equal to that at the positive terminal of the amplifier, even though the current flowing through the electrochemical circuit is varying. The voltage to the cathode was usually controlled at — 0.4 volts by setting of the 1 kΩ potentiometer. The 10 pf feedback capacitor and the 500 Ω input resistance were needed to improve the closed loop stability. The value of the feedback resistance, R_f, varied from 0.4 MΩ to 2.8 MΩ depending on the flow rate through the pipe. The voltage at the output of the 180B amplifier equals the product of the current flowing the electrochemical circuit and the resistance R_f. The time averaged value of the voltage gave the time averaged mass transfer rate. In order to determine the properties of the fluctuating part of the signal the d.c. level of the signal coming from the 180B amplifier was eliminated by supplying a bucking voltage to the positive input of a Nexus SQ 10 A operational amplifier in a gain-of-one position.

For the measurement of local mass transfer fluctuations in a fully-developed concentration boundary layer two electrode circuits were used. One is a very high current circuit for the control electrode which generated the concentration boundary layer. The one for the test electrode is the same as described above. The voltages applied to the test electrode and the reference electrode are the same. The high current amplifier used in the control electrode circuit is the 180B amplifier circuit shown in Fig. 8 to which an additional circuit had been added with a view to handling a current of up to 2 amperes in the feedback loop.

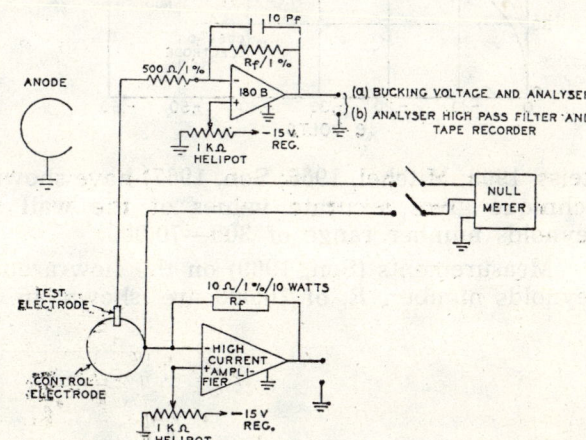

Fig. 8. Fully developed mass transfer electronic circuitry.

7. Results

7.1. VELOCITY GRADIENTS AT THE WALL

Typical polarization curves are shown in Fig. 9. These are plots of average current versus the voltage for a 0.064 in. diameter electrode in the wall of a 1 in. pipe. These show that after a certain voltage a limiting current is obtained whose magnitude is dependent on the flow rate in the pipe. It can be seen that by operating the test electrode at a voltage of —0.4 volts one could be well within the diffusion controlled region for which $C_W \simeq 0$.

The calculation of the velocity gradient at the wall from (8) requires accurate values of the diffusion coefficient. Such measurements have been made for the ferrocyanide, ferricyanide system by Gordon and Tobias (1963).

One of the simplest systems in which to check the accuracy of Eq. (8) is fully developed flow in a circular pipe. For laminar flow the velocity gradient at the wall can be calculated from the measured volumetric flow rate using Poisuille's law. For turbulent flow it can be calculated from the measured pressure drop using a force balance. Tests in a 1 in. pipe

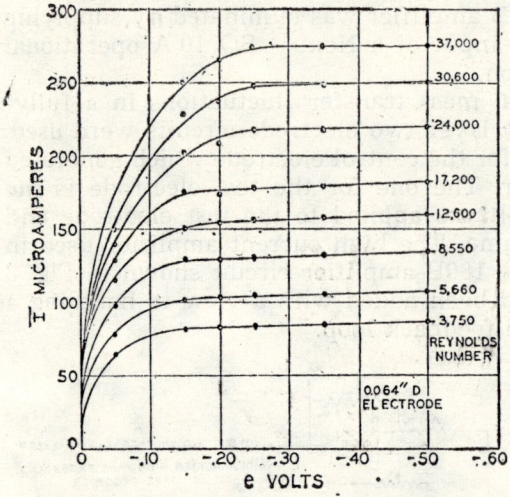

Fig. 9. Reduction of ferricyanide.

(Reiss, 1963; Mitchel, 1966; Son, 1967) have shown that the electrochemical technique gives accurate values of the wall velocity gradient over a Reynolds number range of 300—70,000.

Measurements (Son, 1969) on the flow around a 3/4 in. cylinder at a Reynolds number, R, of 10,000 are shown in Fig. 10. Time averaged

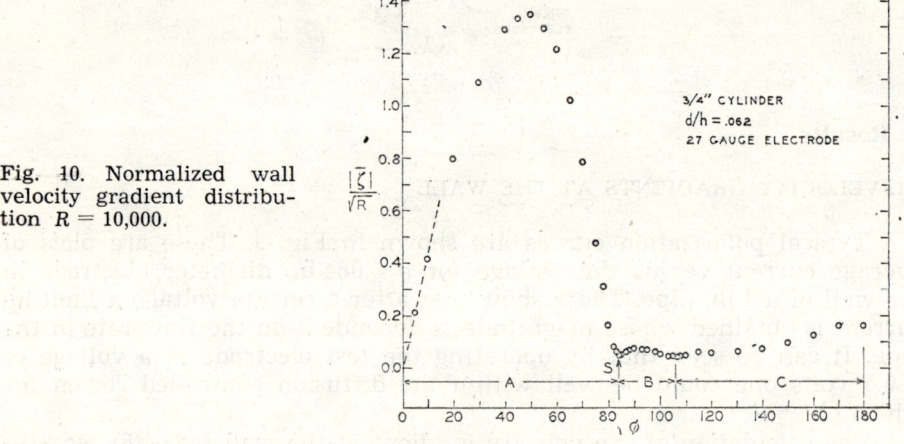

Fig. 10. Normalized wall velocity gradient distribution $R = 10,000$.

absolute values of the velocity gradients, $|\bar{\zeta}|$, at the wall of the cylinder that have been made dimensionless with respect to the free stream velocity, u_∞ and the cylinder radius, $d/2$, are plotted as a function of the angular distance from the front stagnation point. The Reynolds number used in the figure has been defined in terms of the diameter of the cylinder and the free stream velocity. The separation point, as determined by the sandwiched electrodes, is indicated with an arrow and the symbol S. The velocity gradient, ξ, is positive in region A and negative in region B. In the early part of region C it is positive. However, over most of region C it changes direction as time goes on. The region B shown in Fig. 10 was observed over the entire range of Reynolds numbers investigated by Son,

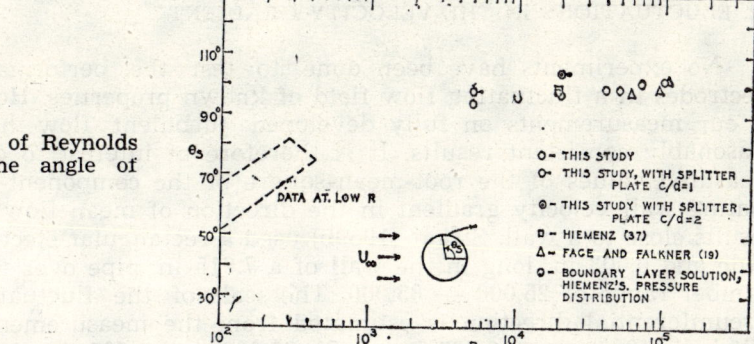

Fig. 11. Effect of Reynolds number on the angle of separation.

5×10^3 to 10^5, and seems to suggest the existence of a separation bubble. Good agreement exists between the separation points determined with the sandwiched electrodes and those previously reported in the literature

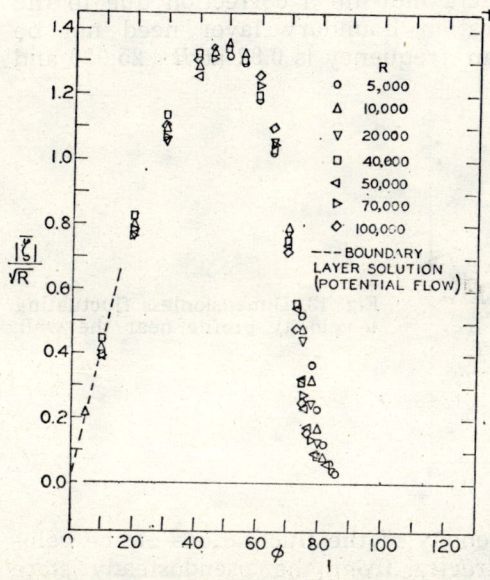

Fig. 12 Effect of Reynolds number on the wall velocity gradient distribution.

(see Fig. 11). Measured velocity gradients prior to separation for a number of flow rates are shown in Fig. 12. As predicted by boundary-layer theory, measurements at different Reynolds numbers collapse on a single curve when the dimensionless velocity gradient is normalized with respect to $\sqrt{R}$. According to the potential flow solution the velocity external to the boundary-layer is given by $U = 4 \times u_\infty / d$ in the neighborhood of the front stagnation point where x is the distance along the cylinder surface. The boundary-layer solution based on this external flow is given by Schlichting and is indicated by the dotted lines in Figs. 10 and 12. It is seen that the measured velocity gradients in the neighborhood of the front stagnation point are in good agreement with this solution.

7.2. FLUCTUATIONS IN THE VELOCITY GRADIENT

No experiments have been done to test the performance of the electrodes in a fluctuating flow field of known properties. However some of our measurements on fully developed turbulent flow have yielded reasonably consistent results. It is therefore of interest to compare our measured values of the root-mean-square of the component of the fluctuating wall velocity gradient in the direction of mean flow to hot wire results close to a wall. Sirkar (1969b) used a rectangular electrode 0.03 in. wide and 0.003 in. long in the wall of a 7.615 in. pipe over the Reynolds number range of 25,000 — 85,000. The scale of the fluctuations in the circumferential direction is estimated from the measurements made by Mitchell (1966) to be 0.067 in. at $R = 25,000$ and 0.027 in. at $R = 85,000$. From Eq. (25) it is found that a correction of less than 2 percent has to be made for spatial averaging at Reynolds numbers less than 50,000. At a Reynolds number of 85,000 the correction is only about $6^{1}/_{2}$ percent. Measurements of the frequency specta indicate a correction due to the capacitance effect of the concentration boundary layer need not be applied. The value $\tilde{n}$ to the median frequency is 0.82 at $R = 25,000$ and

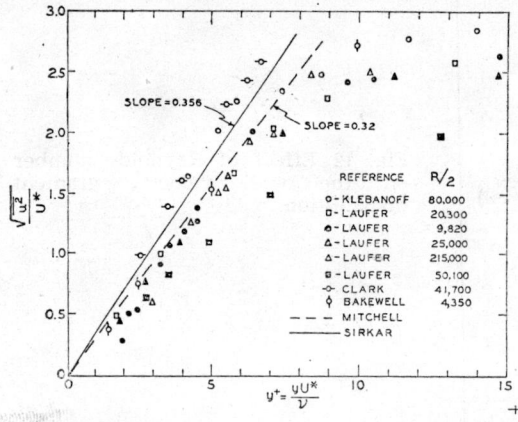

Fig. 13. Dimensionless fluctuating u-velocity profile near the wall.

0.67 at $R = 85,000$. Therefore the intensity of the fluctuations of the velocity gradient can be calculated directly from the pseudosteady state approximation, Eq. (15). It is found that $\sqrt{\overline{s_x^2}}/\bar{S}_x = 0.35$. Mitchell carried out similar measurements in a 1 in. pipe and found that $\sqrt{\overline{s_x^2}}/\bar{S}_x = 0.32$

after applying an appreciable correction factor for spatial averaging. Hot wire measuremets by a number of investigators are plotted as the ratio of the intensity to the local mean velocity in Fig. 13. These should extrapolate to a value of the ordinate equal to $\sqrt{\overline{s^2}/S_x}$ at $y=0$. The large differences among the results of the different investigators probably reflect the difficulty of making anemometer measurements close to a wall.

7.3. LOCAL MASS TRANSFER

Local mass transfer measurements have been made for the fully developed concentration profile in a pipe (Shaw, 1964, Sirkar, 1969b). These are in good agreement with fully developed mass transfer coefficients obtained by examining the effect of the length of the mass transfer section on the rate of mass transfer (Son, 1967). This would seem to indicate that the method of isolating the test electrodes from the rest of the cathode does not disrupt the concentration boundary layer sufficiently to give erroneous results.

Dimopoulos (1868) has studied mass transfer to a 1 in. cylinder over the Reynolds number range of 60—360. His measurements of the local mass transfer rates are shown in Fig. 14. The concentration boundaty layer

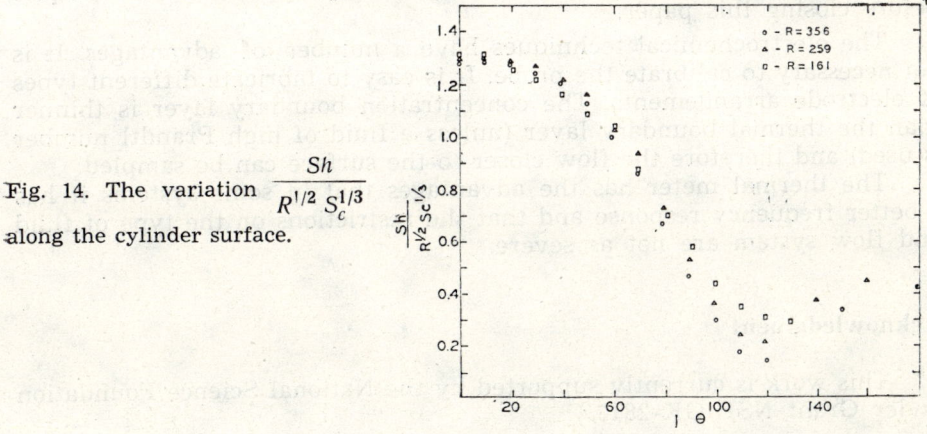

Fig. 14. The variation $\dfrac{Sh}{R^{1/2} S_c^{1/3}}$ along the cylinder surface.

starts at the front stagnation point and thickens as it proceeds around the cylinder. Therefore the mass transfer rate decreases with increasing θ. Dimopoulos also measured the variation of the wall velocity gradient around the surface of the cylinder for $R=247$, $R=293$, and $R=339$. These have been used to calculate the local mass transfer rates shown in Fig. 15. These calculations for the rear of the cylinder must be accepted with some reservation for the reasons cited in the paper by Dimopoulos. The comparison of measured local mass transfer rates on the front half of the cylinder at $R=356$ with the calculations is quite good. This self consistency of electrochemical measurements of local velocity gradients and of local mass transfer rates gives support to the accuracy of the techniques described in this paper.

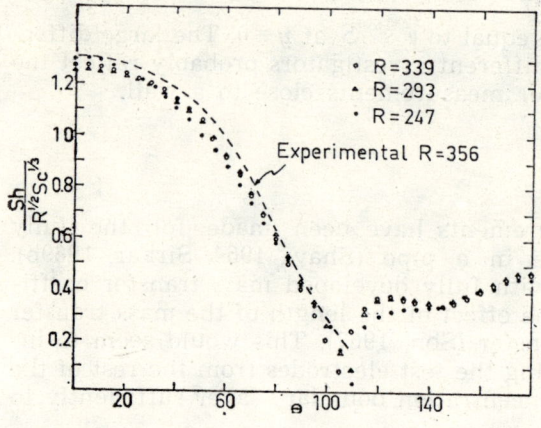

Fig. 15. Mass transfer variation around a circular cylinder predicted from measured shear stress distribution with splitter plate and boundary layer theory.

8. Concluding Remarks

The electrochemical measurement of the wall velocity gradient is the mass transfer analog of the thermal wall meter proposed by Ludwig (1949). It therefore seems to be desirable to compare these two techniques before closing this paper.

The electrochemical techniques have a number of advantages. It is not necessary to calibrate the probe. It is easy to fabricate different types of electrode arrangements. The concentration boundary layer is thinner than the thermal boundary layer (unless a fluid of high Prandtl number is used) and therefore the flow closer to the surface can be sampled.

The thermal meter has the advantages that in some systems it has a better frequency response and that the restrictions on the type of fluid and flow system are not as severe.

Acknowledgment

This work is currently supported by the National Science Foundation under Grant NSF GK-2813X.

REFERENCES

1. Bellhouse, B. J. and Schultz, B. L., J. Fluid Mech., 24, 379 (1966).
2. Dimopoulos, H. G. and Hanratty, T. J., J. Fluid Mech., 33, 303 (1968).
3. Eckelman, L. D., M.S thesis, University of Illinois, Urbana, 1968.
4. Gordon, S. L. and Tobias, C. W., M.S thesis, University of California, Berkeley, 1963.
5. Hanratty, T. J., A.I.Ch.E. J., 2, 359 (1956).
6. Hinze, J. O., *Turbulence*, McGraw-Hill, New York, 1959.

7. Karabelas, A. J. and Hanratty, T. J., J. Fluid Mech., 34, 159 (1968).
8. Ling, S. C., J. Heat Transfer, 685, 230 (1962).
9. Ludwieg, H., Translated as N.A.C.A. Tech. Memo., № 1284, 1950.
10. Mitchell, J. E. and Hanratty, T. J., J. Fluid Mech., 26, 199 (1966).
11. Reiss, L. P. and Hanratty, T. J., A.I.Ch.E. J. 8, 245 (1962).
12. Reiss, L. P. and Hanratty, T. J., A.I.Ch.E. J. 9, 154 (1963).
13. Shaw, P. V., Reiss, L. P., and Hanratty, T. J., A.I.Ch.E. J., 9, 362 (1963).
14. Shaw, P. V., and Hanratty, T. J., A.I.Ch.E. J. 10, 475 (1964).
15. Sirkar, K. K., and Hanratty, T. J., I and Eng. Chem. Fund., May issue, 1969a.
16. Sirkar, K. K., Ph. D. thesis, University of Illinois, Urbana, 1969b.
17. Son, J. S., and Hanratty, T. J., A.I.Ch.E. J., 13, 689 (1967).
18. Son, J. S., and Hanratty, T. J., J. Fluid Mech., 35, 353 (1969).
19. Spence, D. A., and Brown, G. L., J. Fluid Mech., 33, 753 (1968).

OPTICAL MEASUREMENT OF FLUID VELOCITY UTILIZING THE DOPPLER EFFECT

R. J. GOLDSTEIN

Heat Transfer Laboratory, School of Mechanical and Aerospace Engineering, University of Minnesota, Minneapolis, Minnesota, U.S.A.

1. Introduction

The Doppler shift is the change in frequency of radiation associated with relative motion between a source and a receiver of radiation. A similar frequency shift occurs when both the source and receiver are fixed, but the waves are scattered from a body in motion between them. The change in frequency relative to the source frequency is of the order of the speed of the moving object divided by the wave propagation speed.

Until recently the chief utilization of this phenomenon, with light waves, concerned the red shift in the emision spectra of astronomical bodies. This shift to a longer wave length indicates a motion of the source away from the earth. Measurement of the shift permits measurement of the relative motion of the source, and, along with the Hubble hypothesis, has permitted prediction of the distance of stellar bodies from the solar system.

The advent of highly monochromatic light sources, lasers, now permits the measurement of velocities considerably less than astronomical. This is largely due to the monochromaticity of lasers. Thus a gas laser may typically have a minimum bandwidth of about 10 Hz while more common monochromatic light sources have bandwidths of the order of 10^9 Hz. These may be compared to optical frequencies which are of the order of 5×10^{14} Hz.

A recent review of the application of the Doppler effect to measuring fluid velocity is contained in [1]. Optical measurement of fluid velocity offers many advantages over more conventional techniques. Perhaps the most important one is the absence of a probe in the flow, which could distort the flow field. This is often a problem in measurements near a fluid interface, in systems where rotation is present, and where there are time fluctuations in the fluid velocity, particularly those which might cause flow reversal. The absence of a probe is also beneficial in flows which are injurious to contained instruments such as high temperature gases.

Another advantage, as we shall see below, is that use of the Doppler effect enables measurement of the component of velocity in a particular direction. This is in contradistinction to other techniques which measure the velocity vector in a plane. Thus, using the optical method permits measurement of the complete velocity probability function, including the effect of reverse flows which might be encountered when the intensity of turbulence is large.

The Doppler technique permits extreme precision in determining fluid velocity. The accuracy is chiefly dependent on the accuracy of measurement of electrical frequency, which today is highly developed. Thus there have been measurements of velocity to within an accuracy of at least 0.1%. This is not necessarily the limit and thus optical systems should prove useful for calibration of velocity measuring instruments.

The laser-Doppler system gives essentially an absolute measurement of velocity. The output signal is dependent on velocity and not on the properties of the flowing fluid (other than the index of refraction, which can usually be considered constant). Thus it is independent of temperature or density, unlike most other velocity measuring devices. This is often helpful in fluctuating flows where the properties of the fluid may change with time making it quite difficult to separate, for example, fluctuations of temperature and velocity.

2. Frequency Change Associated with Radiation Scattered by a Moving Particle

The classical Doppler shift is easily derived for a fixed source and moving detector of radiation or for a moving source and fixed detector. When waves leave a fixed source and are received by a fixed detector but are scattered by some scattering center in between the relationship is somewhat more complicated. It thus appears advantageous to present a heuristic non-relativistic derivation of this frequency shift.

The derivation can best be understood by reference to Fig. 1 in which radiation is scattered by a moving particle. The impinging radiation is emitted by a stationary source with frequency ν_i, wavelength λ_i, and the speed of the radiation is c. The unit vector $\hat{n}_i$ defines the direction of this incoming radiation which strikes the scattering center (particle), having the velocity $\vec{V}$.

The number of wavefronts passing the scattering center per unit time would be $\nu_i \left(= \dfrac{c}{\lambda_i} \right)$ if the particle were motionless. However, the scattering center appears, from a non-relativistic viewpoint, to recede from the incoming illumination with the speed $\vec{v} \cdot \hat{n}_1$. Thus to the scattering center the incoming waves appear to have a speed $\vec{c} - \vec{v} \cdot \hat{n}_i$. Therefore the number of wavefronts incident upon the scattering center per unit time, which is also the apparent frequency of the radiation to the particle, is

$$\nu_p = \frac{c - \vec{v} \cdot \hat{n}_i}{\lambda_i} \qquad (1)$$

This is also the number of wavefronts per unit time scattered by the moving particle.

Next consider a fixed detector towards which a moving particle is emitting (or scattering) radiation in the direction assigned by the unit

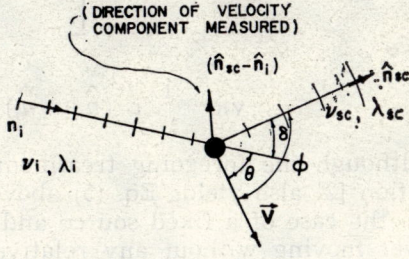

Fig. 1. Light scattering by a moving particle.

$$\nu_p = \frac{c - \vec{v} \cdot \hat{n}_i}{\lambda_i}$$

$$\lambda_{sc} = \frac{c - \vec{v} \cdot \hat{n}_{sc}}{\nu_p} = \lambda_i \left(\frac{1 - \frac{\vec{v} \cdot \hat{n}_{sc}}{c}}{1 - \frac{\vec{v} \cdot \hat{n}_i}{c}} \right)$$

$$\nu_D = \nu_{sc} - \nu_i = c \left(\frac{1}{\lambda_{sc}} - \frac{1}{\lambda_i} \right) = \frac{1}{\lambda_i} \left(\frac{\vec{v} \cdot (\hat{n}_{sc} - \hat{n}_i)}{1 - \frac{\vec{v} \cdot \hat{n}_i}{c}} \right)$$

$$\nu_D \simeq \frac{1}{\lambda_i} \vec{v} \cdot (\hat{n}_{sc} - \hat{n}_i) = \frac{n}{\lambda_0} \vec{v} \cdot (\hat{n}_{sc} - \hat{n}_i)$$

vector $\hat{n}_{sc}$; the number of wavefronts emitted or scattered per unit time being ν_p. After the scattering of one wavefront in the direction $\hat{n}_{sc}$ the particle moves toward that wavefront with the speed $\vec{v} \cdot \hat{n}_{sc}$. Thus when the next wavefront is scattered after a time interval given by $1/\nu_p$ the first wavefront is a distance from the particle given by

$$\lambda_{sc} = \frac{c - \vec{v} \cdot \hat{n}_{sc}}{\nu_p} \qquad (2)$$

This is the apparent wavelength of the scattered radiation to a fixed detector and can be written in the form

$$\lambda_{sc} = \lambda_i \left\{ \frac{1 - \frac{\vec{v} \cdot \hat{n}_{sc}}{c}}{1 - \frac{\vec{v} \cdot \hat{n}_i}{c}} \right\} \qquad (3)$$

The difference in the frequency of this radiation from the initial radiation is called the Doppler shift or Doppler frequency, ν_D

$$\nu_D = \nu_{sc} - \nu_i = c\left(\frac{1}{\lambda_{sc}} - \frac{1}{\lambda_i}\right) \qquad (4)$$

or

$$\nu_D = \frac{1}{\lambda}\left[\frac{\vec{v}\cdot(\hat{n}_{sc} - \hat{n}_i)}{1 - \frac{\vec{v}\cdot\hat{n}_i}{c}}\right] \qquad (5)$$

For $v \ll c$

$$\nu_D \simeq \frac{1}{\lambda_i}\vec{v}\cdot(\hat{n}_{sc} - \hat{n}_i) = \frac{n}{\lambda_o}\vec{v}\cdot(\hat{n}_{sc} - \hat{n}_i) \qquad (6)$$

Although the foregoing treatment is not relativistic, the complete derivation [2] also yields Eq. (5) above. Thus it is interesting to observe that in the case of a fixed source and observer or even for a source and observer moving without any relative motion to one another [3], there is no transverse Doppler effect even considering relativistic effects.

If the directions of the incoming and scattered light beams are fixed, which would normally be the case, the frequency shift indicated by Eq. (6) is proportional to the component of the velocity in the direction given by the difference in the unit vectors of the scattered and incoming radiation. This differs from measurements taken with most types of velocity instrumentation and greatly enhances measurement of velocity fluctuations in a given direction, as well as the direction of the velocity.

The Doppler frequency may be measured in several ways, perhaps most directly using an optical heterodyne receiver. The photocathode of a photomultiplier acts as a square law detector. Thus if two optical signals are superposed on a photocathode, the output includes a term with a frequency equal to the difference between the frequencies of the two signals. Using laser illumination, which is very monochromatic, and putting both the scattered signal and direct rays from the original laser on the photocathode, a beat signal will result with a frequency equal to the difference of the frequencies of these two waves, namely the Doppler frequency. Measurement of this Doppler frequency, the angle of incidence and scattering of the light beam and knowledge of the index of refraction of the fluid and the vaccum wavelength of the light source will permit the specification of the component of the scattering center velocity in a specific direction, the direction of the difference vector of the unit vectors of the scattering and incident beams.

Although the heterodyne principle is the one that has been most widely used for measuring the Doppler frequency, other techniques are possible. Thus a Fabry-Perot interferometer can be used to measure the difference in wavelength or frequency of two optical beams [4]. This is accomplished by changing the length of the cavity of a Fabry-Perot interferometer. Since only resonating waves are effectively transmitted, the measurements of output light signal as a function of cavity length permits measurement of small wavelength or frequency differences.

3. Measurement of Fluid Velocity

As noted above, it is actually the velocity of scattering centers, usually particles, that are measured, utilizing the Doppler effect. The velocity at any point in a flowing fluid can be determined from the motion of scattering particles carried with the fluid. The particles used as scattering centers should be very small, neutrally bouyant, and in relatively dilute concentration. Often the flow disturbances introduced by particles can be made negligible and the particles can almost exactly follow the flow of the fluid. Then all the information concerning the velocity of the fluid at the point observed is contained in the frequency of the scattered illumination. An apparatus can be designed to measure a single velocity component or the full velocity vector by determining components in three different directions.

For studies in water, half-micron polystyrene spheres have often been used as scattering centers. Polystyrene has a specific gravity of approximately 1.04 and if necessary, additives can be placed in solution in the water to increase its density to this point. Usually, however, when the velocities are not high, this is not necessary and the particles will still follow the flow quite accurately. Typical particle concentrations have been of the order of 1:50,000. In air flows dust or smoke particles can be added to serve as scattering centers.

Usually the Doppler shift is determined by recombining the scattered (Doppler-shifted) beam with the reference beam from the same laser on the photocathode of the photomultiplier. The resulting photomultiplier current is amplitude modulated at the Doppler frequency. With suitable electronic processing the average or mean velocity can be obtained quickly, as well as the velocity probability function, the turbulence intensity, and with suitable optics the scale of turbulence, Reynolds stresses, etc.

In one recent study the velocity could be detected audibly. Since the velocities were low, the Doppler frequency shifts were in the audio range. The output from the photomultiplier was amplified and sent to a loudspeaker. Fluid velocity changes could be detected from the changes in pitch.

Equation (6) can be rewritten to contain the Doppler frequency in terms of the angles shown in Fig. 1. Thus,

$$\nu_D = \frac{2nv}{\lambda_o} \sin\frac{\delta}{2} \sin\left(\theta + \frac{\delta}{2}\right) \tag{7}$$

If the direction of the velocity is known, as is often the case, it is usually convenient to set $\theta + \delta/2 = \pm \pi/2$. Then the vector $\hat{n}_{sc} - \hat{n}_i$ is in the same direction as or directly opposite to the velocity. With this arrangement there is also symmetry of the incoming scattered radiation about the velocity vector which facilitates system alignment and also minimizes the signal bandwidth [1].

A typical arrangement of the apparatus is shown in Fig. 2. Light from a continuous wave laser is incident upon a partially coated splitter plate where there is division of amplitude into an illuminating or scattering beam and a reference beam. Both beams are focused on the same region

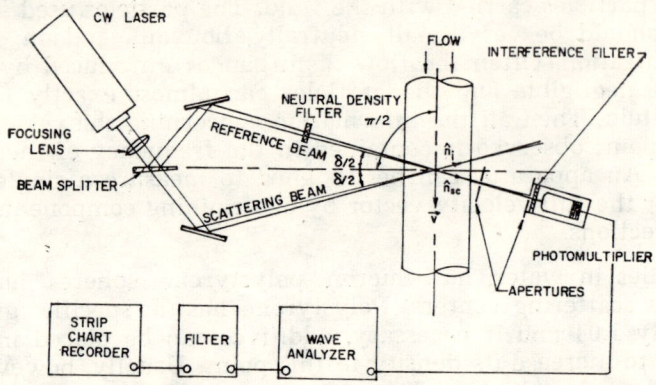

Fig. 2. Arrangement of apparatus for measurement of velocity from Doppler shift of scattered laser radiation

in the flowing fluid by the lens of focal length f. The reference beam is reflected by a mirror, passes through a neutral density filter (to prevent overloading of the photomultiplier), into the test section and then through apertures to the photocathode of the photomultiplier. The illuminating beam is scattered in the test section in all directions. However, only the light scattered from the focus toward the apertures before the photomultiplier will be incident upon the photocathode. The resultant output of the photomultiplier contains a beat signal with a frequency of the Doppler shift. The photomultiplier output can be analyzed directly by a spectrum analyzer. The output of the analyzer in general is a signal which indicates amplitude as a function of frequency which is often directly related to the velocity probability function [5]. For highly turbulent flows at high velocity other electronic measurement techniques can prove useful [6].

The geometry shown in Fig. 2 is one of many optical systems that can be used to superpose the reference and scattered beams onto the photomultiplier. This particular design has certain advantages, including equal path lengths for the two beams and simple visual detection of the point at which velocity measurement is being made. The initial alignment of the beams is simplified by observing the output signal of the photomultiplier on an oscilloscope.

Typical outputs of the spectrum analyzer are shown in Fig. 3. Note the relatively smooth curve obtained for a laminar flow. Ideally, if there were no fluctuation in the velocity and zero velocity gradient, one might expect to get a very narrow signal from the output of the spectrum analyzer. However, in general there is a finite signal width, having a magnitude dependent on the optical geometry. This will be discussed below. If there were a velocity gradient, then the signal width might be increased due to the finite region over which scattering takes place. If the flow were not steady with time, as with transition or turbulent flow, the signal would be expected to represent in some manner the velocity probability function. If a sufficiently long average were taken, these factors could be used to measure, among other things, the intensity of turbulence in a flow. Figure 3 shows typical outputs from the spectrum analyzer for flow in a circular tube. As the flow goes into transition the signal broadens and has a number of large spikes. In fully turbulent flow the signal is broad, representative of the velocity probability function and not smooth as is true for laminar flow.

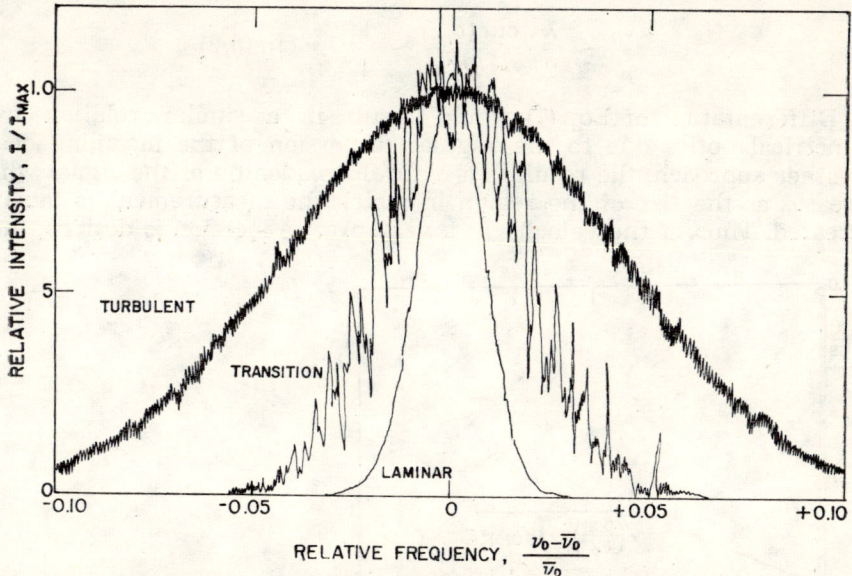

Fig. 3. Spectrum analyzer output for different flow regimes.

The size of the region from which scattering takes place can be estimated by considering the region at the focus of the beam. If the original beam from the laser were of uniform intensity (it actually varies with a Gaussian distribution), then upon focusing through a fixed aperture lens, the result would be an Airy disc at the focal point. If the aperture of the lens is A and its focal length f, then the focal region [7] is represented basically by a depth of field (or length along light beam) of $4\lambda (f/A)^2$ and the diameter of the Airy disc, $d_2 \sim 2.44\lambda (f/A)$. This gives some estimate of the volume of the region from which the velocity measurement is taken, although a full consideration must include the optics from the point of scattering to the photocathode [1].

For a focal length of 5 cm and an aperture (usually the laser beam diameter) of the order of 3 mm, the minimum dimension of the region observed would be approximately 20 μm. The bandwidth of the signal for a steady, uniform flow can be estimated using the dimension d_2 considering the light scattered from a particle as it passes through the region of the focus. Since the particle is present for only a short period of time, it can only scatter a wave train of finite length. A wave train from a monochromatic source having a duration Δt does not appear to be a perfect sine wave but has an effective frequency range of the order of $1/\Delta t$. The time Δt is approximately the time in which a particle passes through the region of the Airy disc.

$$\Delta t \sim \frac{d_2}{v \sin \theta} \tag{8}$$

and using Eq. (7) for $\left(\theta + \dfrac{\delta}{2}\right) = \pi/2$;

$$\frac{\Delta \nu_D}{\nu_D} \sim \frac{\lambda}{2} \frac{\operatorname{ctn}(\delta/2)}{d_2} \sim \frac{1}{4} \frac{A}{f} \operatorname{ctn}(\delta/2) \tag{9}$$

Differentation of Eq. (7) would result in a similar relation from geometrical optics due to the angular dispersion of the incoming beam. By either approach, the result is a natural broadening of the signal which increases as the size of the region in which the measurement is taken is decreased. Thus, if the velocity at a very precise location is desired, there

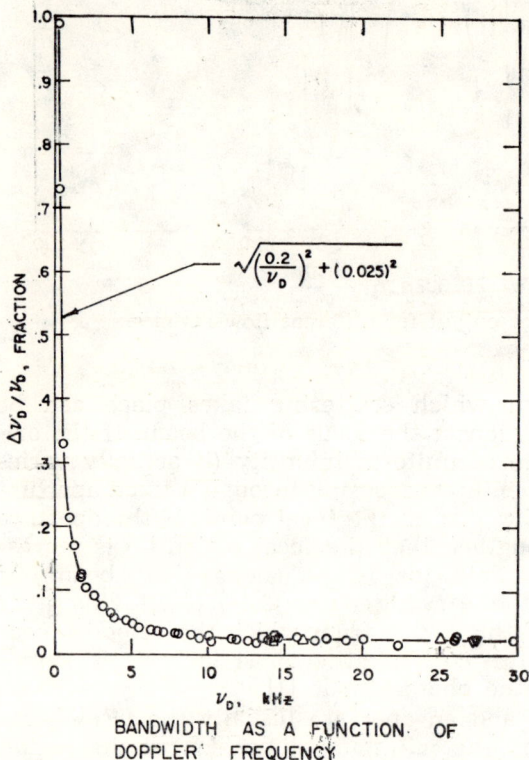

Fig. 4. Doppler signal bandwidth variation.

is an inherent increase in the width of the signal, while conversely, if a narrow bandwidth is required then the dimension of the volume element over which the measurement is made must be made sufficiently large.

Figure 4 shows the variation in signal bandwidth of a spectrum analyzer output for laminar flow in a square duct. At very low velocity and frequency the bandwidth is very large. This is due not only to the broadening described above, but is primarily produced by the frequency bandwidth of the laser and spectrum analyzer. As the velocity (frequency) is increased, the bandwidth decreases to a constant value in good qualitative agreement with the prediction (Eq. 9). In the apparatus used to obtain the figure, $f = 193$ mm, $\delta/2 = 18.5^0$, and the laser beam diameter is nominally 4 mm.

Figure 5 is a photograph of a stored oscilloscope trace of the photomultiplier wave analyzer output. The frequency of the Doppler signal is about 6.5 kHz (velocity $\sim$ 6.5 mm/s) and the volume of fluid from which the signal emanates has, on the average, 525 scattering centers. The time

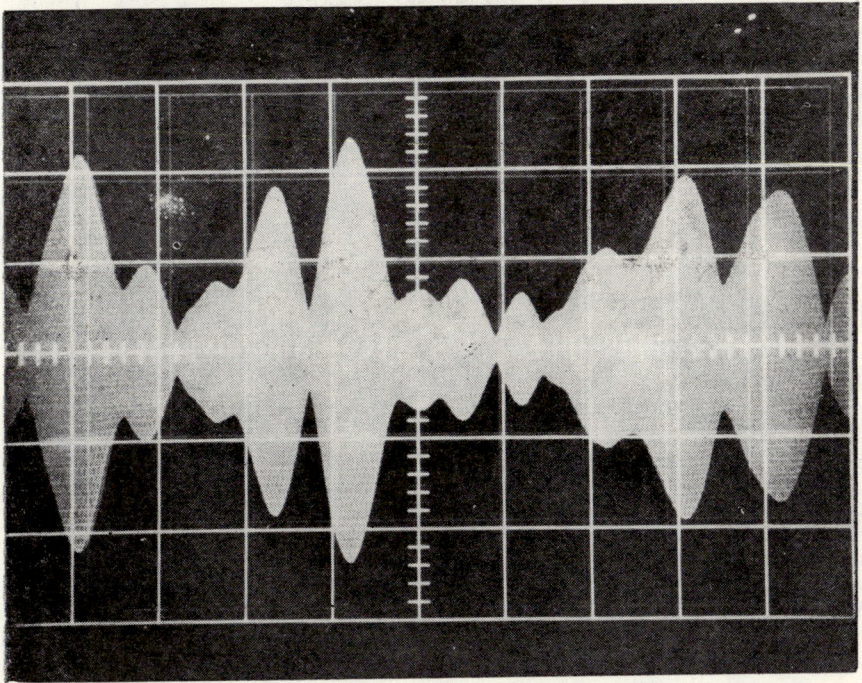

Fig. 5. Oscilloscope trace of amplitude variation of Doppler signal.

scale on the trace is about 2 seconds for a full sweep so that the Doppler frequency itself is not visible, only its amplitude. The signal indicates the effect of the addition of many Doppler signals arising from individual particles.

Figure 6 shows results of measurement of the longitudinal velocity fluctuation for turbulent flow in a circular tube. If a sufficiently long average of the output of the spectrum analyzer is taken the result is essentially a velocity probability function modified by the bandwidth of the instrument. This bandwidth can be obtained either by calculation (of Eq. 9) or by measurement in a steady flow. In the data used to obtain Fig. 6 both the laminar flow and the turbulent flow signals were approximately Gaussian. Thus it is quite easy to subtract the instrument bandwidth from the measured signal distribution to get the turbulence intensity.

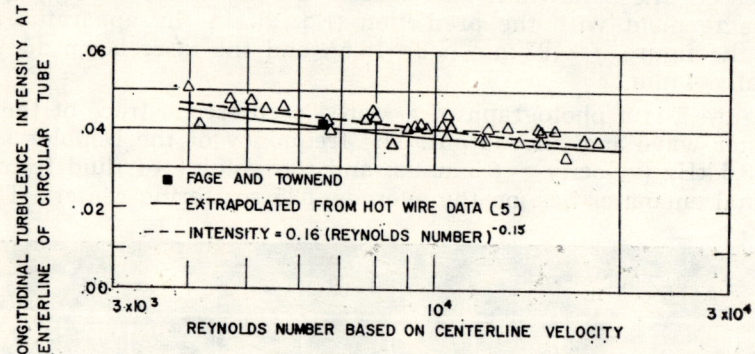

Fig. 6. Turbulence intensity in circular tube.

There are limitations to the velocity range over which the laser-Doppler system can be used. The smallest bandwidth from a cw laser is approximately 10 Hz. Application of Eq. (6) indicates that the minimum velocity that one could reasonably measure would be of the order of 10^{-3} cm/sec. At high velocity there are two difficulties: one is the peak frequency response of the photomultiplier ($\sim$ 500 MHz), and the other is the presence of longitudinal modes in the laser output. A laser has longitudinal modes at a spacing of $c/2L$. For a cavity length, L, of 1 meter the separation would be approximately 150 MHz. This corresponds to a velocity of the order of 100 m/s. Actually, by proper choice of the scattering angle this velocity could be considerably increased.

Another technique which is useful at high speeds involves the frequency shift of the reference light beam by means of an acoustical Bragg--refraction modulator [8]. A typical modulator is shown in Fig. 7. A traveling wave is generated in the fluid by the crystal, the waves being absorbed or scattered at the far end of the cell. The light beam enters and is taken out at the first order Bragg refraction angle with a frequency shift of narrow bandwidth at the frequency of the crystal.

There are many applications in which single sideband modulation of one of the light beams can prove valuable. Consider a fluid whose velocity passes through zero, that is, it reverses in direction. This would be encountered in systems that have a high intensity of turbulence or perhaps in separated flow when measuring the velocity at different positions. It has often proved difficult to distinguish the positive component of velocity in a given direction from the negative component. Equation (6) indicates the sign of the frequency shift and depends on the sign of the velocity component. When measuring frequency changes from zero, the sign of the shift is not detectable. However, if the reference beam frequency is shifted such that at zero velocity there is a frequency difference between the reference and the scattered beam it would be possible to detect positive and negative Doppler shifts (and velocity components).

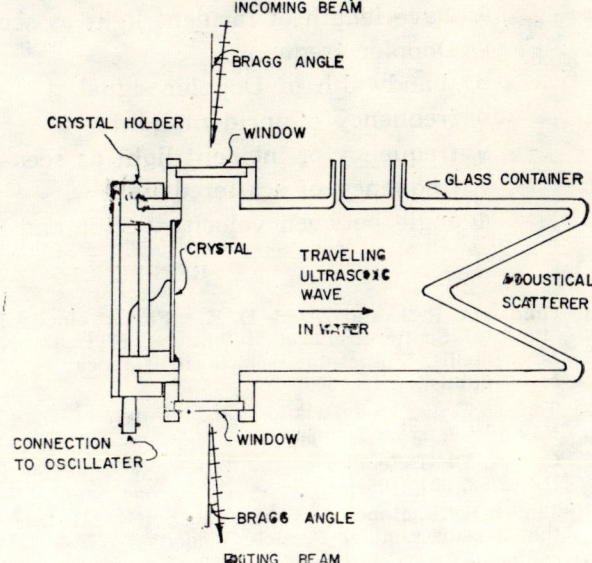

Fig. 7. Ultrasonic wave diffraction cell.

Acknowledgement

R. Adrian and A. Garon aided in the preparation of the figures. Support for the research on which this article is based came from National Science Foundation Grant GK 1737 and Project Themis.

NOMENCLATURE

A aperture diameter
c speed of light
d_2 diffraction limited spot size of a focused beam
f focal length of lens
n index of refraction
$\hat{n}_i$ unit vector in direction of incident light
$\hat{n}_{sc}$ unit vector in direction of scattered light
$\vec{v}$ particle velocity
v magnitude of velocity
δ scattering angle $= \Phi - \theta$
θ angle between velocity vector and incident light beam

λ wave length
λ_i wave length of incident light
λ_0 vacuum wave length of fixed source (laser)
λ_{sc} wave length of incident light as seen by the particle
ν_D Doppler frequency
$\Delta\nu_D$ bandwidth of Doppler signal
ν_i frequency of incident light
ν_p frequency of incident light as seen by the particle
ν_{sc} frequency of scattered light
Φ angle between velocity vector and scattered light beam

REFERENCES

1. Goldstein, R. J. and Kreid, D. K., »Fluid Velocity Measurement from the Doppler Shift of Scattered Laser Radiation«, HTL TR № 85, University of Minnesota, Minneapolis, Minnesota, 1968. (also to appear in *Measurements in Heat Transfer*, to be published in 1969).
2. Temes, C. L., »Relativistic Consideration of Doppler Shift«, IRE Trans. Aero & Navig. Elect., **6**, 37 (1959).
3. Censor, D. »Detection of the Transverse Doppler Effect with Laser Light«, Proc. IEEE, **52**, 987 (1964).
4. James, R. N., Babcock, W. R., and Seifert, H. S., »A Laser-Doppler Technique for the Measurement of Particle Velocity«, AIAA J., **6**, 160 (1968).
5. Goldstein, R. J and Hagen, W. F., »Turbulent Flow Measurements Utilizing the Doppler Shift of Scattered Laser Radiation«, Phys. of Fluids, **10**, 1349 (1967).
6. Rolfe, E., Silk, J. K., Booth, S., Meister, K., and Young, R. M., »Laser Doppler Velocity Instrument«, NASA Cr-1199, 1968.
7. Born, M. and Wolf, E., *Principles of Optics*. Second Edition, Pergamon Press, Oxford, 1964.
8. Cummins, H., Knable, N., Gampel, L. and Yeh, Y., »Frequency Shifts in Light Diffraction by Ultrasonic Waves in Liquid Media«, Appl Phys. Letters, **2**, 62 (1963).

TURBULENCE MEASUREMENT BY OPTICAL METHODS

A. M. TROHAN

Institute of Physico-Technical and Radiotechnical Measurements, Moscow, USSR

1. Introduction

Interest in the application of optical methods in turbulence measurement is increasing steadily. This comes from the fact that optical means offer important advantages over other measurement techniques in turbulence research. In comparison with the methods using probes which are introduced in the fluid, such as; thermoanemometers, inductive and electric discharge anemometers, Pitot tubes, etc., the advantages of the optical means are the absence of the flow disturbances and the possibility of conducting measurements in high temperatures and at high velocities. Compared with the other probeless methods such as ultra-sonic and superhigh frequency oscillation methods and the use of radioactive isotopes, optical techniques offer an important advantage in having high space resolution.

Optical methods for turbulence measurement can be used in liquid and gas flows, flames and plasmas. Although some of them have been known for a long time (e.g. diffusion methods) most of the optical methods have been developed and applied only recently. The optical methods which are now used in turbulence investigations could be divided into three main groups: diffusion methods, densiometric methods and anemometric methods.

2. Diffusion methods

Methods of this group are based on the dependence of the diffusion of the micro or macro particles dispersed in the fluid on the turbulence characteristics. Diffusion methods in turbulence research were first applied by Reynolds. He used the sharp increase of the diffusion rate of paint in water as a means of detecting the transition from laminar to turbulent flow.

In the investigations of gas flows diffusion of heat, generated by electric discharge [1] or heated wires [2], is employed. A heated gas cloud is observed by the corresponding optical means (interferometer, Thoeppler or schlieren method). Instead of by gas heating, the schlieren could be obtained by the introduction of another gas having a different index of refraction as compared to the investigated medium. However, owing to the simplicity of its generation and detection, the most commonly used diffusion »sonde« in gas flows is smoke [3]. For turbulence investigations in water, paints [4], fluorescent mixtures [5], and salt solutions [6] are widely used.

The analysis of the measurements by diffusion methods is usually based on the relation first introduced by Campé de Férié [7] in uniform turbulent flows:

$$\overline{y_2^2}(t) = 2v_2'^2 \int_0^t (t-\tau) R_L(\tau) d\tau \qquad (1)$$

where: $y_2(t)$ — Lagrange coordinate of a marked particle in Cartesian system;
v_2' — Lagrange turbulent velocity fluctuation;
R_L — Lagrange correlation function.

The merits of the diffusion methods in trubulence research are: relative simplicity of the required equipment, intelligibility, and the variety of the measureable characteristics. Disadvantages of these methods are: the complexity of the mathematical evaluation of the measurement results, low space resolution, and poor accuracy, caused by disturbances of the flow by the probe introducing marking substance and by the dimensions of the source.

3. Densiometric methods

These methods are based on the detection of the turbulent density fluctuations and the corresponding variation of the refraction index, on the detection of the density gradient or the second derivative of the density, or the detection of the variations in the concentration of the additives.

One of the unfavorable characteristics of these methods is that the observed optical effects could depend simultaneously on several parameters of the investigated medium which makes the interpretation of the results difficult. For instance, the fraction of the passing light flux, scattered by the elementary medium volume, depends simultaneously on the values of the gradients of the refraction coefficient, concentration and composition of the medium, concentration and dimensions of the marked particles, etc.

Densiometric methods could be divided into three subgroups, with no sharp division between them, namely: a) methods using passing light; b) methods based on the scattered light detection; and, c) methods based on the detection of self-radiation. The methods of the first subgroup allow the detection of the values averaged over the path of observation and the methods belonging to the second and the third subgroups allow the detection of the local values.

3.1. PASSING LIGHT METHODS

This group includes interferometric, schlieren, Thoeppler and absorption methods.

By interferometric methods it is possible to measure integral density fluctuations along the observation line z. The interference band shift S equals the length change of the optical path measured in light wave lengths λ:

$$S = \frac{1}{\lambda} \int_z [n(x, y, z) - n_0] \, dz \qquad (2)$$

where: $n(x, y, z)$ and n_o are refraction indices corresponding to the turbulent and the undisturbed flow, respectively.

The gas density ρ is related to its refraction index by the Gladston-Dale equation:

$$n - 1 = (n_0 - 1)(\rho/\rho_0) \qquad (3)$$

Thus:

$$S = \frac{n_0 - 1}{\lambda \rho_0} \int_z [\rho(x, y, z) - \rho_0] \, dz \qquad (4)$$

As a consequence of the averaging along the beam, pulsation spectrum recorded by an interferometer is narrowed in comparison with the actual spectrum, analogous to the case when the measurement is done by a long hot-wire.

The interference band shift is registered by photographic or photoelectric means.

The advantages of the interferometric methods are the linear dependence of the signal on the fluid density and the high sensitivity of these methods. For instance, detection of the variations of the refraction index of the order of $\Delta n = 10^{-7}$ to 10^{-8} is possible by using an interferometer of the type Fabri-Pero in combination with a laser.

However, it is often convenient in turbulence research to use also the Thoeppler or schlieren methods which detect the second or the first derivative of the refraction coefficient in the direction perpendicular to the light beam. These methods are advantageous because the turbulent density fluctuations are usually small in magnitude. However, the scale of unhomogeneities are also small so that the gradient and the higher order derivatives of the density are significant. In applying these methods, photographic and photoelectric registration is also possible.

If the medium through which the monochromatic light beam is passing absorbs the radiation, then the fluctuations of the radiation intensity leaving the medium give information on the concentration of the absorbing substance. In most cases the Baire absorption law may be used:

$$\Phi = \Phi_o \exp\left[-c \int_0^z \alpha \, dz\right] \qquad (5)$$

where Φ and Φ_o are the entering and the exiting radiation intensity, respectively; c is the concentration of the absorbing substance; α is the absorption coefficient; and z is the length of the light path.

By using corresponding resonance radiation, it is possible to determine concentration fluctuations of the separate components of the medium. In the case of a small correlation scale in comparison with the flow scale, the local fluctuations could be determined by correlating two different light beams which intercept each other in the media.

3.2. SCATTERED LIGHT AND SELF-RADIATION METHODS

In turbulence research use is made also of the scattered waves which appear when electromagnetic radiation is passing through a turbulent flow.

In laminar flow the optical characteristics of the flow are smooth functions of the hydrodynamic parameters. The appearance of instabilities leading to fluctuations results in the fluctuations of the scattered light. The frequency spectrum of the scattered light fluctuations evidently corresponds to the frequency spectrum of the turbulent pulsations of the refraction coefficient or to the spectrum of the other optical unhomogeneities. The ratio of the scattered and the incident light fluxes allows the determination of the density at the given point in the flow in function of the time. The two light fluxes are usually registered by two-beam oscillographs and subsequently analzyed.

Relative simplicity, and high space and time resolutions are advantages of the scattered light methods. However, the interpretation of the measurements is strongly dependent on the conditions of the investigated flow. The method was used by a number of investigators in liquid as well as in gas flows.

In the turbulence investigations of plasmas and flames the self--radiation of the media, visible or infrared, could be used [9]. Corresponding frequency and correlation analysis of the registered signals allows the determination of the frequency spectrum and the correlation scale. In the analysis of infrared radiation the pulsations of its intensity could come from the temperature fluctuations, also. In the analysis of visible radiation, complications could come from the influence of the space fluctuations of the concentration of the radiating substances. This particularly occurs in the case of technical arc plasmas and flames.

The space resolution of the methods using self-radiation is lower in comparison with the space resolution of the methods using scattered light. The space resolution is determined by the sharpness depth of the optical system.

An interesting possibility which might be exploited in the investigations of gas flows at lower pressures is the use of the fluorescence provoked by fast electron beams [10]. The measurement circuit is similar to the one used in scattered light methods. In fact, these two methods are complementary in the sense that the scattered light method is used with high density fluids and the fluorescent methods with low density fluids.

4. Anemometric methods

Anemometric methods in turbulence research are based on the direct measurement of the velocity fluctuations. Unlike the above-mentioned methods, signals obtained by the anemometric methods depend on one parameter only — on the velocity. This is a great advantage in the interpretation of the results. Anemometric methods could be of two kinds, either spectral or kinematic.

4.1. SPECTRAL METHODS

The spectral methods are based on the Doppler:

$$\frac{\Delta \nu}{\nu} = \frac{v}{c} n \cos \varphi \qquad (6)$$

and Fizeau effect:

$$\frac{\Delta \nu}{\nu} = \frac{v}{c} \left(1 - \frac{1}{n^2}\right) \qquad (7)$$

where: ν — observed frequency;
$\Delta \nu$ — difference between observed frequency and the frequency emitted by the source;
v — velocity of the emitting body
c — light velocity;
n — refraction index;
φ — angle between the velocity vector and the direction of observation.

Self-radiation, scattered radiation and passing radiation could be used for the velocity measurement and the determination of its time variation.

On the basis of the frequency shift determination, spectral methods could be subdivided into spectroscopic and optical heterodyning methods. In the first case, a spectral instrument is used in the dispersing element of which radiation frequency shift is transformed into a space shift of the spectral lines. Today, minimum absolute error in measuring frequency shifts by the spectroscopic methods is from 10 to 15 m/sec. Therefore, this method is used for the investigations of high-velocity flows of gases and plasmas.

The optical heterodyning method is based on the recording of beat frequencies of the two coherent light beams, one of which is the reference beam and the other is either passing through the medium or is scattered by it.

The use of the Fizeau effect permits the determination of the integral values of the velocity fluctuations along the observation line [11], as in the case of the passing light methods. The use of the scattered light permits the determination of the local values.

4.2. KINEMATIC METHODS

Kinematic methods in turbulence measurement are based on the time-of-flight analysis of the movement of optical unhomogeneities present in the investigated medium. Natural fluctuations of the refraction coefficient, density or radiance could be employed for this purpose as well as the floating particles introduced into the flow. Two possibilities of measurement exist: one, when the unhomogeneities which are measured are discrete and the other, when they have a continuous character.

The velocity of the floating particles could be measured visually (so-called ultramicroscope method) [13] or by fast movie cameras [14]. Subsequent statistical analysis of the measurement results permits the determination of the turbulence intensity [15]. However, this procedure is extremely laborious and of a low accuracy due to insufficient statistics.

4.2.1. Time-of-flight Method

We have suggested a method of turbulence measurement based on the statistical analysis of the photoelectric signals coming from the two closely situated points in the flow in which discrete optical unhomogeneities are present [16]. These points (small volumes) are displaced from eachother by a known distance in the flow direction, so that a fraction of the particles which passed through the first point (volume) passes also through the second.

The scheme of the experimental set-up is shown in Fig. 1. Radiation coming from a laser (1), having passed through a divider (2) and the lenses (3 and 4), enters the investigated medium (5). Scattered light from the

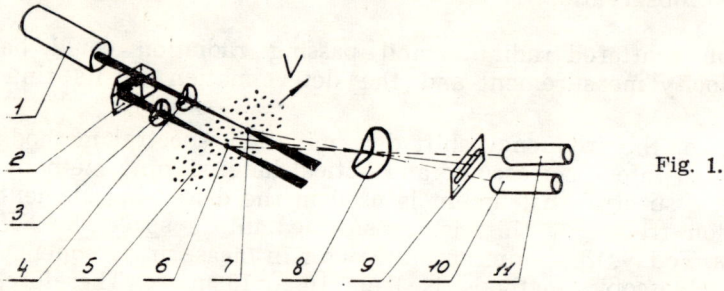

Fig. 1.

areas (6) and (7) when floating particles pass through them, is focussed by an objective onto the diaphragm (9) which determines the size of areas (6) and (7). Subsequently, the light falls on photomultipliers (10) and (11). The signals from the photomultipliers are analyzed.

If a certain particle having passed through the first area subsequently passes through the second, a time shift T results between the impulses provoked by the particle in the two photomultiplier channels. The time shift T is determined by the particle velocity on its way from the first to the second area (measurement basis).

If there are numerous flights over the time basis of the different particles, then for each individual flight:

$$T = \overline{T} + \Delta T \tag{8}$$

The mean time of flight is determined by the mean velocity of the particles; or, if the movement of the marking particles corresponds to the movement of the investigated fluid, the mean-time of flight is determined by the mean flow velocity along the measurement basis:

$$\overline{T} = L/\overline{V} \tag{9}$$

Dispersion of the time of flight is determined by:

$$\overline{\Delta T^2} = 2\delta^2 \int_0^{\overline{T}} (\overline{T} - \tau) R_L(\tau) d\tau \tag{10}$$

where: δ — turbulence intensity;

R_L — Lagrange correlation coefficient.

For small values of τ we have $R_L(\tau) \approx 1$, so that:

$$\overline{\Delta T^2} \sim \delta^2 \overline{T}^2 \tag{11}$$

i.e. if the measurement basis is less than the turbulence scale, time of flight dispersion is proportional to the square of the turbulence intensity. For large τ, when $\overline{T} \gg \tau^*$, where τ^* is a time interval for which $R_L(\tau^*) = 0$, we have:

$$\overline{\Delta T^2} \sim 2\delta^2 \tau_L \overline{T} \tag{12}$$

where τ_L is Lagrange time scale.

Thus, determination of the mathematical expectation and the time of flight dispersion permits the determination of the mean velocity and the turbulence intensity. Variations of the measurement basis and of the position of the second area allow the determination of the turbulence scale and the mean direction of the velocity vector.

It could be considered that the movement of the marking particles corresponds to the movement of the fluid when the characteristic size of the particles is substantially smaller than the turbulence scale and when its density is equal to the flowing fluid density. The first of these conditions could be easily fulfilled. The second could be fulfilled only in the case of the liquid flows when some emulsions having the same density as the flowing fluid is used. Therefore, the accuracy of the measurements in gas flows is determined by the degree of correspondence of particle movement and fluid movement, which is one of the basic problems in gas flow investigations.

The optical time-of-flight method was used for the investigations of flames [17], plasmas and cold gas flows. As was shown in these investigations, the optical time-of-flight method could be used for the measurement of mean velocity and turbulence intensity during investigations of gas flows having a temperature of three to four thousand degrees and a velocity up to 1 km/sec, with a measurement base of the order of 1 mm or less.

For instance, these investigations have led to better understanding of the laws of turbulence generation and dissipation provoked by combustion in homogeneous mixtures. These investigations have also led to the determination of the dependence of these values on the parameters of the mixture. As an example, Fig. 2 gives typical measurement results of the

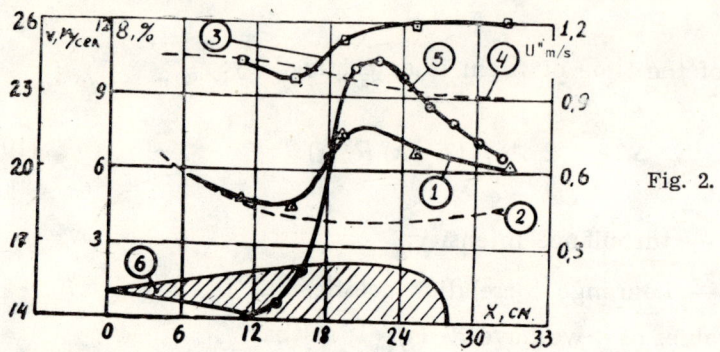

Fig. 2.

flow characteristics in the axis of combustion produced by a mixture of 4.5 percent propane and air flowing in a quadratic cross section channel. The measurements have been made by using the optical time-of-flight method with the measurement basis equal to 1.1 mm (on the abcissa is given the distance along the flow; flow is from left to right).

A comparison of the results obtained by a time-of-flight method with the results obtained by a thermoanemometer, in cases when a thermoanemometer could be used (cold gas), showed a satisfactory agreement. The measurement error in applying optical time-of-flight might be less than one to two percent in a broad range of the flow parameters.

In the investigations of flows of high temperature selfluminous gases, especially in the presence of high accelerations when a minimum size of the tracer particles is required, the background radiation influence becomes significant. In these conditions we have recommended detection of the tracing floating particles by X-ray radiation provoked by illuminating the particles with fast electron beams. Self-radiation of the medium in the X-ray range is absent, except in the mediums with decaying elements, and the absorption of the radiation coming from the particles could be very low because of its intensity. This method of detection is very suitable for the investigation of highly luminous, optically opaque flows as well as for flows inside optically opaque walls.

The use of available electron guns and X-ray detectors made possible reliable determination of the time-of-flight of particles having a sedimentation radius of less than 0.75 microns.

The time-of-flight method is suitable for both liquid and gas flows. In gas flow investigations the upper temperature limit is determined by the possibility of the existance of discrete floating particles and is of the order of three to four thousand degrees. The lower limit is determined by the viscosity drop appearing at temperatures of the order of —200° C. Upper gas pressure limit does not exist. The lower is determined by the viscosity drop at the pressure of the order of 0.1 to 1 mm mercury column. Maximum flow velocity is determined either by the sensitivity of the photoelectrical detection system or by the minimum duration of the impulses permissible by the electronics. Minimum flow velocity corresponds to Brownian motion.

The application range of the optical time-of-flight method is schematically given in Fig. 3 for the measurements of turbulence in air

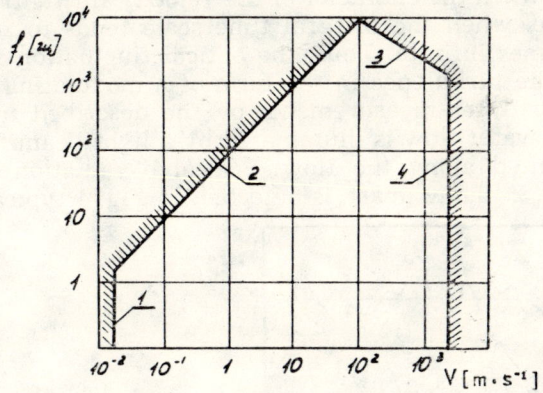

Fig. 3. 1 — Turbulence intensity with combustion; 2 — turbulence intensity in-isothermal flow; 3 — mean velocity with combustion; 4 — mean velocity in isothermal flow; 5 — velocity pulsations caused by combustion; 6 — burning area.

at normal conditions. The scheme corresponds to a measurement basis of 2 mm, an incident light intensity of 40 W/cm^2 with $\lambda = 0.63$ m, and tracing particles which have a spherical form and the density equal to one. The limiting values are determined by a measurement error of the order of 10 percent.

Mean velocity values are plotted on the abscissa and Lagrange frequency of the turbulent fluctuations on the ordinate. Limitations are caused by: 1 — Brownian motion; 2 — integration during the time of flight; 3 — particles inertia; 4 — limiting frequency characteristics of the recording electronics.

4.2.2. Correlation Method

The time-of-flight method could be used for flows in which discrete floating particles are introduced. For the investigation of flows having continuous optical unhomogeneities, we have suggested a method based on the correlation of the signals provoked by the optical unhomogeneities as they pass two or more points in the flow [18].

In flows of an optical unhomogeneity field with variable velocity, an optical fluctuating signal at a point results from the frequency modulation, caused by the velocity pulsations $V(t)$ of some random function $\varepsilon(t)$. This random function could be interpreted as a component caused by the displacement of the optical unhomogeneity field with a constant velocity.

The frequency deviation of each of the elementary harmonics of the process $\varepsilon(t)$ is proportional to the mean square of the velocity fluctuations, and the working frequency modulation is determined by the Eiler interval time scale, the higher the integral time scale the lower the working frequency modulation. A limiting case when the integral time scale tends to infinity corresponds to the absence of modulation, i.e. displacement of the field occurs with a constant velocity which is the case of a laminar flow. In the case when the character of the velocity fluctuations approaches a white noise, i.e. when the integral time scale tends to zero, frequency modulation becomes infinite. Thus, the optical fluctuation signal includes information on the turbulence in the form of a modulating function.

Figure 4 illustrates measurements by the described method. In this particular case a water flow is illuminated by a light beam from a helium-neon laser, oriented along the flow. Scattered radiation from the two areas of the flow, 2.16 mm apart, is detected. In Fig. 4 typical oscillograms

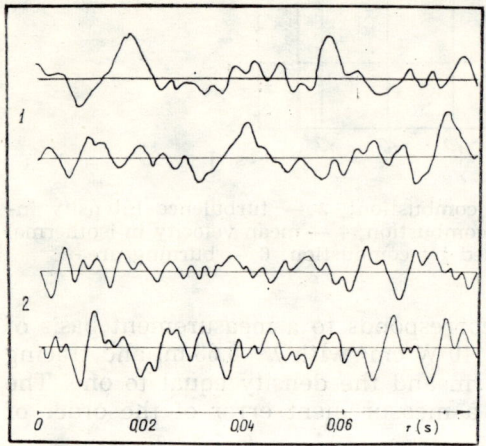

Fig. 4.

corresponding to the flow velocities in the two areas are presented. From Fig. 4 it is seen that the two signals are correlated and that the signal from a downstream area is time-lagged in correspondence to the upstream signal.

Mathematical expectation of the time lag corresponding to the mean flow velocity is calculated from the cross-correlation function:

$$r_{xy}(\tau) = \frac{K_y(\tau)}{K_x(0) \, K_{yy}(0)} \tag{13}$$

In Fig. 5 diagrams of the cross-correlation functions corresponding to the mean velocities of 9.4, 17.6, 18.8 and 35.2 cm/sec are given. It can be seen that the increase of the flow velocity results in the decrease of the time lag corresponding to the maximum of the cross-correlation function. The position of this maximum is used for the determination of the mean velocity.

Curve 3 on Fig. 5 corresponds to a subcritical flow and Curve 4 to a supercritical flow. It is evident from Fig. 5 that in the transition region a small velocity change modifies substantially the frequency of the periodic component of the cross-correlation function.

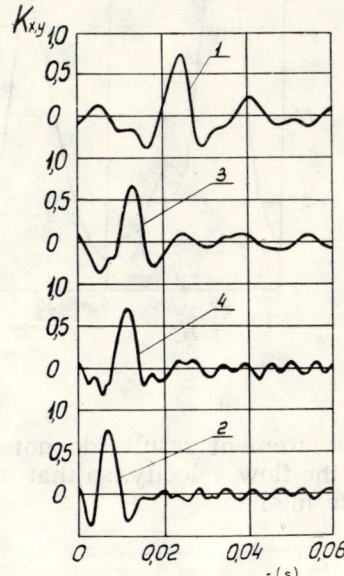

Fig. 5.

The cross-correlation function spectra $S_{xy}(\omega)$ corresponding to flows 1, 3 and 4 from Fig. 5 are presented in Fig. 6 (solid lines). In the same figure corresponding structural function spectra $S(\omega)$ are also plotted (dotted lines). The difference in spectra of the cross-correlation and the structural function allows the determination of the velocity fluctuation spectra. As an example, resulting velocity fluctuation spectra S_v for the flows 1, 3, and 4 are plotted in Fig. 7. As is seen from Fig. 7, low velocity flow (№ 1) has a narrow fluctuation spectra and low pulsation energy. With the increase of the mean velocity, the spectra are broadened although the pulsation energy is still low because the flow is subcritical (flow № 3). In the transition from subcritical to supercritical flow pulsation energy increases substantially and the spectra are correspondingly broadened.

The given method was used not only for turbulence research in liquid flows [19] but also in gas flows [20] and plasma flows [18]. In the investigations of supersonic flows, it is possible to determine the Mach number M from the relation:

$$\frac{\Delta \omega_y}{\omega_y} = \pm \sqrt{\delta^2 + \frac{1}{M^2}} \qquad (14)$$

where δ is turbulence intensity.

The described optical kinematic methods for turbulence measurements make possible measurements with a space resolution of the order of fractions of a millimeter without any perturbations of the flow.

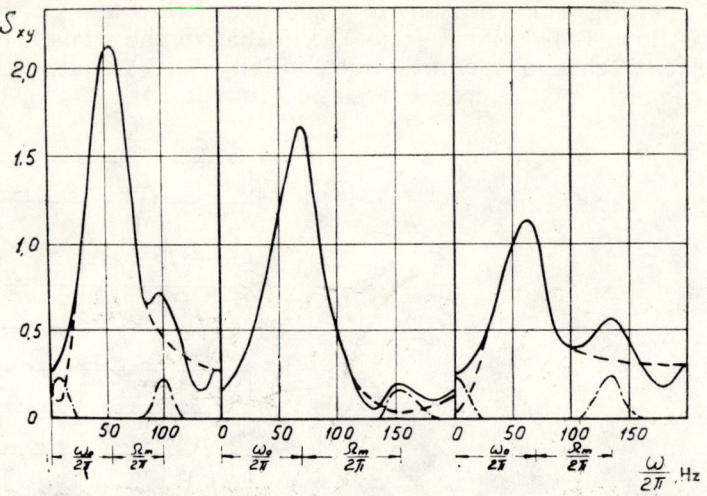

Fig. 6.

Measurement results do not depend on any parameter of the flow except the flow velocity, so that in this respect the described methods are absolute methods.

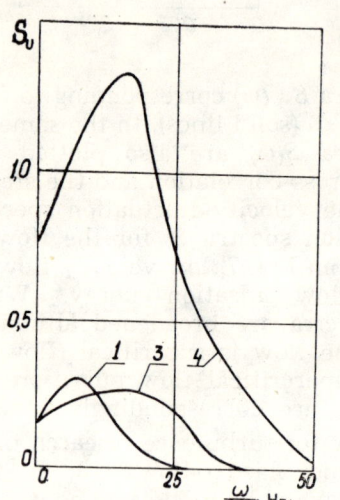

Fig. 7.

These methods permit measurements in flows of liquids, gases and plasmas in practically an unlimited range of velocities and temperatures and with the same instrumentation composed of standard elements. It can be expected with relative confidence that these methods will soon be widely used in the practice of hydrodynamic measurements.

REFERENCES

1. H. C. H. Townend, A Method of Air Flow Cinematography Capable of Quantitative Analysis, J. Aero Sci., **3**, 10, p. 343 (1936).
2. V. Ya. Trubchikov, A Thermal Method of Turbulence Measurement in Aerodynamic Tubes, CPGI Transactions, N 372, 1938.
3. A. G. Prudnikov, A Diffusion Method of Turbulence Intensity and Diffusion Factor Measurements, in Collected Articles about Turbulent Burning of Homogeneous Mixture, M, 1956.
4. L. A. Zhukov, A. V. Maier, G. R. Rechtzamer, The Application of Underwater Photographic Mapping and Filming for the Turbulence Investigation in a Sea, materials of Conference 2 on the problem »Interworking of atmosphere and hydrosphere in the North of the Atlantic Ocean«, LGU publication, pp. 151—155, 1964.
5. G. S. Karabashov, R. V. Ozmidov, Investigation of Turbulence Diffusion in a Sea by Means of Fluorescence Indicators, Izv. Acad. Nauk SSSR, Ser. fiz. atm. and ocean, **2**, pp. 1178—1189, (1965).
6. I. A. Pishchenko, To the Question of the Determination of Fluid Turbulence Characteristics from the Turbulent Diffusion Measurements, Chem. Mechanical Engineering, Mezhvedomstv. resp. sci.-techn. collections, issue 3, pp. 157—161 (1965).
7. I. O. Khintse, *Turbulence,* M. Fizmatgiz, 1963.
8. V. V. Struminsky, V. M Filippov, Experimental Investigations of Light Scattering Phenomena in Laminar and Turbulent Fluid Flows, Izv. Acad. Nauk SSSR, OTN, Mechanics and Mech. Eng., **6**, pp. 10—16 (1962).
9. Dreiper, Measurement of Infrared Radiation of Turbulent Flows, Rocket Techn. and Astronautics, **9**, pp. 120—123 (1966).
10. A. M. Trokhan, Measurement of Gas Flows Parameters with the Help of a Beam of Fast Electrons, PMTF, **3**, pp. 81—94 (1964).
11. Goldstein, Miles, Shabe, Measurements of Light Propagation in Turbulent Atmosphere by a Heterodynizing Method, Trans. Radio-eng. Inst., Russ. transl., 9, pp. 1333—1341, 1965.
12. Foreman, E. W. George, J. L. Jetton, R. D. Lewin, J. R. Thornton, H. J. Watson, Fluid Flow Measurements with a Laser Doppler Velocimeter, JEEE, J. Quantum Electronics, **2**, 8, pp. 260—266 (1966).
13. Fadge, Townend, Investigation of a Turbulent Flow with the Help of an Ultramicroscope, in Collections Turbulence Problems, M. 1936.
14. B. A. Fidman, Application of High Speed Filming to the Investigation of a Turbulent Flow Velocity Field, Izv. Acad. Nauk SSSR, Ser. Geophysical, **12**, 2, pp. 99—106 (1948).
15. V. V. Orlov, Experimental Study of Wall Turbulence in a Channel, PMTF, **4**, pp. 124—126 (1966).
16. A. M. Trokhan, Measurement of a Gas Flow Velocity by Kinematic Methods, PMTF, **2**, pp. 112—121 (1962).
17. A. M. Trokhan, I. L. Kuznetsov, G. P. Baranova, Yu. V. Ignatenko, A Photoelectric Method of Turbulence Measurement of High-Temperature Fluxes, Physics of Burning and Explosion, **1**, pp. 112—116., (1966).
18. N. F. Derevjanko, A. M. Trokhan, About Application of a Correlation Method for Measurements of Plasma Flow Velocity, Measuring Technique, **10**, pp. 24—29 (1966).
19. N. F. Derevjanko, V. M. Latishev, A. M. Trokhan, Investigation of Fluid Flows by an Optical Correlation Method, I. T., **4**, pp. 34—36 (1969).
20. N. F. Derevjanko, I. L. Kuznetsov, A. M. Trokhan, Turbulence Measurement by an Optical Method, Dok. Acad. Nauk SSSR, **180**, 4 (1968).

FLOWS WITH SEPARATION AND REATTACHMENT TO A WALL: ANALYTICAL FRAMEWORK AND EXPERIMENTAL RESULTS

A. F. CHARWAT

University of California, Los Angeles, U.S.A.

Regions of separated flow will appear when the viscous boundary conditions on an attached flow wetting the surface of an immersed body cannot be satisfied. A portion of the boundary is then replaced by a free streamline enclosing a region with closed internal streamlines. The theoretical problem of attached flow can be posed unequivocally, if not always solved. A flow with separations is very much more complex: it involved distinct domains matched along dividing free-streamlines, the shape of which is not known a priori. The uniqueness of these composite structures is demonstrated experimentally, although a theoretical proof has yet to be given.

In spite of much progress, our understanding of flows with separation and reattachment remains often vague and incomplete. The existing models are insufficiently refined to explain many of the important observable phenomena over the full range of Mach and Reynolds number of practical interest, even in the particular geometry for which they are designed. Base pressures and the gross aspect of the flow field can be correlated to an engineering accuracy by relatively simple semiempirical theories. Any improvement of the precision of these semiempirical theories, questions pertaining to the fine details of the pressure distribution, the flow-field at the closure point of the near wake (an important initial condition for the study of the far wake), etc., require a totally different level of analysis involving very extensive numerical work. There seems to be no intermediate step between the elementary model and a virtually complete solution of the entire flow field over the body. A particular deficiency of the existing theories is that they provide no reliable framework for dealing with the important problem of heat-transfer to the walls of the cavity. Heat transfer problems are still mainly a question of judicious interpolation of experimental data.

In this paper, we shall review the problem of flow over notches and grooves, outline the results pertaining to separation — controlled cavities, and the recently developed techniques of numerical analysis of flows with separation and reattachment. Virtually all the theoretical models and most of the experimental work in this field is done for supersonic flows. This is partly because the essential interaction between it and the separated region is much simpler to handle when the external flow is hyperbolic, and partly because practical problems encountered in supersonic aerodynamics gave impetus to the research. We shall therefore speak of supersonic flow but conceptually, the nature of the problems is not different in subsonic flow.

LECTURE

ON SOME PROPERTIES OF REATTACHING LAMINAR AND TRANSITIONAL HIGH SPEED FLOWS

JEAN J. GINOUX

von Karman Institute and Brussels University, Belgium

This paper presents a survey of basic experimental results obtained at the von Karman Institute on separated and reattaching flows at supersonic and hypersonic speeds.

Separation is induced by flaps, steps or cavities. It is shown that periodic patterns of streamwise vortices develop in reattaching laminar and transitional shear layers causing local peaks in heat transfer. It is also observed that a recovery temperature peak exists in the reattachment region when transition is present. The presence of transition is detected by a new method.

It is also shown, in the case of flows over cavities, that static pressure and heat transfer peaks develop at reattachment which can be suppressed by mass injection inside the cavity, light gases being more efficient than air or heavier gases.

SEPARATED SUPERSONIC TURBULENT FLOWS AHEAD OF FORWARD FACING STEPS

H. T. UEBELHACK

von Karman Institute, Bruxelles, Belgium

Flow separation ahead of steps occurs mainly in the following cases:
1. in free flight when sudden enlargements are necessary,
2. interaction between the outer flow and the exhaust plume of an underexpanded jet,
3. jet injection into supersonic flows for control purposes,
4. fuel or gas injection into supersonic streams (supersonic combustion),
5. contraction of a supersonic diffuser (second throat) of an ejector.

The first-mentioned case has been almost always successfully avoided on aircraft designs because of the very large pressure coefficients related to a step. This also explains the lack of interest in obtaining experimental data earlier. The cases mentioned under 2 to 4 are examples of jet interaction which cannot be avoided. In the last-mentioned case the step even has a beneficial effect by reducing the starting pressure and the suction pressure of the secondary stream and therefore increasing the efficiency.

The application mentioned under point 5 has been the purpose of an investigation on flow separation ahead of a forward facing step. Measurements on ejectors with a second throat formed by a ramp or by a step have shown the same positive effect. The separated region in front of a forward facing step forms a »natural« ramp of a minimum length. A minimum length of contraction in such a second throat is required in many cases. The pressure distribution on the ramp or the step forming a second throat must be known in order to solve the momentum equation.

An experimental study has been carried out in order to provide data on the pressure distribution in the separated flow region and particularly on the step face. Literature delivers only few experimental data, since this field was of low interest in the past. Almost all the data presented in the literature refers to supersonic two-dimensional flow at Mach numbers between 1.5 and 3.5. Those experiments show that the peak pressure, as a characteristic pressure, is a function of the upstream Mach number mainly. Reynolds number has little or no influence on this characteristic pressure in fully turbulent supersonic flows. The experiments made by various investigators show always the same tendency (increasing peak pressure with increasing Mach number). The scatter in experimental data increases with Mach number.

The object of this research was to measure and compare the pressure distribution in the separated flow region ahead of forward facing steps,

— in supersonic axisymmetric flow on a cone-cylinder-step model at $M=3.5$,
— in supersonic axisymmetric internal flow at $M=5.0$,
— in two-dimensional flow at $M=3.5$ on flat plate models,
  I) on models which span the wind tunnel completely
  II) on models having $80^0/o$ of the tunnel span.

It was intended to:
— compare data received from axisymmetric flows with those from two-dimensional flows,
— give an explanation for the scatter of the various measurements, particularly at higher Mach number (three-dimensional effects, side walls, angle of attack),
— show the influence of Reynolds number.

Flow visualization was made by schlieren photos and oil flow techniques in order to detect separation lines and to get a more detailed understanding of the flow picture in the separated region.

A STUDY OF SEPARATED FLOWS INDUCED BY TRAILING-EDGE FLAPS ON DELTA WINGS AT HYPERSONIC SPEEDS

D. M. RAO

Aeronautics Dept., Imperial College, London, U.K.

The investigation concerns the control effectiveness of trailing-edge flaps on delta-winged hypersonic vehicles. Flow separations peculiar to this configuration were studied using flow visualisation methods, and their overall effects on the vehicle aerodynamic characteristics were measured. The tests were carried out on two delta wing models (of 70° and 76° leading-edge sweep angle) at M=8.2 in a Gun Tunnel.

While the extent of separation was different on the two wings at the same incidence and flap deflection, producing marked differences in the normal force and patching moment, the separation patterns were qualitatively similar and characteristic of the three-dimensional situation. At low incidence and large flap deflection, a deep »conical« separation was observed, associated with relatively large loss in flap effectiveness. Based on the evidence of flow visualisation, a tentative flow model of this mode of separation is proposed. For only a small increase in incidence (or Reylonds number) the »conical« separation rapidly collapses into a more localised bubble, indicating a strong effect of transition. There is a corresponding improvement in flap loading, which under these conditions is well predicted by two-dimensional shock theory for the smaller flap deflections.

STUDY OF THE THERMAL AND DYNAMIC PHENOMENA AT THE REATTACHMENT OF A SUPERSONIC ROTATIONAL JET ON A CYLINDRICAL WALL

J. DELERY, P. ROUGIER

O.N.E.R.A., Chatillon, France

This communication presents the results of an experimental study of the thermal and dynamic phenomena at the reattachment of a supersonic rotational jet on a cylindrical coaxial surface.

The Mach number of the jet is between 2 and 5, the generating temperature T_i and the generating pressure p_i have been varied from 300 to 1000° K, in case of T_i, and from 5 to 35 bars, in case of p_i.

In the first part we are investigating the influence of the thermal parameters on the purely dynamic phenomena of the reattachment.

The second part is devoted to the study of the heat flux distribution in the region of the reattachment. In particular we are studying the effect of a fluid injection in the separated zone on these distributions.

An attempt of the correlation of the data is presented and discussed.

SHOCK-INDUCED JETS AND QUASI SEPARATED FLOWS

E. EDNEY

Textron's Bell Aerosystems Company, Buffalo, U.S.A.

An extraneous shock impinging on a blunt body in hypersonic flow is observed to alter the flow around the body and increase both the local heat-transfer rate and pressure near the impingement point. Six distinct types of interference pattern can be predicted depending on the strengths of the intersecting shocks. The severest heating is associated with Type III and Type IV interference. In Type III interference a free shear layer, which separates regions of subsonic and supersonic flow, attaches to the surface of the body. The flow in the immediate vicinity of the attachment point is directly analogous to the reattachment of a separated boundary layer in a supersonic flow. In Type IV interference a curved supersonic jet is formed, embedded in a subsonic flow field. This jet may also impinge on the surface of the body, resulting in peak pressures and heat transfer rates an order of magnitude higher than those recorded in the absence of the extraneous shock. The peak pressure and heat transfer rate, associated with these two types of interference, are shown to increase with increasing free stream Mach number and increasing strength of the extraneous shock, reaching a maximum for shock generator angles between 10° and 15°, depending on Mach number. Experimental data obtained at $M=4.6$ and 7 in air and shock generator angles up to 15° confirm these theoretical predictions. The peak pressure is also shown to increase with decreasing ratio of specific heats, indicating that real gas effects will compound the heating problem at very high Mach numbers.

SEPARATED REGIONS IN SOURCE FLOW BETWEEN STATIONARY AND CO-ROTATING, PARALLEL DISKS

F. KREITH and E. BAKKE

University of Colorado, Boulder, U.S.A.

In an investigation of source flow between stationary and co-rotating parallel disks, two kinds of separated flow regions were observed. The first kind occurred near the lip of the entrance to the space between the two disks where the fluid impinged on a plate and had to make a 90 degree turn before commencing out-flow in the radial direction. The flow in this type of separated flow region resembled the conditions at a 90 degree bend in a channel and re-attachment occurred some distance down-stream from the inlet.

The flow between two parallel disks with a source in the center procedes in a direction of increasing pressure due to the increase in flow-cross sectional area with increasing radial distance. Flow separation of the second kind was studied in the system, which has applications to radial diffuser design, when the disks were co-rotating and stationary. The influence of the centrifugal forces introduced into the flow by virtue of the viscous coupling between the fluid and the surface of the rotating disks was investigated and a set of flow parameters which describe the flow was obtained by analysis and verified by experiments. Hot wire measurement very close to the surface and static pressure measurement in the flow field were taken and the results were related to semi-empirical equations. Applications to heat-, mass-, and momentum transfer are discussed.

WALL SHEAR-STRESS IN RECTANGULAR CAVITY

S. N. OKA

Boris Kidrič Institute, Beograd, Yugoslavia

Wall shear-stress measurement in flows with recirculation regions is a difficult experimental problem. Existing measurements in cavities, performed by hot-wire technique, do not give reliable data.

A recently developed electrochemical method was used in this experiment to measure local wall shear-stress in a rectangular cavity.

The height to length ratio H/L was changed from 0,5 to 2,0 and the cavity Reynolds number from 2000 to 40000.

Wall shear-stress distribution shows different flow patterns for different H/L ratios. Dependence of wall shear-stress on Reynolds number is also shown.

SEPARATED FLOWS RESULTING FROM A SUDDEN VARIATION OF THE BOUNDARY

F. SANANES and C. OIKNINE

Institut de Mécanique des Fluides, Université de Toulouse, France

We are presenting here part of the results obtained at the Institut de Mécanique des Fluides of Toulouse during the past three years, following studies on separated flows at subsonic and supersonic speeds to which MM. Robert, Hebrard and Pereysson contributed.

The first studies concerned flows over backward-facing steps. At subsonic speed, wind tunnel tests on a very large scale model permitted us to determine precisely the speed profile of the isobaric zone. This profile can be determined theoretically when the boundary layer at the point of separation is not very thick. At supersonic speed the tests carried out show the existence of a base pressure whose theoretical calculation seems risky within the framework of existing theories.

Numerous tests on a circular wind tunnel involving a sudden widening of the section showed that at subsonic speed the separated zone can be treated in much the same manner in the case of two-dimensional and axially symmetrical flows.

On the other hand, at supersonic speed the flow pattern depends essentially on the ratio of the total upstream and downstream pressures. We give the pressure limits of the supersonic flow for ejector systems. The tests carried out show that the reattachment criterium necessary for the calculation of base pressure still has to be improved. The visualisation of the flow by strioscopic photography permits a better understanding of the phenomenon.

LOCAL HEAT TRANSFER IN A SEPARATION ZONE BEHIND A STEP

O. G. MARTYNENKO and V. A. BAIRASHEVSKY

Heat and Mass Transfer Institute, Minsk, USSR

This paper considers flow dynamics and heat transfer for a turbulent flow within a separation zone in a flat channel and circular tube after sudden expansion of the channel section. The study has been carried out for a wide range of area ratios ($0 < f_o < 1$) before and after the sudden expansion $f_o = \dfrac{F_1}{F_2}$ at large $Re > Re$ of similarity.

The study of flow dynamics behind the step has been carried out with water and air. For a circular tube, symmetry of the separation zone was observed along the circumference, while for a flat chanel the separation zone was three-dimensional and different on the contrary walls of the channel. The region of the main flow contact with the wall changed with time and space. The change in flow parameters before sudden expansion (thickness of the boundary layer and shape of the velocity profiles) did not exert considerable influence on the length of the separation zone.

Local heat transfer in the separation zone behind the step was simulated by an electrochemical method on the basis of the available analogy between heat and mass transfer. The data obtained from the study of local mass transfer coincide quantitatively and qualiatively with those on local heat transfer known from literature. In the work no noticeable influence of Pr on the intensity of heat transfer in the separation zone is found. Although the structure of the separation zone and distribution of local heat transfer along the length were not observed simultaneously, the comparison of the corresponding regimes shows that the maximum local numbers Nu are consistent with the zone of the maximum rate of recirculation of the reverse eddy flow rather than with the zone of contact of the main stream with the channel wall (as assumed in the literature).

WAKE RECIRCULATION FLOWS GENERATED BY FLOWS (M > 1) OVER REARWARD FACING STEPS AND BLUNT BASES

M. G. SCHERBERG

Aerospace Research Laboratories, Wright-Patterson AFB, Ohio, U.S.A.

In the past the wake flows generated by Supersonic and Hypersonic flows over rearward facing steps and blunt bases have been treated as »dead water« regions of constant pressure and the flow surrounding them was studied. Recently much effort has been devoted as well to the wake flows which experiments have shown to be quite complex and in need of attention so that the whole flow structure may be better understood. Previous and current experimental ARN programs at Technion in Israel and at Arnold Engineering and Development Center (AEDC) in Tullahoma, Tennessee have disclosed, [Scherberg, Smith, 1967], [Rom, Seginer, 1966], that the boundary layer turns the step or base corner and separates somewhat downstream of the corner, that generally two recirculating cells may be expected in rearwarding facing step wakes and four such cells in two dimensional wedge base wakes. For axisymmetric case the two cell donut type recirculations would be anticipated. It has also been observed that the prime recirculation flow may be expected to contain streamwise vortex systems of the type under investigation by L. N. Persen, [Persen, 1968], but this effect will not be included in the present analytical investigation.

The investigations of ARN and others have indicated that the recirculation flows are subsonic and the speeds for the most part are such as to make a constant density assumption plausible for preliminary treatment of this flow. Past investigations and current ones have made it clear that the recirculation flows will not be either irrotational or inviscid.

In the light of the above information this paper reports on an investigation undertaken for a rearward facing two dimensional step under the following assumed conditions:

1. That the recirculation flows are constant density flows. Recent experiments have shown that temperature variations are indeed small.

2. That they are composed of two essential parts:

    a. Core flow with constant vorticity. Constant vorticity is used because it is the simplest type of rotational flow.

    b. Thin layer shear flows surrounding the core flow. These are boundary layers when adjacent to fixed walls and interface shear layers when adjacent to interface surfaces.

3. That the geometry of the flow is represented by a geometry in which the free flow corners are used for mathematical expediency and represent in reality smooth sharp turns. Preliminary calculations indicated that reattachment should be at right angles and that there are relatively thin shear layers except at the vertical wall.

LOCAL SEPARATION IN TURBULENT FLOW IN CHANNELS AS A METHOD OF HEAT TRANSFER INTENSIFICATION

E. K. KALININ, G. A. DREITSER, S. A. YARKHO, and W. P. KUSMINOV

Moscow Aircraft Institute, Moscow, USSR

Analysis shows that efficient heat transfer intensification can be accomplished with reasonable hydraulic resistance if an increase in the turbulent thermal conductivity can be obtained and restricted to a region near the wall. The qualitative analysis and visual investigation show that this can be done by appropriately positioning ridges or grooves along the channel wall causing regular distribution of local separation. The shape of the ridges and grooves, their approximate size and distance between them is discussed, as well as the mechanism of increasing turbulent thermal conductivity near the wall.

The second part of the paper is an experimental study of heat transfer and hydraulic losses in tubes having circular ridges inside spaced at certain intervals and in longitudinal flow past the tube bundles having annular grooves or ridges. This study was made over wide ranges of pitch and sizes or ridges and grooves, Pe and Re, including transition from laminar to turbulent flow and influence of variable physical properties under conditions of heating and cooling. Criterial relations for heat transfer and hydraulic resistance were derived from this study. This method of intensifying heat transfer makes it possible to decrease the size and weight of tubular heat exchanges by 1.5 to 2 times.

VELOCITY AND TEMPERATURE DISTRIBUTION NEAR A ROUGH WALL

S. BEĆIRSPAHIĆ

Laboratoire de Mécanique Expérimentale des Fluides, Orsay, France

An experimental investigation has been made of flow through a rectangular channel in which one of the large sides was rough. The roughness was formed by transverse ribs. Two forms were used: square and trapezoidal, but the ratio of pitch to height was the same (2 : 1).

The profiles of mean velocity and the turbulence intensity are determined by a hot wire anemometer. The signal was linearized. The profiles of temperatures were obtained by a thermocouple probe.

Our investigation confirmed the existence of a logarithmic law for velocity and temperature near a very rough wall ($e/k=10$); it remains to determine the origin for y, which is a difficult problem in the case of a large scale roughness. The slope of the velocity profile is about 5.65, which is the value for a smooth wall. The slope of the temperature profile is a little smaller; the ratio A_T/A is about 0.95.

The measurements of turbulence intensity and spectrum have shown that outside the very thin layer near the wall, the fine structure of turbulence in the rough and smooth channels is probably the same.

FLOW FIELD BETWEEN TWO ROUGHNESS ELEMENTS IN TURBULENT CHANNEL FLOW

S. N. OKA

Boris Kidrič Institute, Beograd, Yugoslavia

The results of measurements of mean velocity and turbulence field between two two-dimensional roughness elements are presented. Velocity profiles outside vortex region, in vortex region and in viscous sublayer are measured. Turbulence characteristics $\overline{u'^2}$, $\overline{v'^2}$ and $\overline{u'v'}$ between two roughness elements are also given.

Change in flow pattern with the change in distance between elements is evident.

CIRCUMFERENTIAL DISTRIBUTION OF FRICTION STRESSES IN A CLUSTER OF CORRUGATED RODS PARALLEL TO THE DIRECTION OF FLOW

J. GEFFROY, G. PAUMARD, M. JUDE

Section de Thermique C.E.N., Saclay, France

Experimental results are presented concerning the distribution of friction stresses in a cluster of 19 corrugated rods placed inside a circular channel, parallel with the direction of flow.

Local friction stresses were determined from the velocity profiles by the use of a universal law of velocity distribution along normals to the contours.

The integral of the friction stresses is compared with the pressure drop measurements, and the flow rate distribution in the different subchannel of the cluster is given.

DISTRIBUTION OF TEMPERATURE AND HEAT TRANSFER COEFFICIENTS ON THE PROFILE OF A CORRUGATED SURFACE

J. GEFFROY and J. C. MOURGUES

Section de Thermique C.E.N., Saclay, France

Performances are often limited by the maximum temperature attained locally on an exchange surface. For this reason a study was made of the heat transfer coefficient variation along several corrugated surface profiles.

The experiments were based on the analogy between mass transfer and heat transfer, using forced sublimation of naphthalene.

Experimental results relating mainly to sinusoidal profiles are given. The influence of machining defects on the heat transfer coefficient distribution and on conduction effects are studied.

AEROTHERMAL PROPERTIES OF CORRUGATED SURFACES

G. PAUMARD

Section de Thermique C.E.N., Saclay, France

Experimental results on heat transfer from corrugated surfaces are presented. The experiments were carried out on internally corrugated circular tubes with constant-temperature wall heating.

Cooling is obtained by means of a flow of pressurized CO_2 with Reynolds numbers between $2 \cdot 10^5$ and $3 \cdot 10^6$, and for average heat flux values between 15 and 90 kW/m².

The corrugated surfaces examined here have continuous or broken sinusoidal contours or consist of asymmetrical wave forms with different profiles (about 50 geometries). The influence of height and spacing is studied.

Based on the aerodynamic and thermal similarities, a method of correlating the experimental results is proposed.

LOCAL HEAT TRANSFER OF AN ELLIPTICAL CYLINDER IN CROSS-FLOW OF FLUID

J. ZHIUGZHDA

Institute of Physico-Technical Problems of Energetics, Vilnius, USSR

The results dealing with the investigation of local heat transfer of an elliptical cylinder in the flow of various fluids at the boundary condition of constant heat flux are analyzed in the report. The influence of kind of liquid, temperature difference and angle of attack on heat transfer are determined.

Experiments are performed in the flow of air, water and transformer oil in the range of Re_{fx} numbers from 10^2 to 10^6, Re_{fd} to 10^5, and Pr_f numbers from 0.7 to 300.

HEAT TRANSFER FROM A CYLINDER IN THE WAKE OF ANOTHER CYLINDER

Ž. KOSTIĆ

Boris Kidrič Institute, Beograd, Yugoslavia

The results of the measurements of local heat transfer and pressure distributions on the surface of a cylinder in the wake of another cylinder in cross flow are presented.

The variation of average heat transfer and drag coefficients with Reynolds number is discussed. Influence of the distance between two cylinders on heat transfer and drag coefficients is also presented. Measurements are carried out for distances $L/D = 1$ to 10, and Reynolds numbers 10^4 to $6 \cdot 10^4$.

INFLUENCE OF SURFACE ROUGHNEFS ON THE CROSS FLOW AROUND A CIRCULAR CYLINDER AND THROUGH A STAGGERED TUBE BUNDLE

E. ACHENBACH

Kernforschungsanlage Jülich, Federal Republic of Germany

In the range of Reynolds-numbers $5 \cdot 10^4 < Re < 10^7$ investigations on the cross flow around a circular cylinder and through a staggered tube bank with rough surfaces have been made. It was intended to find out how roughness influences the boundary layer and thereby the flow resistance of bluff bodies. For this purpose the local pressure and skin friction distribution around the single cylinder, and around the tubes in different positions of the bundle, was measured at two roughness conditions. In addition, the total pressure loss of the heat exchanger model was experimentally determined.

As the rough surface of the tubes was realized by glued grinding paper, it was easy to construct a pipe with the same roughness conditions. In this pipe with inline flow the skin friction probe could be calibrated. Moreover, this test channel enables one to classify the investigated roughness according to Nikuradse's friction loss diagram for pipes. In this way the drag coefficient ξ of the bundle can be given as a function of Reynolds-number and Nikuradse's sand grain roughness K/D. It is shown, that an increase of the roughness height causes a decrease of the critical Reynolds-number and a rise of the drag coefficient ξ in the supercritical flow régime. Comparing, for example, a bundle of the roughness $K/D = 2 \cdot 10^{-3}$ with a surface polished tube bank, one finds a decrease of the critical Reynolds-number by the factor 5 and an increase of the ξ — value up to nearly 2.5. In the subcritical flow régime the influence of the surface roughness seems to be negligible.

The experimental local pressure and skin friction distributions give information as to how the position of the transition point laminar-turbulent and the separation point of the boundary layer varies depending on Reynolds-number, roughness height and turbulence level. Whereas in the first two rows of a staggered tube bundle the boundary phenomena are quite similar to those ones of a single cylinder in cross flow, these effects are observed no longer in the following rows because of the high turbulence level.

The tube diameters of the single cylinder and of the heat exchanger model was 150 mm. The staggered bundle consisted of 5 rows with a transversal pitch of $s_t/D = 2.0$ and a longitudinal pitch of $s_l/D = 1.4$. Up to $Re = 5 \cdot 10^5$ the experiments were conducted under atmospherical conditions. The high Reynolds-numbers have been achieved in a high pressurized wind tunnel at a static pressure of 40 bar.

HEAT TRANSFER BY PULSATING AIR FLOW AROUND CYLINDERS

M. MARTIN and J. GOSSE

Faculté des Sciences, Nancy, France

For the last few years, the Research Laboratory has been interested in dynamical and thermal responses of boundary layers and separated regions on cylinders to periodical fluctuations in the stream velocity (pulsating flow but not reversed flow).

Our purpose is to compare the efficiency of heat transfer in pulsating and steady flows for the same mean Reynolds number.

A study about cylinders in an air cross-flow shows that the mean heat transfer is increased by velocity pulsations. Moreover, a comparison between measured values of local heat transfer for the forward and backward portion of the cylinder proves that the pulsating effect is not the same on every region of the cylinder surface. It seems to result from our experiments that velocity pulsations have no action on the convection phenomena in steady boundary layers, as is proved by a theoretical study in laminar boundary layer. However the heat transfer in unsteady flow regions is greatly increased by velocity pulsations. It is experimentally observed on the cylinder backward face an increase of the local heat transfer coefficient which rises to 40 or 50 per cent when $Re = 30,000$.

Several parameters have a beneficient effect on thermal convection: mean flow velocity, pulsating rate, frequency, and also surface condition.

The heat transfer efficiency can be determined by the quality index J: the ratio of heat power convected to spended mechanical power. The quality index is studied in relation to a single parameter: the pulsating rate — and an unfavourably influence of velocity fluctuations on the heat transfer efficiency is always noticed. Experiments show that the quality index is a decreasing function of the pulsating rate.

HEAT TRANSFER FLUCTUATIONS IN SEPARATED FLOWS

PETER D. RICHARDSON

Brown University, Providence U.S.A.

Measurements of Parnas (1965) and Fabula (1966) of overall heat transfer from cylinders in crossflow showed that it was possible to observe fluctuations in heat transfer at the shedding frequency of the Kármán street. Local measurements of fluctuating heat transfer have been made more recently, using a new technique involving a small semi-conductor film. A silicon wafer, 0.13 inch long, 0.02 inch wide, and 0.001 inch thick was attached to the surface of an aluminium cylinder, 0.75 inch diameter, and connected to a resistance-measuring bridge. The sensor had a sensitivity of 14.8 ohms/deg C. The cylinder could be rotated to put the sensor in any desired position relative to the crossflow of air. The sensor acts essentially as a resistance thermometer of low thermal inertia, and measures the fluctuations in surface temperature consequent upon fluctuations in heat transfer coefficient. A frequency correction is necessary to allow for transient conduction attenuation through the small thickness of the sensor.

Measurements were made with the cylinder in a steady air stream in the range $5.500 < Re < 11,000$. Hot-wire anemometry was employed in the same experiment to measure some details of the associated flow pattern. At the of the cylinder the dominant fluctuations in heat transfer coefficient occur sharply at the shedding frequency of the Kármán street; the same is seen in the signal of the hot-wire when it is placed near the boundary layer at the front of the cylinder. The r.m.s. Nusselt number for the oscillating heat transfer increased approximately as Re^3 in the range tested. At the back of the cylinder the fluctuations in heat transfer occur over a much wider spectrum of frequencies, but the maximum still seems to be at the shedding frequency. The contributions at other frequencies are still large for a factor of three on each side of the Kármán street frequency. At the back of the cylinder the fluctuating Nusselt number varies very approximately as Re^2. The level of fluctuating heat transfer can be increased at both front and rear by introducing disturbances into the flow to which it is sensitive (Peterka and Richardson 1969).

There is no adequate means for predicting the heat transfer in separated regions on bluff bodies. However, some inferences can be drawn from the present results that are useful in examining theoretical models for the convection process.

THE TURBULENCE CHARACTERISTICS OF ANNULAR WAKE FLOW

T. W. DAVIES and J. M. BEÉR

Dept. of Chemical Engineering, University of Sheffield, U.K.

Experimental results for the flow around bluff bodies situated in a nozzle are reported in two parts.

a) The influence of initial flow boundary conditions on the characteristics of the mean flow field are given. The spatial distributions of mean stream function and static pressure, axial distributions of static pressure, reverse mass flow rate and mass entrainment rate are presented as functions of bluff body blockage ratios in the range 0 to 0.54 and bluff body divergent half-angles in the range 0° to 90°.

b) The turbulence characteristics of the flow in the immediate wake of a »disc-in-nozzle« arrangement are given; for example, the spatial distributions of turbulent shear stress, turbulence intensities and turbulence kinetic energy. These were determined using a new method of interpreting anemometer responses developed for use in highly turbulent flows.

When correlated with the mean flow characteristics the turbulence measurements show that simple models of turbulence are inadequate for the evaluation of momentum exchange coefficients in highly turbulent flows. The correlation between the spatial distribution of mean stream function and local kinetic energy of turbulence, both calculated from measurement data, show good agreement with that predicted for a similar flow system using the Prandtl — Kolmogorov model of turbulence. This evidence adds support to the use of this model of turbulence for the evaluation of the coefficients of momentum exchange in highly turbulent flows, and provides the data from which it is possible to extract information on the empirical constants of the model.

LECTURE

THE CALCULATION OF HEAT AND MASS TRANSFER IN SEPARATED FLOWS UTILIZING BOUNDARY LAYER THEORY METHODS

A. I. LEONT'EV

Institute of High Temperatures, Moscow, USSR

Results of the analysis of the existing experimental data on convective heat transfer and flow hydrodynamics in separated flows are presented. A physical model of the fluid flow in separated zone is proposed, based on the hypothesis of the constant vorticity value. It is proposed that the calculations of the heat and mass transfer in this region are based on the development of the boundary layer created by the vortex flow of the fluid. The range of the application of the proposed calculation method is discussed on the basis of the comparisons of the calculation results with the experimental data. The following flow configurations are analysed: heat and mass transfer in regions of the boundary layer separation in cross flows, in vortex flows in fields of volumetric forces, in clearances between rotating coaxial cylinders, in bubble boiling of liquids, in flows along very rough surfaces. Trends in the further development of the heat and mass transfer calculation methods in separated regions are discussed.

LECTURE

STATISTICAL TRANSFER THEORY IN TURBULENT SHEAR FLOWS

B. A. KOLOVANDIN

Heat and Mass Transfer Institute, Minsk, USSR

The report deals with the problem of transfer of momentum and scalar substance in turbulent shear flows. The problem is formulated in terms of statistical moments of fluctuations of hydrodynamic and scalar substance fields.

At present numerous attempts are made to develop the theory of momentum transfer in turbulent shear flows, which is based on the use of the equations for the Reynolds stresses and the equations for higher-order moments (for example, works by P. Y. Chou, 1945; J. Rotta, 1951; B. I. Davydov, 1959). The essence of this theory is as follows: The Reynolds equations involving double correlations $R_{ij} = \overline{u_i u_j}$ are supplemented by the equations for the double and triple correlations.

In order to assure that the system of equations with respect to second moments be closed, the statistical hypotheses on the fourth moments (M. D. Millionshchikov, 1941) and on the terms involving the decay of turbulent momentum fluxes (A.P. Kolmogorov, 1941; P. Y. Chou, 1945; J. Rotta, 1951) should be employed. With the aid of the Poisson equation the terms containing pressure fluctuations are expressed in terms of two-point correlations of velocity fluctuations and velocity gradients of the mean motion.

Finally, the equations for two-point correlations of velocity fluctuations are used, on the basis of which the equation for the velocity correlation scale is obtained. As a result, it is possible to obtain the closed system of the equations with respect to the first and one-point second moments of velocity fluctuations. These equations include some constants, whose values may be determined either from the conditions of local isotropy or from experiments.

Similar approach may be applied to scalar substance transfer in turbulent shear flows. Here, we proceed from the possibility to close the mean equation for transfer of scalar substance with the system of the equations for higher order moments of velocity and scalar substance fluctuations.

In order to assure that the system of equations with respect to the second moments be closed, the hypotheses on the fourth moments (M. D. Millionshchikov, 1941) and on the terms involving the decay of turbulent fluxes of scalar substance are employed. The pressure fluctuations are expressed, as before, with the help of the Poisson equation in terms of two-point correlations between velocity and scalar substance as well as in terms of velocity gradients of the mean flow.

Finally, with the help of the equations for two-point correlations of the velocity and scalar substance the equation for correlation scales between the velocity and scalar substance is derived. As a result, we have the closed system of the equations with respect to an averaged value of scalar substance and one-point correlations between the velocity and scalar substance. This system of the equations also contains the coefficients which may be determined, for example, by statistical relations of axisymmetrical turbulence. Such relations may be found from the theory of axisymmetrical turbulence (G. K. Batchelor, 1946; S. Chandrasekhar, 1950).

LECTURE

THE CALCULATION OF STEADY, TURBULENT, RECIRCULATING FLOWS

D. V. SPALDING

Imperial College, London, U.K.

In order that heat- and mass-transfer processes may be predicted, for the flows indicated by the title, two conditions must be fulfilled: — (I) the turbulent-exchange process must be described by a sufficiently realistic set of differential and algebraic equations; and (II) these equations together with those for momentum, concentration, and stagnation enthalpy, must be rendered capable of sufficiently exact solution. To be valuable to engineers, and interesting to scientists, both the model of turbulence and the solution procedure for the equations must be generally applicable.

The lecture will describe the obstacles which have hitherto prevented the complete fulfillment of these conditions, and will show that prospects of removing them are good. Already a sufficiently general mathematical procedure exists. This calculation procedure is undergoing continuous development and improvement by the author and his colleagues.

The attainment of a satisfactory turbulence model is less complete. Computations of turbulent recirculating flows have been made, so far, with models which employ only one differential equation, for the kinetic energy of turbulence; the second turbulence quantity, the length scale, has to be specified as a function of position by way of an empirically-derived algebraic equation. The lecture will describe a new model of turbulence which deduces the length scale from a second differential equation; that for the time-averaged square of the vorticity fluctuations. It will be shown that this is capable of generating satisfactory predictions for some boundary — layer flows. The next step will be to incorporate it into the general solution procedure for the elliptic equations of recirculating flow.

PREDICTIONS AND EXPERIMENTAL RESULTS FOR FLOW PAST SUDDEN ENLARGEMENT IN A PIPE

A. K. RUNCHAL

Mechanical Engineering Dept., Imperial College, London, U.K.

The paper will deal with the theoretical and experimental results for heat/mass transfer in the separated and reattaching zones of flow immediately downstream of a sudden enlargement in a circular pipe.

The predictions are obtained from a numerical method based on a special, and to some extent, unconventional, flow-oriented finite-difference scheme which has been devised by Runchal, Spalding and Wolfshtein (1968). The numerical method is general enough to enable almost any of the current models of turbulence to be incorporated. The model employed is based on the turbulence-energy hypotheses of Kolmogorov (1942) and Prandtl (1945).

The experimental results are obtained from an electro-chemical technique — sometimes referred to as the diffusion-controlled-electrolysis technique. The data obtained are for Schmidt numbers of 1400 and 2500, and for Reynolds number between 2500 and 89000.

The predictions and experimental results are compared and it is concluded that a theoretical procedure is now available which can yield reliable predictions for such flows.

COMPUTATION OF TEMPERATURE DISTRIBUTION AND RADIANT HEAT FLUX IN A CYLINDRICAL COMBUSTION CHAMBER

W. M. PUN

Mechanical Engineering Dept., Imperial College, London, U.K.

This communication describes some computations made for a preliminary study of the temperature distribution and radiant heat flux in gas fired »radiant panels«.

The combustion chamber chosen for this study is of cylindrical shape. A mixture of fuel gas and air is admitted through a small, concentric opening at one end of the chamber; the products of combustion exit through the other end. The burning of the fuel heats the walls of the chamber, mainly by conduction and convection. Radiation from the walls passes out of the chamber, either directly or after multiple reflections.

The chemical reaction is regarded as proceeding by way of a single step, at a rate described by an expression of the Arrhenius type. The walls are treated as grey bodies with emissivities that are independent of temperature.

The governing equations are solved by the general, finite-difference procedure for recirculating flows developed in the Mechanical Engineering Department of Imperial College.

The computations that are described in the communication are chosen so as to exhibit the influence on the maximum wall temperature and radiant efficiency of the geometry of the chamber and the flow rate of the combustible gas.

PREDICTION OF THE HYDRODYNAMIC AND COMBUSTION PROPERTIES OF THE CONICAL STIRRED REACTOR

A. D. GOSMAN

Mechanical Engineering Dept., Imperial College, London, U.K.

The conical stirred reactor consists of a hollow cone, open at one end and closed at the other by a hemispherical cap. At the open end, a pre-mixed stream of fuel and oxidant is introduced through a nozzle into the reactor, where combustion occurs. The products of combustion exit through the annular gap between the nozzle and the reactor wall. This configuration results in a flow in which there are large zones of recirculation, and high turbulence levels: these are features which enhance mixing, and hence efficiency of combustion.

This paper describes a theoretical study of the conical reactor, the object of which was to obtain predictions of the hydrodynamics of the flow (i.e. the flow pattern, distribution of turbulence energy) and of the combustion properties (i.e. distributions of temperature, concentrations of reacting species). The predictions were obtained by solving numerically the set of simultaneous elliptic partial differential equations which describe two-dimensional turbulent flows with recirculation and chemical reaction. The numerical solution procedure employed was one which has been developed by a group of workers in the Mechanical Engineering Department of Imperial College.

Results are presented for a range of operating conditions (i.e. mass flow rate, inlet concentrations): for each set of conditions, solutions are presented with and without combustion, so as to show how the latter influences the flow pattern.

RECIRCULATION IN CONFINED VORTEX FLOWS — PREDICTIONS AND EXPERIMENTAL RESULTS

N. A. CHIGIER and N. SYRED

Department of Chemical Engineering, University of Sheffield, U.K.

Effective viscosities have been determined for recirculating flows in confined vortex chambers by matching profiles of velocity measured by hot wire anemometer with predictions obtained from the finite difference numerical analysis of Spalding et al.

Measurements have been made in a cylindrical vortex chamber with tangential and axial inlets, with the aid of a DISA hot wire anemometer for flow conditions over a range of swirl numbers. Solid body rotation is obtained when all flow is introduced tangentially and free vortex flow is approximated with high axial and low tangential flow rates. Recirculation is found between the central jet and chamber walls at low swirl numbers and internal recirculation along the axis is found at higher swirl numbers.

Predictions have been made by applying the finite difference solution of the full Navier Stokes equations after Spalding et al. The turbulent kinetic energy model of turbulence has been used for the confined swirling flow case. By matching the predicted and measured profiles of axial, radial and tangential velocity distributions, values of effective viscosity are determined at each point in the flow field for the range of flow conditions.

NUMERICAL SOLUTION OF THE PROBLEM OF A TURBULENT IMPINGING JET

M. WOLFSHTEIN

Mechanical Engineering Dept., Imperial College, London, U.K.

The problem of impinging jets is very often met in many branches of engineering. In separated flows impinging jets are found in the reattachment region.

The prediction of turbulent impinging jet flow is difficult for two reasons. Firstly, because the boundary layer approximation cannot be made, and the full elliptic Navier-Stokes equations have to be solved. Secondly, near the stagnation point, the mixing length hypothesis is entirely unrealistic, because convection of turbulence becomes a phenomenon of overriding importance in this region.

The first of the above difficulties may be overcome by the use of finite-difference equations and numerical methods. For the second difficulty a model of turbulence, which incorporates a transport equation (or equations) of the turbulence, is necessary.

In the present paper the application of both the above ideas to a plane turbulent impinging jet flow is described. The differential and difference equations which control the phenomenon are presented, and the way by which the boundary conditions are incorporated in the numerical technique is discussed.

Results of the computations include contour plots and profiles of velocity, pressure and turbulence. It is also shown that the exact pattern of the recirculating, entrainment flow has no significant influence of the main jet flow. Comparison with experimental data is favourable.

EXPERIMENTAL RESULTS ON HEAT TRANSFER RATES OF SINGLE AND MULTIPLE JETS IMPINGING ON WALLS

W. KOSCHEL

Institut für Luftfahrttriebwerke, Technische Universität Berlin

In the Institute of Aero Engines at the Technical University of Berlin an experimental work on the heat transfer of impinging jets has been carried out since 1967 in order to applicate impinging jets for the inside cooling of turbine blades. The experiments are performed with cold air for nozzles with diameters of 3.1 mm; 3.8 mm and 4.4 mm. The air jets are impinging normal to a plane wall and to the concave sides of cylinders, the inside diameters of the cylinders ranging from 20 mm to 80 mm. The study involves the effects on heat transfer of single jets as well as of an array of jets with variable number, size and spacing of the nozzles. The air flow rates are measured by a special orifice. Parameters in the experiments are the distance nozzle-to-wall and the exit velocity of the air with Mach-numbers from 0 to 0,5.

The data, which are obtained by the measurements, are the pressure distribution and the local heat transfer coefficients. The measurements of the heat transfer are made possible by a small heat-flow sensing device, which allows the determination of the heat transfer rate on the surface of the wall within an area of 0.8 mm in diameter.

The purpose of this study is to find the combination of size and spacing of nozzles in an array of jets for a given air flow rate, by which the best distribution of the heat transfer coefficients on the inside wall of a cylinder could be obtained.

LECTURE

GAS FLOWS ALONG POROUS WALLS UNDER INTENSIVE BLOWING CONDITIONS

V. P. MOTULEVICH

University, Moscow, USSR

One cause of the flow separation from a wall could be the intensive mass blowing through porous elements of the surface. These phenomena are of special interest when it is necessary not only substantially to decrease dynamic and thermal interference of the flow with the wall but also to create a region in which the concentration of the blown-in materials is around 100%.

Results of an experimental investigation carried in a air-tunnel and using Mah-Zender interferometer with a laser light source, allowing for the substantial increase in measurement accuracy, are presented. For the determination of the velocities in the binary zones a hot-wire anemometer specially calibrated for various mixture concentrations was used. Velocity vector direction on the flow separation line was determined by optical methods.

The structure of the boundary layer was investigated. Velocity and concentration profiles have been measured in a wide range of the blowing parameter. All these profiles have a characteristic S-form with the inflection point in the zone of the viscous interference. With intensive blowing a flow region near the wall is created in which the velocity profile is nearly linear with a small gradient and in which the concentration of the blown-in material is practically 100%.

Separation of the turbulent boundary layer appears for substantially larger blowing parameters than in the laminar flow case. Preceeding the laminar-turbulent transition is a zone of regular transverse fluctuations. It was found that the dependence of the critical Reynolds number on the blowing parameter has a minimum. With the increase of the blowing intensity the principal parameters of the perturbance motion, wave length and amplitude, are substantially changed.

LECTURE

INFLUENCE OF ROUGHNESS AND NON-ISOTHERMICITY ON HEAT AND MASS TRANSFER IN TURBULENT WALL JETS

E. P. VOLCHKOV

Institute of Thermophysics, Novosibirsk, USSR

Results of the investigation of effectiveness and heat transfer in the turbulent wall jets are presented. Experimental data concerning of jet flow on smooth and rough (with regular tubular roughness) surfaces under quasi-isothermal conditions are also presented. Experimental results for the smooth surface are in agreement with the suggested method of calculation.

The effectiveness (heat-insulated wall temperature) of the wall jets is found to be weakly depending on the roughness and this dependence with an increase of the injection parameter. The heat transfer in the wall jet on the rough surface is circumscribed by the same relations as in the case of a homogeneous gas flow over rough surface if a heat flux is defined by the actual wall temperature and heat-insulated wall temperature difference.

Results of heat and mass transfer investigation in the wall jet under considerable non-isothermal conditions are also presented.

Experiments were performed with a hollow graphite cylinder, which has been heated to the temperature in the range of 1600 — 2000°K by means of a high-frequency induction heater. The jet of the nitrogen or air has been injected through the tangential slot at the entrance of the graphite section. Experimental data for the intensity of burning of the graphite surface in the wall jet are in good agreement with the results of the suggested theoretical method.

The microamperemeter is connected with the electric circuit of the electromagnetic pump. The readings of this microamperemeter give the possibility to determine the value of measuring pressure drop.

The given mercurial differential pressure-gauge for measuring of small pressure determining of pressure drops in the range of $0.1 - 100 \frac{N}{m^2}$ with accuracy about 1%.

EFFICIENCY OF GAS SCREENS IN THE PRESENCE OF TURBULENT EXCITATIONS

J. V. BARISCHEV, V. I. ROZHDESTVENSKY

University Lomonosov, Moscow, USSR

The research was aimed at gauging gas screens in the presence of a projection of differing configuration before the thermal insulated surface.

A calculation method for gas screen effects was advanced whereby of vortex currents in the area of the projection were considered. The calculation results were then compared with experimental data.

LECTURE

HEAT AND MASS TRANSFER IN THE VORTEX JET FLOWS

J. IVANOV and V. HENDRIKSON

Institute of Thermal Physics, Tallin, USSR

The task of the creation of the high efficiency constructions in the different branches of technics establishes the problem of finding the methods of intensification of the processes of heat and mass transfer. One of such methods of intensifying heat and mass transfer between gas and surface is the way of provoking vortex jet flow. It appeared to be possible to force the same kind of interaction between two gas flow without any rigid surface. One of the flows (as a rule with a smaller total mass flow) is divided into several jets which are injected normally into the main stream. Vortex flows which are produced behind the jets intensify very effectively the heat and mass transfer between two gas media.

The present report deals with a study of the process of heat transfer between hot jet and cold cross wind or vice versa. Although this method is technically simply applied and is widely used we know quite little about the process itself.

In this complicated pattern of flow the only reasonable way of research would be experimentation, with a further approximate analytical solution on the basis of experimental data.

The experimental study was made with various constructive and flow parameters. The jet path, distribution of velocity and temperature in the jet were studied. The approximate flow similarity over studied range was found, on the basis of which an analytical solution was proposed. The results of the analysis are in satisfactory agreement with the experimental data.

HEAT TRANSFER IN THE VORTEX FLOW OF COMPRESSED GAS IN THE FIELD OF BODY FORCES

G. B. PETRAZITSKY, V. I. POLEZAEFF

The Moscow Higher Technical School, Moscow, USSR

The results of numerical study of flow and temperature fields for the case of laminar motion of viscous heat conducting compressible gas caused by body forces are presented. The study was carried out for rectangular regions with different side-to-side ratios and with different orientation of the outer force.

The investigation was conducted on the basis of the numerical solution of the total system of two-dimensional, non-stationary equations of motion, continuity and energy with variable physical constants and with compressibility and dissipation terms.

Stationary solution was obtained using the method of setting. Numerical calculations were used for obtaining velocity, density and temperature fields and flow functions as well as the local and mean values of heat flows at the boundaries of the region. These calculations were conducted for different time values in the wide range of 8 determining dimensionless complexes corresponding to the problem.

The main calculations were carried out for the condition of the two opposite boundaries of the rectangular region being at constant but different temperatures, with two other boundaries being heat-insulated.

TRANSITION FROM TURBULENT MOTION TO LAMINAR ONE FOR NON-LINEAR STOKES FLUID

V. E. AEROV and B. A. KOLOVANDIN

Heat and Mass Transfer Institute, Minsk, USSR

This study considers the motion of fluids which obey the Stokes postulates, and for which viscous stresses are in non-linear relation with the applied deformations. In particular, for incompressible pure viscous fluid an approximate relationship describing the dependence of the stress tensor on the deformation velocity tensor may be found in the form

$$\tau_{ij} = -p\delta_{ij} + (\mu - \mu_2 I_2)\dot{e}_{ij} \tag{1}$$

where p is hydrostatic pressure; δ_{ij}, Kronecker symbol; μ and μ_2, constants; I_2, second invariant of the deformation velocity tensor.

To find the critical Reynolds number depending on nonlinear properties of the fluid, a method proposed by V. B. Levin is used, according to which at some value of the dynamic Reynolds number the equation of energy for fluctuational motion has no positive solution.

By using this condition from the simultaneous solution of the equations of mean motion and fluctuational motion energy, the critical values of the parameters necessary for determination of the critical Reynolds number may be found.

As an example, a fluid flow in a plane-parallel channel is considered. The computations performed on the digital computer Minsk-22 revealed the effect of non-linear fluid properties on the critical value of the Reynolds number.

USE OF THE ELECTRON GUN FOR THE INVESTIGATION OF RAREFIED GAS FLOWS

J. C. LENGRAND

Laboratoire d'Aérothermique du C.N.R.S. Meudon, France

The use of the electron beam probe for investigations in rarefied gas flows has been developed in these last few years. The present paper gives some examples of applications of the probe, and more precisely its use in the »Laboratoire d'Aérothermique du C.N.R.S.«

When a beam of energetic electrons is passed through a rarefied gas flow, a fluorescence of the gas is observed, which is essentially due to the creation of excited molecules or ions.

If a preliminary calibration is made, the local density is obtained by measuring the fluorescence of a small volume excited by the beam.

The flow may be visualized by means of the beam displacement across the flow field.

Rotational and vibrational temperature of a diatomic gas as nitrogen may be determined by an analysis of the emitted light, provided a theoretical description of the excitation-emission path is postulated.

Concentrations in gas mixtures may be obtained after a preliminary study for a given mixture by measuring relative intensities of lines or bands.

Line-broadening produced by Doppler-effect is a means of determining translational temperature. The shift of the line gives a measurement of the flow velocity. This measurement has already been conducted with helium.

The equipment used in the Laboratoire d'Aérothermique is composed of an electron gun and an optical device including essentially an objective, and a photomultiplier for density measurements, or an objective, a spectrometer and a photomultiplier for temperature measurements. The whole system is inside the vessel of a rarefied gas wind-tunnel. The electron gun and the optical device are fastened with the same support, which may be displaced in the three directions. Then it is easy to explore any point of the flow. The support displacement and operations concerning both the gun and the spectrometer are remotely controlled.

The influence of the beam properties and of the dimension of the viewed volume on the correspondence between measured light intensity and density has been experimentally studied, as well as the influence of the density of the adjacent points on the light emitted by a given volume.

Wakes of cylinders and spheres at Reynolds numbers from 3.2 to 320 and Knudsen numbers from 0.1 to 10 as well as the boundary layer on a flat plate have been studied. The obtained densities will be presented.

The determination of rotational temperatures requires a theoretical model for the excitation. Such a model being selected, different operating ways are possible:

— one may measure the relative intensities of all the rotational lines of a given vibrational band. This eventually permits discovering a non--Boltzmann distribution of gas molecules.

— one may compare the whole intensities of two sets of lines in the same band. Such a set of lines is selected for instance with bandpass filters. This method is faster.

— we prefer to use both preceding methods with a spectrometer. A narrow exit slit permits recording the spectrum. A wide exit slit permits getting out a well-defined set of lines.

The electron beam probe is found to be an interesting means of investigation by its multivarious applications for studying rarefied gas flows. However, it is necessary to examine carefully the conditions of validity of the different formulas and theories in order to obtain an accurate knowledge of the gas parameters.

LECTURE

RECENT DEVELOPMENTS IN HOT WIRE ANEMOMETRY

P. O. A. L. DAVIES

Institute of Sound and Vibration Research, Southampton University, U.K.

Hot wire or hot film anemometry provides the most highly developed technique for dynamic measurements at a point within a fluid system. This lecture first reviews the present state of the art, both from a fundamental as well as a practical point of view. The lecture then goes on to discuss the difficult problem of statistical measurements and the interpretation of hot wire signals in terms of fluctuating flow properties.

THE PERFORMANCE OF NORMAL AND YAWED WIRES

H. H. BRUUN

Institute of Sound and Vibration Research, Southampton, U.K.

The heat transfer for normal and yawed wires has been studied for flows with very low turbulence intensities. Using the hot-wire probes and the hot-wire anemometer developed at the Institute of Sound and Vibration Research, it was experimentally established that the calibration laws for normal and yawed wires with the same nominal geometry could be expressed in universal forms.

The reliability of this method, for low intensity turbulence flows, was successfully shown by Reynolds stress measurements in a pipe with fully developed turbulent flow.

Due to non-linearities in the heat transfer relationship, great uncertainty was found when applying the steady state calibration law to high turbulence intensity flows.

HOT WIRE ANEMOMETER TECHNIQUE FOR TURBULENCE MEASUREMENT ALONG AND ACROSS THICK BOUNDARY LAYERS

D. L. ZORIĆ

Department of Traffic Engineering, University of Beograd, Yugoslavia

A large scale turbulent boundary layer with no pressure gradient, developed on a flat plate 95 feet long, has been investigated. Artificial thickening of the boundary layer allowed development of the boundary layers up to 3 feet thick. To obtain turbulence data, measurements with the hot wire anemometer were used. Measurements covered the full length of the flat plate in continuous runs, and all three components of turbulence were measured. A special technique was therefore developed to ensure continuity, reliability and precision of measurements. Single rotating wire probes were used and uncertainty intervals calculated for the results. The investigation took place at the Colorado State University, Ft. Collins, U.S.A.

AUTO-CORRELATION MEASUREMENTS OF TURBULENT VELOCITY FLUCTUATIONS IN THE VISCOUS SUBLAYER USING REAL TIME DIGITAL TECHNIQUES

NGUYEN VAN THINH

Laboratoire de Méchanique Expérimentale des Fluides, Faculté des Sciences, Université de Paris, France

The hypothesis of an instantaneous linear profile of the longitudinal turbulent velocity fluctuation in the viscous sublayer and its consequence on the correlation and spectra functions has been examined on the basis of recent experimental results. It is shown that a similarity exists for the correlation curves and as a consequence, the spectra function, normalized with suitable wall parameters, is only a function of the Strouhal number. A program of autocorrelation measurements using real time digital techniques is then developed to verify these results.

Both linearized and unlinearized signals have been processed, showing à difference in the correlation results. Finally, the distribution of non--dimensional time micro-scale and time derivative in the viscous sublayer is presented.

LECTURE

STATISTICAL INTERPRETATION OF THE INSTANTANEOUS VELOCITY AND TEMPERATURE SIGNALS IN HIGH INTENSITY TURBULENT SHEAR FLOWS

Z. ZARIĆ

Boris Kidrič Institute, University Beograd, Yugoslavia

A method of measurement and statistical analysis of the instantaneous velocities and temperatures with a single probe in the form of a microne--thick wire is presented. Velocity and temperature probability density distributions in wall layers of flows with various turbulence intensities, obtained by this method, are given.

These results are discussed on the basis of theoretical distribution functions depending on several variables. It follows that probe signals contain more information on turbulence characteristics than what is normally supposed. The possibilities of exploiting this fact for the obtainment of additional information on turbulence structure in the wall vicinity and for the elimination of measurement error due to high intensity turbulence, are further discussed.

THE ANALYSIS OF ANEMOMETER RESPONSES IN HIGHLY TURBULENT FLOWS

R. G. SIDDALL and T. W. DAVIES

Dept. of Chemical Engineering, University of Sheffield, U.K.

During a study of the turbulence characteristics of separated flows a new method of interpreting the electrical responses of an anemometer was developed which allowed the instrument to be used in highly turbulent flows.

The method is based on an improved form of the steady state heat transfer relationship for a wire situated normally to a laminar flow. This may be written as

$$E^2 = A + BU_e^{1/2} + CU_e$$

where E is the indicated wire voltage and U_e is the total effective wire cooling velocity. This equation alone represents a considerable improvement in the accuracy attainable by normal anemometry techniques.

The analysis is developed in a general form so that the three mean velocity components and the six turbulent stresses at a point may be determined easily without prior knowledge of the flow field and regardless of turbulence level. The assumptions made in analysis are more flexible and less restrictive than the assumptions made in conventional analyses. Binomial expansions are not used and velocity terms are not neglected. Instead, assumptions are made about the characteristics of the fluctuating component of the wire voltage. These are that the fluctuation takes the form of a square wave, the period and phasing of which is independent of the orientation of the wire at a given point within the flow field. Calculations have shown that the square wave assumption is justifiable and measurements are available, which show that the assumption about period and phasing is also valid.

The square wave assumption enables the mean value of the total instantaneous wire cooling velocity ($\bar{U_e}$) to be expressed as a simple function of the measured anemometer responses ($\bar{E}$) and $(\overline{E'^2})^{1/2}$ which are the mean d.c. and r.m.s. values of the fluctuating bridge voltage respectively.

i.e. $\bar{U_e} = 1/2 \{ f[(\bar{E} + \overline{E'^2})^{1/2}] + f[\bar{E} - (\overline{E'^2})^{1/2}] \}$

The 'period and phasing' assumption permits the calculation of the mean velocity components and hence the values of the six turbulent stresses.

The anemometry technique associated with the new analysis requires the determination of six separate anemometer responses at a point. The six wire positions may be arbitrarily chosen on the grounds of experimental convenience and without reference to the mean flow direction at the point. The measurements may easily be made with one or two wires.

The new analysis has been tested by using it to determine the turbulence characteristics of a free jet. In regions where the turbulence intensity $(\overline{U'^2})^{1/2}/U$ was below 20% the new method gave results which agreed well with the results given by conventional analyses. In the highly turbulent outer regions of the jet conventional analyses indicated apparent turbulence intensities which were much higher than those indicated by the new method.

The new method of analysis has been used in the determination of the turbulence characteristics of a region of separated flow (free annular jet). The measured correlation between the spatial distributions of mean stream function and local kinetic energy of turbulence show good agreement with the correlation prediction for a similar flow system using the Prandtl-Kolmogorov model of turbulence.

HOT-WIRE MEASUREMENTS IN ROTATING IMPELLERS FLOW AND THE CALCULATION OF INTENSITIES

J. D. VAGT

Herman Föttinger-Institut für Strömungstechnik, Berlin, FR of Germany

In the turbulence level high, the equations of Champagne for the calculations of intensities and correlations can no longer be used. Champagne only considered the first order velocity terms and the deviation from the cosine law up to the fourth order. The calculation was extended by taking into account velocity fluctuation terms up to the fourth order as well as the deviation from the cosine law.

One of the basic demands for the calculation of the intensities is — if the simplified equations are used-that the inclination angles of the two hot wires (X-array) are equal to 45°. To assure this, a special hot wire probe was developed. With this probe it was possible to adjust the two wires to the exact position.

HOT-WIRE MEASUREMENTS IN ROTATING IMPELLERS

K. GIESE

*Hermann Föttinger Institut Technische Universität Berlin,
Federal Republic of Germany*

The influence of the yaw-angle between the flow and a line perpendicular to the hot-wire axis can be used for the determination of direction and magnitude of the velocity by making two measurements at each point of the flow field with either two crossed hot wires simultaneously or with one hot-wire which is turned between the two measurements. The second technique which can give results only for the mean values, was used for measurements in a rotating radial impeller. Problems of this measuring method related to the special conditions in rotors of turbomachines are discussed as influence of the centrifugal forces on the hot-wire calibration, influence of axial velocity-components and high turbulence level on the accuracy of the measurements. Some results of the measurements in a rotating impeller will be shown.

A NEW LINEARIZER UNIT FOR HOT WIRE ANEMOMETRY

E. FROEBEL

Institut für Turbulenzforschung, Berlin, Federal Republic of Germany

The purpose of a hot-wire anemometer is the measurement of flow velocities including the velocity fluctuations. Due to new electronic elements it is now possible to improve hot-wire anemometers with respect to noise level and frequency band width. Especially in turbulence measurements one needs high quality linearisation stages in order to obtain an output signal proportional to the momentary flow velocity. The built-in linearisation stage of the DVL-CT-hot-wire anemometer converts the bridge voltage into a polygonal function of it by means of a switched resistor-network. This switched resistor-network can be replaced by a recently developed unit with a continously curved characteristic made with integrated circuits. Advantages and disadvantages of both units will be discussed.

LECTURE

REMARKS ON AVERAGING IN TURBULENT FLOWS

S. J. KLINE and T. B. MORROW

Mechanical Engineering Dept., Stanford University, U.S.A.

Recent investigations of the turbulent boundary layer have shown two characteristically dissimilar types of flow structure in the layers near the wall. The dual structure has been examined through the use of the hydrogen bublle visualization technique in a water channel and is found to consist of (1) the unstable growth of disturbances near the boundary which accounts for the production of turbulent fluctuations, and (2) turbulent dissipation including the cascade of energy to smaller scales.

Conventional statistical space-time averages formed by long time averaging, which have been used succesfully to reveal the flow structure of decaying turbulent flows, are appropriate only to flows with a single type of structure and fail to reveal the dual structure of a turbulent boundary layer flow.

This situation, of dual structure, may well exist in other complicated flow patterns. Some ways for determining when this is so are discussed. Some more appropriate statistical procedures for such dual structures are suggested.

FLOW VELOCITY MEASUREMENT USING AN ELECTROCHEMICAL METHOD

J. L. BOUSGARBIES

Laboratoire de Dynamique des Fluides, Faculté des Sciences de Poitiers, France

In the presented measurement technique, the liquid in which the velocity field is to be determined, is an electrochemical solution of an electrochemical cell. The electrodes of the cell are made of different metals and one of them is the probe for the flow velocity determination. Electrolytical liquid used is a water solution of iodine of a very low concentration and the potassium-iodide of a high concentration. Measuring electrode is made of platinum, its active part only being in contact with the solution. The active part is in the form of a half-sphere of very small dimensions (0.2 to 1 mm in diameter). The opposite electrode is made of copper and its area is large in comparison with the area of the probe.

When the cell is connected to a resistance R a current is produced; consumption of the iodine on the surface of platinum and the oxidation of the copper start. It is easy to determine the current intensity value I in the circuit in function of the voltage difference at the end of the resistance, measured by a digital voltmetre. For a suitable value of the R the iodine concentration C is zero on the probe surface. Current intensity is then limited by the transport velocity of the iodine towards the platinum surface. If the liquid around the probe flows with a velocity V, there is a transport of the reacting substance by convection and a relation between the flow velocity V and the current intensity exists.

The platinum probe is placed on the axis of a cylindrical tube in which the solution is flowing. Experiments show an increase of the voltage U with the increase of the mass flow rate. The probe could be calibrated as the the velocity on the axis could be calculated from the known flow rate. The calibration curve allows to determine the flow velocity in any point in the flow from the measurement of the current intensity. Measurements are reproducible with an error of less than 3%. By this method very low velocities of the order of 1 mm/second could be measured.

ON THE APPLICATION OF AN ELECTROLYTIC PROBE AS LIQUID VELOCITY DETECTOR

C. PISONI

Istituto di Fisica Tecnica, Università di Genova, Italia

An experimental procedure for the determination of the velocity fluctuations induced in a stagnant liquid by injected air bubbles, is presented.

For the small agitation amount occurring in the liquid, a measuring probe operating on the basis of electrolytic effects was found suitable; this device presents a high sensitivity at the lowest values of the liquid relative velocity.

The experimental results obtained and their reproducibility in several test conditions, at different air bubble rates, show that the test procedure adopted is particularly suitable to the study of the fluodynamic problem here investigated.

DYNAMIC RESPONSE OF AN ELECTROLYTIC PROBE FOR VELOCITY MEASUREMENTS IN LIQUIDS

G. GUGLIELMINI

Istituto di Fisica Tecnica, Università di Genova, Italia

The present communication refers on the performance set-up of an electrolytic velocity detector.

The influence of the geometric characteristics of the probe (as: shape, dimensions and spacing of the electrodes) both on the steady state response and on the dynamic response has been investigated at different liquid velocity values.

A method which allows the obtainment of the calibration curve and the evaluation of the time response is presented and discussed.

It has been observed that the time response of the probe strongly depends upon the liquid velocity value.

POLAROGRAPHY, ELECTROCHEMICAL MEANS FOR THE STUDY OF THE MOTION OF LIQUIDS

P. PY and J. GOSSE

Faculté des Sciences, Nancy, France

Polarography subject to certain precautions like the choice of chemically inert electrodes and the use of a fast red-ox electrolyte, allows a sensible and reproductible study of mass transfer by convection up to high velocities. Hanratty and co-workers have showed that microelectrodes inserted in the wall permit an evaluation of the local and instantaneous characteristics of the flow. In this case, the mass transfer coefficient is mainly determined by the velocity gradient at the wall.

When frontal or tangential microelectrodes are inserted in a probe, which is located in the flow, the reduction current is a function of the main component of the velocity on a level with the boundary layer. The laws of transfer are dissimilar according to whether the flow is laminar or turbulent, and according to whether or not the boundary layer which develops on the probe becomes turbulent on the microelectrode.

The polarographic method permits the determination of the velocity profile in a channel. It has been particularly adequate for detecting the apparition of various modes of instability in a Couette flow and for studying their developments. Its advantages from the point of view of sensibility, reproducibleness and easy application, allow its use for experimentantion about turbulents or non-steady flows.

ELECTROCHEMILUMINISCENCE METHOD FOR MEASURING VELOCITY FIELDS IN TRANSPARENT COMPLEX LIQUIDS AND ITS APPLICATION TO KARMAN STREET

A. V. LUIKOV and Z. P. SHULMAN

Heat and Mass Transfer Institute, Minsk, BSSR, USSR

At present the methods of pneumometry and thermoanemometry are mainly used for measuring flow velocities. Both methods are well-knwon and widely used. They are characterized by such defects as low sensitivity, especially at low velocities, and complicated calibrating in unhomogeneous velocity fields. They are of no use for velocity measurements in low--molecular solutions with high-polymer additions, in suspensions, emulsions, etc.

The electrochemiluminiscence (ECL) anemometer is based on the phenomenon of spontaneous luminescence of ions of special luminophore under the action of an electric field. The brightness of electrochemiluminescence with fixed voltage in the electrochemical cell is determined only by the intensity of diffusion mass transfer from an active electrolyte to the anode surface. ECL-luminescence is generated in the region near the anode, in a layer a few light wavelengths thick, i.e. at distances much less than the thickness of the diffusional boundary layer. For small potential differences between the anode and the solution there exists almost linear dependence of the current on the applied voltage. When the potential difference increases up to 0.45—0.65 V, the current attains its maximum value and does not depend any longer on the applied e.m.f. The maximum value of the current corresponds to zero concentration of ions discharging on the anode surface, i.e. the ions approaching the anode surface discharge instantaneously and then the current through the cell may be controlled only by the intensity of ion diffusional transport to the anode. Under certain conditions the ECL-method may be used for the study of bulk effects (eddies, wakes, cavities, stall, etc.).

The effect of polymer additions was studied by this method. The experimental results allowed a region to be found beyond which the presence of a polymer has no effect on the flow. Polymers affect the flow region where dimensions of vortices are comparable with assemblies of polymer macromolecules and solvent molecules in the solutions.

A different possibility may be offered by the study of a vortex structure of stalled boundary layers. To establish the regularities of formation of Karman vortices in stalled flows around bodies und the role of rheological factors is of great interest and importance. Special experiments using circular cylinders in a transverse flow have been carried out with this aim.

MECHANISM OF THE TRANSITION BETWEEN CONDUCTION AND CONVECTION

C. THIRRIOT and S. BORIES

Institut de Mécanique des Fluides, Université de Toulouse, France

By interferometry and then, by photography of particles in suspension in a fluid, we investigated the transition regime conduction-convection in a two-dimensional Hele-Shaw cell system.

The growth of the disturbances which, after development, will give the eddies of convection, is investigated in particular.

Velocities and temperatures are determined from measurements using photographs of particles and interference fringes.

ASTIGMATIC USE OF A DIFFERENTIAL INTERFEROMETRIC FITTING WITH A WOLLASTON DOUBLE PRISM FOR THE STUDY OF THE THERMAL BOUNDARY LAYER

G. GONTIER and A. MARTINOT-LAGARDE

Institut de Mécanique des Fluides, Université de Lille, France

When, to observe and measure the density gradient in a moving gas, an interferometric device of the differential type is used, with a spherical mirror and a Wollaston double prism for instance, the optical pieces are usually set up to realise at best the stigmatism conditions. If the question is to study a narrow field such as that of a boundary layer along a flat plate, the best sensitivity is obtained when the fringes are parallel to the plate; but then, a few number of fringes are observed in the boundary layer and consequently the measurements are not accurate.

The method of measurement may be improved a great deal when the spherical mirror is used in an astigmatic manner, but not too much so that the images remain clear. Then, keeping the best sensitivity, one can give the fringes any direction with respect to that of the displacement which occurs between the two separated beams. In such a case, the device shows a suppleness of adjustment similar to that of the conventional Mach-Zehnder fitting. The fringe direction can be kept normal to the plate outside the boundary layer and, according to the fringe deformations observed in the boundary layer, one can measure the normal gradient of the density with good accuracy.

In order to study a phenomenon of heat conduction, this procedure has been used to determine the temperature distribution in the neighbourhood of a vertical plate heated at one end. The results given by interferometry agree with that given by a thin thermal gauge.

AN INTERFEROMETRIC STUDY OF COMBINED FORCED AND FREE CONVECTIVE MASS TRANSFER ON A VERTICAL FLAT PLATE

G.G. HARDEL and M.M. EL-WAKIL

University of Wisconsin, Madison, U.S.A.

Local mass transfer coefficients and concentration profiles for pure and combined forced and free convective mass transfer of a heavier-than--air vapor from a vertical flat plate to a vertical air stream were measured using a Mach-Zehnder interferometer. Results for flows in which the buoyancy forces both aid and oppose the free stream were obtained. The latter included the regions before and after separation.

The experimental model consisted of a vertical falling liquid film of evaporating n-heptane. An unwetted starting length insured a sharp leading edge and reduced undesirable liquid build-up at the film entrance and exit. A uniform velocity, low turbulence air stream was blown either upward or downward as desired. The experiments were taken under near isothermal conditions, and a constant concentration boundary condition existed at the liquid-vapor interface.

Experimental mass transfer coefficients and concentration profiles were compared to analytical solutions where possible. There were good agreements for pure forced and free convection. In combined convection, there was good agreement with an existing approximate analytical technique in aiding flow, but increasing disagreement in opposing flow as the point of separation was approached. Data in the region beyond separation indicated that the mass transfer coefficients rapidly approach those for pure free convection.

The data is presented in the form of concentration profiles as well as plots with Sherwood, Reynolds and Grashoff numbers as parameters.

LECTURE

THE PHOTOCHROMIC VISUALIZATION OF FLOWS WITH AN ILLUSTRATION OF PIPE JET FLOW

A. de P. IRIBARNE and R. L. HUMMEL

Dept. of Chemical Engineering, University of Toronto, Canada

The technique described in this lecture was proposed by R. L. Hummel and developed by A. T. Popovich, F. Frantisak, the co-author and others.

Photochromic indicators like: 2- (2, 4-dinitrobenzyl) pyridine and nitrospiropyran dissolved in appropriate solvents produce a dye streak when irradiated with an intense beam of UV-light. A giant pulse ruby laser with frequency doubler was used to induce the tautomeric reaction in the dye and measurements were made from high-speed movies. Optimal experimental conditions for a trace of as little as 30 µ in diameter and length of 2.5 cm have been determined. This non-disturbing flow visualization technique has been used in a number of related subjects, including the flow patterns in an axisymmetrical pipe jet with diametrical expansion ratio of 1:2. Reynolds numbers based on the larger diameter from 50 to 1000 were investigated. A wave-like flow instability was discovered in laminar flow even at the lowest Reynolds numbers studied.

Instantaneous and mean velocity profiles were determined at various down-stream positions, as well as other flow properties. Because of the flow instability, the reattachment point of the separated region was confined to a wide band for any Reynolds number.

THE APPLICATION OF PHOTOCHROMIC VISUALIZATION TO STUDIES OF FLOW IN ROUGH PIPES AND OTHER SEPARATED FLOWS

J. W. SMITH and R. L. HUMMEL

Dept. of Chemical Engineering, University of Toronto, Canada

The technique described previously by Dr. Iribarne has been applied to a wide variety of fluid flow situations by the students of the authors. Accurate velocity profiles and other turbulent properties have been obtained very near the rough pipe walls. The nitrospiran dye was used in a solution in which refractive indices of fluid and pipe wall were carefully matched. Velocity gradients and shear stress distributions have also been obtained in liquid fluidized beds using a similar technique. Separated and laminar flows around submerged objects, such as spheres, have also been studied. The technique is particularly useful in studies very close to solid surfaces or in flows in which a probe disturbs the flow pattern unduly. Included in the latter category is a study of viscoselacticity, of falling films, and of the pipe jet near the reattachment point, previously described. These various programs are illustrated by films obtained with high-speed motion picture cameras.

SMOKE VISUALIZATION OF WAKES BEHIND A GROUP OF CYLINDERS

M. M. ZDRAVKOVICH

Department of Mechanical Engineering, University of Beograd, Yugoslavia

A smoke visualization technique was used to study the laminar and transitional wakes behind a group of cylinders. It was conducted in the Aerodynamics Laboratory of the Engineering Department of the Cambridge University (U.K.) in a vertical low speed wind-tunnel with 30×15 cm working section. The wake was made visible by the introduction of smoke filaments, actually liquid kerosene in the form of a fog of fine droplets produced by a Cambridge smoke generator of the Preston and Sweeting type.

The modes of interaction of three cylinders and their wakes are shown for different spacings and velocities. The following series of mechanisms was revealed and made visible:
a) Origination of an instability and its development,
b) Formation of a new Kármán vortex street,
c) Crossing of vortex rows,
d) Transition to turbulence,
e) Origination and development of three-dimensionality in a two-dimensional flow in wake.

Items (a) to (d) serve as an illustration of the method because they are published in the Journal of Fluid Mechanics. Ithem (e) is new and still unpublished.

DETERMINATION OF MICROTHERMOCOUPLE DYNAMIC CHARACTERISTICS

V. PIŠLAR, N. AFGAN, M. STEFANOVIĆ and LJ. JOVANOVIĆ

Boris Kidrič Institute, Beograd, Yugoslavia

A microthermocouple probe was developed for the temperature fluctuation measurements in two-phase flows. The microthermocouple was made from 0.02 mm copper and konstantan wires. Assuming the step function of heat flux at the surface of thermocouple, response of the thermocouple was calculated. Experimental verification was done by chopped light source focused on thermocouple and photodiode. The signals from microthermocouple probe and photodiode were fed directly to the two-channel oscilloscope. By measurement of transfer function for different chopper speed the frequency characteristics of microthermocouple were determined. Measurements have been performed in vacuo where the convection heat transfer can be neglected so that the Biot number is infinite. Some measurements have been done with the finite values of the Biot number. Comparing the calculated response of microthermocouple to the step function of heat flux at $Bi=\infty$ with the measurements it has been shown that the determination method of microthermocouple dynamic characteristics gives reproducible results.

A pH METHOD FOR THE TEMPERATURE MEASUREMENT IN WALL LAYERS

E. I. NEVSTRUEVA, I. M. ROMANOVSKII

Institute of High Temperatures, Moscow, USSR

A contactless method for temperature distribution measurements, based on the property of some substances — indicators to change colour at a certain value of the hydrogen ion concentration (pH) for a given temperature — is presented. When the pH value does not depend on the temperature, which is the case for pH values less than 5, for a given pH value, a separation boundary in the region of the constant temperature corresponding to the chosen pH value is visualized. For a different pH value, also less than 5, another separation boundary, displaced in relation to the first, corresponding to the second pH value is visualized. By gradually changing pH, the isotherms of the temperature distribution near a heated wall could be obtained on a film.

Using another indicator type for which the colour change does not depend on the temperature but on the pH values only, it is possible by the colour change of the indicator to determine the concentration field in the wall layer created by the acid solution evaporation and the local hydrogen ion concentration increase. This method could present some advantages in comparison to the salt method of the evaporation rate determination in the wall layers, the accuracy of which is low due to the difficulties of the liquid phase temperature determination in the wall layers.

A STUDY OF TEMPERATURE FIELD USING A VISUALISATION METHOD

J. HUETZ

C.N.R.S., Cachan, France

For the reduced variables u^+ y^+ T^+, the boundaries of the different flow zones vary with Prandtl's number. A method is here proposed for a direct measurement of temperature profile in a transparent and a high Prandtl number liquid.

This method consists of creating a solution or a suspension of luminescent products and using the variations of luminescence with temperature. The accuracy is of the order of 2—3°, which is sufficient for low conducting fluids for which the T variations are easily 100° or more between the wall and the mixing temperature.

A more accurate process consists of using the falling of phosphorescence with T for ZnS. The field is thus separated into a luminescent and a dark zone, the boundary being at the falling luminescence temperature.

The light discontinuity surface at a constant temperature moves towards the heating wall with thermal level increase: few measurements at different mixing temperatures allow the plotting of T against r.

The method is especially suitable for zones having a high gradient temperature, i.e. an interesting zone very close to the wall for which anemometric methods cannot easily be used.

LARGE POSSIBILITIES OF APPLICATION OF THE ADSORPTION METHOD FOR THE STUDY OF BOUNDARY LAYER SEPARATION

S. KONČAR-ĐURĐEVIĆ, M. MITROVIĆ, G. POPOVIĆ

Faculty of Technology and Metallurgy, University of Beograd, Yugoslavia

A critical review of the adsorption method technique is presented. The method uses adsorption of dyes from flowing dye solutions along the surfaces coated by white film of adsorbens.

Different coloration intensities of the coloured surfaces reflect various phenomena in the boundary layer, especially in the zone of the separation.

In the paper are presented different modifications of the techniques used for the study of the boundary layer separation. The influence of the roughness of the surface on the separation of the boundary layer was studied for different and complicated geometries of the system. The experiments have proved that the adsorption method can be used for qualitative as well as for quantitative studies.

THE APPLICATION OF THE ADSORPTION METHOD ON THE MASS TRANSFER BY MIXED KINETIC WITH SIMULTANEOUS HEAT TRANSFER

S. KONČAR-ĐURĐEVIĆ and S. CVIJOVIĆ

Faculty of Technology and Metallurgy and Institute of Chemical, Technological and Metallurgical Researches, Beograd, Yugoslavia

The already published papers that refer to the determination of the coefficients of the mass transfer by the adsorption method have been analysed. The condition of the application of this technique requested mainly the diffusional kinetics to influence the phenomenon of the adsorption.

Through these results and through the original measurements as well. the possibility of widening the adsorption method application on the mixed kinetic mass transfer, determined by the rate of diffusion and adsorption, was considered. It is shown that this method can be applied to the determination of the coefficients of the mass transfer as well as the concentrational disturbances in the boundary layer having the heat and mass transfer at the same time.

SEMINAR COMMITTEE

Chairman: Z. ZARIĆ, University of Beograd, Boris Kidrič Institute of Nuclear Sciences
Vice-chairman: J. P. HARTNETT, University of Illinois, Chicago Circle
Members: J. GINOUX, Bruxelles University, von Karman Institute
J. GOSSE, University of Nancy
A. I. LEONT'EV, Institute of High Temperatures Moscow
R. G. TAYLOR, Imperial College, London
R. WILLE, Technical University, Berlin
Executive Secretary: B. JOVIĆ, Boris Kidrič Institute of Nuclear Sciences

LIST OF PARTICIPANTS

ABADŽIĆ Esref, Linde A. G., Höllriegelskreuth, F. R. Germany.

ACHENBACH Elmar, Kernforschungsanlage, Jülich, F. R. Germany.

AEROV V. E., Heat and Mass Transfer Institute, Minsk, U. S. S. R.

AFGAN Naim, Boris Kidrič Institute of Nuclear Sciences, Beograd, Yugoslavia.

ANASTASIJEVIĆ Predrag, Boris Kidrič Institute of Nuclear Sciences, Beograd, Yugoslavia.

APELQUIST, Dept. of Applied Thermo and Fluid Dynamics, Chalmers University of Technology, Göteborg, Sweden.

ARSIĆ Branislava, Boris Kidrič Institute of Nuclear Sciences, Beograd, Yugoslavia.

BARAT M., Laboratoire de Mécanique Expérimentale des Fluides, Faculté des Sciences, Orsay, France.

BARYCHEV JU. V., University Lomonosov, Moscow, U. S. S. R.

BEĆIRSPAHIĆ Sulejman, Laboratorie de Mécanique Expérimentale des Fluides, Faculté des Sciences, Orsay, France.

BERTELA Manlio, Università di Bologna, Bologna, Italy.

BORIES S., Institut de Mécanique des Fluides, Université de Toulouse, Toulouse, France.

BOŠNJAKOVIĆ Fran, Technische Hochschule, Stuttgart, F. R. Germany.

BOUSGARBIES Jean-Louis, Laboratoire de Dynamique des Fluides, Poitiers, France.

BREVI Roberto, C. S. N. Casaccia, Roma, Italy.

BRUN Edmond-Antoine, Université de Paris, Paris, France.

BRUUN H. H., Institute of Sound and Vibration Research, University of Southampton, Southampton, U. K.

CHARWAT A. F., School of Engineering, University of California, Los Angeles, U. S. A.

CHEVALIER P., Centre National de la Recherche Scientifique, Cachan, France.

CHIGIER N. A., Dept. of Fuel Technology and Chem. Eng., University of Sheffield, Sheffield, U. K.

CREMER Helmut, Inst. für Technische Thermodynamik der Rhein. — Westf. Techn. Hochschule, Aachen, F. R. Germany.

CVIJOVIĆ Mihailo, Boris Kidrič Institute of Nuclear Sciences, Beograd, Yugoslavia.

CVIJOVIĆ Svetomir, Institute of Chemical, Technological and Metallurgical Researches, Beograd, Yugoslavia.

DAVIES P. O. A. L., Institute of Sound and Vibration Research, Univ. of Southampton, Southampton, U. K.

DAVIES T. W., Dept. of Fuel Technology, Univ. of Sheffield, Sheffield, U. K.

DE COMELLI Giovanni, Istituto di Fisica Tecnica, Trieste, Italy.

DELERY J., Office National d'Etudes et de Recherches Aérospatiales, Chatillon--Sous-Bagneux, France.

DELGADO Domingos, Instituto Superior Técnica, Universitada Técnica de Lisboa, Lisboa, Portugal.

ĐORĐEVIĆ Bojan, Faculty of Technology and Metallurgy, Beograd, Yugoslavia.

ĐORĐEVIĆ Radivoje, Faculty of Mechanical Engineering, Beograd, Yugoslavia.

DURST Franz, Dept. of Mech. Eng. Imperial College, London, U. K.

ECKERT E. R. G., Dept. of Mech. Eng., University of Minnesota, Minneapolis, U. S. A.

EDNEY Barry, Bell Aerosystems, Research Dept., Buffalo, U. S. A.

EL-WAKIL M. M., Dept. of Mech. Eng., University of Wisconsin, Madison, U. S. A.

FISSAN Heinz, Inst. für Techn. Thermodynamik der Rhein. — Westf. Techn. Hochschule, Aachen, F. R. Germany.

FONTANA Donato, Istituto di Fisica Tecnica, Roma, Italy.

FROEBEL Eberhard, Deutsche Forschungs- und Versuchsanstalt für Luft- und Raumfahrt e. V., Berlin, Germany.

FRÖSSLING Nils, Dept. of Applied Thermo and Fluid Dynamics, Chalmers University of Technology, Göteborg, Sweden.

GELIN Paul, Centre d'Etudes Nucléaires, Gif-sur-Yvette, France.

GIESE Klaus, Hermann Föttinger-Institut, Technische Universität, Berlin, Germany.

GINOUX Jean, von Karman Institute for Fluid Dynamics, Rhode-Saint-Genèse, Belgium.

GOLDSTEIN R. J., Dept. of Mech. Eng. University of Minnesota, Minneapolis, U. S. A.

GOSMAN A. D., Dept. of Mech. Eng. Imperial College, London, U. K.

GOSSE Jean, Ecole Nationale Supérieure d'Electricité et de Mécanique, Université de Nancy, Nancy, France.

GRIGULL U., Lehrstuhl und Institut für Technische Thermodynamik der Technischen Hochschule, München, F. R. Germany.

GRÜTTER Rudolf, Brown, Boveri and Co, Ltd., Baden, Switzerland.

GUGLIELMINI Giovanni, Istituto di Fisica Tecnica, Università di Genova, Genova, Italy.

HANJALIĆ K., Dept. of Mech. Eng. Imperial College, London, U. K.

HANRATTY T. J., Dept. of Chemistry and Chem. Eng. University of Illinois, Urbana, U. S. A.

HARTNETT J. P., Dept. of Energy Engineering, University of Illinois, Chicago, U. S. A.

HARTZUIKER J. P., Technical University, Delft, and National Aerospace Laboratory, Amsterdam, Netherlands.

HEINECKE Jochen, Internationale Atomreaktorbau GmbH, Bensberg/Köln, F. R. Germany.

HIRT Cyril, Los Alamos Scientific Laboratory, Los Alamos, U. S. A.

HUMMEL R. L., Department of Chemical Engineering, University of Toronto, Toronto, Canada.

IRIBARNE Agripina, Department of Chemical Engineering, University of Toronto, Toronto, Canada.

IRVINE T. F., Dept. of Engineering, State University of New York at Stony Brook, New York, U. S. A.

IVANOV JU. V., Institute of Thermal Physics, Tallin, U. S. S. R.

JOVANOVIĆ Milan, Faculty of Technology and Metallurgy, Beograd, Yugoslavia.

JOVAŠEVIĆ Vladimir, Boris Kidrič Institute of Nuclear Sciences, Beograd, Yugoslavia.

JOVIĆ Larisa, Boris Kidrič Institute of Nuclear Sciences, Beograd, Yugoslavia.

KALININ E. K., Moscow Aircraft Institute, Moscow, U. S. S. R.

KJELLSTRØM Bjørn, AB Atomenergi, Studsvik, Sweden.

KOBUS Helmut, Institut für Hydromechanik, Universität Karlsruhe, F. R. Germany.

KOLOVANDIN B. A., Heat and Mass Transfer Institute, Minsk, U. S. S. R.

KONCAR-ĐURĐEVIĆ Slobodan, Faculty of Technology and Metallurgy, Beograd, Yugoslavia.

KORST Helmut, Dept. of Mech. Eng., Universaty of Illinois, Urbana, U. S. A.

KOSCHEL Wolfgang, Lehrstuhl und Institut für Luftfahrttriebwerke, Berlin, Germany.

KOSTIĆ Života, Boris Kidrič Institute of Nuclear Sciences, Beograd, Yugoslavia.

KREITH Frank, Dept. of Chemical Engineering, Universty of Colorado, Boulder, U. S. A.

LECKNER Bo, Dept. of Applied Thermo and Fluid Dynamics, Chalmers University of Technology, Göteborg, Sweden.

LEDINEGG Max, Institut für Techn. Warmelehre, Technische Hochschule, Wien, Austria.

LENGRAND Jean-Claude, Laboratoire d'Aérothermique du C. N. R. S., Meudon, France.

LIND Leif, Den Polytekniske Laereanstalt Afdelingen for Fluid Mekanik, København, Denmark.

LUIKOV A. V., Heat and Mass Transfer Institute, Minsk, U. S. S. R.

MAKSIMOVIĆ LJiljana, Boris Kidrič Institute of Nuclear Sciences, Beograd, Yugoslavia.

MALIĆ Dragomir, Faculty of Technology and Metallurgy, Beograd, Yugoslavia.

MARKOVIĆ Slavimir, Boris Kidrič Institute of Nuclear Sciences, Beograd, Yugoslavia.

MARKOVSKI Mile, Faculty of Mechanical Engineering, Beograd, Yugoslavia.

MARTIN Maurice, Ecole Nationale Supérieure d'Electricité et de Mécanique, Université de Nancy, Nancy, France.

MARTINENKO O. G., Heat and Mass Transfer Institute, Minsk, U. S. S. R.

MARTINOT-LAGARDE André, Faculté des Sciences, Lille, France.

MILINČIĆ Dobrosav, Faculty of Mechanical Engineering, Beograd, Yugoslavia.

MIRKOVIĆ Zdravko, ITEN — Energoinvest, Sarajevo, Yugoslavia.

MITROVIĆ Jovan, Institute for Technology of Nuclear Raw Materials, Beograd, Yugoslavia.

MOTULEVICH V. P., Institute of Energetics, Moscow, U. S. S. R.

NENADOVIĆ Miroslav, Faculty of Mechanical Engineering, Beograd, Yugoslavia.

NEVSTRUEVA E. J., Institute of High Temperatures, Moscow, U. S. S. R.

OBROVIĆ Branko, Faculty of Mechanical Engineering, Kragujevac, Yugoslavia.

OBSIEGER Vilka, Faculty of Mechanical Engineering, Rijeka, Yugoslavia.

OIKNINE C., Université de Besançon, France.

OKA Simeon, Boris Kidrič Institute of Nuclear Sciences, Beograd, Yugoslavia.

OPSENICA Jovo, Aircraft Institute, Beograd, Yugoslavia.

PANDOLFI Maurizio, Politecnico di Torino, Istituto di Macchine e Motori per Aeromobili, Torino, Italy.

PATTEN T. D., Dept. of Mech. Eng., Heriot-Watt University, Edinburgh, U. K.

PAVLOVIĆ Pavle, Boris Kidrič Institute of Nuclear Sciences, Beograd, Yugoslavia.

PETRAZITSKY G. B., The Moscow Higher Technical School, Moscow, U. S. S. R.

PETRIE A. M., Paisley College of Technology, Paisley, U. K.

PEUBE Jean-Laurent, Laboratoire de Dynamique des Fluides, Faculté des Sciences, Poitiers, France.

PISONI Claudio, Istituto di Fisica Tecnica, Università di Genova, Genova, Italy.

PIŠLAR Vladislav, Boris Kidrič Institute of Nuclear Sciences, Beograd, Yugoslavia.

POPOVIĆ Mladen, Faculty of Mechanical Engineering, Beograd, Yugoslavia.

PUN W. M., Dept. of Mech. Eng. Imperial College, London, U. K.

PY Bernard, Ecole Nationale Supérieure d'Electricité et de Mécanique, Université de Nancy, Nancy, France.

RAO D. M., Aeronautics Dept. Imperial College, London, U. K.

RICHARDSON Peter, Division of Engineering, Brown University, Providence, U. S. A.

ROUGIER P., Office National d'Etudes et de Recherches Aérospatiales, Chatillon, France.

RUNCHAL A. K., Dept. of Mech Eng., Imperial College, London, U. K.

SALJNIKOV Viktor, Faculty of Mechanical Engineering, Beograd, Yugoslavia.

SANANES F., Ecole Nationale Supérieure d'Electronique, d'Informatique et d'Hydrolique, Université de Toulouse, Toulouse, France.

SANTARELLI Francesco, Istituto Impianti Chemici, Università di Bologna, Bologna, Italy.

SCHERBERG Max, Aerospace Research Laboratories, Dayton, U. S. A.

SCHNELLER Jiri, Statny Vyskumni Ustav pro Stavby Stroi, Bechovice, Czechoslovakia.

SCHWEFEL Hans-Paul, AEG Forschungsinstitut.

SHABEER Ahmed, Dept. of Mech. Eng. University of Waterloo, Waterloo, Canada.

SMITH J. W., Dept. of Chemical Eng. and Applied Chemistry, University of Toronto, Toronto, Canada.

SMYTH RAYMOND, Dept. of Mech. Eng., University of Sheffield, Sheffield, U. K.

SPALDING D. B., Dept. of Mech. Eng. Imperial College, London, U. K.

SPASOJEVIĆ Dušan, Boris Kidrič Institute of Nuclear Sciences, Beograd, Yugoslavia.

SPEE B. M., National Aerospace Laboratory, Amsterdam, Netherlands.

STOJANOVIĆ A., Institute of Space Mech. Beograd, Yugoslavia.

STOJANOVIĆ Dragutin, Faculty of Mech. Eng., Beograd, Yugoslavia.

STUDOVIĆ Milovan, Boris Kidrič Institute of Nuclear Sciences, Beograd, Yugoslavia.

SYRED N., Dept. of Chem. Eng. University of Sheffield, Sheffield, U. K.

SANDOR Mario, ITEN — Energoinvest, Sarajevo, Yugoslavia.

ŠAŠIĆ Mane, Faculty of Mech. Eng., Beograd, Yugoslavia.

ŠEL Jovan, Faculty of Mech. Eng., Beograd, Yugoslavia.

ŠIKMANOVIĆ Slobodan, Boris Kidrič Institute of Nuclear Sciences, Beograd, Yugoslavia.

TACCOEN Lionel, Electricité de France, Chatou, France.

TASIĆ Aleksandar, Faculty of Technology and Metallurgy, Beograd, Yugoslavia.

TAYLOR R. G., Dept. of Mech. Eng., Imperial College, London, U. K.

TOMŠIČ Mihael, Nuclear Institute »Jožef Stefan«, Ljubljana, Yugoslavia.

TOŠIĆ Dragica, Boris Kidrič Institute of Nuclear Sciences, Beograd, Yugoslavia.

TRIFUNOVIĆ R., Faculty of Mechanical Engineering, Beograd, Yugoslavia.

TROHAN A. M., Institute for Physical-Technical and Radiotechnical Measurements, Moscow, U. S. S. R.

UEBELHACK Helmut, von Karman Institute for Fluid Dynamics, Rhode-Saint--Genese, Belgium.

VAGT Jorg-Dieter, Hermann Föttinger Institut für Strömungstechnik, Technische Universität, Berlin, Germany.

VAN THINH Nguyen, Laboratoire de Mécanique Expérimentale des Fluides, Faculté des Sciences, Orsay, France.

VEHAUC Aleksandar, Boris Kidrič Institute of Nuclear Sciences, Beograd, Yugoslavia.

VOLCH'KOV E. P., Institute of Thermophysics, Novosibirsk, U. S. S. R.

VORONJEC Dimitrije, Faculty of Mechanical Engineering, Beograd, Yugoslavia.

VORONJEC Konstantin, Faculty of Mechanical Engineering, Beograd, Yugoslavia.

WHITTAKER Maurice, Rolls Royce Ltd., Bristol Eng. Division, Bristol, U. K.

WOLFSHTEIN M., Dept. of Mech. Eng. Imperial College, London, U. K.

ZARIĆ Zoran, Faculty of Mechanical Engineering and Boris Kidrič Institute of Nuclear Sciences, Beograd, Yugoslavia.

ZDRAVKOVIĆ Momčilo, Faculty of Mechanical Engineering, Beograd, Yugoslavia.

ZEMANEK Jan, Statny Vyskumni Ustav pro Stavby Stroi, Bechovice, Czechoslovakia.

ZHIUGZHDA J., Institute of Physico-Technical Problems of Energetics, Vilnius, U. S. S. R.

ZHUKAUSKAS A., Institute of Physico-Technical Problems of Energetics, Vilnius, U. S. S. R.

ZORIĆ Dušan, Faculty of Traffic Engineering, Beograd, Yugoslavia.

REPRESENTATIVES

SERBIAN ACADEMY OF SCIENCES, Beograd, *Veličković Dušan*.

FEDERAL COMISSION FOR NUCLEAR ENERGY, Beograd, *Guzina Vojin*, President.

INTERNATIONAL ATOMIC ENERGY AGENCY, Vienna, *Ristić Milorad*.

AUTHOR INDEX

Achenbach, E., 978
Aerov, V. A., 990
Afgan, N. H., 761, 1003

Baev, V. C., 475, 489
Bairashevsky, V. A., 973
Baker, R. J., 321
Bakke, E., 971
Barat, M., 887
Barishev, J. V., 988
Bećirspahić, S., 975
Beér, J. M., 980
Bobrovich, G, I., 663
Bories, S., 1000
Bousgarbies, J. L., 997
Broise, B., 715
Brun, E. A., 867
Bruun, H. H., 992
Burdukov, A. P., 669

Carlson, L. W., 177
Charwat, A. F., 967
Chigier, N. A., 985
Cvijović, S., 1005

Davies, P. O. A. L., 992
Davies, T. W., 980, 994
Davis, M. R., 577
Delery, J., 970
Delhaye, J. M., 715
Devold, I., 529
Dreitser, G. A., 975
Druker, I. G., 677
Dumas, R., 157
Dvorak, F. A., 257

Edney, E., 971
El-Wakil, M. M., 1001
Escudier, M. P., 239

Favre, A., 157
Fedorov, B. I., 151
Fedorovich, E., 683
Fiore, A., 715
Fortier, A., 199
Froebel, E., 996
Fulachier, L., 157

Geffroy, J., 976
Giese, K., 995
Ginoux, J. J., 968
Gleb, L. K., 419
Goldshtik, M. A., 39, 42, 437
Goldstein, R. J., 941
Gontier, G., 1000
Gosman, A. D., 984
Gosse, J., 979, 998
Guglielmini, G., 998

Hahne, E., 523
Hanjalić, K., 393, 411
Hanratty, T. J., 919
Hardel, G. G., 1001
Hartnett, J. P., 97
Head, M. R., 257
Hedberg, S., 593
Hendrikson, V., 988
Hirt, C. W., 827
Huetz, J., 1004
Hummel, R. L., 1001, 1002

Iribarne, A. de P., 1001
Ivanov, J., 988

Jones, W. P., 307
Jonsson, V. K., 321
Jovanović, LJ., 753, 1003
Jovašević, V., 227
Jude, M., 976

Kalinin, E. K., 975
Khabakhpasheva, E. M., 573
Kjellstrøm, B., 593
Kline, S. J., 996
Kolovandin, B. A., 359, 981, 990
Končar-Đurđević, S., 1005
Korst, H. H., 781
Koschel, W., 986
Kostić, Ž., 425, 451, 977
Kovalev, S. A., 657
Kreith, F., 971
Kumar, R. N., 141
Kusminov, W. P., 975
Kutateladze, S. S., 939
Kuzma-Kichta, Ya. A., 657

Launder, B. E., 307, 393, 411
Lengrand, J. C., 990
Leont'ev, A. I., 51, 981
Lockwood, F. C., 339
Luikov, A. V., 999

Martin, M., 979
Martinot-Lagarde, A., 1000
Martynenko, O. G., 973
Mathieu, J., 379
Maye, J.-P., 629
Mironov, B. P., 71
Mitrović, M., 1005
Morrow, T. B., 996
Motulevich, V. P., 987
Mourgues, J. C., 976

Nakoryakov, V. Ye., 619
Nicoll, W. B., 239
Nevstruyeva, E. I., 651, 1003

Ogorodnikov, V. P., 657
Oiknine, C., 972
Oka, S. N., 425, 451, 972, 976
Ong, P. H., 339

Paumard, G., 976, 977
Petrazitsky, G. B., 989
Petukhov, B. S., 495
Pisoni, C., 997
Pišlar, V., 1003
Kolezaeff, V. I., 989
Polyakov, A. F., 517
Popov, V. P., 419
Popović, G., 169, 1005
Pun, W. M., 984
Pustyntsev, G. N., 351
Py, P., 998

Rao, D. M., 970
Richardson, P. D., 979
Rohsenow, W. M., 683
Romanovskii, I. M., 1003

Rougier, P., 970
Rozhdestvensky, V. I., 988
Rubtsov, N. A., 87
Runchal, A. K., 983

Sananes, F., 972
Sastri, V. M. K., 97
Scherberg, M. G., 974
Semeria, R., 701
Shulman, Z. P., 999
Siddal, R. G., 994
Smith, J. W., 1002
Spalding, D. B., 1, 269, 982
Stefanović, M., 743, 753, 1003
Struminskii, V. V., 459
Styrikovich, M. A., 641
Syred, N., 985

Taccoen, L., 217
Talmor, E., 177
Thirroit, C., 1000
Tretjakov, C. P., 475

Uebelhack, H. T., 968

Vagt, J. D., 995
Van Thinh, N., 585, 993
Vasil'ev, L. L., 125
Vernier, Ph., 715
Verollet, E., 157
Volchkov, E. P., 987

Whitelaw, J. H., 289
Wolfshtein, M., 985

Yarkho, S. A., 975
Yasakov, V. A., 489

Zarić, Z., 555, 993
Zdravkovich, M. M., 447, 1002
Zhiugzhda, J., 977
Zhukauskas, A., 843
Zhukov, V. M., 657
Zorić, D. L., 992